Second Edition

FACILITY LAYOUT AND LOCATION: AN ANALYTICAL APPROACH

Richard L. Francis
Leon F. McGinnis, Jr.
John A. White

PRENTICE HALL INTERNATIONAL SERIES IN INDUSTRIAL AND SYSTEMS ENGINEERING

W. J. FABRYCKY and J. H. MIZE, editors

Facility Layout and Location

PRENTICE HALL INTERNATIONAL SERIES
IN INDUSTRIAL AND SYSTEMS ENGINEERING

W. J. Fabrycky and J. H. Mize, Editors

ALEXANDER *The Practice and Management of Industrial Ergonomics*
AMOS AND SARCHET *Management for Engineers*
ASFAHL *Industrial Safety and Health Management,* 2/E
BABCOCK *Managing Engineering and Technology*
BADIRU *Expert Systems Applications in Engineering and Manufacturing*
BANKS AND CARSON *Discrete-Event System Simulation*
BLANCHARD *Logistics Engineering and Management,* 4/E
BLANCHARD AND FABRYCKY *Systems Engineering and Analysis,* 2/E
BUSSEY AND ESCHENBACH *The Economic Analysis of Industrial Projects,* 2/E
CANADA AND SULLIVAN *Economic and Multi-Attribute Evaluation of Advanced Manufacturing Systems*
CHANG AND WYSK *An Introduction to Automated Process Planning Systems*
CHANG, WYSK, AND WANG *Computer Aided Manufacturing*
CLYMER *Systems Analysis Using Simulation and Markov Models*
ELSAYED AND BOUCHER *Analysis and Control of Production Systems*
FABRYCKY AND BLANCHARD *Life-Cycle Cost and Economic Analysis*
FABRYCKY AND THUESEN *Economic Decision Analysis,* 2/E
FRANCIS, MCGINNIS, AND WHITE *Facility Layout and Location: An Analytical Approach,* 2/E
GIBSON *Modern Management of the High-Technology Enterprise*
HALL *Queing Methods: For Services and Manufacturing*
HAMMER *Occupational Safety Management and Engineering,* 4/E
HUTCHINSON *An Integrated Approach to Logistics Management*
IGNIZIO *Linear Programming in Single- and Multiple-Objective Systems*
KUSIAK *Intelligent Manufacturing Systems*
OSTWALD *Cost Estimating,* 3/E
PULAT *Foundations of Industrial Engineering*
TAHA *Simulation Modeling and SIMNET*
THUESEN AND FABRYCKY *Engineering Economy,* 7/E
TURNER, MIZE, AND CASE *Introduction to Industrial and Systems Engineering,* 2/E
WOLFF *Stochastic Modeling and the Theory of Queues*

2nd Edition

Facility Layout and Location

An Analytical Approach

Richard L. Francis
Department of Industrial and Systems Engineering
University of Florida

Leon F. McGinnis, Jr.
School of Industrial and Systems Engineering
Georgia Institute of Technology

John A. White
School of Industrial and Systems Engineering
Georgia Institute of Technology

PRENTICE HALL
Englewood Cliffs, New Jersey 07632

Library of Congress Cataloging-in-Publication Data

FRANCIS, R. L.

Facility layout and location : an analytical approach / Richard L. Francis, Leon F. McGinnis, Jr., John A. White. — 2nd ed.

p. cm.

Includes bibliographical references and index.

ISBN 0-13-299231-0

1. Plant layout. 2. Factories—Location. 3. Operations research I. McGinnis, Leon F. (Leon Franklin) II. White, John A. III. Title.

TS178.F7 1992

658.2'3—dc20 91-12931

CIP

Acquisitions editor: *Elizabeth Kaster*
Editorial/production supervision
and interior design: *Kathleen Schiaparelli*
Copy editor: *Zeiders & Associates*
Prepress buyer: *Linda Behrens*
Manufacturing buyer: *Dave Dickey*
Supplements editor: *Alice Dworkin*
Editorial assistant: *Jaime Zampino*

Printed in the United States of America

10 9 8 7 6 5 4 3 2

ISBN 0-13-299231-0

Prentice-Hall International (UK) Limited, *London*
Prentice-Hall of Australia Pty. Limited, *Sydney*
Prentice-Hall Canada Inc., *Toronto*
Prentice-Hall Hispanoamericana, S.A., *Mexico*
Prentice-Hall of India Private Limited, *New Delhi*
Prentice-Hall of Japan, Inc., *Tokyo*
Simon & Schuster Asia Pte. Ltd., *Singapore*
Editora Prentice-Hall do Brasil, Ltda, *Rio de Janeiro*

This book is dedicated

to the memory of

David F. Baker

Contents

Preface

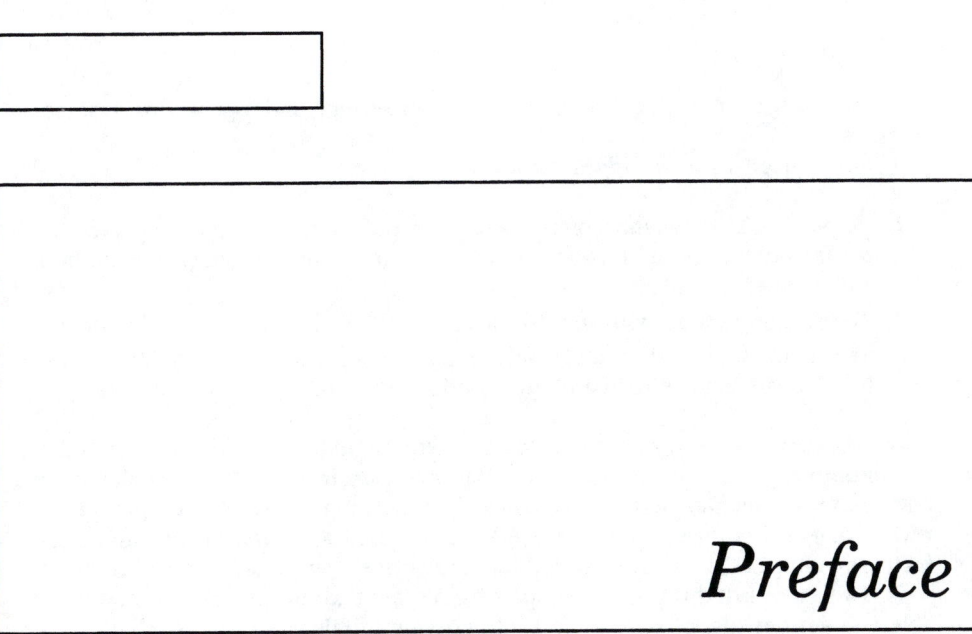

The more we learn about the process of facility and system design, the more we appreciate how difficult these problems are. This book, which is a major revision of the first edition, was written because we believe that quantitative models are essential for complex facility design and location decisions. Our intended audience is students in courses which address facility layout and location. While these courses will be found primarily in curricula in industrial engineering, manufacturing systems, management science, and operations research, the book also should be of interest to students in systems engineering, transportation science, and urban and regional studies.

Our approach to the subject of facility layout and location is a quantitative one. We view these problems as optimization problems, i.e., what decisions will optimize some measure of performance, subject to some constraints. To fully appreciate and master the advanced material will require prior work in several advanced topics. However, we have been careful to present the basic material so that it is accessible to a student who has completed the standard engineering sequence in calculus. Some of the material in Chapter 5 will employ basic concepts of probability theory, some of the material in Chapters 6 and 8 will use linear programming results, and an understanding of discrete optimization will facilitate the study of Chapters 8 and 9. We have taught the material in the book at both the undergraduate and graduate levels to industrial engineering, systems engineering, and operations research students.

Our objectives in writing the book were:

1. To provide a reasonably comprehensive introduction to quantitative methods for facility layout and location, in contrast to the mostly qualitative methods widely used in practice
2. To provide students with an introduction to the basic tools and methodologies
3. To provide advanced students and beginning researchers a reasonably comprehensive view of the field and an introduction to the issues and literature

Given the explosive growth of the research literature in the past fifteen years, these are ambitious goals. We have been forced to make choices regarding which models and methods to include in the book, so we do not claim to provide a comprehensive review of the literature. However, we have included a section on recommended further reading at the end of each chapter to facilitate access to additional literature.

Two aspects of this book set it apart from other texts in the field. First, we take a very quantitative approach to the facility design problem. This is especially clear in Chapter 5, dealing with warehouse design. Second, we treat the facility layout and location problems together. We view the facility location problem as a "layout problem in the large." The problems of configuring a single facility and of configuring a system of facilities have much in common, and that commonality is best seen when the problems are formulated as decision problems.

This book can support a number of different courses. At the undergraduate level, we have covered Chapters 1, 2, 3, 4, and 5 in a course on facility design. We have covered Chapters 4, 6, 7, 8, and 9 in a graduate course on facility location. Chapters 4, 5, 6, 7, and 8 can be used to support a dual-level course on distribution systems.

No textbook reaches the presses without the contribution of many people. Our respective academic departments at Georgia Tech and the University of Florida have encouraged and supported the preparation of the manuscript. Our students have patiently suffered through rough drafts, diplomatically pointing out our notational inconsistencies and logical confusion. John White III provided several of the figures in Chapter 2. Even in this age of personal computers, we owe a debt of thanks to the able staff support provided by Joene Owen, Jo Funk and Robin Vanlue. Any errors that remain are ours alone.

Excerpts appearing in Chapter 6 are reprinted with the permission of The Institute of Management Sciences.

Richard L. Francis
Leon F. McGinnis
John A. White

Facility Layout and Location

Chapter 1

Introduction

1.1 BACKGROUND

This chapter is written in order for the book to have a first chapter. Since few faculty ever read the first chapter of the assigned text for a course, the chapter is not written for faculty. Instead, it is written for those students who have been told to read it (and do so!), as well as for those who have an interest in facilities layout and location and want to read a brief introduction to the subject.

Our treatment of facility layout and location problems, to say the least, is nontraditional, as well as unconventional. Traditionally, the subject of the book is referred to as *plant layout*. The more general term of facility layout and location is used here, rather than plant layout and location, for four reasons. First, we did not want to have the book categorized by a librarian as a book on botany or horticulture; even though one can interpret facility to include plant life, the book will be of limited use to botanists and horticulturists. Second, a very broad facility location literature has developed and is so identified. Third, in a sense, facility layout and location problems include plant layout problems as a subclass, since many facility location problems do not require a layout decision; further, while the plant layout problem should include "laying out" facilities other than industrial plants, its name implies otherwise. Fourth, the field of plant layout (no pun intended) has become identified as one that is predominately qualitative and depends heavily on the use of checklists, principles, rules of thumb, heavy doses of experience, and pragmatism; as you will see, beginning with Chapter 4 our treatment of the subject is quite different from that typically accorded the plant layout problem.

Facility layout and location problems have been the subject of analysis for centuries. As an example, a special case of a location problem treated in Chapter 4 was solved as early as the seventeenth century. Even though facility layout and location problems have received considerable attention over the years, it was not until the emergence of the interest in operations research and management science that the subject received renewed attention in a number of disciplines. Currently, there exists a strong interdisciplinary interest in facility layout and location problems.

Architects, computer scientists, economists, engineers from several disciplines, home economists, management scientists, mathematicians, operations researchers, regional scientists, technical geographers, transportation systems designers, and urban planners have discovered a commonality of interest in a concern for the location and layout of facilities. Each brings different interpretations and different solutions to the problem. Indeed, each defines "facility" differently.

As suggested in the book's title, we take an analytic approach to the subject, one in which mathematical models are proposed as an aid to solving layout and location problems. Creating useful mathematical models of real layout and location problems always involves abstraction and frequently requires simplifying assumptions. Thus the results obtained by analyzing the models are not necessarily solutions to problems. The degree to which results from the model are useful depends on the degree to which the model captures the important aspects of the real problem. At best the model solves the problem directly; at worst the model provides a *benchmark* against which operationally accepted solutions are compared. As such, *the facility layout and location models considered in the book should be viewed as design tools*, in the same way that models are viewed by other disciplines.

1.2 MODEL CLASSIFICATION

Reference has been made to the development and use of models in solving facility layout and location problems. Perhaps it is worthwhile to review one classification of models and contrast the approach taken in the past with that taken in this book with regard to the analysis of facility layout and location problems.

Models can be classified as iconic, analog, and symbolic. *Iconic models* are scalar representations of objects in that they look like the objects represented. Iconic models may be two- or three-dimensional representations and closely maintain the visual effects of the situation under study. *Analog models* substitute one property for another. After the problem is solved in the substituted state, the solution is translated back to the original dimensions or properties. *Symbolic* or *mathematical models* are an abstract representation of a system.

Prior to the development of interest in analytical approaches, facility layout and location problems were solved primarily using iconic models. Templates and scale models were maneuvered on a floor plan of the facility until a number of alternative solutions were obtained. The generation of these alternatives was largely dependent on the subjective criteria of the analyst. The alternatives were judged on the basis of visual effects as well as other qualitative objectives. In some cases, checklists and rules of thumb were employed in an attempt to reduce the degree of subjectivity involved. Gradually, such approaches have been replaced by a greater degree of reliance on a quantitative analysis.

In this book our concern is with the development of symbolic or mathematical models. There are basically two types of mathematical models: descriptive and prescriptive (normative). In the former, the model is used to *describe* the behavior of the system. Normally, as an example, queueing models are descriptive. Normative models are used to *prescribe* a course of action that, in some sense, is optimal. Thus normative models require a measure of effectiveness against which alternative solutions can be judged. Classical optimization models are examples of typical normative models.

1.3 CRITERION SELECTION

The facility layout and location models developed in the book are normative models. For this reason the problem of selecting an appropriate criterion must be considered. The choice of the criterion to be used in choosing the "best" solution from among several alternative solutions is not an easy one in the case of facility location and design problems.

Perhaps the most popular criterion used is to minimize some function of distance traveled. Within an industrial setting, it is argued that minimizing distance will minimize material handling cost. However, it may be the case that reducing distance creates congestion in a concentrated area and material handling costs increase. Also, it is often desirable to maintain some minimum separation between facilities.

A number of attempts to model facility layout and location problems have made some rather bold assumptions regarding data and criteria. In particular, it is often assumed that material handling cost is the only significant factor and that material handling costs are linear functions of distance and flow volumes; furthermore, static and deterministic conditions are assumed; and the impact of the layout/location decision on other aspects of the business is ignored.

Assuming the performance measure is a linear function of distance can be invalid for a number of reasons. First, the functional relationship might not be linear, but a more complicated function of distance. Second, performance might not be related to distance at all. Third, true distances can be quite different from those computed using a particular travel metric. In the latter case, for example, the topology of the situation might preclude straight-line distances between two points; after all, crossing a river by truck usually requires a bridge or ferry. Similarly, one-way aisles and streets can complicate the computation of distances traveled.

Returning to the matter of whether or not performance is a function of distance, one should not blindly assume, for example, that minimizing total distance traveled is an appropriate objective. As an absurd illustration, with such an objective, locating a school such that 1500 students must each travel 4 miles to school is as desirable as locating the school such that either 1000 students travel zero miles and 500 students travel 12 miles each or 1400 students travel zero miles and 100 students travel 60 miles each.

Instead of the criterion of minimizing total distance traveled, one may wish to minimize the maximum distance traveled. Such a solution is called a *minimax location*. In the case of the location of a school, a minimax location would guarantee that the largest distance traveled by a student would be as small as possible. Such an objective is encountered in the design of stadiums and theaters, as well as the location of fire stations, hospitals, and civil defense units.

Yet another criterion that might be appropriate is maximization of the minimum distance to an "offensive" facility. The location of solid waste disposal sites, nuclear fuel waste, prisons, and other such "facilities" is clearly one in which everyone wants to be as far away from the facility as possible.

Although many of the public-sector location problems are amenable to analysis using one of the criteria given above, it is also the case that most such problems are "solved" through a political process rather than on the basis of analytical models. However, it is also the case that an objective assessment of the situation, based on the use of analytical models, can serve to defuse a potentially explosive political issue.

1.4 COST ELEMENTS

We require a number of costs in the various models we develop in the book. More specifically, we require incremental costs. As an illustration, suppose that trucks are used to perform all material handling between stores in a city. A new store is to be located in the city, and there are two possible locations, P and Q, for the new store. If the new store is located at P, there will be 50 miles of travel per day between the new store and other existing stores; if located at Q, there will be 70 miles of travel per day. If the number of trucks and employees required for material handling is the same whether the new store is located at P or Q, the only incremental costs are those resulting from the additional travel of 20 miles per day, plus the costs of site preparation and constructing the store at each site. If the new store is actually an existing store that is to be relocated, relocation costs must be considered. Finally, if costs have a nonlinear relationship with total distance, traveling 20 additional miles per day must be evaluated in the context of the total travel distance involved.

In some cases the choice of location affects the type of material handling equipment to be used. For example, suppose that there exists a machine at location A. A new machine, B, can be located at either X or Y. If located at X, a roller conveyor will be installed between A and X; if located at Y, a lift truck will transport material between A and Y. The incremental costs to be considered are:

1. Cost of purchasing, installing, and maintaining a roller conveyor if located at X
2. Cost of operating the lift truck if located at Y
3. Cost of in-process inventory differences resulting from different material handling systems
4. Other costs produced by the differences in material handling systems

In the case of lift trucks, the incremental annual cost of item movement between machines A and B is typically the product of the incremental cost per mile, number of round trips between A and B per year, and round-trip distance per trip between A and B. If a conveyor connects machines A and B, the incremental annual cost is the incremental cost of owning, operating, and maintaining a conveyor of sufficient length to connect A and B and of sufficient capacity to meet the yearly material handling requirements. Normally, this cost is assumed to be a linear function of conveyor length.

1.5 MODEL VALIDATION

In addition to an appropriately selected criterion, the validation of the model is a very important step in the analysis of facility location and design problems. By validation we mean the process of verifying that the model does indeed accurately represent the physical system under study. One measure of the validity of our models is whether or not they lead to reliable predictions of the system's performance and subsequent improvements in the system. Normally, it is quite difficult to validate a facility location and design model.

In practice, model validation usually consists of a verification of the assumptions of the model. In a number of cases this approach is justified. However, as will be seen in later discussion, the model can be relatively insensitive to a number of the assumptions. Consequently, it does not necessarily follow that a model based on inaccurate assumptions will yield inaccurate solutions. The true test of the validity of the model is whether or not it serves as an aid in obtaining solutions that are better than would otherwise be obtained.

There exist trade-offs between the degree of realism in the model, its ease of manipulation, expense of solution, and clarity of cause-and-effect relationships. Normally, very simple models are developed in the text. Many simplifying assumptions are made. As a result, the accuracy of the assumptions is sometimes questionable. On the other hand, it is often possible for simple models to yield answers that agree closely with those obtained from a highly sophisticated model closely approximating the "real world." Furthermore, more elaborate, higher-fidelity models generally require more data and more structured assumptions about the underlying problem; the data for the "more realistic model" are usually more expensive to collect and more susceptible to errors and uncertainties.

Of the several approaches suggested in the literature for model validation, one that can be employed with the type of models considered here is to compare the answers obtained from different models of the problem. For example, the solutions obtained using an iconic model, an analog model, a simulation model, a simple mathematical model, and a complex mathematical model could be compared. Obviously, such an approach could prove to be quite expensive. In practice, one might choose two models and compare their solutions. If the answers agree, there is a higher probability that the model is valid. If the answers disagree, one model could still be valid or, for that matter, both could be invalid. In the latter case, a third model might be developed, and its solution compared with the other models. However, one can never be assured that a facility layout and location model is valid in the strictest sense.

In general, the notion of starting with the simplest models first, then bringing to bear more elaborate models, and looking for consistency of solutions is an important concept. Unfortunately, most faculty, students, and design engineers pay too little attention to model validation; instead, they become enamored of the model and lose sight of the problem. Due to the difficulties associated with performing model validation, the process is often ignored.

It is frequently the case that model validation cannot be performed because no hard data are available. As an example, suppose that the location of a new plant to produce a new product is to be determined in the face of high degrees of uncertainty concerning customer locations and customer demands. Without hard data it would be difficult to validate the location model. However, once experience is gained, it should be possible to perform a retrospective validation of the model. This would be especially useful if repetitive location decisions are to be made.

By performing a retrospective audit of the location decision, we are able to refine the model, identify the data that will be important for similar decisions in the future, and assess the need for future relocations. One should continually ask the question: If a different location had been chosen, would costs have been reduced? It is this evolutionary property of analytical models that provides strong support for their use, as opposed to a strict reliance on qualitative factors.

1.6 THE DESIGN PROCESS

Throughout the book we formulate a variety of layout and location problems as optimization problems. It will be important for you to put such formulations in the proper context; namely, the formulations are abstract representations of the true layout and location problem and the solution process is a part of an overall design process. Too often, the location problem and layout problem are considered to be entirely different kinds of problems. For example, Simon [12] considers the facility layout problem to be a *design problem*, whereas the facility location problem is considered to be an *optimization problem*. As noted by Simon, "in ordinary language, however, we apply the term 'design' only to problem solving that aims at synthesizing new objects. If the problem is simply to choose among a given set of alternatives, e.g., to choose the location or site for a plant, we do not usually call it a design problem, even if the set of available alternatives is quite large, or possibly infinite [12, p. 295]."

Typically, design problems do not have well-defined, unique, optimum solutions. Instead of striving to obtain an optimum layout design, we are generally interested in obtaining a satisfactory layout design. Simon characterizes the solution process used in design as satisficing rather than optimizing. He also noted: "If Frederick Taylor had been correct in insisting that there is one right way of doing anything, there would be an inevitable clash between function and style—we could exercise stylistic freedom only by sacrificing function" [12, p. 287].

In discussing the impact of style on design, Simon pointed out:

> There is a proverb to the effect that the best is the enemy of the good. A little reflection shows that the proverb can be read in two diametrically opposed ways, and that it is not clear which reading is intended—or is the more defensible. By one interpretation, the proverb means that if we are willing to settle for the good-enough—to "satisfice"—we will never attain the best. By the other interpretation, it means that our striving to reach our unattainable best may prevent us from reaching an achievable good-enough. It is this latter interpretation that describes the stern reality of the design process in the world we know—in any world, for that matter, that is even moderately complex. [12, p. 288]

Simon contrasted design problems and optimization problems in the following way: "There are two reasons, one negative and one positive, why we call the layout problem, but not the maximising problem, a design problem. The negative reason is that we do not have a simple finite algorithm for going directly to a solution of the layout problem. The positive reason is that the processes we do have for attacking it involve synthesizing the solution from component decisions that are *selective, cumulative,* and *tentative*" [12, p. 296].

We have considerable empathy with Simon's contrast in design and optimization problems, but we take exception with any characterization of the location problem that precludes it being a design problem. The rendering of a location problem as an optimization problem is simply an abstraction of the overall location problem. It has been our experience that both layout and location problems are design problems. Further, we have found that abstract representations of both types of problems can be rendered as optimization problems.

In the case of the facility layout problem, we seek the "best" layout design; in solving the facility location problem, we wish to find the "best" locations for the facilities. In the case of the former, neither the objective function nor the constraints are usually well formulated; as a result, the process becomes one of satisficing. In the latter case, it is usually the case that the objective function and constraints are better, if not perfectly, formulated; as a result, the process becomes one of optimizing. Yet, in both contexts, the formulation itself is an imperfect representation of the complex interactions that generally accompany any facility layout and/or location problem.

Even in the case of facility location problems, the solution obtained is a solution to the model, which may or may not be an accurate representation of the real-world location problem. Qualitative, rather than quantitative, considerations may dominate the decision process. (For example, a number of location decisions are made solely on the basis of where the decision maker wants to live. As a result, we offer the following theorem: *Location decisions are based on objective, quantitative criteria if and only if the decision maker and spouse are not affected directly by the outcome of the decision*. Our proof procedure is the popular *proof by assertion*.)

In solving facility layout or facility location design problems, it is important that the general procedure advocated for solving design problems be used. Specifically, the following steps are recommended:

1. Formulate the problem.
2. Analyze the problem.
3. Search for alternative solutions.
4. Evaluate the design alternatives.
5. Select the preferred design(s).
6. Specify the solution(s).

Subsequently, we treat each step of the design process. To facilitate the discussion, we will concentrate on the layout design problem. However, the discussion is also applicable to location design problems and the design of a system of facilities (the number, locations, and interactions of facilities). Before treating each of the six steps embedded in the design process, it is important to note that the procedure is seldom followed exactly as given. In particular, "in the course of the design process very few complete designs are generated, compared and evaluated—in the typical case, only one" [12, p. 297].

In practice, an iterative process is used. First, a macro-level approach is used to gain acceptance of an overall design concept. After agreeing on the concept, the designer might perform detailed design on subsets of the overall design and gain acceptance component by component. Finally, the overall design is developed by synthesizing the design components generated previously.

Furthermore, it would not be unusual to find the six-step design process used at the concept level but not at the detailed design level. In designing a new distribution center for a retail firm, as an example, one might develop three or four alternative concepts at a macro-design level. One concept might centralize receiving and shipping, whereas another concept might decentralize the two functions. Yet another concept might depict the use of high-rise automated storage and retrieval, whereas alternative concepts might depict palletized product being block stacked or stored in alternative configurations of storage rack.

In designing a distribution center, the overall design effort is usually broken into a number of components or segments. Typically, the design components represent the major functions performed: receiving, in-bound inspection, large-item storage, medium-item storage, small-item storage, order accumulation, packing, shipping, offices, maintenance, personnel services, parking, and so on.

Each component is then designed using the overall design process, with changes made to components depending on the choices made among other components. Finally, the overall layout is designed by synthesizing the various design components. As Simon noted, "the problem space through which the designer searches is not a space of designs, but a space of design components and partially completed designs" [12, p. 297].

Whether designing facilities or designing systems of facilities, the objective is to organize the space. The result of the organization is a structure within which many specific decisions must be made. For example, in designing a facility, we decide first which departments or activities will be located in the facility; next, we arrange them relative to one another. Although we do not have useful models to guide our decisions regarding the departments and activities to include in the facility, we do have a number of models to assist us in determining their relative locations. Similarly, in designing a distribution system, we must decide how many levels will be used in the overall distribution hierarchy; next, we determine how many facilities will be provided at each level and where they will be located. Again, the former decision regarding the overall organization of the distribution system is not one that is easily modeled, whereas the determination of the number and locations of facilities at each level of the hierarchy is amenable to modeling as a location problem.

1.6.1 Formulate the Problem

In some cases the formulation of the problem is obvious. In fact, it may be formulated for you: for example, "Go over to the personnel building and determine the best location for a new data-processing machine." In such a case, care should be taken to assure that the proper problem is formulated. In the present case, a change in the design of the paperwork system might eliminate the need for the new data-processing machine!.

Rather than having the problem correctly formulated for you, more commonly you will be confronted with a situation, and you must proceed from there. Typically, the situation is the present solution to the problem. This is especially true in re-layout problems. There are rather serious dangers associated with such a formulation. Specifically, it is rather easy to become biased and lose one's objectivity when confronted with an existing layout.

The formulation of the problem is aided by taking a *black-box* approach. As Krick [8] describes the black box, there exists an originating state of affairs (state A) and a desired state of affairs (state B). Also, a transformation must take place in going from state A to state B, as shown in Figure 1.1. There is more than one method of accomplishing the transformation from state A to state B, and there is unequal preferability of these methods (otherwise, no problem exists). The solution to the problem is visualized as a black box of unknown, unspecified contents, having input A and output B.

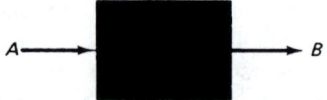

Figure 1.1 Black-box approach.

The black-box approach facilitates a proper identification of states *A* and *B* during problem formulation. As an illustration, consider the problem of locating a lathe in department *X*. The problem could be formulated in terms of states *A* and *B* as follows:

State A	State B
1. Lathe to be located in department *X*.	Lathe located in department *X*.
2. Lathe to be located.	Lathe located.
3. Machine tool to be located.	Machine tool located.
4. Operations to be performed.	Operations performed.
5. Customer having need for a product.	Customer's need satisfied.

The first formulation is very narrow. Using it, we would concentrate on determining the best location for the lathe in department *X*. The second formulation allows us to consider location sites outside department *X*. The third formulation suggests the best machine tool might not be the lathe initially specified. The fourth formulation allows us to develop the process best suited for the performance of the set of operations and to design the overall layout. The fifth formulation involves a total systems design. No doubt, there are other formulations having varying degrees of breadth.

Just how broadly should the problem be formulated? In general, we recommend that the problem be formulated as broadly as the economics of the situation, time constraints, and the organizational boundaries permit. In the example just considered, it is not likely that the last formulation given would be justified. The point to be made is that the narrower the formulation the fewer solutions there are to the problem. Thus, by formulating a problem narrowly you might miss a solution that would produce significant cost savings for the total system. In the case of the lathe, it may well be that the product mix has changed sufficiently over time to the point that a complete re-layout is justified. In fact, adding another lathe might be counter-productive for the total system.

1.6.2 Analyze the Problem

The *analysis of the problem*, which has been formulated previously, consists of a relatively detailed phrasing of the characteristics of the problem, including restrictions. This phase of the design process involves considerable *fact gathering*. Be careful to separate the real restrictions from the fictitious restrictions. Just because something has been done the same way for the past 20 years does not justify it as a real restriction. The analysis of the problem also involves the identification of the appropriate criteria to be used in evaluating alternative solutions to the problem.

The process of fact gathering brings you face to face with the present solution to the problem. Consequently, it is wise to make every effort to avoid becoming biased in your thinking. Preoccupation with the present solution often results in solutions to the problem that are only slight modifications of the present solution.

If original and creative designs are to be developed, it is essential for the layout designer to concentrate initially on the correct problem formulation and then to analyze the true problem. Although charts and diagrams can be quite useful in layout design, it should be remembered they are aids to, not substitutes for, a sound analysis of the problem.

1.6.3 Search for Alternative Solutions

The *search for alternative solutions* to the layout problem consists of the specification of alternative contents of the black box employed in formulating the problem. Remember, the quality of the final solution can be no better than the quality of the alternative solutions generated during this phase of the design process. Since the final solution to the problem must come from the set of alternatives generated, we suggest that you attempt to maximize the number, quality, and variety of alternative solutions.

In the search for alternative solutions, strive to be creative, and make every effort to divorce your thinking from the present solution. There exist a number of aids to creativity that can improve your ability to generate more and better alternative designs. Krick [9] has summarized a number of these. We shall draw on his discussion.

1. *Exert the necessary effort*. Set a time limit and force yourself to concentrate on the problem during the allotted time.

2. *Do not get bogged down in details too soon*. Think big initially; the details can be considered later. Of course, you can overdo it. We are reminded of the operations research analyst during World War II who supposedly suggested that the German submarine force could be destroyed by boiling the ocean. When questioned further about the feasibility of the plan, the analyst responded, "I'm supposed to develop solutions, not implement them. I leave the details to others."

3. *Make liberal use of the questioning attitude*. The questions what, who, when, where, which, how, and why should be applied to the problems.

4. *Seek many alternatives*. Establish a goal for the number of ideas to be generated and work to achieve it. Again, it is emphasized that the preferred solution will come only from the set of alternative solutions generated.

5. *Avoid conservatism*. Do not restrict yourself to simple variations of the present solution. Think big.

6. *Avoid premature rejection*. This phase of the design process emphasizes the generation of alternative solutions, not the evaluation of these alternatives. Consequently, do not reject possible solutions, regardless of their apparent feasibility.

7. *Avoid premature satisfaction*. Do not terminate the search for alternative solutions when a seemingly good solution has been generated. Save your evaluation of the solutions until later.

8. *Refer to analogous problems for ideas*. Consult trade magazines, look around you, refer to recent architectural literature, and be aware of the large variety of layout solutions that surround you. Ideas for your layout can come from the layout of communities, offices, libraries, plants, stores, restaurants, homes, and schools, to list but a few.

9. *Consult others*. Actively seek information and suggestions from urban planners, engineers, workers, supervisors, and others. The more people actively involved in the generation of ideas, the greater the chance of (a) getting new and different solutions, (b) selling the final solution, and (c) stimulating your own thought processes.

10. *Attempt to divorce your thinking from the existing solution*. Employ the black-box approach repeatedly. Many of today's layouts are the product of evolution rather than careful design. As an illustration, consider the following anonymous poem given in [10]:

Path of the Calf

One day through the primeval wood
A calf walked home as good calves should
But made a trail all but bent askew,
A crooked trail as all calves do.
Since then three hundred years have fled,
And I infer the calf is dead.
But still he left behind his trail,
And thereby hangs my moral tale.
The trail was taken up next day
By a lone dog that passed that way;
And then a wise bell wether sheep
Pursued the trail o'er vale and steep,
And drew the flock behind him, too,
As good bell wethers always do.
And from that day o'er hill and glade
Through those old woods a path was made.

The years passed on in swiftness fleet,
The road became a village street;
And this, before men were aware,
A city's crowded thoroughfare.
And soon the central street was this
Of a renowned metropolis;
And men two centuries and a half
Trod in the footsteps of that calf.

Each day a hundred thousand rout
Followed this calf about
And o'er his crooked journey went
The traffic of a continent,
A hundred thousand men were led
By one calf near three centuries dead.
They followed still his crooked way,
And lost one hundred years a day;
For thus such reverence is lent
To well established precedent.

For men are prone to go it blind
Along the calf-paths of the mind,
And work away from sun to sun
To do what other men have done.

11. *Try the group approach*. This method, popularly known as brainstorming, involves a group of four or five or preferably more persons assembled for the stated purpose of generating layout solutions. No evaluation takes place. Instead, one attempts to maximize the number of solutions generated. The problem should be stated in a broad, unrestricted manner. For example, the actual problem may be that of devising a new and effective way of transporting emergency patients to surgery in a hospital. In this case the initial instructions to the group could be, "generate many ways of moving objects." After a period of brainstorming, this group could be instructed, "now generate many ways of moving emergency patients." By brainstorming a very general formulation first, followed by a more specific formulation, we can improve the benefits of the group approach.

12. *Remain conscious of the limitations of the mind in this process of idea generating*. If you are constantly aware of the tendency to impose fictitious restrictions, to be overly conservative, to accept prematurely and reject, and so on, you have already made an important step toward overcoming the tendencies that stifle creativity.

1.6.4 Evaluate the Design Alternatives

Once the set of alternative designs has been generated, an evaluation of the alternatives is performed. This phase of the design process consists of the measurement of the alternative solutions, using the criteria as the basis for comparison. If the criteria are easily quantified, the process of measurement is easily performed. However, in the following discussion we list a number of objectives that are qualitative rather than quantitative. Realistically, there usually exist multiple criteria that are to be used in evaluating alternative layouts. Some of the criteria probably will be quantifiable, but inevitably some will not be. Thus the process of evaluating alternative layout designs is not a simple one.

Among the various techniques used to evaluate alternative facility layouts are the following:

1. *List of pros and cons*. The simplest way to evaluate alternative layouts is by listing the advantages (pros) and disadvantages (cons) of each.
2. *Ranking*. Select the factors or considerations felt to be important to the layout, list them, and rank the alternatives in numerical order for each factor.
3. *Factor analysis*. Each factor is assigned a numeric weight and then each alternative is ranked against each factor. The weighted rankings are totaled for each alternative, and the alternative having the best score is chosen.
4. *Cost comparison*. All costs associated with each alternative are identified as well as cost savings produced, and that alternative which is most economical is chosen.

The listing of pros and cons is probably the easiest way to evaluate alternative layouts. However, it also is probably the least accurate. Its primary purpose is to allow an initial screening of those alternatives that have major deficiencies.

The ranking procedure has the property that all alternatives are compared against the same set of factors. The shortcomings of this method are that some important factors might be overlooked and that a final selection of the preferred design is not easily made. After using the ranking procedure, you still have to combine the rankings in some way so that a choice can be made. The factor analysis technique makes this process explicit by assigning weights to each factor. Since quantitative and qualitative factors can be included and the preferred solution is made explicit, factor analysis is a very popular method of evaluating alternative layouts.

Both the ranking procedure and the factor analysis procedure involve a comparison of layout alternatives for each factor. It is important that this comparison be made as objectively as possible. One approach that is used is the paired comparison technique, where all combinations of two layout alternatives are ranked for each factor. As an illustration, suppose that there are five layout alternatives (A, B, C, D, and E), and the paired comparison results for a given factor are

$$\begin{array}{ll} A > B, & B < D \\ A > C, & B < E \\ A < D, & C < D \\ A > E, & C < E \\ B < C, & D > E \end{array}$$

where $X > Y$ means X ranks higher than Y, and $X < Y$ means X ranks lower than Y. Combining these comparisons indicates that

$$D > A > E > C > B$$

for the factor considered. If there are n alternatives and m factors, then

$$\frac{mn(n-1)}{2}$$

comparisons are required.

The paired comparison technique is quite beneficial in testing for inconsistencies. As an example, suppose it is found that someone has provided the rankings

$$A > B, \qquad B > C, \qquad C > A$$

Obviously, there is an inconsistency in the rankings. Such inconsistencies might occur due to imprecise definitions of the factors or to a selection of factors that are too broadly defined.

Some of the most commonly involved factors or considerations are [1]

1. Ease of future expansion
2. Flexibility of layout
3. Material handling effectiveness
4. Space utilization
5. Safety and housekeeping
6. Working conditions
7. Ease of supervision and control
8. Appearance, promotional value, public or community relations
9. Fit with company organization structure
10. Equipment utilization
11. Ability to meet capacity or requirements
12. Investment or capital required
13. Savings, payout, return, profitability

A potential weakness of both the ranking procedure and the factor analysis procedure is the possibility of the *halo effect*. The halo effect refers to the situation in which the analyst ceases to act objectively and lets high rankings on a few factors for a particular layout alternative influence the rankings on a number of other factors. One can detect this by looking for an alternative that clearly dominates all others. If the search process was truly effective, there should not exist a dominant alternative. The design ultimately chosen will normally represent a compromise among the important factors.

A cost comparison of the layout alternatives involves a consideration of investment, operating, and maintenance costs. In making this comparison, one should make explicit the planning horizon (number of years) over which the alternatives are to be compared. An economic analysis can be performed using the time value of money and considering income tax effects. There are a number of excellent references that can be used as guides in performing this analysis. For example, see [3], [7], and [13].

1.6.5 Select the Preferred Design(s)

After evaluating the design alternatives, one or more are selected for recommendation to management. Depending on the management style of the decision maker, the designer might recommend only one design for consideration by management.

However, because of the multiplicity of objectives held by various members of management, it is often the case that the two or three "best" designs are presented, with the designer's recommendation given as to the one that should be selected.

Since the designer's value system might differ from management's value system, the designer's recommendation might not be selected. Hence *the designer should ensure that each alternative presented to management is acceptable in terms of the functions it is to support*. To minimize the risks of a technically unacceptable alternative being selected, the designer must understand management's selection process.

Unfortunately, management's selection process is best described as "the one that looks best!" Of course, what looks best to one person does not necessarily look best to another. At times, it appears that management believes the members of the board of directors will be flying over the plant in the corporate jet just at the time a tornado rips off the roof of the building. When the board members look down on the plant (sans roof), management wants to ensure that the plant "looks good!"

To facilitate the selection process, the designer might prepare three- or two-dimensional scale models of the layout. Alternatively, a computer might be used to display computer-aided design (CAD) representations of the layout alternatives. Examples of two- and three-dimensional representations of layout alternatives are given in Figures 1.2 and 1.3.

1.6.6 Specify the Solution(s)

The final phase in the design process involves specifying the preferred solutions. Actually a two-step process is involved. First, a management presentation is developed for selling the preferred solution. After the layout design has been chosen for implementation, a detailed specification of the design is prepared. Depending on the scope of the layout problem, a detailed set of drawings is used to specify the location of electric, sewage, gas, and water lines. These carefully prepared, detailed, dimensional drawings are often the main vehicle for documenting and communicating the layout solution to the contractors responsible for its implementation.

1.6.7 The Design Cycle

The specification of the solution might be the final phase of the design process, but it seldom represents the end of the designer's association with the layout problem. Still ahead are selling the solution, installing the design, observing and evaluating the design in use, and detecting the need for a redesign. These functions form the design cycle depicted in Figure 1.4. Implementation consists of *selling* all persons involved in the solution, *documenting* the procedures to be followed, *training* personnel in the performance of their duties, and *monitoring* the operation of the newly designed system.

It should be noted that Figure 1.4 is an idealized representation of the design cycle. The process seldom occurs sequentially without interruption or backtracking.

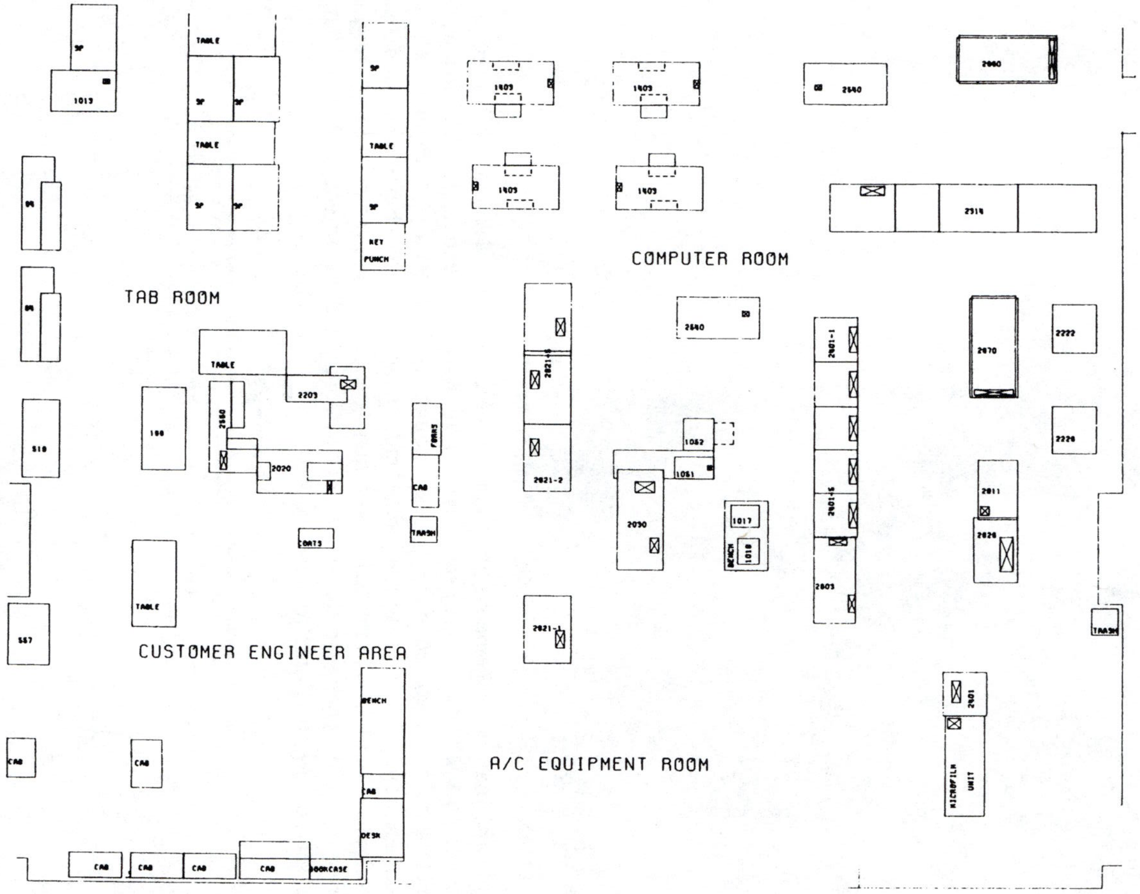

Figure 1.2 Two-dimensional iconic model. (Courtesy of AT&T.)

Figure 1.3 Three-dimensional iconic model.

In particular, management's involvement should not be delayed until after the preferred solution(s) is(are) specified. To the maximum extent possible, management should be involved in the process throughout. Furthermore, the step of selling management generally results in one or more new alternatives being generated and evaluated. In fact, the interaction with management at the time of "selling" might even result in a change in the mission of the facility and the objective of the layout; in such a case, it is time to start over!

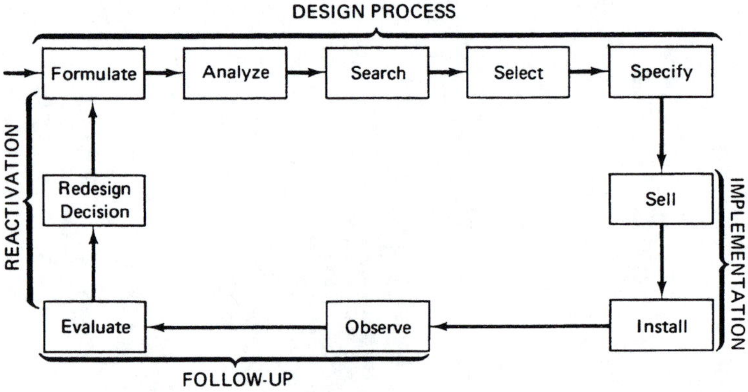

Figure 1.4 Design cycle.

Many excellent layout designs have been developed that were never implemented. Undoubtedly, there are many reasons for this. Layout designers tend to be problem solvers. As such, many bring a great deal of enthusiasm to the analysis until, to them, the problem is solved. Typically, such persons turn in their recommendations to management with a statement to the effect that they will be available to answer any questions concerning the proposed layout; until then, they will be "solving the next problem."

Another reason that layout solutions are not implemented is a poor job of selling. Not only must top, middle, and lower management be sold on the solution, but more critically, the individual employee must be sold. Management may *make* the decisions, but the employee can *break* the decision if he or she is not sold on it.

A third reason that implementation fails is that a good layout designer is often in high demand and will have a backlog of requests to be met. As such, management may pressure the designer to let someone else handle the implementation in order that the designer be free to tackle other problems. Such an approach is very shortsighted and will eventually lead to dissatisfaction with the work of the designer (and a subsequent perusal of the want ads).

One of the major aids in selling a layout is the iconic model of the layout. Admittedly, we have deemphasized the role of the iconic model in developing alternative solutions, but its importance in aiding visualization of the new layout cannot be overemphasized. A two- or three-dimensional representation of the proposed layout can go a long way toward selling the manager on the design. Along with the iconic model, clear plastic overlays are often used to show product flow lines, alternative designs, expansion plans, service lines, and so on.

The documentation of the procedures to be followed with the new layout is essential for successful implementation. Recommended flow lines should be specified, the material handling procedures must be detailed, and any other information pertinent to the new layout design must be documented.

If new procedures are developed to facilitate the layout design, the personnel affected should be properly trained to carry out these procedures. New rules for moving items in and out of storage, new methods of handling materials, and new flow patterns can result from a change in the layout. Unless the employees affected are able to adjust to these changes, the potential benefits to be derived from the new layout may never be realized.

In addition to seeing that the design is installed, control procedures should be devised to monitor the production system. Management information systems should be designed to indicate when a re-layout study should be conducted. One should make periodic checks for the presence of the problem indicators cited earlier. Also, one should compare the performance of the layout with that which was predicted. If serious discrepancies exist between the two, adjustments should be made in the methods of predicting system performance.

If you compare the steps of the design process with the sequence of steps suggested in a number of the more traditional texts on plant layout, it is apparent that traditional approaches often begin with the analysis of the problem. Consequently, when the traditional approach is followed, one assumes that the correct problem has been formulated. The dangers inherent in this approach should be obvious. We strongly recommend that careful consideration be given to the formulation of the problem before analysis begins.

Recall, the problem should be formulated as broadly as the organizational boundaries and economics permit. One of the major prerequisites to successful layout design continues to be an understanding of how the organization functions and how to work within the organization to bring about desired changes.

In much of our subsequent discussion we employ analytical approaches in solving facility layout and location problems. It is emphasized that analytical approaches are used to stimulate the search phase of the design process. Even though normative models are employed, a number of relevant nonquantifiable factors have been ignored. Consequently, in many situations the solutions obtained from the models must undergo separate evaluations by taking into consideration all relevant criteria.

1.7 CLASSIFICATION OF FACILITY LAYOUT AND LOCATION PROBLEMS

As noted previously, facility layout and location problems arise in a variety of contexts. They also differ significantly in their characteristics. Because of the multiplicity of facility layout and location problems that have been addressed in the research literature, it is useful to define a number of categories that can be used to classify facility layout and location problems. In classifying facility location problems, six major elements are to be considered: new facility characteristics, existing facility locations, new and existing facility interactions, solution space characteristics, distance measure, and the objective. These elements are depicted in Figures 1.5 to 1.10.

Facility location problems can be classified according to the number of new facilities involved. Furthermore, the new facilities can be considered to occupy either point locations or area locations. In a number of facility location problems, the number of new facilities is a decision variable, rather than being given. Finally, the location of a new facility can be either dependent on or independent of the locations of the remaining new facilities. If area locations are to be considered for both new and existing facilities, the facility location problem is often classified as a facility layout problem, in which case the size and the configuration of the area location are frequently decision variables. In the facility layout problem, the facilities might be plants, offices, warehouses, and so on.

The location of an existing facility can be either static or dynamic, as well as deterministic or probabilistic. Additionally, depending on the size of an existing facility, it can be considered to have a point location or an area location. Again, a special class of facility location problems involving area locations is the facility layout problem. When existing facilities are included in the layout problem a re-layout problem can exist, in which case the configuration and locations of the existing facilities

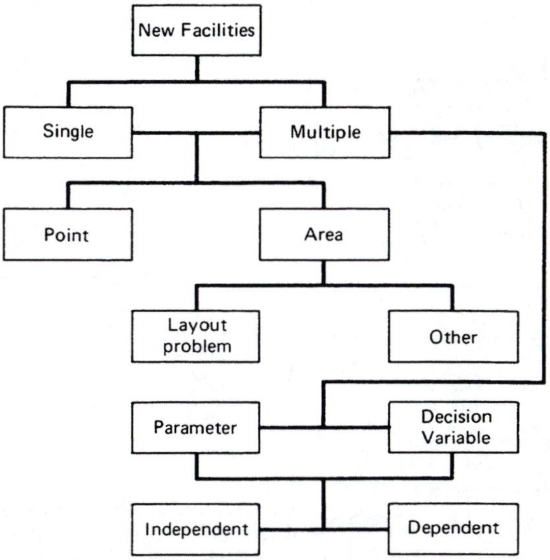

Figure 1.5 Classification of facility layout and location problems—new facility characteristics.

must be determined. A *re-layout* or *relocation* problem arises when the location of an existing facility is a decision variable.

When there exists a quantitative or qualitative relationship between facilities, we consider the facilities to interact. In some cases the degree of interaction is a function of the locations of the facilities. Of course, there are a number of situations in which such interaction is independent of the facility locations. Furthermore, the magnitude of the interaction can be either static or dynamic, either deterministic or probabilistic, and either a parameter or a decision variable.

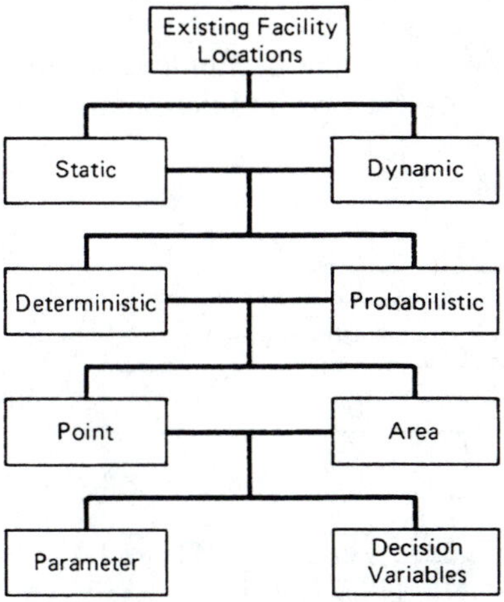

Figure 1.6 Classification of facility layout and location problems—existing facility locations.

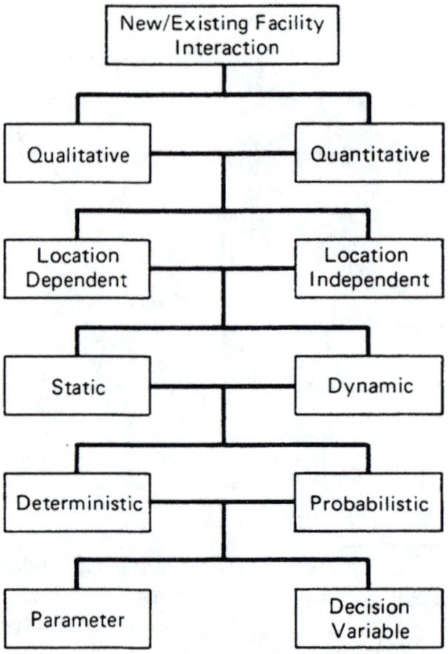

Figure 1.7 Classification of facility layout and location problems—new and existing facility interactions.

The fourth category considered in classifying a facility location problem concerns the solution space for the location problem. In some facility location problems the solution space is one dimensional; however, more commonly a two- or three-dimensional solution space exists. Additionally, the solution space can be either discrete or continuous. Typically, the discrete solution space consists of a finite number of possible locations, whereas a continuous solution space consists of an infinite number of possible locations. In either case, the solution space can be further restricted by one or more constraints. A discrete layout involves discrete modules, such as bays in a warehouse, whereas a continuous layout considers the modules to be points.

The distance measure involved in the facility location problems provides another basis for classifying facility location problems. In subsequent discussions,

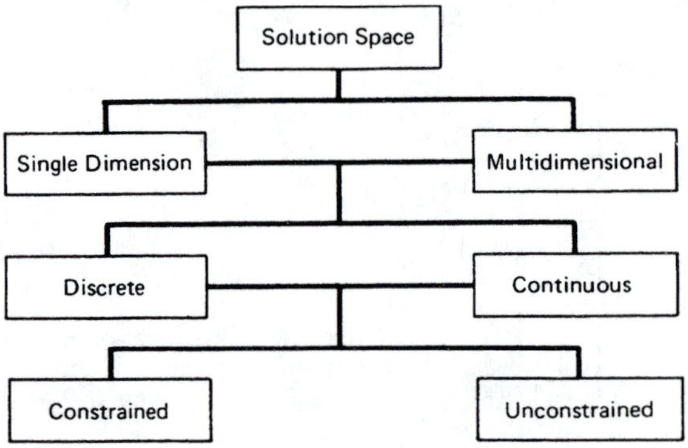

Figure 1.8 Classification of facility layout and location problems—solution space characteristics.

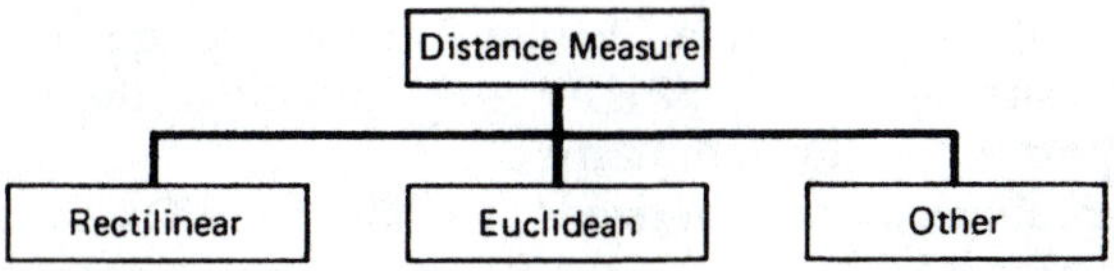

Figure 1.9 Classification of facility layout and location problems—distance measure.

rectilinear and Euclidean distances are commonly used. However, there do exist a number of facility location problems in which the distances between facilities cannot be reasonably represented as either rectilinear or Euclidean distances. An illustration is an urban location problem in which streets do not form a rectangular grid.

The final category commonly used to classify facility location problems concerns the objective function employed to evaluate alternative solutions. In the case of facility *layout* problems a number of qualitative objectives are commonly used, whereas quantitative objectives are used to solve facility *location* problems. The quantitative objectives normally encountered in facility location problems are the minimization of some total cost function or the minimization of the maximum cost between pairs of facilities. Of course, there can be other quantitative objectives, but these two occur most frequently in the literature.

Although some facility location problems might not be completely classified using the categories we have suggested, the classification scheme serves to unify a growing body of literature treating the subject of facility layout and location. It might be instructive to refer to Figures 1.5 to 1.10 as we consider in detail in subsequent chapters a number of different facility layout and location problems.

1.8 OVERVIEW OF THE BOOK

As an introduction to facility layout and location, we examine the traditional subject of plant layout in Chapter 2. The discussion in Chapter 2 is based on the systematic layout planning (SLP) approach. Using the steps of SLP as a base, we treat the

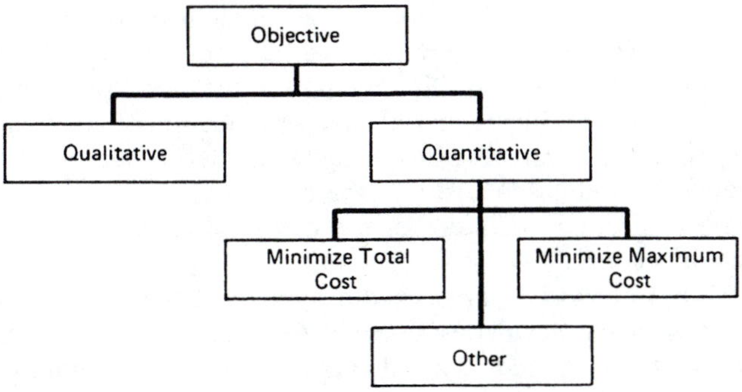

Figure 1.10 Classification of facility layout and location problems—the objective.

subject of plant layout briefly. The discussion in Chapter 2 emphasizes qualitative considerations in plant layout. Although Chapter 2 is focused primarily on the design problem, it does identify opportunities for optimization.

In Chapter 3 we treat the subject of computerized layout planning. The chapter is organized into two major sections: one describing the efforts to develop normative computer-aided design (CAD) models and one that addresses descriptive CAD models.

Chapter 4 treats single-facility location problems. We assume that the single new facility is to be located in a continuous region; we further assume that the new facility has distance-related relationships with one or more existing facilities. We consider a number of distance metrics, as well as several different criteria, in the chapter. The chapter treats the situation in which a particular device or resource is to be located in a plant or a particular facility is to be located within a region.

Chapter 5 provides a coverage of storage systems design. Specifically, optimization models are presented that are appropriate in designing warehouse layouts and in sizing warehouses. The treatment of storage systems design addresses the determination of the amount of storage space to provide, the assignment of products to storage locations, and the storage configuration to be used.

We treat a number of other location problems in Chapter 6. All of the problems involve the location of one or more new facilities in a continuous region. By and large, the models presented are generalizations of the models presented in Chapter 4. Additionally, some simple location-allocation problems are considered. Thus Chapter 4 is a prerequisite for Chapter 6.

The location of one or more new facilities on planar networks and trees is the subject of Chapter 7. In a sense, the problems treated are constrained versions of those treated in Chapters 4 and 6.

Discrete location problems are covered in Chapter 8. We consider a number of integer programming formulations of location problems, including set covering problems and the plant location problem.

Our coverage of discrete location problems is continued in Chapter 9, with a consideration of discrete location assignment problems. Among the discrete location problems we address is the quadratic assignment problem.

In case you have not noticed, we recommend strongly the use of analytical approaches to "solving" facility layout and location "problems." We believe that solutions to such problems should have a foundation of analysis. Although intuition and experience are valuable in solving facility layout and location problems, we do not believe they should constitute the entire basis for a solution. Furthermore, we advocate you follow the steps of the design process in solving facility layout and location problems.

We have already emphasized a number of times that many of the analytical approaches which we will describe will yield benchmark designs against which other designs can be compared. We will make this point many more times throughout the text, since we believe that it is quite important for you to adopt this view of analytical design aids. Since a number of qualitative, as well as quantitative, factors are not normally included in the analytical model, the solution obtained from the model will usually be modified based on a number of considerations not accounted for explicitly in the model.

In a sense, the design obtained from analysis is an ideal solution. Our philosophy is that it is better to begin with an idealized solution and modify it based on real-world considerations than to attempt to develop the preferred solution directly. In subsequent discussions we shall repeat our philosophy, since it is important that you place in proper perspective the use of analytical approaches in solving facility layout and location problems.

REFERENCES

1. Apple, J. M., *Plant Layout and Material Handling*, 2nd ed., The Ronald Press Company, New York, 1963.
2. Canada, J. R., and J. A. White, *Capital Investment Decision Analysis for Management and Engineering*, Prentice Hall, Englewood Cliffs, NJ, 1980.
3. DeGarmo, E. P., and J. R. Canada, *Engineering Economy*, 5th ed., The Macmillan Company, New York, 1973.
4. Elmaghraby, S. E., *The Design of Production Systems*, Van Nostrand Reinhold Company, Inc., New York, 1966.
5. Elmaghraby, S. E., "The Role of Modeling in IE Design," *Journal of Industrial Engineering*, Vol. 19, No. 6, 1968, p. 292.
6. Francis, R. L., and J. A. White, *Facility Layout and Location: An Analytical Approach*, Prentice Hall, Englewood Cliffs, NJ, 1974.
7. Grant, E. L., and W. G. Ireson, *Principles of Engineering Economy*, The Ronald Press Company, New York, 1970.
8. Krick, E. V., *An Introduction to Engineering and Engineering Design*, John Wiley & Sons, Inc., New York, 1965.
9. Krick, E. V., *Methods Engineering*, John Wiley & Sons, Inc., New York, 1962.
10. Lehrer, R. N., *Work Simplification (Creative Thinking about Work Problems)*, Prentice Hall, Englewood Cliffs, NJ, 1957.
11. Nadler, G., *Work Design: A Systems Concept*, Richard D. Irwin, Inc., Homewood, IL, 1961.
12. Simon, H. A., "Style in Design," in C. M. Eastman, ed., *Spatial Synthesis in Computer-Aided Building Design*, John Wiley & Sons, Inc., New York, 1975.
13. Thuesen, H. G., W. J. Fabrycky, and G. J. Thuesen, *Engineering Economy*, 4th ed., Prentice Hall, Englewood Cliffs, NJ, 1971.
14. Tompkins, J. A., and J. A. White, *Facilities Planning*, John Wiley & Sons, Inc., New York, 1984.
15. Vollmann, T. E., and E. S. Buffa, "The Facilities Layout Problem in Perspective," *Management Science*, Vol. 12, No. 10, June 1966, pp. 450–468.
16. White, J. A., M. H. Agee, and K. E. Case, *Principles of Engineering Economic Analysis*, 3rd ed., John Wiley & Sons, Inc., New York, 1989.

PROBLEMS

1.1. Suppose that you are asked to recommend the location for a hospital that is to be placed in a major city within the state. What criteria would be appropriate in evaluating alternative sites?

1.2. Suppose that you are asked to recommend the number, sizes, and locations for storerooms in an industrial plant. What criteria would be appropriate in evaluating alternative solutions to the problem?

1.3. List six facility location problems and classify them using the scheme provided in Figures 1.5 to 1.10.

1.4. Using the black-box approach, provide alternative formulations of states A and B for Problems 1.1 to 1.3.

1.5. Perform a literature search and obtain the titles, authors, and other bibliographic data for 10 recent facility layout and location papers not included in the list of references given at the end of Chapters 2 to 10.

1.6. Locate ten professional journals which contain articles dealing with facility layout and location problems.

1.7. Prepare a written critique of the Vollmann-Buffa paper [15].

1.8. Suggest alternative methods of monitoring the operation of a newly designed system to ascertain when a new design is justified.

1.9. Suppose that you are asked to re-layout the library at your university, plant, or town. Formulate the problem using the black-box approach, giving alternative formulations having varying degrees of breadth. List the criteria to be used in evaluating alternative layout designs.

1.10. List a number of common iconic, analog, and symbolic models that are used in your daily activity.

Chapter 2

The Plant Layout Problem

2.1 INTRODUCTION

In this chapter we treat the classical plant layout problem. Although our focus is on the development of layouts for manufacturing facilities, the discussion is easily extended to such nonmanufacturing contexts as offices, banks, hospitals, shopping centers, post offices, schools, restaurants, warehouses, recreation centers, amusement parks, libraries, and military bases, among others.

Perhaps a more appropriate title for the chapter would be "facility layout." However, the term *plant layout* is so widely used in practice that we continue to use it. As expected, the name is derived from the act of designing *layouts* for production *plants*.

The process of developing plant layouts contains elements of both art and science. The artist's dependence on creativity, synthesis, and style is very evident in designing plant layouts. Similarly, the scientist's use of *analysis*, *reduction*, and *deduction* are essential in designing plant layouts.

As noted in Chapter 1, because the plant layout problem is a *design problem*, it is fundamentally different from an optimization problem. Furthermore, solutions to the plant layout problem depend heavily on the use of *synthesis*, rather than resulting directly from *analysis*. These distinctions are important and should not be treated lightly.

Since in this book we emphasize the use of analysis in solving facility layout and location problems, why include a coverage of plant layout, which depends more heavily on synthesis than analysis? In a sense, analysis is a process of dissecting, whereas synthesis is a process of combining or creating. Where analysis might generate numbers, synthesis looks between and beyond the numbers. Although synthesis is different from analysis, the process of synthesis can be enhanced through the use of analysis. Through the use of synthesis, the results of analysis can be used to "build" a solution that has never before been built.

There are many "fuzzy" criteria involved in selecting the preferred layout design. Both qualitative and quantitative considerations enter into the selection process. As Simon [49] would put it, designers tend to "satisfice" rather than "optimice"; in particular, a design is sought that satisfies the decision maker's expectations (i.e., it performs at or above minimum acceptable levels of all criteria).

In Chapter 1 we noted that both the facility layout problem and the facility location problem are design problems. Further, in some contexts the layout problem can be formulated as an "area location problem," where the relative shapes and locations of individual "departments" are to be determined. Recall that we defined the facility layout problem to include the joint determination of the locations of multiple activities (departments, manufacturing cells, workstations, etc.), as well as their sizes and configurations. However, it should be noted that the configuration aspects of the layout problem are such that few similarities exist in the way in which layout problems and location problems are solved.

Unfortunately, no algorithm exists for solving the plant layout problem per se. Instead, the final plant layout is built up from a large number of selective, cumulative, tentative decisions. In making these decisions, models can provide insight, suggest solutions, or provide assessments of alternatives. The plant layout problem embodies both design and analysis components, as do problems involving the design of distribution systems, transportation systems, and health care delivery systems.

In this chapter we focus specifically on the *design* aspects of the layout problem, including the emphasis placed on qualitative considerations. Consequently, the material presented in this chapter will be quite different from that presented in subsequent chapters. Specifically, a stronger emphasis is placed on analytical modeling in subsequent chapters. However, the results obtained from analytical models can be used in synthesizing the final plant layout. (While the content of the chapter varies from that in subsequent chapters, note that the plant layout problem is not the only location/layout problem for which qualitative considerations apply. Much of the focus of this chapter applies to a broad range of location/layout design problems.)

Here we provide a qualitative overview of the entire subject of plant layout. Our reasons for doing so are twofold. First, as we emphasized in Chapter 1, qualitative considerations are very important in solving real-world problems. Consequently, we want to recognize those qualitative aspects of facility layout and location that we believe to be worthy of consideration. Second, we wish to place the remainder of the book in context within the overall design process.

As a member of the family of design problems, the plant layout problem relies heavily on the use of synthesis and is amenable to the use of the engineering design process. Thus the layout design process consists of:

1. Formulating the layout design problem
2. Analyzing the design problem
3. Searching for alternative layout designs
4. Evaluating the layout design alternatives
5. Selecting the preferred design
6. Specifying the layout design to be installed

The synthesis aspect of design is an essential component of the search phase of the design process. Many of the analytical approaches described in subsequent chapters can be useful to the designer in searching for solutions to facility layout design problems.

In addition to the primary objective of treating the plant layout problem, in this chapter we relate the plant layout problem to a number of other problem areas in an industrial setting. Also, we indicate a variety of analytical approaches that are applicable to each problem area.

In striving to achieve our stated objectives, the chapter is organized as follows. First we discuss, in very general terms, the overall plant layout problem. Then we consider a number of different approaches that have been presented for solving plant layout problems. Next, we provide a more in-depth treatment of one particular approach, systematic layout planning, developed by Muther [36]; it serves as the basis for the balance of our discussion in this chapter. Our consideration of the plant layout problem concludes with a recommendation concerning the position of analytical approaches in layout design.

Depending on the objectives of the course, as well as your objectives, the subject matter of this chapter might not be emphasized as heavily as subsequent chapters. If such is the case, we recommend that you read at least the following sections: 2.4, 2.5, 2.6.3, 2.6.4, 2.7, and 2.9. In particular, the material presented on systematic layout planning and from–to charts is referred to a number of times in later discussions.

2.2 OVERVIEW OF THE PLANT LAYOUT PROBLEM

Defining the boundaries on a plant layout problem is no easy undertaking. In fact, the boundaries are different from one problem to the next. For example, the layout analyst might be called on to determine the location for a new machine, and then the next assignment might be to develop the layout for a new production plant. Consequently, a tremendous variety in the types of layout problems can be encountered. This variety is due, in part, to the number of ways in which plant layout problems develop. For example, a layout problem might arise because of a change in the design of a product, the addition or deletion of a product from a company's product line, a significant increase or decrease in the demand for a product, changes in the design of the process, the replacement of one or more pieces of equipment, the adoption of new safety standards, organizational changes within the company, or a decision to build a new plant. Plant layout problems may also develop because of gradual changes over time that finally manifest themselves in terms of bottlenecks in production, crowded conditions, unexplainable delays and idle time, backtracking, poor housekeeping, excessive temporary storage space, obstacles to materials flow, failure to meet schedules, and a high ratio of material handling time to production time.

Thus we see that the plant layout designer might be called on to interact with the product designer, the process designer, and the schedule designer. Furthermore, it is clear that the layout problem can be a very complex systems design problem requiring the most sophisticated systems analysis and design tools in order to develop satisfactory layout solutions.

Plant layout problems can occur in a large number of ways and can have significant effects on the overall effectiveness of the production system. Therefore, it is highly desirable that the optimum plant layout be designed. Unfortunately, the magnitude of the problem is so great that true system optimization is beyond our current capabilities. The approach normally taken in solving the plant layout problem is to try to find a "satisfactory" solution employing the *component approach*. We define the overall system as a collection of components or subsystems and attempt to obtain "optimum" solutions for the subsystems. Granted, the resulting system will be a suboptimum solution rather than an optimum solution. However, the solution is believed to be better than that which would otherwise be obtained.

In using the component approach there is the danger of developing a solution for one component that is detrimental to the overall system. To minimize this possibility, it is necessary to coordinate the product, process, schedule, and layout design decisions, as depicted in Figure 2.1. In searching for a plant layout design that "satisfices," one must agree on the basis for evaluating alternative designs. As mentioned in Chapter 1, one of the criteria commonly used to evaluate alternative layouts is that of material handling cost. It was also emphasized that for many situations this may not be an appropriate criterion. Typically, one has a number of goals that are important. For example, some of the objectives of the plant layout study may be to:

1. Minimize investment in equipment.
2. Minimize overall production time.
3. Utilize existing space most effectively.
4. Provide for employee convenience, safety, and comfort.
5. Maintain flexibility of arrangement and operation.
6. Minimize material handling cost.

7. Minimize variation in types of material handling equipment.
8. Facilitate the manufacturing process.
9. Facilitate the organizational structure.

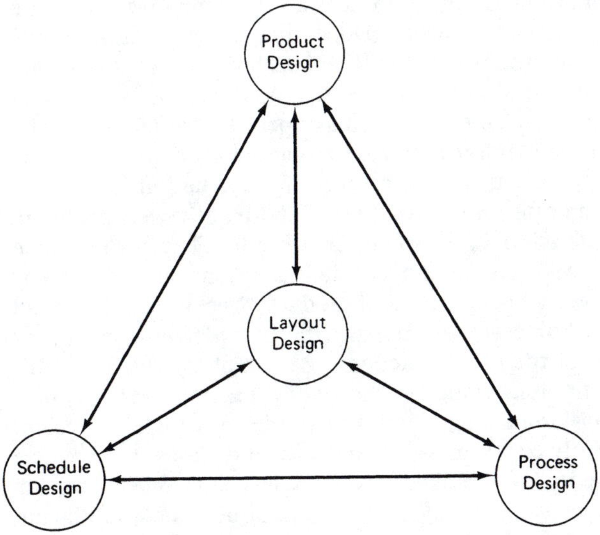

Figure 2.1 Communication links among product, process, schedule, and layout design.

In addition to there being a variety of objectives that can be used to guide the analyst in solving the plant layout problem, there may also exist a number of constraints on the solution. For example, there exist local, state, and federal governmental restrictions on allowable noise levels. This, along with other standards concerning aisle widths, ventilation, temperature, and lighting, can affect the solution to the layout problem. Building geometry can serve as a restriction on the layout. In the case of an existing building, the layout might be required to fit within the present building. In the case of a new building, the building site can restrict the shape of the building and, consequently, the layout. When there exists a building within which the layout is to be designed, there can exist a large number of restrictions on the solution. For example, the layout solution will be affected by the present location of walls and columns, equipment, footings to support heavy equipment, loading docks, windows, lights, ventilating equipment, storage and office areas, and sewage, water, and power lines. Some of these can be relocated. However, it should be remembered that whenever an existing layout is to undergo a revision, the analyst must consider the costs of relocating existing facilities together with the advantages derived from the relocation.

Over the years a number of rules of thumb have been developed as guides to a good plant layout. A number of these have been assembled in the form of checklists. Checklists serve the useful purpose of forcing the designer to examine a number of important aspects and details pertinent to the design of a successful plant layout. A sample checklist is given in Figure 2.2. Apple [1] and Muther and Hales [37,38], among others, provide a number of checklists.

It is rarely the case that one person is involved in carrying out all of the analyses involved in solving the plant layout problem. Rather, different people at different levels within the organization are involved. Also, it is not uncommon to find that a "layout team" is formed to design the layout. Depending on the organizational structure of the company, it may well occur that the plant layout analyst does not have the freedom to participate in a portion of the decision-making process that establishes the boundaries within which he or she must operate. Not uncommonly, the product design, process design, and schedule design decisions have been made previously, and the layout analyst must make the best of the situation.

In this chapter we concentrate on the solution of the plant layout problem, assuming certain preliminary decisions have already been made. That is, we assume that a number of product, process, and schedule design decisions have previously been made. As shown in Figure 2.1, the functions of product, process, schedule, and layout design should interact. Some authors portray the plant layout function as encompassing a large portion of the other functions. We admit that we are sufficiently biased in our view of the importance of the layout function to lobby for a larger role in the overall decision-making process in production systems design. However, we are also sufficiently cognizant of organizational realities to recognize the facts of life concerning the role of the plant layout analyst. It is unusual for the plant layout analyst to influence significantly product design decisions; the layout analyst can normally exert some influence on process design decisions; and the layout analyst can usually have a significant influence on schedule design decisions. Furthermore, it is quite common to find that a person who is responsible for facilities planning and design is also responsible for carrying out other duties.

2.3 LAYOUT PROCEDURES

A number of procedures have been developed to facilitate the design of plant layouts. We consider briefly a number of the more significant contributions to the field. Specifically, we contrast the ideal systems approach by Nadler [39] with the layout planning steps recommended by Immer [20], Apple [1], and Reed [43].

2.3.1 Nadler's Ideal Systems Approach

Although it was developed initially for designing work systems, Nadler's ideal systems approach [39] is also applicable for designing plant layouts. Although more a philosophy than a procedure, the ideal systems approach is based on the following hierarchical approach toward design:

1. *Aim* for the "theoretical ideal system."
2. *Conceptualize* the "ultimate ideal system."
3. *Design* the "technologically workable ideal system."
4. *Install* the "recommended system."

Worksheet No. 3 — Sheet 1

MATERIALS HANDLING AUDIT CHECK SHEET

Dept. ______________ Building ______________ Plant ______________

Date ______________ Surveyed by ______________

Conditions indicating possible productivity improvement opportunities	Condition exists here (✓)	To correct this, we need: Supervisor attention (✓)	Management attention (✓)	Analytical study (✓)	Capital Investment (✓)	Other (for comments)
1. Delays in material moving						
2. Excessive material on hand						
3. Production equipment idle for material shortage						
4. Long hauls						
5. Cross traffic						
6. Manual handling						
7. Outmoded handling equipment						
8. Inadequate handling equipment						
9. Insufficient handling equipment						
10. Unbalanced sequence of operations						
11. Idle handling equipment						
12. Obstacles to materials flow						
13. Materials piled directly on floors						
14. Poor work place layout for materials						
15. Disorderly storage						
16. Cluttered aisles						
17. Cluttered work space						
18. Crowded dock space						
19. Motor truck and railroad car tieup						
20. Manual loading techniques						
21. Excessive wasted "cube" in storage						
22. Excessive aisles						
23. Operations unduly scattered						
24. Poor locations of service areas						
25. Lack of in-plant container standardization						
26. Lack of unit load technique						
27. Excessive MH equipment maintenance cost						
28. Rehandling						
29. Handling done by direct labor						
30. Operators traveling for supplies, materials						
31. Supplies moved by poor techniques						
32. High indirect payroll						
33. Materials waiting for papers						
34. Excessive demurrage						
35. Unexplained delays						

Figure 2.2 Sample checklist. (From [61] with permission.)

Worksheet No. 3 — Sheet 2

MATERIALS HANDLING AUDIT CHECK SHEET (cont'd)

Conditions indicating possible productivity improvement opportunities	Condition exists here (✓)	To correct this, we need: Supervisor attention (✓)	Management attention (✓)	Analytical study (✓)	Capital Investment (✓)	Other (for comments)
36. Idle labor						
37. Inspection not properly located						
38. Excessive scrap						
39. Hazardous lifting by hand						
40. Misdirected materials						
41. Clumsy, dangerous "home made" handling rigs						
42. Lack of standardization on handling equipment						
43. Long travel distances for materials, equipment, and personnel						
44. Backtracking of materials						
45. Non-standard process routing						
46. Opportunity for group technology layout						
47. Opportunity for product layout						
48. Opportunity for process layout						
49. No real time dispatching of equipment						
50. No modular MH system						
51. No modular work stations						
52. Automatic identification system not used						
53. No one-way aisles						
54. MH equipment running empty						
55. Different things treated same						
56. Excessive trash removal						
57. Centralized storage						
58. Decentralized storage						
59. No incentive system for MH labor						
60. Low usage of automated MH equipment						
61. Variable path equipment used for fixed path handling						
62. System not capable of expansion and/or change						
63. Low usage of industrial robots						
64. No parts preparation performed prior to manufacturing						
65. No pre-kitting of work						
66. Lack of automated loading/ unloading of trailers						
67. Poor MH at the work station						
68. Lack of industrial truck attachments and below hook lifters						
69. Equipment capacity not matched to load requirement						

Figure 2.2 *(Continued)*

As shown in Figure 2.3, the ideal systems approach is a top-down approach. The focus is the opposite of that typically used in practice, where the layout designer concentrates initially on the present method rather than aiming for the theoretical ideal. We believe it important to focus initially on "what can be" instead of "what has been."

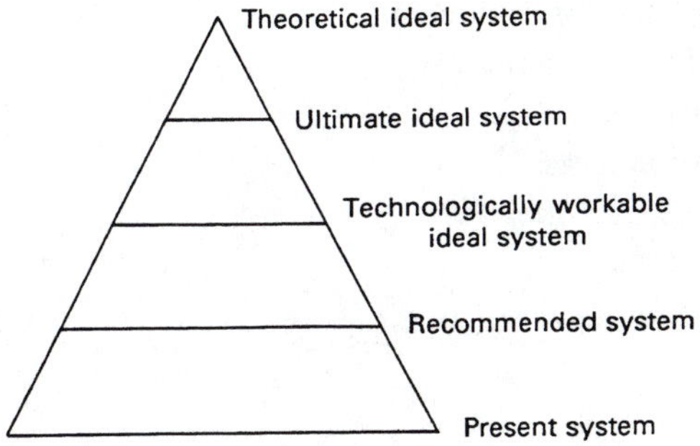

Figure 2.3 Ideal systems hierarchy. (From [39].)

2.3.2 Immer's Basic Steps

In an early book on the subject, Immer [20] described the analysis of a layout problem as follows: "This analysis should be composed of three simple steps, which can be applied to any type of layout problem. These steps are:

1. Put the problem on paper.
2. Show lines of flow.
3. Convert flow lines to machine lines" [20, p. 14].

Immer quoted Mallick and Sansonetti as follows: "Good plant layout, as defined at Westinghouse Electric Corporation, means (a) placing the right equipment, (b) coupled with the right method, (c) in the right place, (d) to permit the processing of a product unit in the most effective manner, (e) through the shortest possible distance, and (f) in the shortest possible time. The importance of good layout as a factor in insuring low-cost production is thus well established, and its need is well understood" [25, p. 102].

2.3.3 Apple's Plant Layout Procedure

Apple [1] recommended that the following detailed sequence of steps be used in designing a plant layout.

1. Procure the basic data.

2. Analyze the basic data.
3. Design the productive process.
4. Plan the material flow pattern.
5. Consider the general material handling plan.
6. Calculate equipment requirements.
7. Plan individual work stations.
8. Select specific material handling equipment.
9. Coordinate groups of related operations.
10. Design activity relationships.
11. Determine storage requirements.
12. Plan service and auxiliary activities.
13. Determine space requirements.
14. Allocate activities to total space.
15. Consider building types.
16. Construct master layouts.
17. Evaluate, adjust, and check the layout with the appropriate persons.
18. Obtain approvals.
19. Install the layout.
20. Follow up on implementation of the layout.

2.3.4 Reed's Plant Layout Procedure

In "planning for and preparing the layout," Reed [43] recommended that the following steps be taken in his "systematic plan of attack":

1. Analyze the produce or process to be produced.
2. Determine the process required to manufacture the product.
3. Prepare layout planning charts.
4. Determine work stations.
5. Analyze storage area requirements.
6. Establish minimum aisle widths.
7. Establish office requirements.
8. Consider personnel facilities and services.
9. Survey plant services.
10. Provide for future expansion.

The layout planning chart incorporates the following: operations, transportations, storage, and inspection; standard times for each work element; machine selection and balance; labor selection and balance; and material handling requirements.

2.4 SYSTEMATIC LAYOUT PLANNING

The layout planning procedures considered in Section 2.3 vary in their degree of specificity. However, they are alike in their emphasis on the design aspects of layout planning. Taken together, and coupled with the engineering design approach, a comprehensive layout planning approach would emerge.

Over the years, the most popular approach used in designing plant layouts has been the *systematic layout planning* approach developed by Muther [36]. Referred to as SLP, the procedure has been applied to production, transportation, storage, supporting services, and office activities, among others. The SLP procedure is depicted graphically in Figure 2.4. We see that once the appropriate information is gathered, a flow analysis can be combined with an activity analysis to develop the relationship diagram. Space considerations, when combined with the relationship diagram, lead to the construction of the space relationship diagram. Based on the space relationship diagram, modifying considerations, and practical limitations, a number of alternative layouts are designed and evaluated. In comparison with the steps of the design process, we see that SLP begins after the problem is formulated. The first five steps of SLP involve the analysis of the problem. Steps 6 through 9, including the generation of alternative layouts, constitute the search phase of the design process. The selection phase of the design process coincides with step 10 of SLP.

Our subsequent discussion will be based on the steps of SLP depicted in Figure 2.4. In particular, in Section 2.5 we examine the data-gathering step of SLP. In Section 2.6 we discuss flow analysis and activity analysis. The development of the relationship diagram is treated in Section 2.7. A consideration of space requirements and space availability takes place in Section 2.8. In Section 2.9 we consider the design of the layout. The discussion in Section 2.9 includes steps 6 through 9 of SLP. In Section 2.10 we consider the selection, specification, implementation, and follow-up steps in the design cycle. Finally, in Section 2.11 we take a look at the remainder of the book as it relates to SLP.

2.5 INFORMATION GATHERING

For the plant layout designer to perform effectively, certain information is required pertaining to the product, process, and schedule. Data requirements may not always coincide with data availabilities. However, we shall assume that such data exist; otherwise, the necessary management information system to obtain the data must be developed.

Data regarding product design decisions can affect the layout significantly. However, the major effect of product design decisions is felt by the process designer. Whether a part is made of aluminum or plastic will influence processing decisions, which ultimately affect the layout. Thus product design decisions can indirectly affect the layout. Product design decisions can also have a direct effect on the layout, since

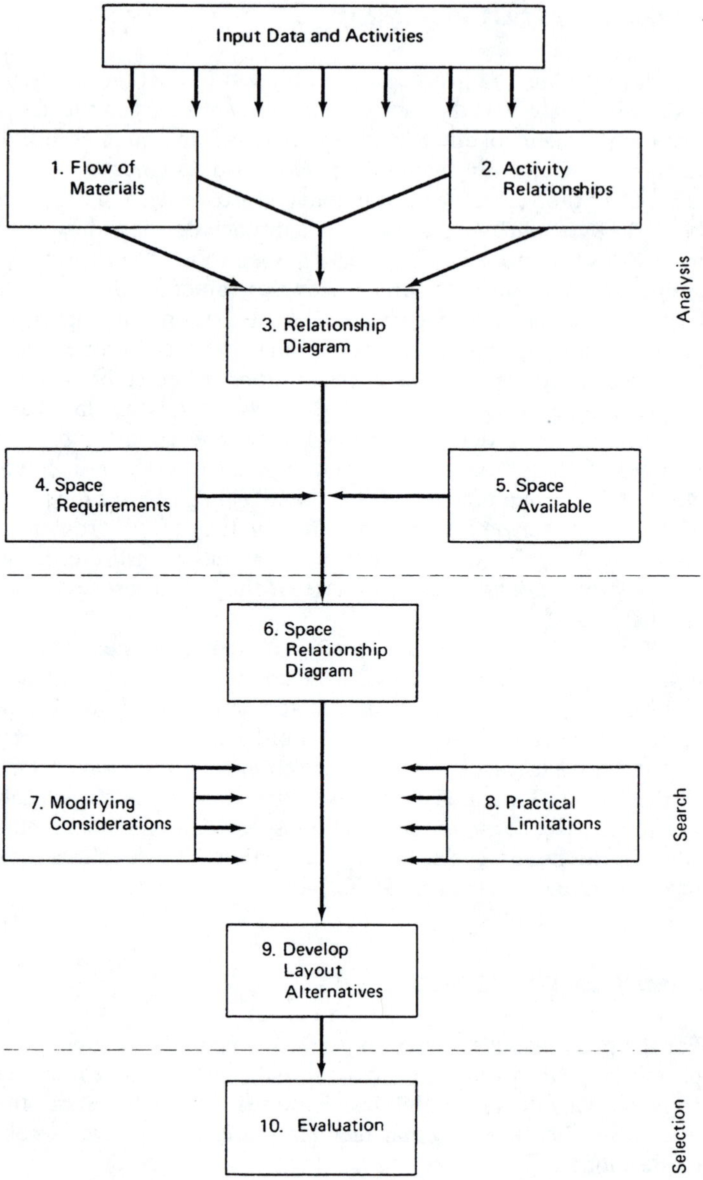

Figure 2.4 Systematic layout planning procedure.

the design of the product affects the sequence of assembly operations and the sequence can influence the layout. Thus it is important to have data available concerning the design of the product.

Basic product design data can be obtained from photographs of the product (Figure 2.5), "exploded" drawings (Figure 2.6), engineering drawings of individual parts (Figure 2.7), parts lists (Figure 2.8), bills of materials (Figure 2.9), and assembly charts (Figure 2.10). Also, when they are used, prototypes of the product can be useful in obtaining data that will influence the design of the layout.

The *assembly chart* is an analog model of the operations and inspections required in assembling a product. Circles represent assembly operations and squares represent inspections; the time sequencing of activities is represented by the vertical relationships among operations and inspections. The assembly chart is also called a "gozinto" chart, since it shows graphically which parts must go into (or be assembled to) other parts. The assembly chart often is the basis for the layout of the assembly department or assembly line. Also, it influences the relative placement of other departments and activities. Hence it is of considerable benefit to the layout designer.

To construct an assembly chart for an existing product, disassemble the product completely. Then, in reverse order, depict the disassembly process and insert the appropriate cleaning operations and inspections. For the gate valve depicted in

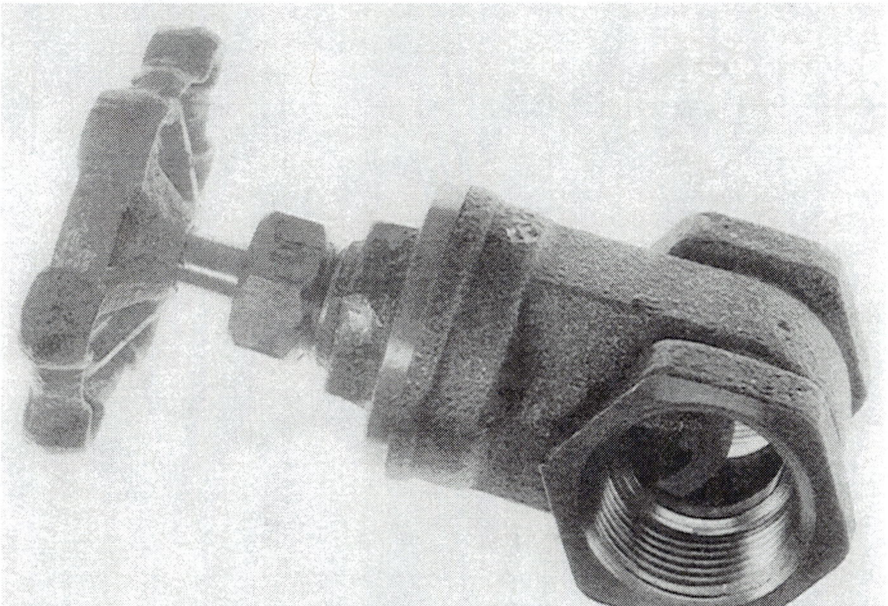

Figure 2.5 Gate valve.

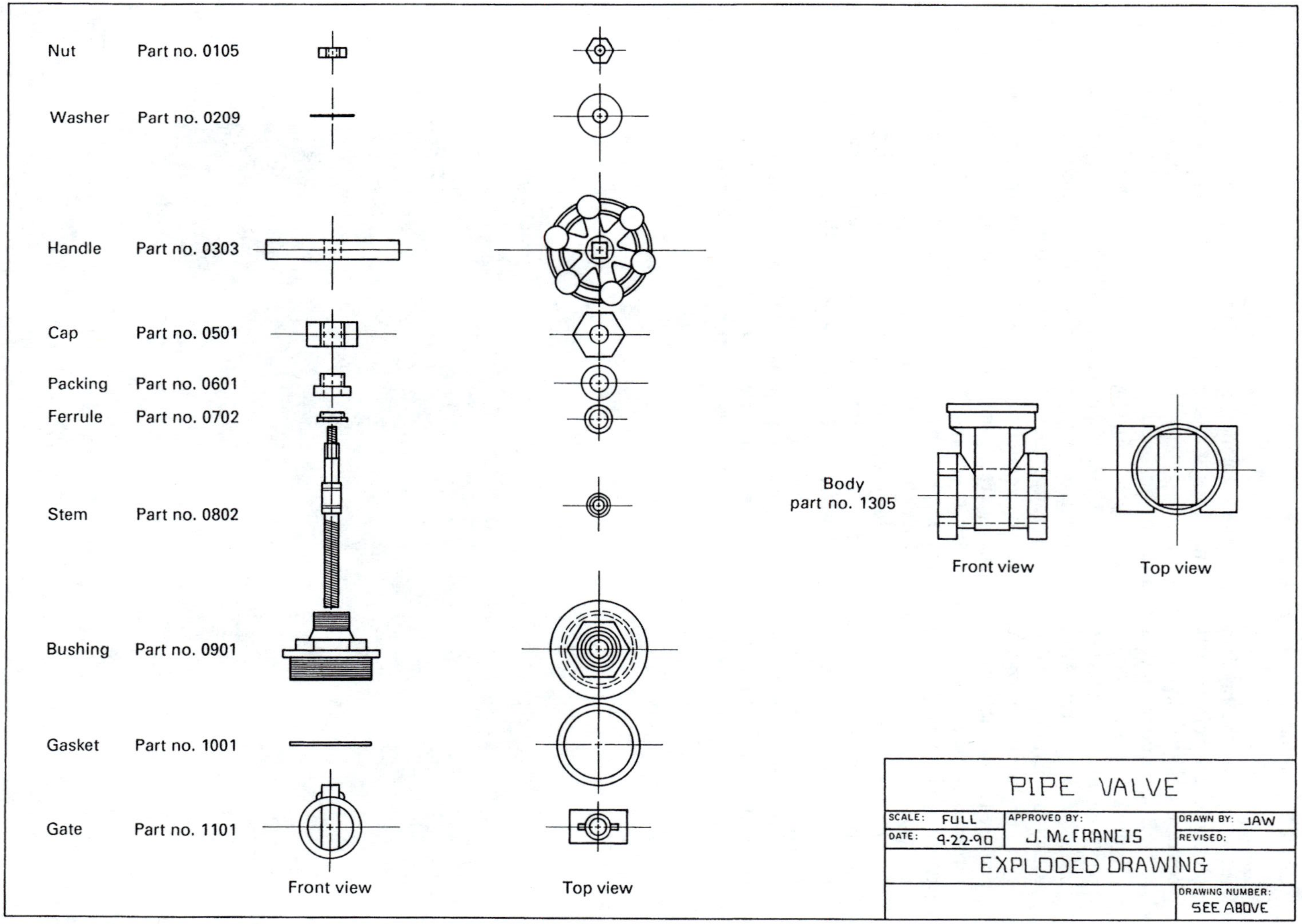

Figure 2.6 "Exploded" drawing of the gate valve.

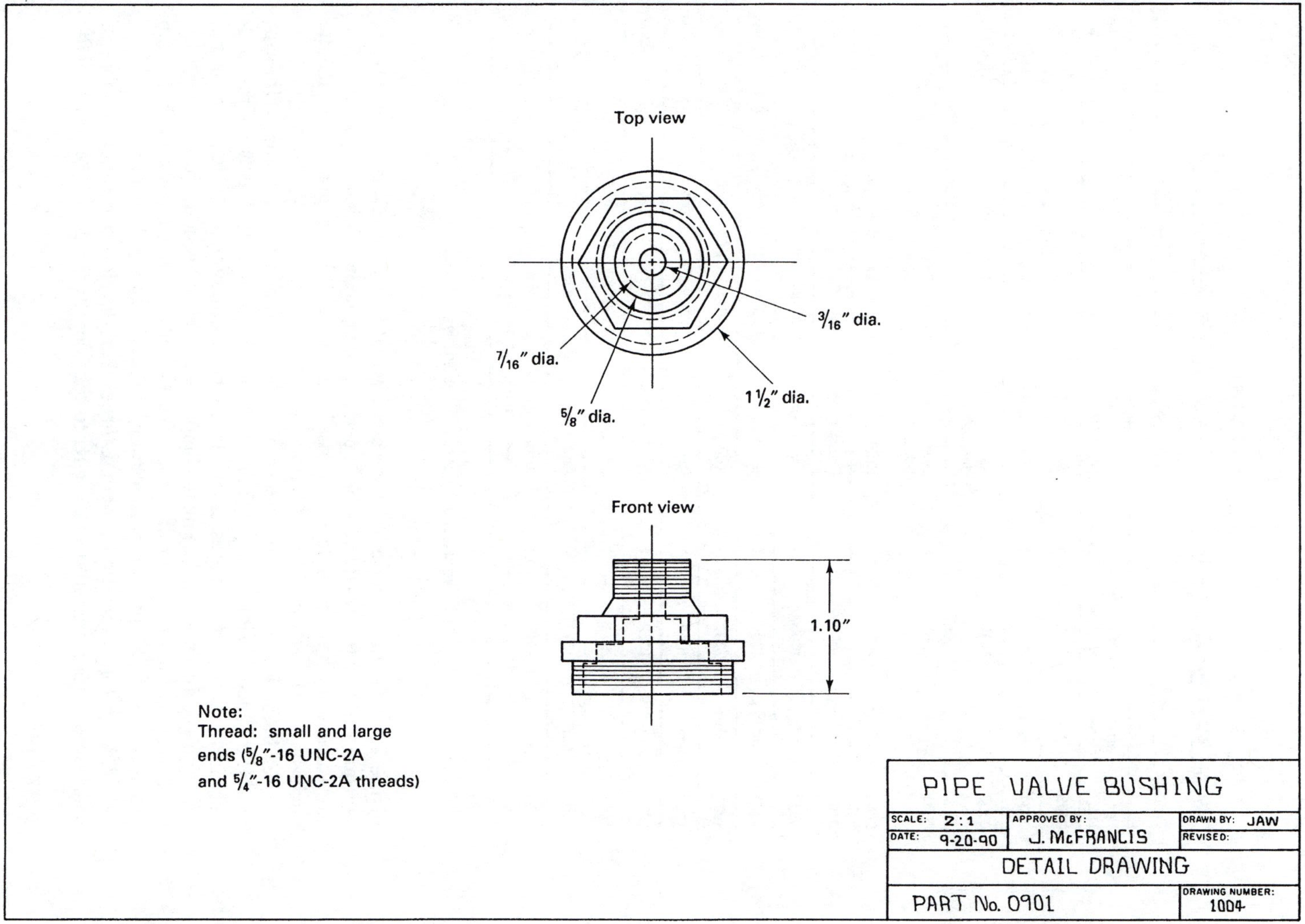

Figure 2.7 Engineering drawing of one part of the gate valve.

PARTS LIST

Company *Open-Close Valve Company* Prepared by *John McFrancis*

Product *Gate valve* Date *October 31, 19xx*

Part No.	Part Name	Drawing Number	No./ Unit	Material	Size	Make Buy
0105	Nut	2005	1	Brass		Buy
0204	Washer	2003	1	Aluminum	0.8" od x 0.22" id x 0.01" thick	Buy
0303	Handle	1010	1	Zinc, die casting	2.0" x 0.6"	Make
0501	Cap	1008	1	Brass, hex bar stock	0.75" x 0.40"	Make
0601	Packing	2001	1	Rubber	0.58" od x 0.25" id x 0.25"	Buy
0702	Ferrule	1007	1	Brass, tube	0.4" x 0.2"	Make
0802	Stem	1000	1	Brass, rod	3.0"" x 0.375"	Make
0901	Bushing	1004	1	Bronze	1.10" x 1.50"	Make
1001	Gasket	2007	1	Rubber	1.40" od x 1.17" id x 0.025"	Buy
1101	Gate	1006	1	Bronze	1.20" x 0.96" x 0.55"	Make
1305	Body	1002	1	Bronze	1.40" x 1.90" x 2.125"	Make
1402	Box	2011	1	Cardboard	2.125" x 2.125" x 4.25"	Buy
1503	Staple	2012	4	Aluminum		Buy
1603	Label	2013	1	Adhesive backed paper	2.0" x 1.0"	Buy

Figure 2.8 Parts list for the gate valve.

Figure 2.10, the first disassembly operation would be to unpack the gate valve; a pressure and leak test would have been performed immediately before packing. The gate valve would have been cleaned before being tested. Prior to cleaning, the handle might have been attached to the stem.

A limitation of the assembly chart is its two-dimensional format. Unfortunately, it is not possible to capture in two dimensions the true directional relationships involved in performing assembly. From the assembly chart it might appear that everything is assembled to the main component by entry from the left, as the flows are depicted. Such is seldom the case. The exploded drawing given in Figure 2.6 provides a better pictorial representation of the "assembly directions."

Design for automation programs have been developed within a number of companies to force the product designer to consider the impact of the design of the

BILL OF MATERIALS

Company *Open-Close Valve Company* Prepared by *John McFrancis*

Product *Gate valve* Date *October 31, 19xx*

Level	Part No.	Part Name	Drawing Number	No./ Unit	Make Buy	Comments
0	0010	Pipe valve	0010	1	Make	
1	0100	Bottom assembly	0020	1	Make	
2	0200	Top assembly	0030	1	Make	
4	0105	Nut	2005	1	Buy	
3	0204	Washer	2003	1	Buy	
2	0303	Handle	1010	1	Make	
3	0403	Paint	—	0.010 oz.	Buy	
4	0501	Cap	1008	1	Make	
5	0601	Packing	2001	1	Buy	
3	0702	Ferrule	1007	1	Make	
1	0802	Stem	1000	1	Make	
2	0901	Bushing	1004	1	Make	
3	1001	Gasket	2007	1	Buy	
2	1101	Gate	1006	1	Make	
2	1201	Grease	—	0.001 oz	Buy	
3	1305	Body	1002	1	Make	
6	1402	Box	2011	1	Buy	
7	1503	Staple	2012	4	Buy	
7	1603	Label	2013	1	Buy	

Figure 2.9 Bill of materials for the gate valve.

product on the assembly process. The primary thrusts of such programs have been threefold: *dimensional reduction*, *parts elimination*, and *parts standardization*.

With respect to the former, it is generally the case that assembly occurs with insertions from below, above, the front, the back, the right, and the left, as well as from various angles and involving a variety of contortions on the part of the assembly operator. The cost of assembly has been reduced significantly by designing products so that assembly can occur in a single dimension (e.g., along the z-axis). It has been

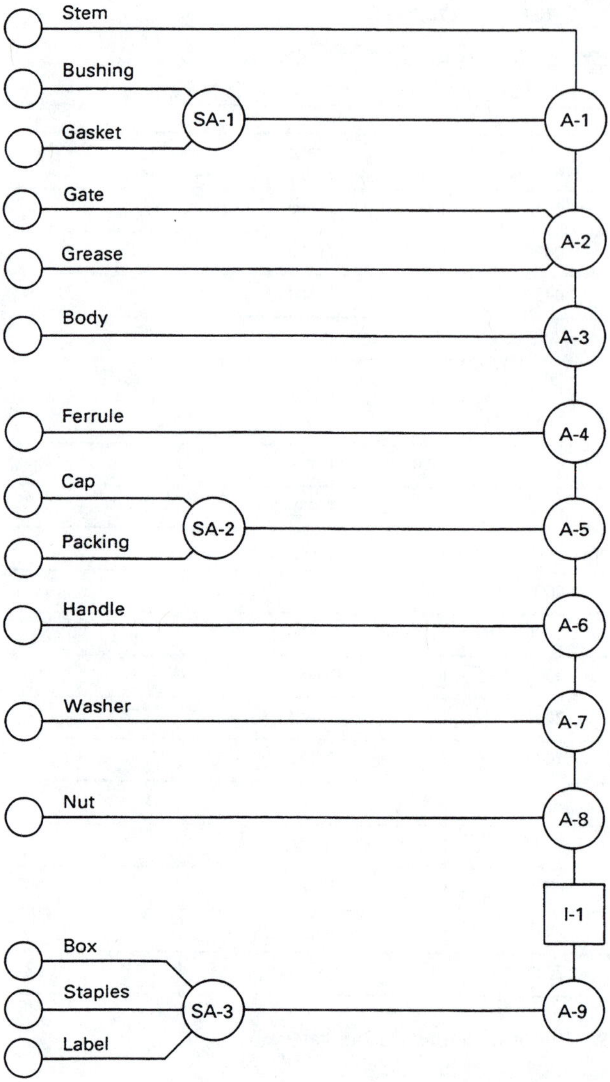

Figure 2.10 Assembly chart for the gate valve.

found, for example, that robotic assembly is much easier to justify when assembly occurs along one axis.

Not surprisingly, the complexity of programming a robot increases geometrically with the number of assembly dimensions. However, simplifying the process of assembly can have surprising results; for example, several firms have found that after redesigning a product to make it easier to assemble with a robot that it was also easier to assemble manually and robotic assembly could not be justified.

By redesigning the product some firms have reduced the number of component parts by more than 50%. For example, snap fasteners have been used to eliminate screws, rivets, bolts, and nuts. Forming processes have been used to create complex shapes rather than creating them via assembly or joining operations.

Although dimensional control and parts reduction efforts have produced dramatic results, perhaps the greatest improvement has come from parts standardization. Numerous firms have found their inventory to include as many as a dozen identical screws, but with different part numbers. Gaskets, washers, bolts, nuts, and flanges, for example, are often specified by the product designer by consulting a supplier's catalog, rather than a firm's listing and description of its own inventory. Product designers tend to "put their signature on the product" by specifying unique parts, when standard parts are available. As a result, many firms have experienced tremendous growth in the number of part numbers in inventory.

The *bill of materials* is also referred to as a *structured parts list* since it includes all of the information typically included in the parts list, as well as information concerning the structure of the product. Typically, a hierarchical approach is taken to indicate the assembly structure for the product. As shown in Figure 2.9, the bill of materials for the gate valve includes an indication of the assembly level for each part. The main component is the stem; level 2 parts are assembled directly to the stem; to level 2 parts are assembled level 3 parts; and so on.

Continuing with our consideration of information gathering, process design decisions determine whether a part will be purchased or produced, how the production of a part will be achieved, what equipment will be used, and how long it will take to perform each operation. Such information is typically summarized on an *operation sheet* or *route sheet*, such as that shown in Figure 2.11. A separate route sheet is normally required for each part.

The *operation process chart* is often used to supplement the route sheet; in a sense, it is a pictorial representation of the route sheet. As shown in Figure 2.12, the operation process chart is also an analog model. It is an expansion of the assembly chart in that it includes all operations and inspections performed on the product. Vertical flow lines depict sequential relationships; horizontal flow lines depict points in time when secondary flows merge with the primary vertical flows. The operation process chart also serves as the basis for the layout of a manufacturing plant.

Since the assembly chart and the operation process chart often are the foundation for the layout, care must be taken to ensure that their construction does not inhibit the design process. We have seen a number of situations where the wrong assembly chart and wrong operation process chart were used to determine the layout. It was not a matter of them being incorrect; rather, they were poor choices from among the multitude of possible assembly charts and operation process charts.

To motivate the discussion, consider the precedence diagram given in Figure 2.13 for assembling the gate valve treated previously. A *precedence diagram* establishes the prerequisite assembly steps that must be completed before performing a given assembly step. In the case of the gate valve, the packing must be placed in the cap before the cap is attached to the stem; similarly, the cap must be placed on

ROUTE SHEET

Company *Open-Close Valve Company* Prepared by *John McFrancis*

Product *Gate valve* Date *October 31, 19xx*

Part Name *Bushing* Part Number *0901*

Material *Bronze casting* Production Quantity *1500*

Op. No.	Operation Description	Machine Type	Tooling and Supplies	Setup Time (hrs)	Oper. Time (hrs)
05	Cast: using green sand mold with 50 bushings per mold; pour, cool, cut-off gates and risers	Bench mold Band saw	T-shaped dry sand core for each bushing	0.50	2.00
06	Clean: grind, as needed to remove flashing, and tumble	Bench or pedestal grinder; tumbling barrel	A-24-M-8-V type grinding wheel	0.25	0.75
07	Drill hole: 3/16" x 1/2"	Turret lathe	3/16" drill	0.50	2.50
08	Bore hole: 7/16" x 1/4"	Turret lathe	7/16" bore	—	2.75
09	Turn: small end to 5/8"; large end to 3/2" and then large end to 5/4"	Turret lathe	Round-nose turning tool, cut-off tool, and right-hand facing tool	—	10.50
10	Thread: small and large ends (5/8" – (16 UNC-2A and 5/4" – 16 UNC-2A threads)	Turret lathe	Stationary, self-opening die heads for 5/8" and 5/4"	—	9.25
			TOTALS	1.25	27.75

Comments: Operations 7 through 10 are performed on the same turret lathe, using one setup. The operating times given include rechucking the workpiece. The operations are performed as follows: chuck large end; drill, bore, turn, thread, and cutoff small end to length; chuck small end, grip hexagonal surface; turn large end to 1.50" over 0.375" length, cutoff to length, turn large end to 1.25" over 0.25" length, retaining 1.50" diameter over 0.125" length.

Figure 2.11 Route sheet for one component of the gate valve.

the stem before the handle is assembled to the stem; the washer must be placed on the stem after the handle and before the nut is assembled to the stem.

In parallel, it is possible to slide the stem through the bushing. The bushing must be placed on the stem before the gate is attached to the stem. Also, it is more convenient to attach the gasket to the bushing before assembling the gate to the stem. Similarly, the gate must be attached to the stem before inserting the stem into the body of the gate valve. However, it does not matter when the stem/bushing/gate/body assembly occurs relative to the stem/packing/cap/handle/washer/nut assembly.

The precedence diagram is a *directed network* representation of the assembly operations. In general, the assembly chart and the operation process chart are tree

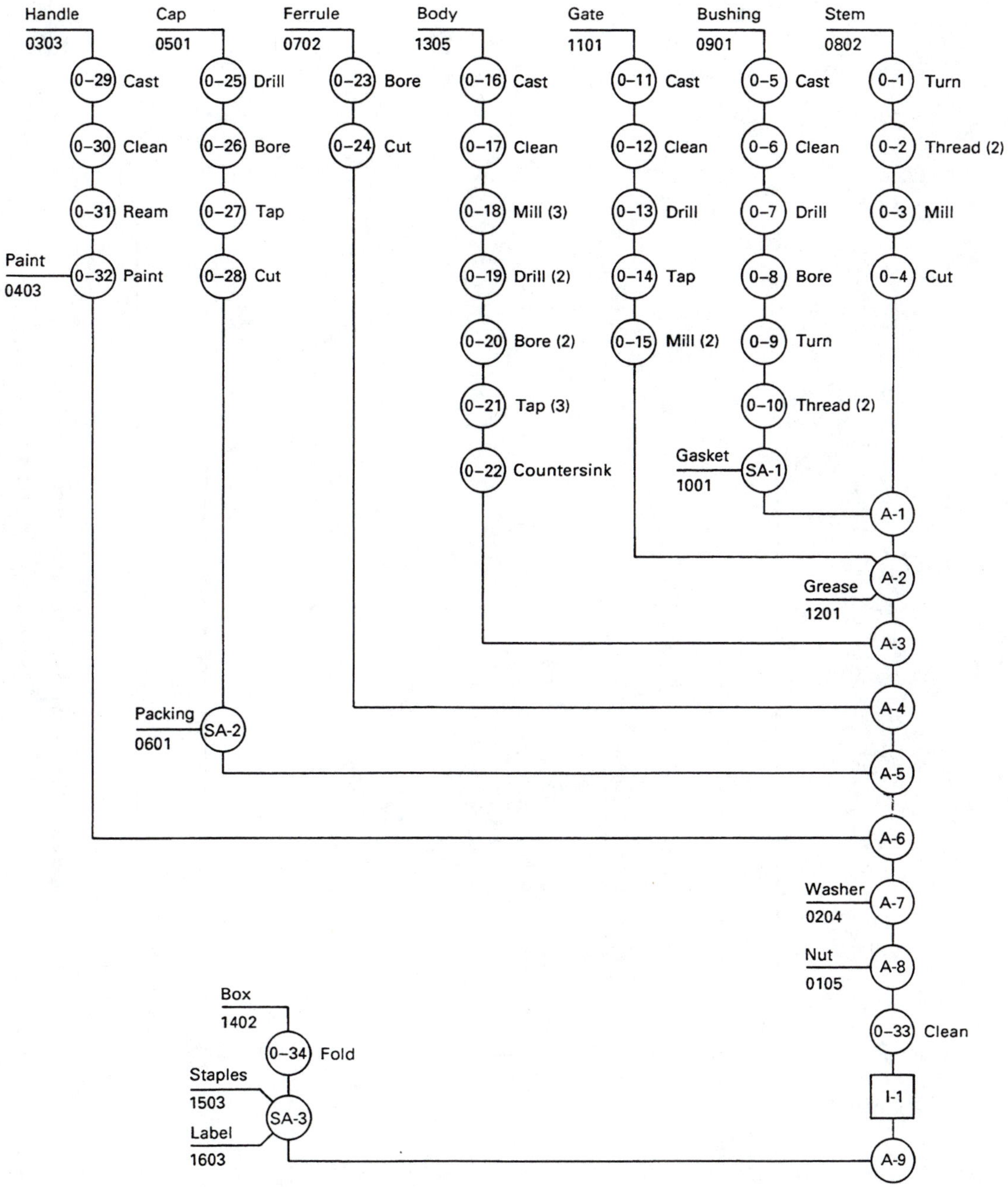

Figure 2.12 Operation process chart for the gate valve.

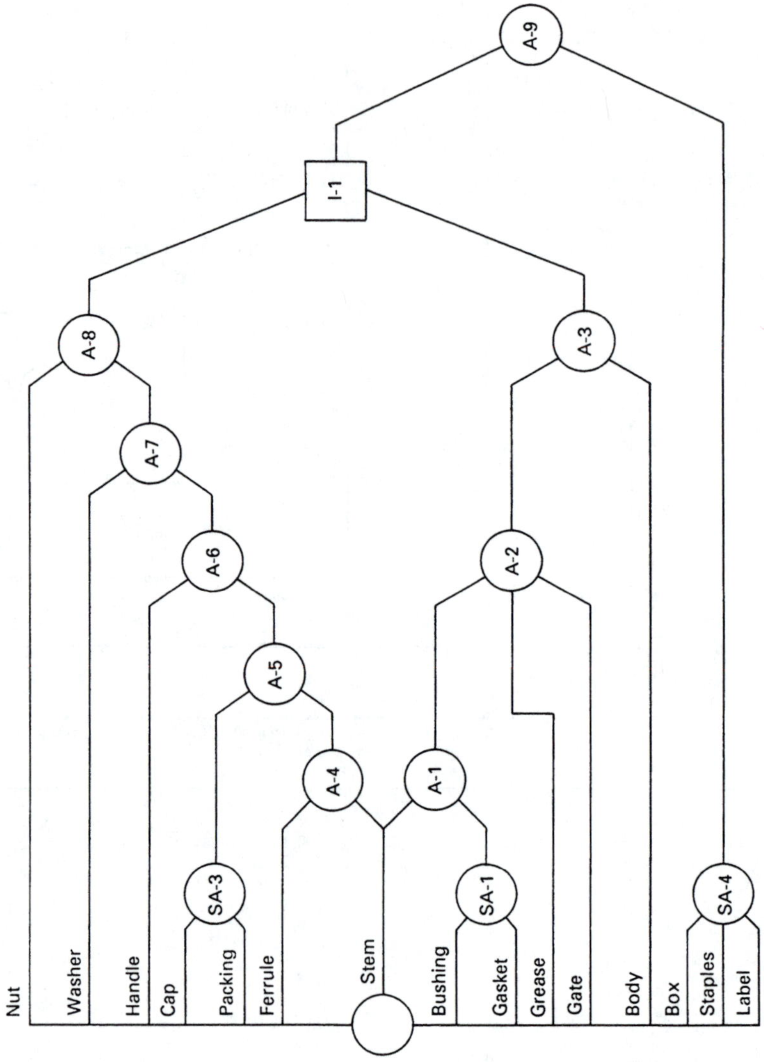

Figure 2.13 Precedence diagram for assembling the gate valve.

representations of the production process. From the precedence diagram it is clear that more than one assembly chart can be developed and still satisfy all of the precedence relationships. As an example, an alternative assembly chart for the gate valve is given in Figure 2.14. Yet, in describing how to construct an assembly chart, we said: "*To construct an assembly chart for an existing product, disassemble the product completely. Then, in reverse order, depict the disassembly process and insert the appropriate cleaning operations and inspections.*"

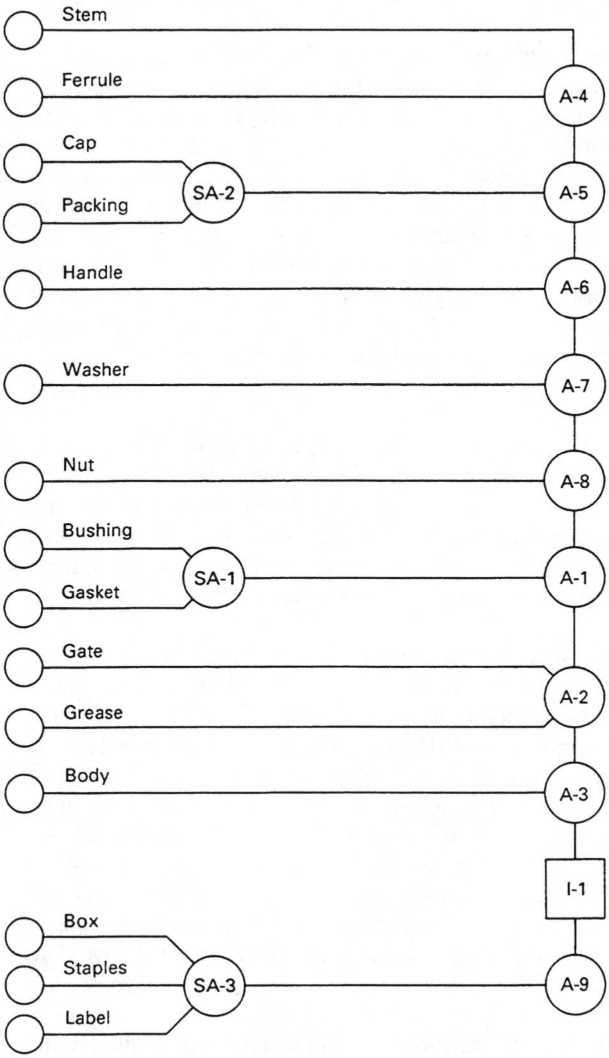

Figure 2.14 Alternative assembly chart for the gate valve.

Just as there are many possible ways to disassemble a product, so are there multiple ways to assemble it. Potentially, each way can have a significant impact on the material handling system and the layout. For this reason, the assembly chart and operation process chart should be constructed with care. In fact, we recommend constructing a number of different versions of each and assessing their impact on the layout.

Having considered the impact of product and process design decisions on the layout, it remains to consider the role played by schedule design relative to plant layout design. In brief, schedule design decisions tell us how much to produce and when to produce.

The market forecast is converted into production demand, and decisions are made concerning the production rate. Associated with these decisions is the determination of the number of machines of each type required to meet the production volume. Depending on the product mix (number of different products) for the firm and the required production rate, the decision may be made to produce each product on a continuous basis or an intermittent basis. The plant layout designer is often involved in this decision. The degree of involvement is dependent on the organizational policies of the firm. In any case, the layout will be affected by decisions concerning the number of machines and the production schedule. Production schedules are also sometimes given using an analog model called the *Gantt chart* (Figure 2.15). Many decisions made in designing the production schedule are dependent on layout decisions. Thus the layout designer must work closely with the schedule designer.

Throughout the information-gathering phase, remember to avoid a preoccupation with the present solution. Most information sources available portray yesterday's and, possibly, today's data. We are interested in *tomorrow's* data, since we are developing a solution to be used tomorrow. Thus available information might need to be modified. Remember, the product, process, and schedule data available from the sources we have listed are probably representative of the present solution, not the preferred solution.

If you have taken a quantitative course in production management or production planning and control, you are aware that analytical approaches are available to the product, process, and schedule designer in making many of the decisions which affect the plant layout. We have not treated these approaches, since our concern is with the design of the layout. Regardless of how product, process, and schedule design decisions are made, the layout designer should be aware of the decisions.

Our discussion of data sources emphasizes the interrelationships among product, process, schedule, and layout design. The layout must support the objectives of the facility. However, the effective arrangement of personnel and equipment depends on accurate information regarding product, process, and schedule requirements. When such requirements change rapidly, it is important for the layout to accommodate changes in requirements. Such layouts, called *flexible layouts*, are discussed in Section 2.8.

Before ending our treatment of information gathering, and while on the

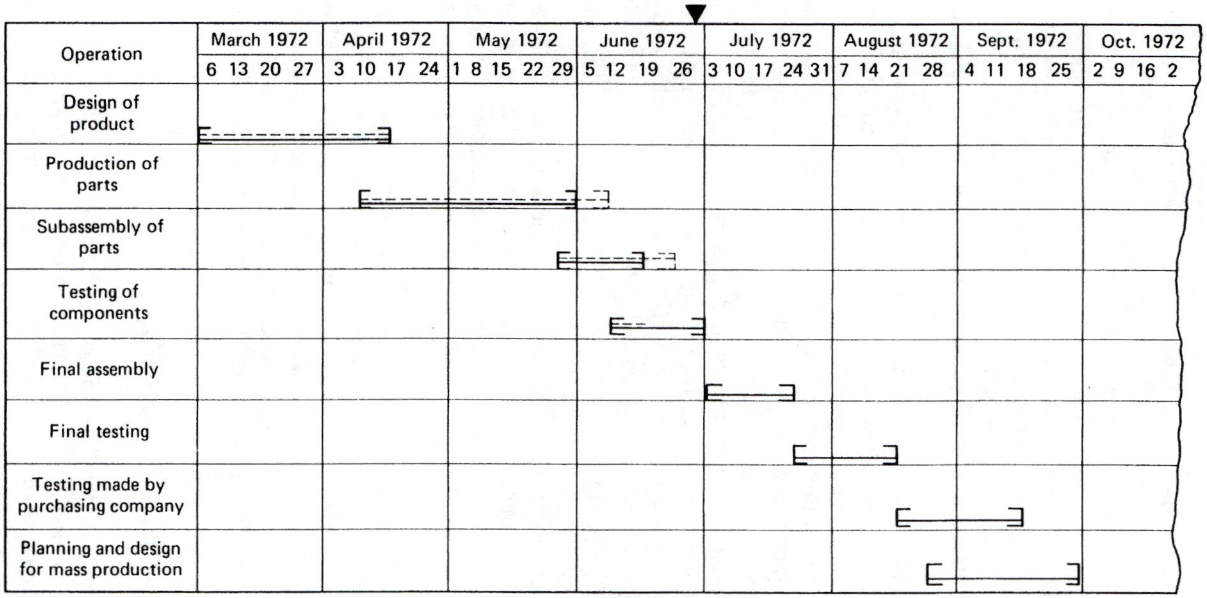

Figure 2.15 Gantt project planning chart. (Indicates current week of operation, the estimated amount of time a particular operation will take, and the actual amount of time that the particular operation has taken.) We see that the project is 1 week behind schedule.

subject of product, process, schedule, and layout design, it should be noted that many organizations are giving considerable emphasis to strengthening the communication links among the design functions. The design of the product determines the way the product is manufactured, assembled, and packaged. Also, it affects the way the product is handled, stored, inspected, and identified throughout production, as well as packed, shipped, and maintained. As a result, considerable emphasis has been given to designing the product to facilitate subsequent activities; the integrated consideration of product, process, and schedule design is called *concurrent engineering*. In the past, marketing considerations have governed all aspects of product design. More recently, other functions have had a greater voice in the design of the product.

In a sense, the thrust of computer-integrated manufacturing (CIM) is to provide the information links that will connect all design and operating functions. Traditionally, the product designer designed a product without consulting the process designer; the completed product design was "thrown over the transom" to the process designer who was required to design a process to produce the product. The process designer would then design a process for producing the product without consulting the schedule designer or layout designer; the completed process design was "thrown over the transom" to the schedule designer and the layout designer. In turn, independently they would design production schedules and layouts for producing the product. As a result, considerable inefficiencies resulted. Each designer in the serial process made decisions that reduced the "design degrees of freedom" for designers "downstream."

To counteract the "transom design" phenomenon, some firms form design teams with each design function represented; others require all four design functions' approval of each function's design; still others transfer personnel among the design functions to improve the designer's understanding of the impact on others of the design decisions.

Other attempts to strengthen the communication links among the design functions have included the development of "computer aids." For example, computer-aided design (CAD), computer-aided process planning (CAPP), computer-aided scheduling (CAS), computer-aided layout (CAL), and computer-aided logistics system (CALS) have been developed to help the product, process, schedule, and layout designers. (In Chapter 3, we address further computer-aided layout.)

2.6 FLOW ANALYSIS AND ACTIVITY ANALYSIS

In this section we describe flow analysis techniques and then describe activity analysis techniques. Flow analysis concentrates on some quantitative measure of movement between departments or activities, whereas activity analysis is primarily concerned with the nonquantitative factors that influence the location of departments or activities.

Once certain basic data have been obtained concerning the process, product, and schedule, the plant layout analyst is in a position to analyze the flow of materials, equipment, and personnel. Since the layout is designed to facilitate the flow of the product, from raw material to the finished product, we are primarily concerned with the flow of materials.

Some of the factors that affect the flow pattern are given by Apple [1] as:

1. External transportation facilities
2. Number of parts in product
3. Number of operations on each part
4. Sequence of operations on each part
5. Number of subassemblies
6. Number of units to be produced
7. Necessary flow between work areas
8. Amount and shape of space available
9. Influence of processes
10. Types of flow patterns
11. Product versus process type of layout
12. Location of service areas
13. Production department locations
14. Special requirements of departments
15. Material storage
16. Desired flexibility
17. The building

2.6.1 Types of Flow Patterns

Most of the factors listed require little additional discussion. However, we shall examine two of these factors in greater detail. First, consider the various types of flow patterns. We can classify flow patterns as being either *horizontal* or *vertical*. There are at least five basic types of horizontal flow patterns, as shown in Figure 2.16. A number of other flow patterns can be developed by combining these basic flow patterns. [In fact, we can visualize A, B, C, D, E, F, G, H, J, K, M, N, P, Q, R, T, V, W, X, Y, and Z flow patterns. No doubt, other alphabetic systems would suggest even more flow patterns (e.g., π, λ, ϕ, and γ).]

Straight-line flow is the simplest form of flow. However, when employed in a plant, separate receiving and shipping crews are normally required. The L-shaped flow pattern is usually adopted when either straight-line flow cannot be accommodated in an existing facility or construction costs do not permit straight-line flow. The U-shaped flow pattern is very popular, since it is simple to administer and facilitates a combination of receiving and shipping activities. Circular flow is applicable when it is desired to terminate the flow very near the point where the flow originated. The

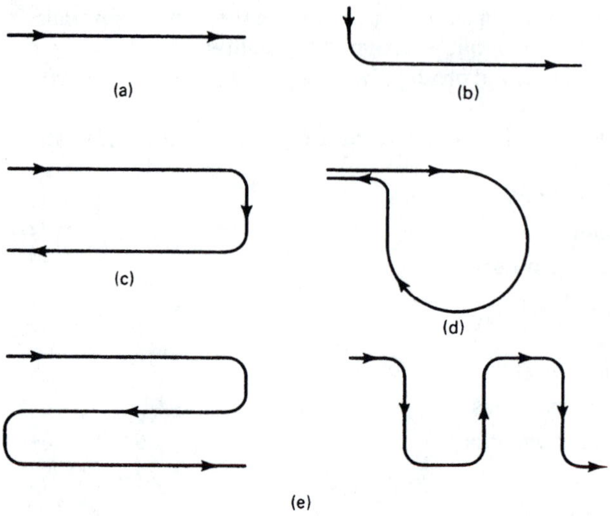

Figure 2.16 Basic horizontal flow pattern: (a) straight, or I flow; (b) L flow; (c) U flow; (d) circular, or O flow; (e) serpentine, or S flow.

serpentine flow pattern is used when the production line is so long that zigzagging on the production floor is required.

Vertical flow patterns exist both in single-story and multistory buildings. Utilization of overhead space with conveyorized material flow has focused greater attention on the design of vertical flow patterns. We shall examine vertical flow patterns as they commonly exist in multistory structures. It should be recognized that similar patterns also occur in single-story buildings. Figure 2.17 presents six different vertical flow patterns in a three-story building.

Vertical flow pattern (a) is often used when there is flow between buildings and there exists an elevated connection between the buildings. Pattern (b) is used when ground-level ingress (entry) and egress (exit) are required. In (c), ground-level ingress and egress occur on the same side of the building. In (a), (b), and (c), decentralized elevation is present, since travel between floors can occur on either side of the building. Centralized elevation exists in (d), since travel between floors occurs on the same side of the building. Inclined flow occurs in (e); some bucket and belt conveyors and escalators result in inclined flow. Backtracking occurs in (f) due to the return to the top floor.

2.6.2 Types of Layout

The second factor affecting the flow pattern is the choice of layout type. There exist four general layout categories: fixed product, product, group, and process layouts. The *fixed* or *static product layout* is used when the product is too large or cumbersome to move through the various processing steps; consequently, rather than take the product to the processes, the processes are brought to the product. Examples include the shipbuilding industry, aspects of the aircraft industry, and the construction

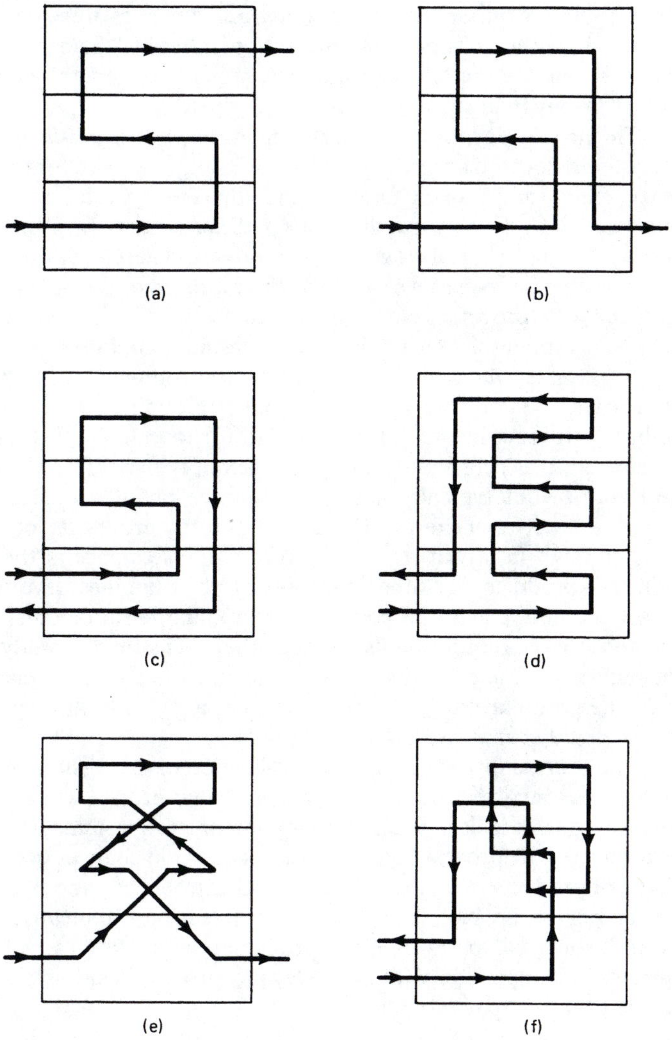

Figure 2.17 Vertical flow pattern.

industry. The fixed product layout is developed by locating workstations or production centers around the product in the appropriate processing sequence. Considerable logistics are involved in ensuring that the right processes are brought to the product at the right times and are located in the right places. Although the fixed product layout is typically found when the product is large, it has been used in the electronics industry to support computer systems integration and test; materials, subassemblies, components, peripherals, and housings are brought to a single workstation, where the finished computer system is "built." Since the workpiece remains stationary, equipment and personnel movement are generally higher with the fixed product layout than for any of the other types.

The *product layout*, also referred to as the *production-line layout*, results when processes are located according to the processing sequence for the product. Material flows directly from a workstation to the adjacent workstation. Product layouts are used when high-volume production conditions exist. In a strict product layout, machines are not shared by different products. Therefore, the production volume must be sufficient to achieve satisfactory utilization of the machines. Because of the ability to locate processes close together in the production sequence, it is usually the case that a product layout minimizes the distances between processing operations.

A *group layout* is used when production volumes for individual products are not sufficient to justify product layouts, but by grouping products into logical product families, a product layout can be justified for the family. The groups of processes are referred to as cells; hence the group layout is also referred to as *cellular layout*. The group layout typically has a high degree of intradepartmental flow; it is a compromise between the product layout and the process layout.

Although in layout design the processing sequence is the basis for forming families of products, for other purposes product families have been formed based on part geometry, material composition, tooling requirements, and material handling and storage requirements, among others. The field of study that addresses the grouping of products into families is referred to as *group technology* (GT). As a result, the group layout is also referred to as a group technology or GT layout and the individual groups are called GT cells.

The *process layout* consists of a collection of processing departments or cells. All machines involved in performing a particular process are grouped together in a process layout. It is used when there exist many low-volume, dissimilar products to be produced. The process layout is used when rapid changes occur in the mix and/or volume of products to be produced, as well as when conditions are such that neither product nor group layout approaches are feasible. Typically, job shops employ process layouts due to the variety of products manufactured and their low production volumes. In comparison with group layout, process layout is characterized by high degrees of interdepartmental flow.

Example 2.1

To illustrate the differences in fixed product, product, group, and process layouts, consider a situation in which four parts (A, B, C, and D) are to be produced and assembled into a single product. The processing sequence for part A is saw, turn, mill, and drill; for part B it is saw, mill, drill, and paint; for part C the processing sequence is grind, mill, drill, and paint; and for part D the sequence is weld, grind, turn, and drill. All parts go to a central assembly department. The daily production rate of products is to be 100, with each product consisting of one of part C, two of parts B and D, and three of part A.

TABLE 2.1 DATA FOR EXAMPLE 2.1

		Equipment requirements						
Part	Daily production rate	Weld	Grind	Saw	Turn	Mill	Drill	Paint
A	300	—	—	0.5	0.5	0.3	0.2	—
B	200	—	—	0.4	—	0.5	0.3	0.2
C	100	—	0.4	—	—	0.3	0.5	0.3
D	200	0.3	0.5	—	0.3	—	0.2	—

The daily production rates and equipment requirements for each part are given in Table 2.1. Based on the data provided, if a product layout is used, two saws and four drills will be required, whereas only one saw and two drills are required with group layout, and one saw and one drill are required with process layout. Depending on how production is scheduled, the equipment requirements for fixed product layout might be identical to those for process layout. In Figure 2.18 we have provided illustrations of each type layout for the example problem.

Fixed product, product, group, and process layouts are compared in Table 2.2. The decision to use a fixed product layout is generally dictated by a particular characteristic of the product. However, the decision to use either a product, group, or process layout generally depends on the volumes of production and variety of products being produced.

You might think that it would be relatively straightforward to determine the variety of products being produced. However, it is not always obvious how to proceed. As an example, part numbers or product numbers should not be used arbitrarily as the basis for measuring variety. The real measure of processing variety is the "degree of sameness" among products in their processing requirements. For instance, if 50 different products differ only in their color and if painting is the last processing operation, the products could be treated as a single product prior to the painting process. The differences in the colors of the products might not affect the layout.

The volume of production is also subject to interpretation. Tonnage produced annually, annual sales volumes, number of units produced annually, and cubic volume of materials consumed in production are alternative measures of volume. Yet none might capture the real issue of volume from a layout perspective. The layout type depends on an assessment of the impact of the layout selection on material handling and the utilizations of *space*, *equipment*, and *personnel*.

From a material handling perspective, the number of units produced annually is not nearly as important as the number of loads moving through the plant. The material flow requirement depends on whether the product is moved individually, in totebox loads, in pallet loads, or in truckload quantities. The continuity of flow affects the need for in-process storage, which affects space utilization. Duplication of machines due to decentralized processing in product and group layouts affects machine and personnel utilizations.

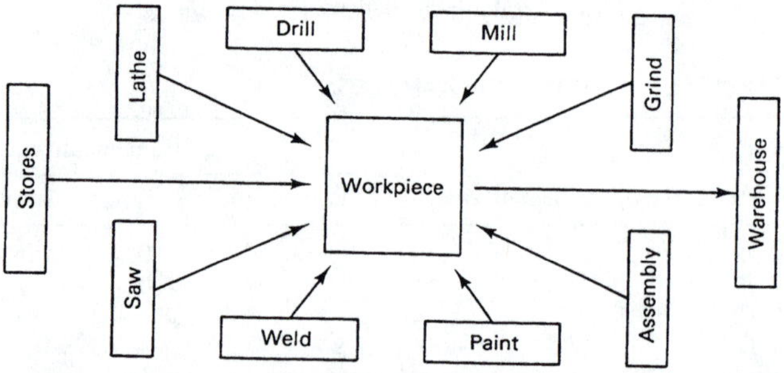

(a) Fixed product layout

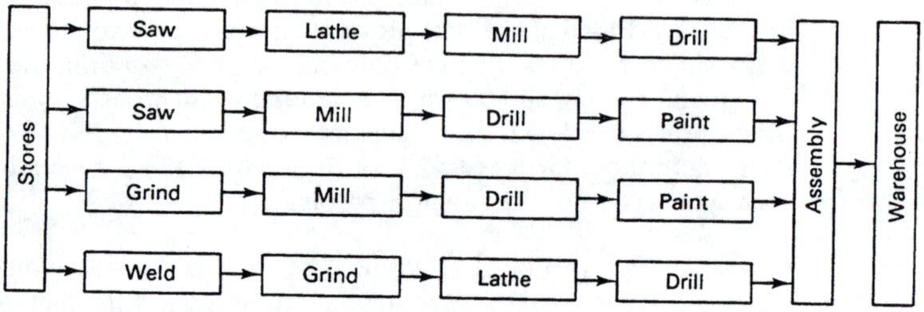

(b) Product layout

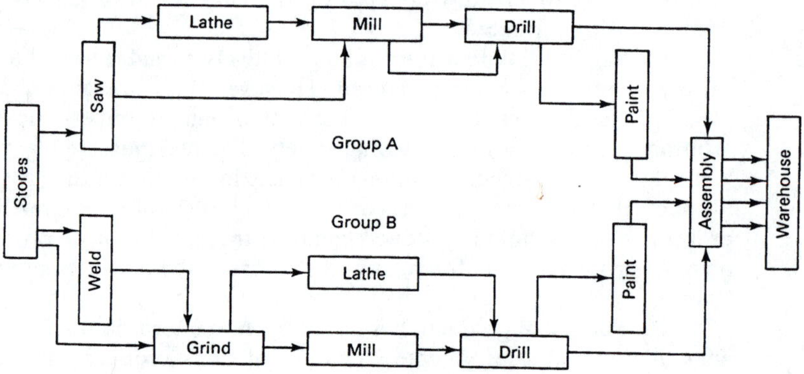

(c) Group layout

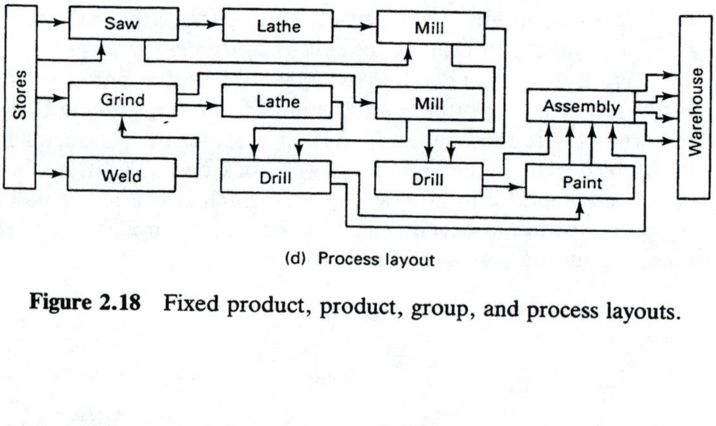

(d) Process layout

Figure 2.18 Fixed product, product, group, and process layouts.

TABLE 2.2 ADVANTAGES AND LIMITATIONS OF FIXED PRODUCT, PRODUCT, GROUP, AND PROCESS LAYOUTS

Fixed Product Layout

Advantages

1. Material movement is reduced.
2. Promotes job enlargement by allowing individuals or teams to perform the "whole job."
3. Continuity of operations and responsibility results from team.
4. Highly flexible; can accommodate changes in product design, product mix, and production volume.
5. Independence of production centers allows scheduling to achieve minimum total production time.

Limitations

1. Increased movement of personnel and equipment.
2. Equipment duplication may occur.
3. Higher skill requirements for personnel.
4. General supervision required.
5. Cumbersome and costly positioning of material and machinery.
6. Low equipment utilization.

Product Layout

Advantages

1. Since the layout corresponds to the sequence of operations, smooth and logical flow lines result.
2. Since the work from one process is fed directly into the next, small in-process inventories result.
3. Total production time per unit is short.
4. Since the machines are located so as to minimize distances between consecutive operations, material handling is reduced.
5. Little skill is usually required by operators at the production line; hence, training is simple, short, and inexpensive.
6. Simple production planning and control systems are possible.
7. Less space is occupied by work in transit and for temporary storage.

Limitations

1. A breakdown of one machine may lead to a complete stoppage of the line that follows that machine.
2. Since the layout is determined by the product, a change in product design may require major alterations in the layout.
3. The "pace" of production is determined by the slowest machine.
4. Supervision is general, rather than specialized.
5. Comparatively high investment is required, as identical machines (a few not fully utilized) are sometimes distributed along the line.

Group Layout

Advantages

1. Increased machine utilization.
2. Team attitude and job enlargement tend to occur.

(Continued on next page)

TABLE 2.2 ADVANTAGES AND LIMITATIONS OF FIXED PRODUCT, PRODUCT, GROUP, AND PROCESS LAYOUTS—*Continued*

3. Compromise between product layout and process layout, with associated advantages.
4. Supports the use of general-purpose equipment.
5. Shorter travel distances and smoother flow lines than for process layout.

Limitations

1. General supervision required.
2. Higher skill levels required of employees than for product layout.
3. Compromise between product layout and process layout, with associated limitations.
4. Depends on balanced material flow through the cell; otherwise, buffers and work-in-process storage are required.
5. Lower machine utilization than for process layout.

Process Layout

Advantages

1. Better utilization of machines can result; consequently, fewer machines are required.
2. A high degree of flexibility exists relative to equipment or manpower allocation for specific tasks.
3. Comparatively low investment in machines is required.
4. The diversity of tasks offers a more interesting and satisfying occupation for the operator.
5. Specialized supervision is possible.

Limitations

1. Since longer flow lines usually result, material handling is more expensive.
2. Production planning and control systems are more involved.
3. Total production time is usually longer.
4. Comparatively large amounts of in-process inventory result.
5. Space and capital are tied up by work in process.
6. Because of the diversity of the jobs in specialized departments, higher grades of skill are required.

As shown in Figure 2.19, product layouts are appropriate for high-volume, low-variety conditions; process layouts are appropriate for low-volume, high-variety conditions; and group layouts are appropriate for "in-between" conditions. There do not exist rules of thumb as to where the divisions occur. In fact, many layouts tend to be be a combination of product, group, and process layouts. The volume–variety mix among products can be such that a few products are produced using product layouts, others use group layouts, and the remainder use process layouts. Similarly, it might be appropriate to use either process or group layouts for the production of individual components and to use a product layout for the assembly of the components. From the foregoing it should be obvious that the choice of layout type should be based on the requirements of the situation rather than determined arbitrarily.

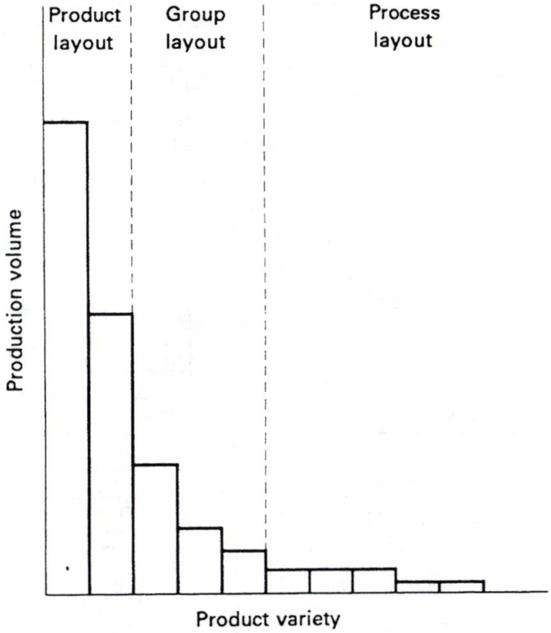

Figure 2.19 Volume–variety chart indicating that different layouts are justified.

2.6.3 Flow Analysis Approaches

The preceding discussion has provided motivation for the importance of flow analysis in layout design. It remains to consider some techniques that are commonly used in analyzing the flow of materials, people, and equipment. The most popular method of analyzing flow is to use charts and diagrams. These descriptive, analog models are used to assist the layout designer's visualization of the required flow. The assembly chart and operation chart were described in Section 2.5. Additional charts and diagrams that have been found to be useful in flow analysis are:

1. Flow process charts
2. Multiproduct process charts
3. Flow diagrams
4. From–to charts

The *flow process chart* is an analog model that substitutes circles for operations, squares for inspections, arrows for transportations, triangles for storages, and the capital letter D for delays. Vertical flow lines connect these symbols in the sequence in which they are performed. A sample flow process chart is given in Figure 2.20.

The *multiproduct process chart* is used to combine conveniently the operation process charts for more than one product. A sample multiproduct process chart is given in Figure 2.21.

Job	Assemble Slab – wooden pencil
Follow the	☐ Product ☐ Main ☐ Material ☐ Form
Chart begins	Slabs in storeroom
Chart ends	Assembled and clamped
Charted by	P.O.E. Date 9/29

Summary — No. ___ Page ___ of ___

600 Assemblies	Present No.	Present Time	Proposed No.	Proposed Time	Differance No.	Differance Time
○ Operations	7	304.8	7	70.0	–	234.8
⇨ Transportations	10	4.2	4	0.5	6	3.7
□ Inspections	–		1	–	+1	–
D Delays	–		2		+2	
▽ Storages	3	✓	1	✓	-2	
Totals	20	309	15	70.5	-5	238.5
Distance travelled	417 ft		80 ft		337 ft	

Details of Present / Proposed Method	Operation / Transport / Inspection / Delay / Storage	Distance (ft)	Quantity	Est. time (min.)	Notes
1. Stored in storeroom	○ ⇨ □ D **▽**				
2. To slotter-groover by hand truck	○ **⇨** □ D ▽	25	1,200	0.25	Finished stock thinner one box contains 1,200 four-stock slabs (2,400)
3. Slot cut in bottom and four grooves in top	**○** ⇨ □ D ▽		1,200	30.00	One pass thru tandem set machines
4. To lead-laying machine (one-half lot – see 9)	○ **⇨** □ D ▽	25	600	0.13	Hand truck
5. Wait for lead layer	○ ⇨ □ **D** ▽		600	✓	Stock delay between lots all four-grove run be-fore starting next size
6. Loaded in machine magazine	**○** ⇨ □ D ▽		600	–	Loaded during machine operation
7. Lead layed in slab	**○** ⇨ □ D ▽		600	20.00	Push-bar mach. pushes slabs from bottom of mag. under lead hopper.
8. Inspected for full leads. Moved to topper (see 12)	○ ⇨ **□** D ▽				Inspected by machine tender on steel bench slide on way to topper.
	~~○ ⇨ □ D ▽~~				During machine time
9. To glue topper (one-half lot – seee 4)	○ **⇨** □ D ▽	30	600	0.15	Hand truck
10. Wait for glue topper	○ ⇨ □ **D** ▽		600	✓	Refer to 5
11. Loaded in glue machine magazine	**○** ⇨ □ D ▽		600	2.40	Glue topper loads 25 slabs at time into mag. = 24 loads @ 10 min/load
12. Glued	**○** ⇨ □ D ▽		600	–	Push-bar mach. pushes slab over glue wheel into topping position
13. Topped and turned	**○** ⇨ □ D ▽		600	11.60	Topper places glued slab on leaded slab and turns on edge
14. Assembled slabs Clamped by topper	**○** ⇨ □ D ▽		600	6.00	Topper clamps unit of 25 assem. slabs = 24 units (topper paced by layer)
	○ ⇨ □ D ▽				
	○ ⇨ □ D ▽				

Figure 2.20 Flow process chart of a proposed method of assembling pencil slabs. (After G. C. Close, *Work Improvement*, John Wiley & Sons, Inc., New York; by permission of the publisher.)

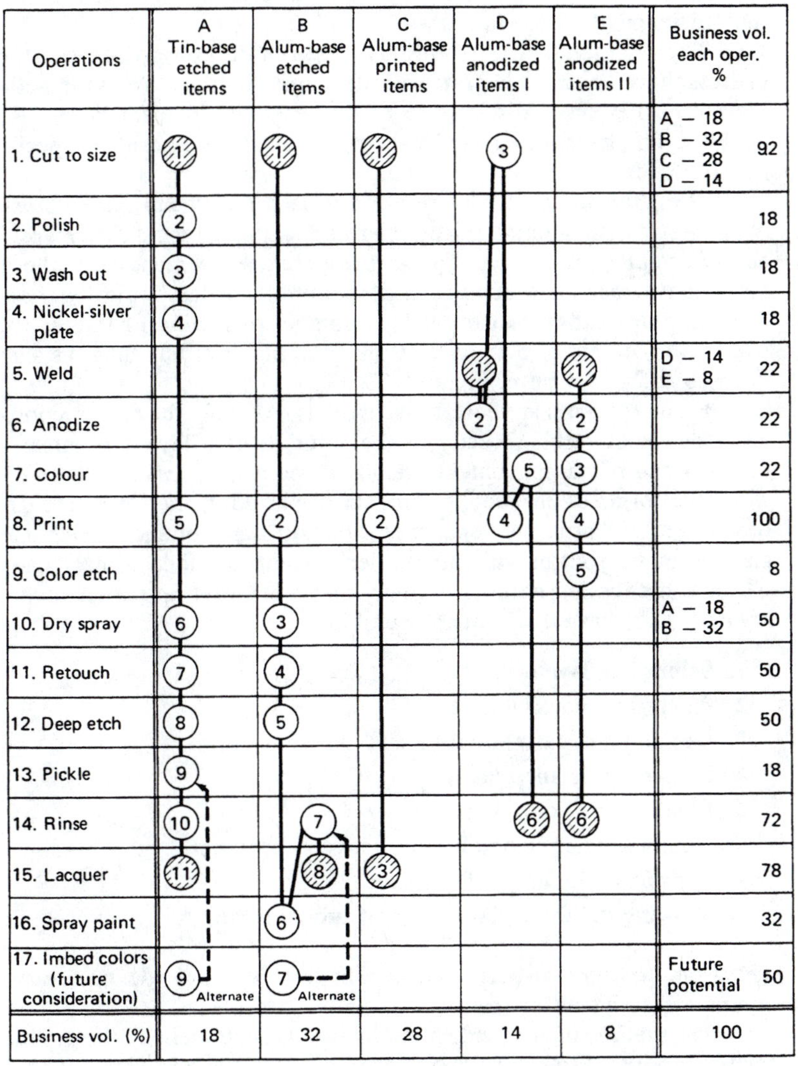
Operations
A Tin-base etched items
B Alum-base etched items
C Alum-base printed items
D Alum-base anodized items I
E Alum-base anodized items II
Business vol. each oper. %
1. Cut to size
2. Polish
3. Wash out
4. Nickel-silver plate
5. Weld
6. Anodize
7. Colour
8. Print
9. Color etch
10. Dry spray
11. Retouch
12. Deep etch
13. Pickle
14. Rinse
15. Lacquer
16. Spray paint
17. Imbed colors (future consideration)
Business vol. (%)
A – 18
B – 32
C – 28
D – 14
92
18
18
18
D – 14
E – 8
22
22
22
100
8
A – 18
B – 32
50
50
50
18
72
78
32
Future potential
50
18
32
28
14
8
100
Alternate
Alternate

Figure 2.21 Multiproduct process chart for five products. (After Richard Murther, *Practical Plant Layout*, McGraw-Hill, Inc., New York; by permission of the publisher.)

A *flow diagram* depicts the probable movement of materials by corresponding lines superimposed on the floor plan of the area under study. An example of a flow diagram is given in Figure 2.22. The flow diagram is quite beneficial in evaluating the efficiency of an existing layout.

The assembly chart, operation process chart, flow process chart, multiproduct process chart, and flow diagram were developed initially as aids for methods analysis. However, they have also proved to be beneficial in plant layout as aids in flow analysis. The from–to chart, on the other hand, was designed specifically for the purpose of facilitating layout design.

The *from–to chart*, also referred to as a *travel chart* and a *cross chart*, is an adaptation of the familiar mileage chart appearing on most road maps. The from–to chart normally contains numbers representing some measure of the material flow between two machines, departments, buildings, or sites. Typically, the from–to chart provides information concerning the number of material handling trips made between two centers of activity and the total material handling distance. A sample from–to chart is given in Figure 2.23.

From–to charts are descriptive models. As such, the construction of a from–to chart does not result directly in the solution of a layout problem. Rather, the from–to chart is a convenient means of reducing a large volume of data into a workable form. By inspecting the data displayed in the from–to chart, the layout designer can identify the departments having large volumes of item movement and can design a layout in which these departments are located near one another.

As mentioned earlier, from–to charts are used in conjunction with process layouts. They have been found useful in:

1. Selling the layout
2. Analyzing material flow
3. Designing departmental block plans
4. Designing detailed layouts
5. Evaluating layout alternatives
6. Demonstrating dependencies between activities
7. Improving the use of floor space
8. Showing the interrelationship of product lines

The from–to chart also is used in Chapter 3 to provide necessary input for a computerized layout program.

The method of analyzing material flow is dependent on the production volume for the product. From the volume–variety chart given in Figure 2.19, high-volume products are candidates for a product layout. Therefore, the assembly chart, operation process chart, and flow process chart are used to analyze material flows for high-volume products. When there are several high- and medium-volume products to be produced, the multiproduct process chart is used to examine the interrelationships among products. The from–to chart is used in designing both group and process

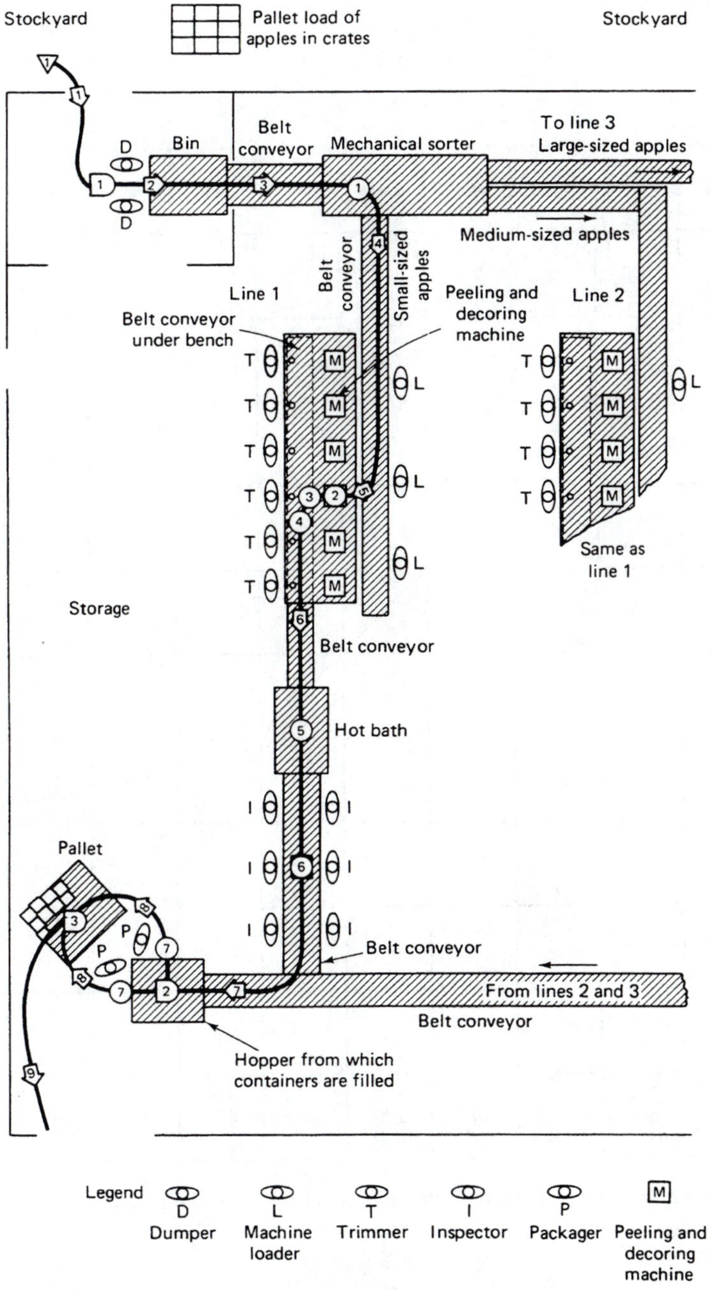

Figure 2.22 Flow diagram for process of preparing apple sections for freezing. (After E. V. Krick, *Methods Engineering*, John Wiley & Sons, Inc., New York; by permission of the publisher.)

To / From	Stores	Saw	Grind	Weld	Lathe	Mill	Drill	Paint	Assemble	W'house	Total
Stores		500	100	200							800
Saw					300	200					500
Grind					200	100					300
Weld			200								200
Lathe						300	200				500
Mill							600				600
Drill								300	500		800
Paint									300		300
Assemble										800	800
W'house											–
Total	–	500	300	200	500	600	800	300	800	800	

Figure 2.23 From–to chart showing flows of parts for the layouts given in Figure 2.18.

layouts. Consequently, it is used to analyze the material flow for low-volume products and product families. The flow diagram is used in product, group, and process layout, as well as their variations.

2.6.4 Activity Relationship Analysis

Flow analysis tends to relate various activities on some quantitative basis, and as mentioned, the relationship is often expressed as some function of material handling cost. We emphasized in Chapter 1, as well as in an earlier discussion in this chapter, that a number of factors other than material handling cost might be of primary concern in layout design. The *activity relationship chart*, or *REL chart*, was designed to facilitate a consideration of qualitative factors. As developed by Muther [36], the REL chart replaces the numbers in a from–to chart by a qualitative *closeness rating*. A typical REL chart is shown in Figure 2.24. All pairs of relationships are evaluated, and a closeness rating (A, E, I, O, U, or X) is assigned to each pair. When evaluating activity relationships for N activities, there are $N(N-1)/2$ such evaluations. With the exception of a U rating, the reason for the assigned rating is indicated using a numeric code.

Closeness ratings represent an ordered preference for "closeness." Specifically, an A rating and an X rating are considered to be the most important ranking. Hence any layout must satisfy the A and X ratings. An E rating is second ranked, and most, if not all, E ratings should be satisfied by the layout. An I rating is third ranked, and as many as possible I ratings should be satisfied by the layout without sacrificing A, E, or X ratings. In the same sense, O ratings are fourth ranked and they should be satisfied by the layout only if A, E, I, and X ratings are not jeopardized. U ratings are neutral ratings; hence they can be ignored in designing the layout. Thus A and $X > E > I > O > U$, where $>$ means "more important or higher ranking than."

The process of developing the activity relationships can be quite valuable; it can generate considerable insight regarding the attitudes, biases, and preferences of the people involved. However, to be most useful in layout design, the closeness ratings should reflect a high degree of discrimination. As an example, very few A and X relationships should be assigned. One rule of thumb is to allow no more than 5% of the closeness ratings to be an A or an X, no more than 10% to be an E, no more than 15% to be an I, and no more than 20% to be an O. Hence at least 50% of the closeness ratings would be a U if the rule of thumb is followed.

Layouts are generally developed using a hierarchical approach. Namely, *block plans* or *block layouts* are developed first by determining the sizes, shapes, and relative locations of departments or other designated activities. Next, detailed layouts are designed for each department. Thus different REL charts are needed for designing block plans and detailed layouts. In the case of a block plan, the activity relationship chart is constructed as follows:

1. List all departments or activities to be included.
2. Obtain closeness ratings by interviewing or surveying persons involved in performing functions within each activity, as well as those responsible for managing one or more activities.
3. Determine reasons used for closeness ratings and record on the REL chart.
4. Assign a closeness rating to each pairwise combination of activities and record the code for the reason behind the rating.
5. Review the REL chart with those providing input in step 2 and make appropriate adjustments in the ratings.

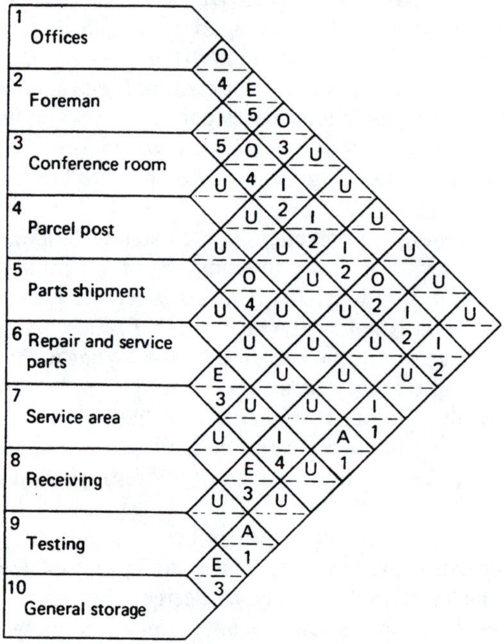

Code	Reason
1	Flow of materials
2	Ease of supervision
3	Common personnel
4	Contact necessary
5	Convenience
6	
7	
8	
9	
10	

Rating	Definition
A	Absolutely necessary
E	Especially Important
I	Important
O	Ordinary closeness OK
U	Unimportant
X	Undesirable

Figure 2.24 Activity relationship chart.

The process of constructing an activity relationship chart can be complicated by the presence of multiple relationships that will influence the design of the layout. For example, material flow, personnel flow, equipment flow, information flow, monetary flow, organizational issues, control requirements, process relationships, environmental considerations, and product characteristics, among others, can impact significantly the design of the layout.

Because of the multiplicity of relationships that might exist, we recommend that the *major* bases for relationships be identified and separate REL charts be constructed for each. As examples, different REL charts might be constructed for material flow, personnel flow, and equipment flow.

2.7 RELATIONSHIP DIAGRAM

Our consideration of the activity relationship diagram (REL diagram) begins with a brief treatment of its purpose. Next, we consider the process of developing an activity relationship diagram.

2.7.1 Purpose, Premise, and Proximity

The purpose of the activity relationship diagram is to depict spatially the relationships of the activities. A sample activity relationship diagram is given in Figure 2.25 for the REL chart given in Figure 2.24. Underlying the development of the REL diagram is the premise that geographic proximity can be used to satisfy particular relationships. For example, when the activity relationships reflect the magnitudes of material flows, pairs of activities having the greatest pairwise flow are located next to one another; the relative locations of pairs of activities having the smallest flows are of little importance. Similarly, pairs of activities having an A rating are located adjacently and pairs of activities having an X rating are located far apart.

We must be careful to avoid thinking that proximity is the only way to satisfy "closeness" ratings. For example, if a production machine is noisy, emits offensive odors, and generates considerable heat, it is a likely candidate for an X rating with several other activities. However, the use of acoustical panels and a ventilating system might eliminate the offensive attributes of the machine.

A second example involves two activities with communication needs sufficient to justify an A rating. One method of satisfying the need for communication is to locate the two activities adjacently to allow face-to-face contacts. Alternatively, live video links, computer ties, intercoms, and pneumatic-tube delivery systems might be used to satisfy the communication requirement.

A conclusion that can be drawn from the two examples is that closeness ratings can be satisfied without measuring closeness by distance separation alone. In particular, it might be less expensive to incorporate special features in the facility to satisfy certain closeness ratings than it is to satisfy the rating with distance separation or proximity.

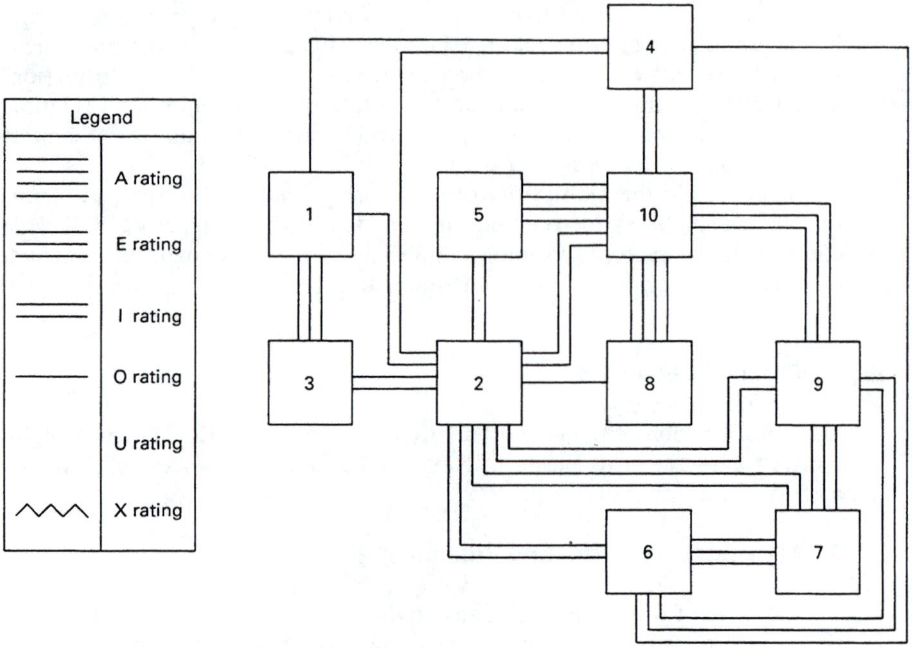

Figure 2.25 Activity relationship diagram.

2.7.2 Traditional Process

The process of constructing an activity relationship diagram often requires compromises, especially when not all closeness ratings can be satisfied. The process is complicated by the lack of a binary (yes or no, 0 or 1) ruling on whether or not a rating is satisfied. For instance, if a 25-ft separation satisfies a closeness rating, does a 50-ft separation also satisfy the closeness rating? Judgment and discretion are required.

Traditionally, relationship diagramming consisted of a manual "cut and try" process in which equal-sized squares were used to represent activities; in turn, the squares were moved around until the designer judged that the closeness ratings had been satisfied adequately.

Example 2.2

To illustrate the traditional process used in developing the REL diagram, consider the REL chart given in Figure 2.24. The following combination of activities have *A* relationships: (5,10) and (8,10). No *X* relationships exist. The combinations of activities having an *E* relationship are (1,3), (6,7), (7,9), and (9,10). The *I* combinations are (2,3), (2,5), (2,6), (2,7), (2,9), (2,10), (4,10), and (6,9). Finally, the *O* combinations are (1,2), (1,4), (2,4), (2,8), and (4,6).

The process of representing relationships spatially is depicted in Figure 2.26. Throughout the process, every effort should be made to avoid having connecting lines cross other connecting lines. Such an approach reduces clutter. Also, if it is possible to eliminate "crossings," a useful property of the problem is established. We consider the property in the next section.

Note that the *A* relationships are considered first and are identified by connecting the activities with four lines. Next, the *E* relationships are incorporated in the partial REL diagram using three connecting lines. The *I* relationships are incorporated using two connecting lines. (To eliminate crossings we change the relative positions of activities 8 and 9.) Finally, the REL diagram is completed by considering *O* relationships, which have a single connecting line. The completed REL diagram appears in Figure 2.25.

Since a manual approach is used to develop the REL diagram, some feel it unnecessary to capture all of the requirements in the REL chart. As an example, suppose it is desirable to ensure that the offices (department 1), parts shipment (department 5), and receiving (department 8) be located on the exterior of the facility. The layout designer could ensure that the REL diagram was developed in such a way that the requirement is satisfied.

However, since the designer might forget the "fuzzy" requirement of placing departments 1, 5, and 8 on the outside of the facility, we might choose to designate the "exterior" as department 11 and assign it an *A* relationship with departments 1, 5, and 8. However, such an approach still requires the designer to remember to keep department 11 on the periphery of the REL diagram, as shown in Figure 2.27.

2.7.3 Graph-Based Process

A number of attempts have been made to apply results from graph theory in developing relationship diagrams and block layouts. To facilitate the process, "close" is interpreted to mean adjacent. If two activities had a common border, they are judged to be close; otherwise, they are not. By adopting adjacency as the definition of closeness, a graph-based approach can be used to aid the designer in determining if a layout can be designed to satisfy the closeness ratings.

To facilitate the process, each activity is represented by a *circle, node,* or *vertex.* Activities that must be adjacent are denoted by connecting the respective nodes or vertices with *lines*, *links*, or *edges*. The resulting representation of the activities and relationships is the activity relationship diagram or *graph*, depicted in Figure 2.28.

A requirement for the existence of a layout satisfying the activity relationships depicted in the graph is that the graph be *planar*. A graph is planar if it can be drawn so that its vertices are points in the plane and each edge can be drawn so that it intersects no other edges and passes through no other vertices (hence our earlier encouragement to eliminate crossings in constructing the REL diagram).

Examples of *nonplanar graphs* are the Kuratowski graphs shown in Figure 2.29. An equivalent condition for a graph to be nonplanar is that it contain either of the Kuratowski graphs; thus a graph will be planar if it contains neither Kura-

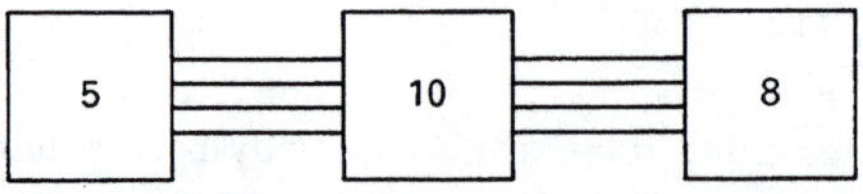

(a) A relationships

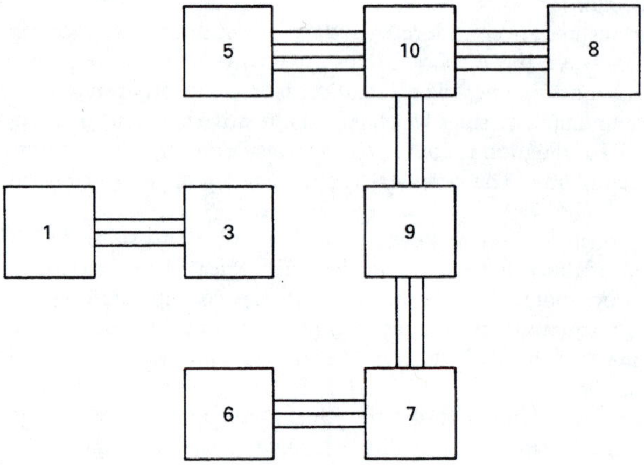

(b) A and E relationships

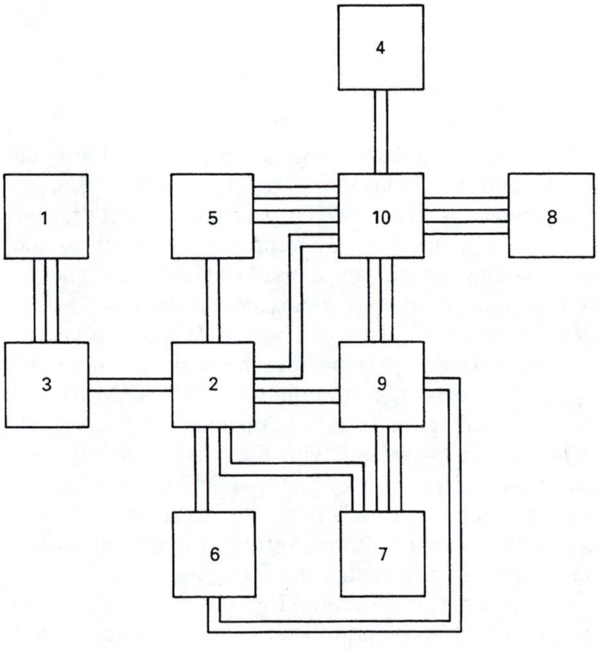

(c) A, E, and I relationships

Figure 2.26 Activity relationship diagram construction.

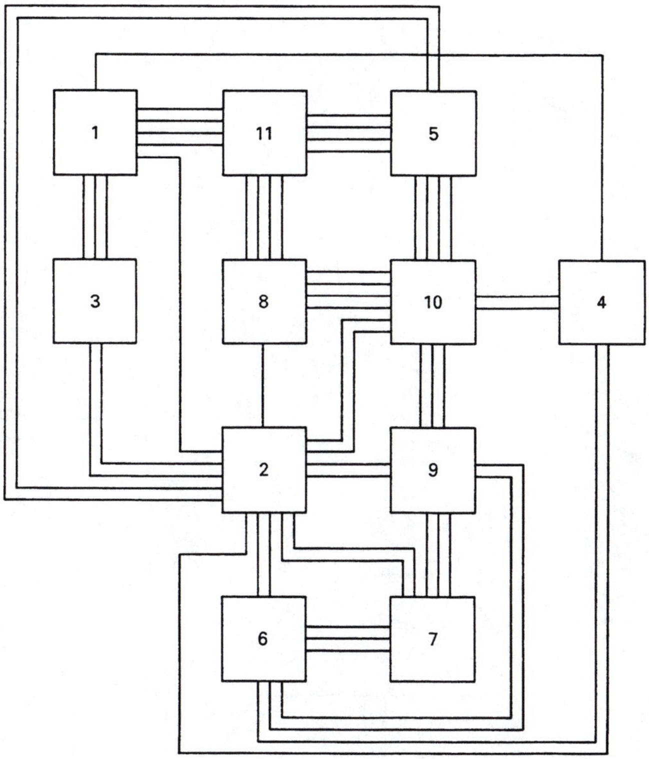

Figure 2.27 Explicit incorporation of "exterior" department 11 in the activity relationship diagram, with a single "crossing."

towski graph. Examples of nonplanar graphs and planar graphs are shown in Figures 2.30 and 2.31, respectively.

Regions defined by a graph are referred to as *faces*; the unbounded outside region is the *exterior face*. In the case of Figure 2.32, the graph has two faces, A and B, with B being the exterior face. Two faces are said to be *adjacent* if they share a common edge. In Figure 2.33, faces A and B are adjacent and faces A and C are nonadjacent.

An additional aspect of a graph is its *dual*. To construct the dual of a planar graph, place a dual node in each face of the primal planar graph. Whenever two faces share an arc e in their common boundary, join the nodes of the corresponding faces by an edge e^* crossing only e. If the primal graph is planar, the dual graph will be planar. Primal and dual graphs are depicted in Figure 2.34, with the edges for the dual graphs shown with dashed lines.

If the REL diagram is a planar graph, its dual graph will be planar. Notice that the faces of the dual graph in Figure 2.34 correspond to the space assigned to the

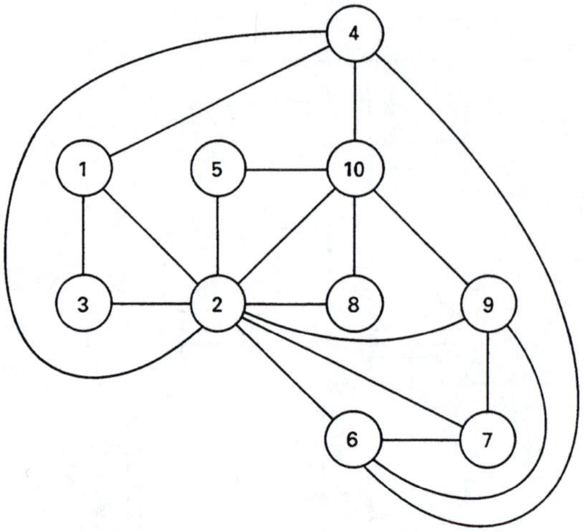

Figure 2.28 Activity relationship graph, with 10 vertices and 19 edges.

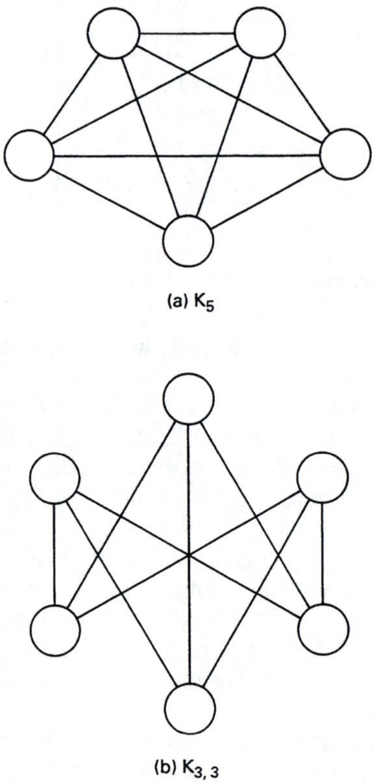

Figure 2.29 Kuratowski graphs.

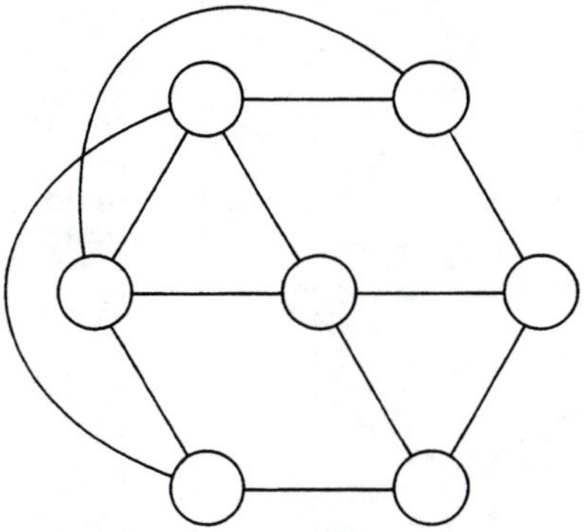

(a)

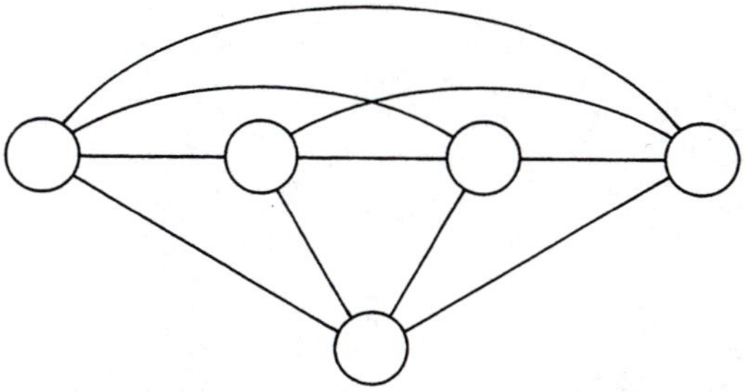

(b)

Figure 2.30 Nonplanar graphs.

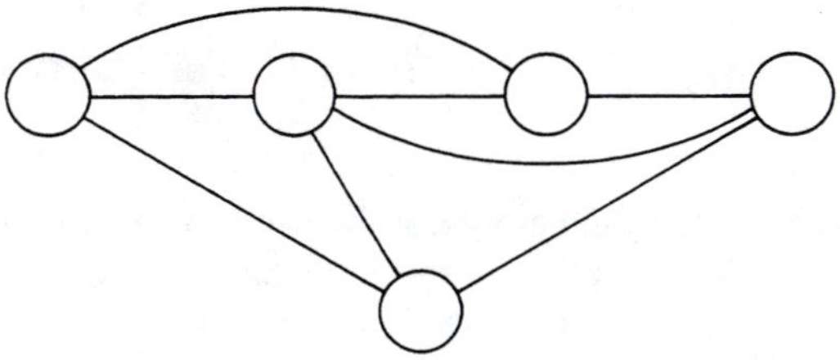

(a)

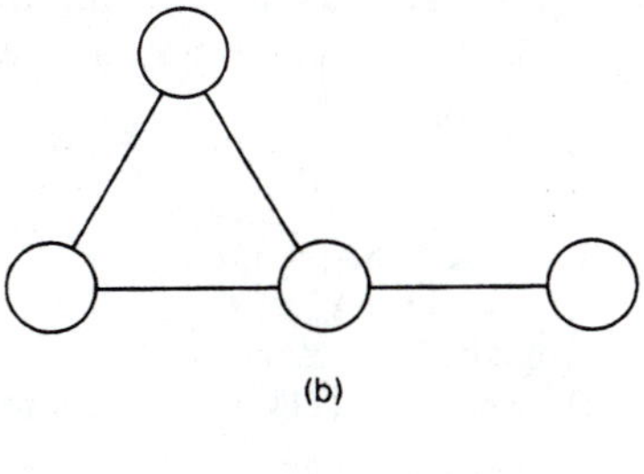

(b)

Figure 2.31 Planar graphs.

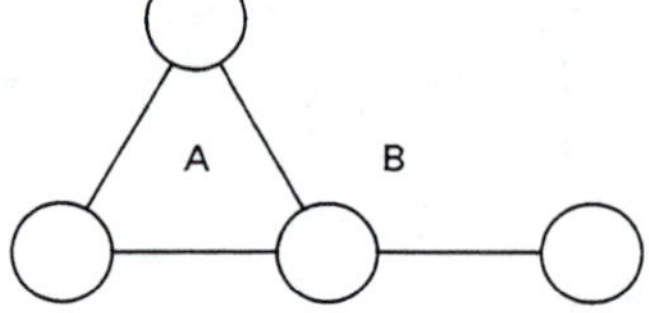

Figure 2.32 Interior (A) and exterior (B) faces of a graph.

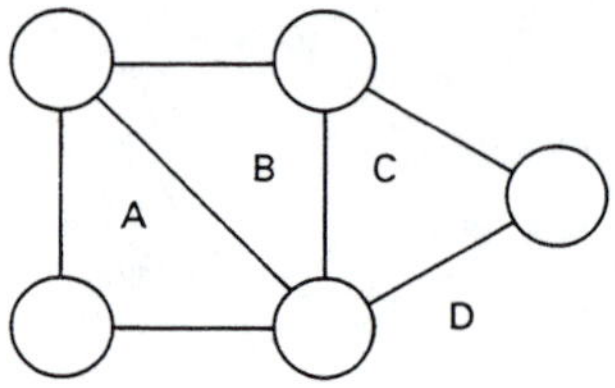

Figure 2.33 Adjacent (A and B) and nonadjacent (A and C) faces of a graph.

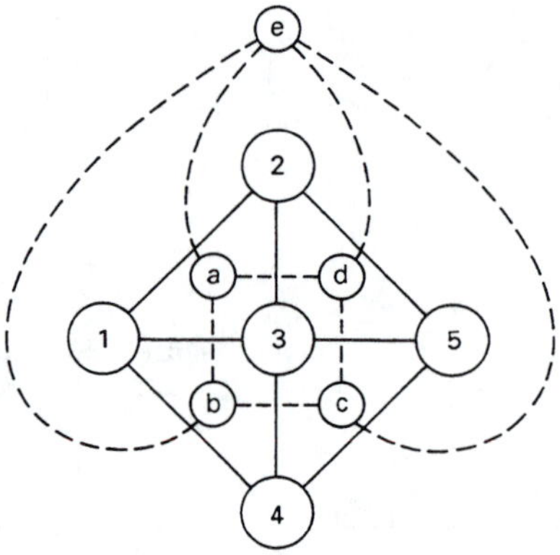

Figure 2.34 Primal and dual graphs.

activities in the primal graph. Since the dual graph can be used to develop the block layout, if a planar dual graph exists, it is possible theoretically to construct a block plan that will satisfy the adjacency requirements. However, in general, it is not straightforward to convert the dual graph to a block layout since the faces of the dual graph do not necessarily enclose sufficient area to satisfy the space requirements of the primal activities. On the other hand, both the dual graph and the planar graph representation of the REL chart can be of assistance in developing the corresponding block layout.

Example 2.3

To illustrate the relationships among the REL diagram, dual graph, and block layout, consider a situation involving five departments (A, B, C, D, and E), plus the "exterior," called department F. Suppose that it is necessary for A to be adjacent to B, C, and D; for B to be adjacent to A, D, and E; for C to be adjacent to A and D; for D

to be adjacent to A, B, C, and E; for E to be adjacent to B and D; and for all departments to be adjacent to F. A REL diagram for the situation is shown in part (a) of Figure 2.35; the dual graph is constructed in part (b) and shown with dashed lines; the block layout corresponding to the dual graph is given in part (c).

Example 2.4

To illustrate the converse situation, consider the block layout given in part (a) of Figure 2.36 for departments A, B, C, and D, with the exterior department labeled E. To

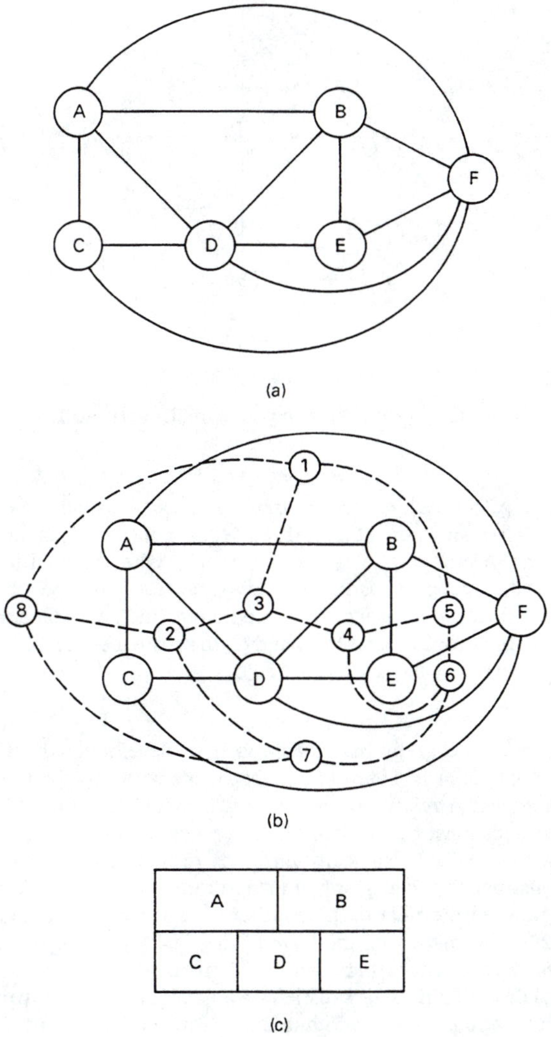

Figure 2.35 Graph representation of REL diagram, dual, and block layout for Example 2.3.

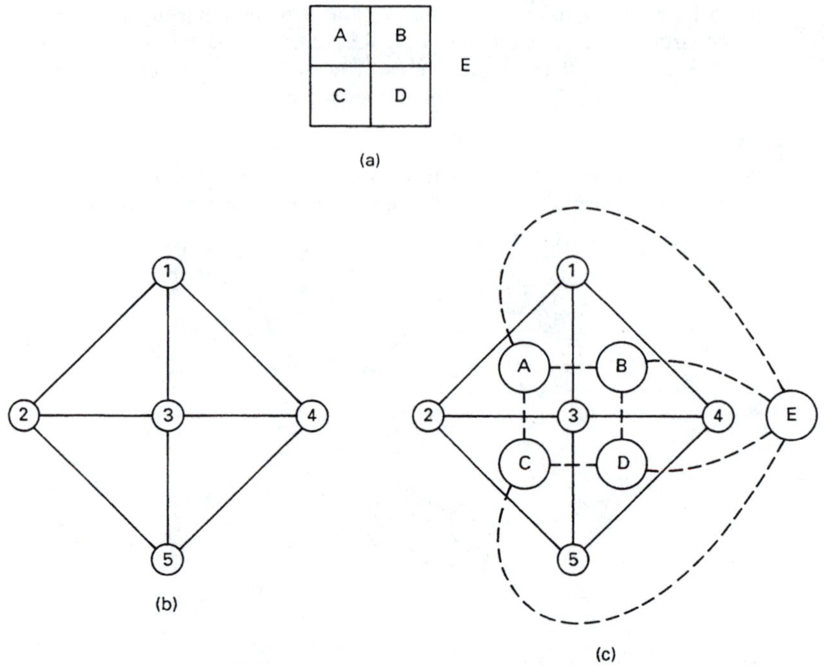

Figure 2.36 Graph representation of block layout, dual, and REL diagram for Example 2.4.

convert the block layout to a graph each *corner point* in the layout becomes a vertex of the graph; a corner point is a point where at least three departments meet, including the exterior department. The primal graph is given in part (b), with the dual graph shown in part (c). Note that the dual graph is the REL diagram (i.e., all departments are adjacent to one another) since we consider any shared point (the common corner point for *A* and *D* and the common corner point for *B* and *C*) to satisfy the requirements of adjacency.

Example 2.5

To illustrate the application of the graph-based approach in developing the REL diagram, recall the REL chart given in Figure 2.24. Department 11 is added as an exterior department and assigned *A* relationships with departments 1, 5, and 8.

If only the *A* relationships must be satisfied via adjacency, a primal graph and a dual graph are given in Figure 2.37. Radically different results can be obtained depending on how you construct the dual graph. Furthermore, the unit block plan follows directly from the graph of the relationships rather than from the dual graph. Also, as shown in Figure 2.38, depending on the form of the dual graph, department 10 can become the exterior department rather than department 11.

Now, if all *A* and *E* relationships must be satisfied via adjacency, a possible primal graph and dual graph combination is shown in Figure 2.39. Notice that the unit block

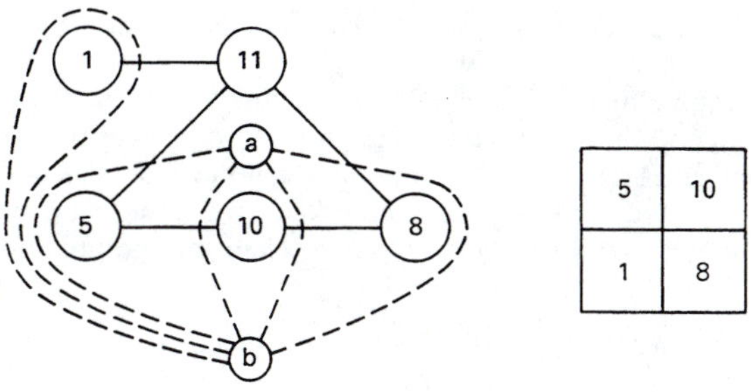

Figure 2.37 Primal and dual graphs of the *A* relationships, with the unit block plan.

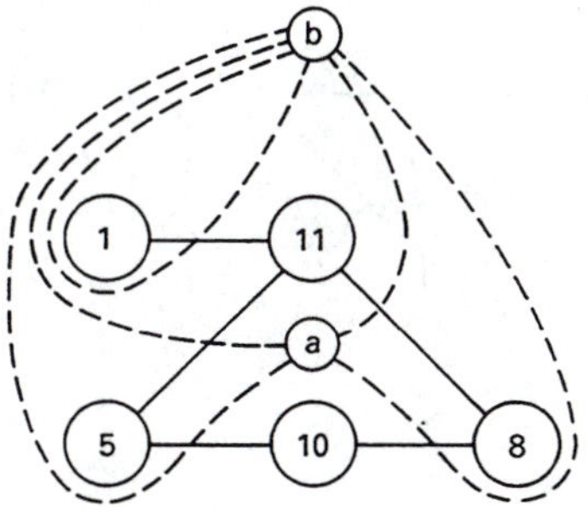

Figure 2.38 Dual graph with department 10 being the exterior department, rather than department 11.

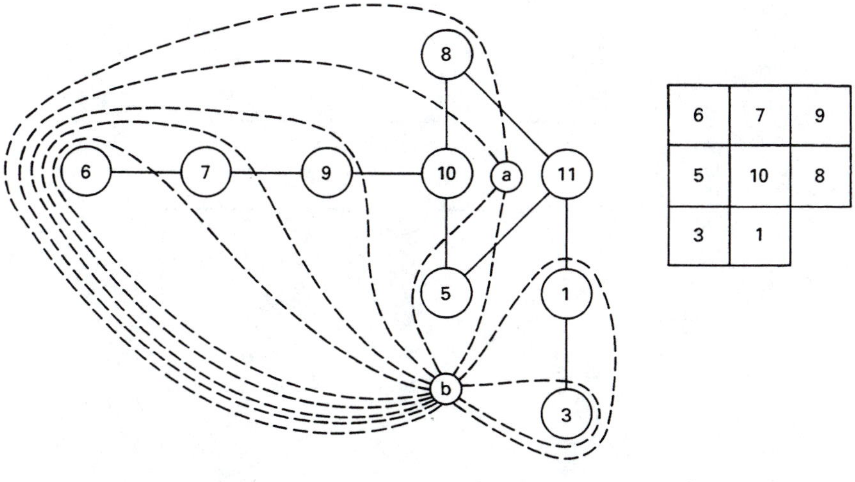

Figure 2.39 Primal and dual graphs of the *A* and *E* relationships, with the unit block plan.

plan followed directly from the primal graph, not its dual. The dual suggests nesting departments (e.g., 3 in 1 and 6 in 7 in 9 in 10).

Next, suppose we require that all A, E, and I relationships be satisfied via adjacency. A possible graph and dual graph combination is given in Figure 2.40.

Finally, suppose that all A, E, I, and O relationships must be satisfied via adjacency. As shown by the boldfaced edge in Figure 2.41, the resulting graph is nonplanar. However, the graph would be planar either by changing the relationship

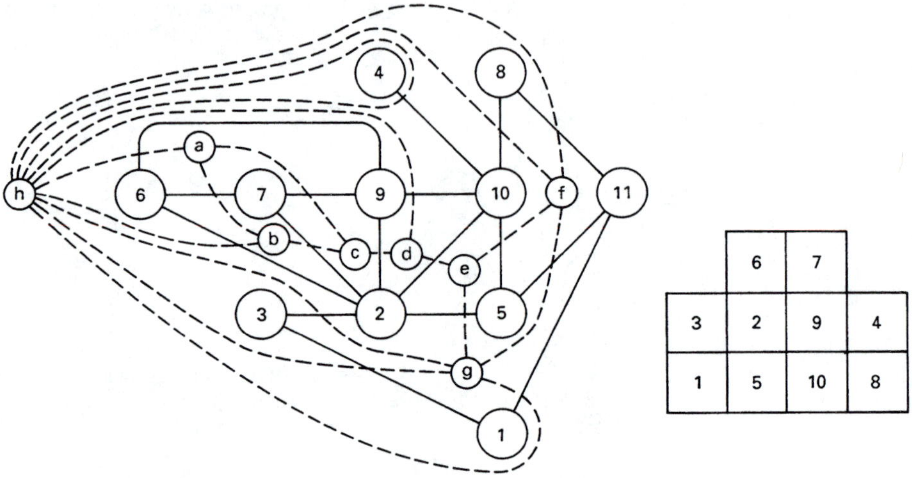

Figure 2.40 Primal and dual graphs of the A, E, and I relationships, with the unit block plan.

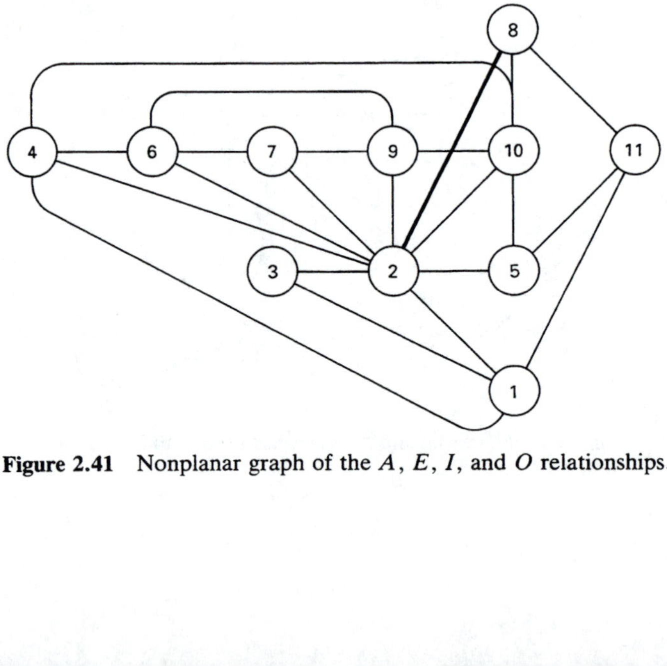

Figure 2.41 Nonplanar graph of the A, E, I, and O relationships.

between departments 2 and 8 from an *O* to a *U* or by relaxing the requirement that the offices be adjacent to the exterior. Figure 2.42 illustrates the latter.

In this case, although the graph is planar, we cannot develop a *unit block plan*. Department 2 has adjacency requirements with nine departments. However, there are only eight adjacent positions available using unit squares—hence the strange-looking block plan.

From the foregoing we see that it is possible to generate a dual graph with the exterior (department 11) being inside the building. Also, "nesting" occurs with high

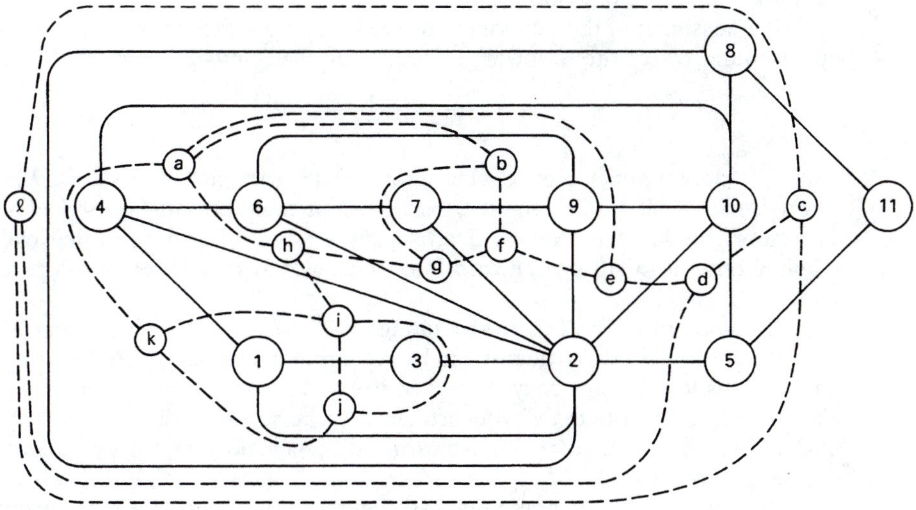

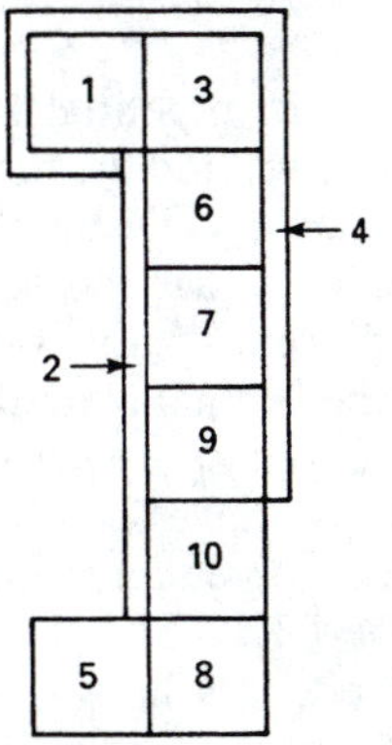

Figure 2.42 Primal and dual graphs of a relaxation of the (1,11) *A* relationship and all other *A*, *E*, *I*, and *O* relationships, with a modified unit block plan.

frequency using the dual graph. However, except for the last case, nesting was unnecessary, as indicated by the unit block plans. Similarly, depending on where we place the vertex for the exterior face and/or how we draw one edge of the primal graph, the dual graph can be turned "inside out."

Although it is not a simple matter to establish the planarity of a graph for a reasonably large layout problem, computer codes do exist for establishing the planarity of a graph [19]. Of interest, an upper bound exists on the number of pairs of adjacent activities. Namely, if there are N activities, no more than $3N - 6$ pairs of activities can be adjacent. Hence if the graph of adjacency relationships has more than $3N - 6$ edges, it cannot be planar.

If at least half of the relationships are U ratings, then for planarity to exist, the upper bound places the following limitation on the number of activities:

$$3N - 6 > \frac{0.50\, N(N-1)}{2}$$

or $N \leq 10$ departments. Hence if more than 10 departments are involved, planarity will not exist if all A, E, I, and O relationships must be satisfied via adjacency. Therefore, for large problems, if adjacency is the basis for satisfying closeness requirements, it might be the case that only A or A and E relationships can be satisfied.

If the distribution of closeness ratings is 5% A, 10% E, 15% I, and 20% O, then as many as 118 departments can be accommodated if only the A relationships must be satisfied via adjacency. However, if A and E relationships are to be satisfied via adjacency, no more than 38 departments can be accommodated. Similarly, if we must satisfy all A, E, and I relationships via adjacency, no more than 18 departments can be accommodated.

Finally, it should be noted that satisfying the upper bound will not necessarily assure the existence of planarity. For the example with 11 departments, nonplanarity resulted when 22 relationships were considered, which is less than the upper bound of 27.

We have noted a number of potential shortcomings with the graph-based approach. Among those, the one we find the most difficult to accept is the interpretation of closeness to mean adjacency. Distance itself can be a questionable determinant of relationship satisfaction, much less adjacency. Furthermore, adjacency is binary, activities are either adjacent or not, whereas there are various degrees of closeness expressed through the ordinal scale used (A and $X > E > I > O > U$).

In summary, the graph-based approach provides a structured approach for developing the REL diagram. We have found the graph-based approach useful in developing activity-based block layouts. It emphasizes the importance of constructing a planar graph of the REL chart if a block plan is to be constructed to satisfy the relationships. Indeed, it is known that an *equivalent condition* for all the adjacencies of the REL diagram to be achieved is that the REL diagram be a planar graph.

Despite its usefulness in developing activity-based block layouts, there are limitations to its use; we have noted several, including the interpretation of "closeness" to mean adjacency and the resulting peculiar shapes of departments to satisfy adjacency requirements. Additionally, there is no well-defined algorithm for drawing the planar graph; hence a wide variety of planar and dual graphs can be developed for the same set of activity relationships. Finally, the planarity requirement itself can be a limitation. Just because a set of relationships cannot be satisfied via a planar graph does not mean that a block layout cannot be developed; instead, it means that it is not possible to satisfy all relationships with adjacency.

As with other analytical techniques, the objective in using the graph-based approach is to aid the layout designer. To the extent that it allows the designer to be more effective and efficient in designing layouts, the graph-based process can be a valuable extension to traditional layout design.

2.7.4 Relationship Diagramming Process

A number of attempts have been made to develop an algorithm for constructing the activity relationship diagram, including a number of computer algorithms [17]. Although we treat the subject of computerized layout in Chapter 3, it is instructive to consider one possible algorithm for generating the activity relationship diagram; the algorithm we describe is similar to the CORELAP algorithm considered in Chapter 3 and is easily applied manually.

In general, layout algorithms are of two types: improvement and construction. The *improvement algorithm* begins with an initial layout and searches for improved solutions, whereas the *construction algorithm* adds departments to the layout, one by one, until all departments have been placed. (Of course, it is possible to develop an algorithm that constructs a layout and then seeks to improve it.)

In this chapter we consider a simple construction algorithm for generating the REL diagram. Before presenting the algorithm, we note that any construction algorithm for REL diagrams must address two issues:

1. *The order of placement* of departments in the REL diagram
2. Their *relative locations*

Further, it should be obvious that many variations to the following algorithm are possible. We make no claim that our algorithm is better than one you might develop.

Example 2.6

To illustrate our algorithm, consider the REL chart given in Figure 2.24. Our algorithm depends strongly on maintaining the ordered ranking of the closeness rankings. To facilitate the process, the following numerical values are assigned to the closeness ratings: A = 10,000; E = 1000; I = 100; O = 10; U = 0; and X = −10,000. (You might choose to assign different numerical values in your algorithm!) The sum of the absolute values for the relationships with a particular department is referred to as the *total closeness rating* (TCR). The total closeness ratings are given in Table 2.3.

TABLE 2.3 TOTAL CLOSENESS RATINGS (TCRs) FOR EXAMPLE 2.4

	Department										Summary					
Dept.	1	2	3	4	5	6	7	8	9	10	*A*	*E*	*I*	*O*	*U*	TCR
1	—	*O*	*E*	*O*	*U*	*U*	*U*	*U*	*U*	*U*	0	1	0	2	6	1,020
2	*O*	—	*I*	*O*	*I*	*I*	*I*	*O*	*I*	*I*	0	0	6	3	0	630
3	*E*	*I*	—	*U*	*U*	*U*	*U*	*U*	*U*	*U*	0	1	1	0	7	1,100
4	*O*	*O*	*U*	—	*U*	*O*	*U*	*U*	*U*	*I*	0	0	1	3	5	130
5	*U*	*I*	*U*	*U*	—	*U*	*U*	*U*	*U*	*A*	1	0	1	0	7	10,100
6	*U*	*I*	*U*	*O*	*U*	—	*E*	*U*	*I*	*U*	0	1	2	1	5	1,210
7	*U*	*I*	*U*	*U*	*U*	*E*	—	*U*	*E*	*U*	0	2	1	0	6	2,100
8	*U*	*O*	*U*	*U*	*U*	*U*	*U*	—	*U*	*A*	1	0	0	1	7	10,010
9	*U*	*I*	*U*	*U*	*U*	*I*	*E*	*U*	—	*E*	0	2	2	0	5	2,200
10	*U*	*I*	*U*	*I*	*A*	*U*	*U*	*A*	*E*	—	2	1	2	0	4	21,200

The first department placed in the layout will be the one with the greatest TCR value. If ties exist, the one with the greatest number of *A* relationships is placed first. If ties still exist, we break the tie randomly. For the example, department 10 is placed in the layout first, because of its TCR value of 21,200.

Next, if a department has an *X* relationship with the first department placed, it will be placed in the layout last and labeled the "last placed department." If more than one department has an *X* relationship with the first one placed, the one having the smallest TCR will be placed in the layout last. (No department has an *X* relationship with department 10, so no department has been labeled the "last placed department.")

The second department placed in the layout will be one with an *A* relationship with the first department placed and having the greatest TCR value; again, if a tie exists, we break the tie randomly. In the example, departments 5 and 8 have an *A* relationship with department 10, but department 5 has the greatest TCR value, so it is placed in the layout second.

If a department has an *X* relationship with the second department placed and no department had an *X* relationship with the first department placed, it will be placed in the layout last and labeled the "last placed department." If more than one department has an *X* relationship with the second one placed, the one having the smallest TCR will be placed in the layout last. If a department had an *X* relationship with the first department, any department having an *X* relationship with the second department placed will be labeled the "next-to-last placed department" and placed in the layout accordingly. (No department has an *X* relationship with department 5, so no department has been labeled the last placed department or the next-to-last placed department.)

The third department placed in the layout is to be the one having the greatest TCR from among those having an *A* relationship with one of the placed departments. If none exist, we choose the department with the greatest TCR from among those having an *E* relationship with one of the placed departments. If none exist, we use the same process, but considering *I* relationships, followed by *O* relationships. At this point, if the next department to be placed has not been identified, we choose one randomly from the list of unplaced departments, excluding the department designated as the last placed department. (For the example problem, department 8 is placed third, due to its *A* relationship with department 10.)

In a manner similar to that used to designate the last placed department and the next-to-last placed department, a check is made to identify departments having an X relationship with the third department placed. If any such departments exist, one will be selected for placement in the layout either last, next to last, or second from last. (Since no X relationships exist in the example, no such designations are made.)

The fourth department is identified using the same procedure as used to identify the third department. (For the example, department 9 is placed in the layout, due to its E relationship with department 10.)

The process continues until all departments have been placed in the layout. In the example, the following sequence of placements occurs: 7, due to an E relationship with 9; 6, due to an E relationship with 7; 2, due to its I relationships with the placed departments and its TCR value; 3, due to its I relationship with department 2; 1, due to its E relationship with 3; and 4, because it is last.

Having generated the placement sequence, we now determine the relative locations of the departments. Since each department is represented by a unit square, its placement will be determined by its *weighted placement value*. Drawing on the use of adjacency in the graph-based procedure, the weighted placement value is determined by summing the numerical values for all pairs of adjacent departments.

Our algorithm for determining the placement of the "new" activity begins at the "western edge" of the partial layout and evaluates all possible locations in counterclockwise order; we then assign the new activity to the location with the greatest weighted placement value. If ties exist, we assign it to the first location encountered having the greatest weighted placement value. The sequential placement of the departments is illustrated in Figures 2.43 to 2.52.

In Figure 2.43, locations 1, 3, 5, and 7 are "fully adjacent" and locations 2, 4, 6, and 8 are "partially adjacent" to activity 10. Hence locations 2, 4, 6, and 8 receive half the weight given to locations 1, 3, 5, and 7. Since department 5 has an A relationship with department 10, locations 1, 3, 5, and 7 have a weighted placement rating of 10,000 and locations 2, 4, 6, and 8 have a weighted placement rating of 5000. Thus department 5 is placed in location 1, as shown in Figure 2.44.

Since department 8 has an A rating with department 10 and a U rating with department 5, location 4 in Figure 2.44 is the preferred location for department 8, as shown in Figure 2.45. Similarly, department 9's E relationship with department 10 and

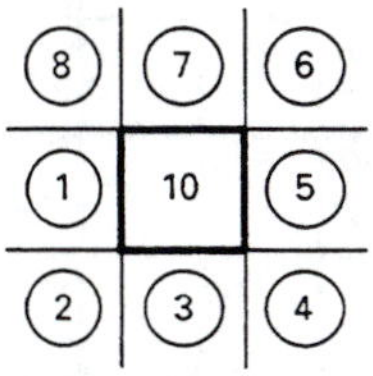

Figure 2.43 Eight alternate "adjacent" locations for the second activity placed.

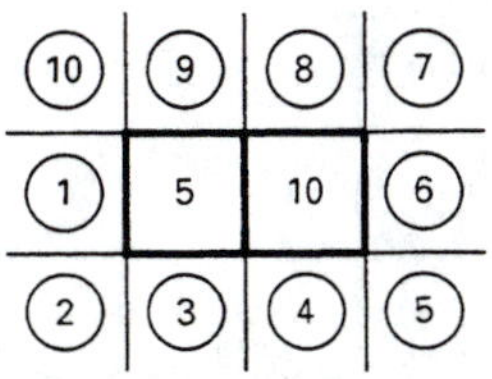

Figure 2.44 Ten alternate "adjacent" locations for the third activity placed.

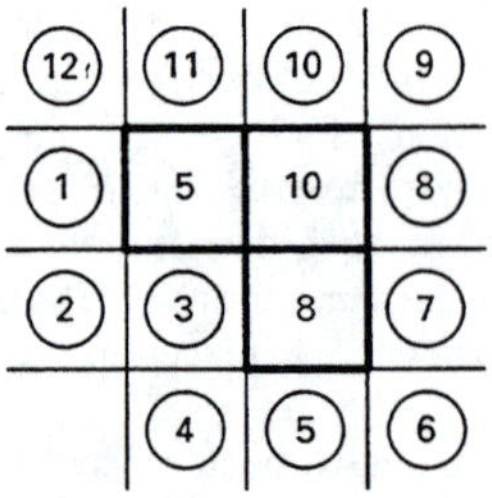

Figure 2.45 Twelve alternate "adjacent" locations for the fourth activity placed.

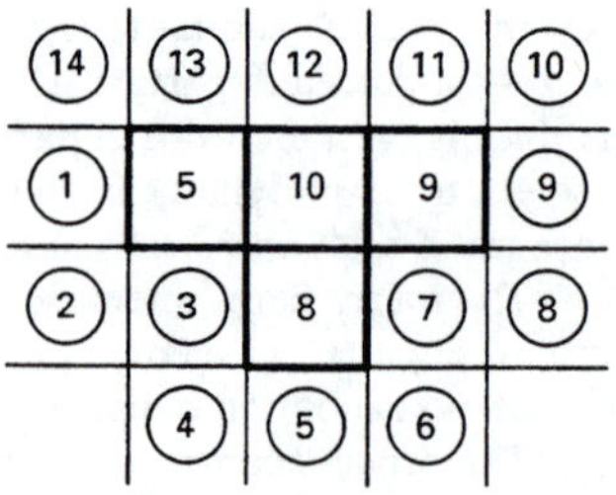

Figure 2.46 Fourteen alternate "adjacent" locations for the fifth activity placed.

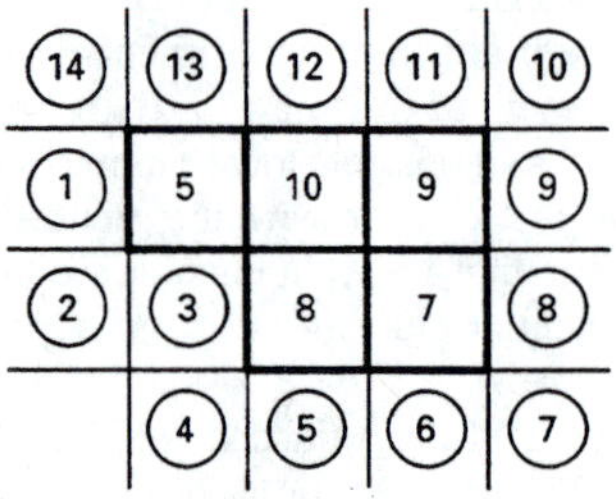

Figure 2.47 Fourteen alternate "adjacent" locations for the sixth activity placed.

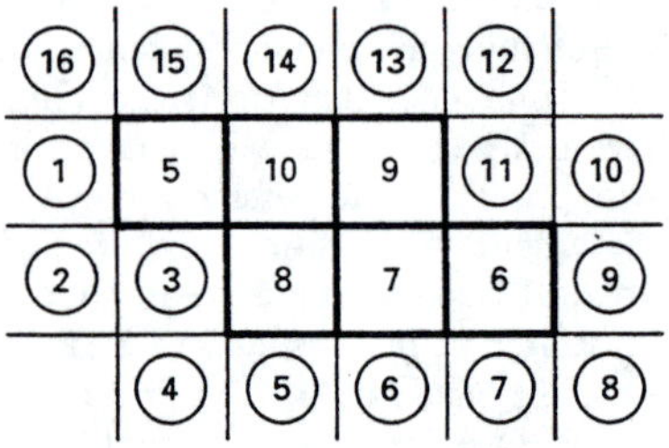

Figure 2.48 Sixteen alternate "adjacent" locations for the seventh activity placed.

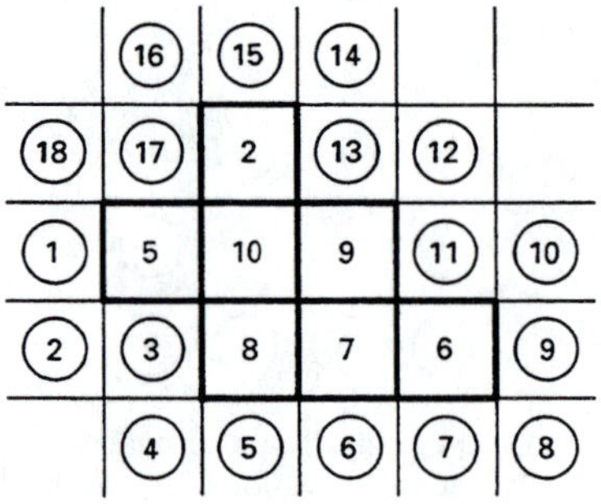

Figure 2.49 Eighteen alternate "adjacent" locations for the eighth activity placed.

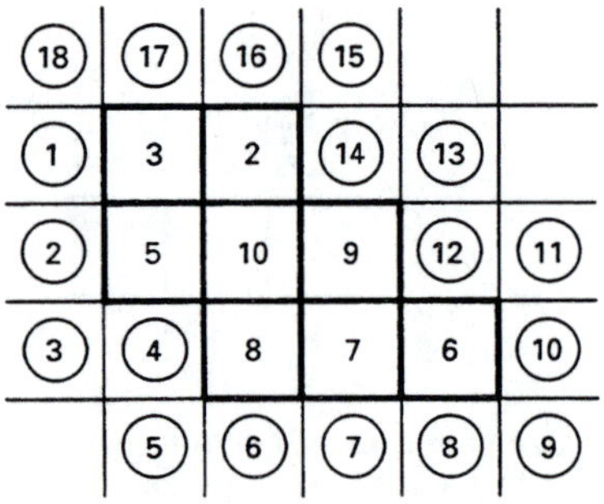

Figure 2.50 Eighteen alternate "adjacent" locations for the ninth activity placed.

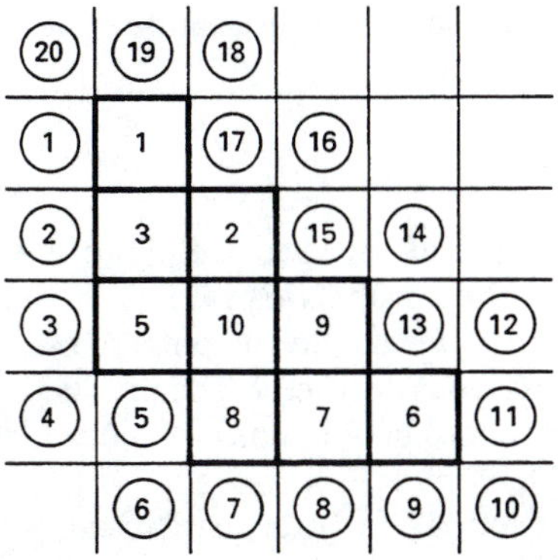

Figure 2.51 Twenty alternate "adjacent" locations for the tenth activity placed.

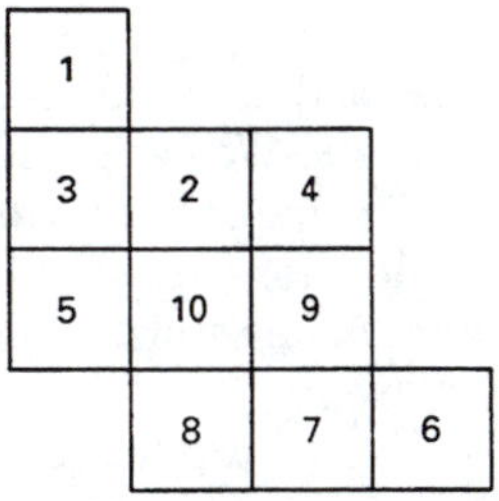

Figure 2.52 Activity relationship diagram for Example 2.6.

U relationship with departments 5 and 8 result in it being placed in location 8 in Figure 2.45, as depicted in Figure 2.46.

As shown in Figures 2.47 to 2.52, the process continues in like manner until all departments are placed. To illustrate the computation of the weighted placement rating when more than one "adjacent" department has a nonzero rating with respect to the next department to enter the layout, consider the entry of department 6 into the layout. Since department 6 has an *E* relationship with department 7 and an *I* relationship with department 9, location 8 in Figure 2.47 has a weighted placement rating of 1050, due to 1000 for the fully adjacent position with department 7 and the partially adjacent position with department 9.

Relationship diagramming is an important step in layout design. In Chapter 3 we describe a number of computer-aided layout approaches to designing layouts. Some of these are based on the relationship chart and generate relationship diagrams. In fact, the algorithm we presented for relationship diagramming is based on one such computer program, CORELAP [24].

2.8 SPACE REQUIREMENTS AND AVAILABILITY

Once consideration has been given to the flow of materials and the relationship of activities, and the appropriate relationship diagram has been constructed, we are in the position to evaluate the space requirements for the layout. Ideally, we would like to develop the layout and then construct the building around the layout. However, from a practical viewpoint, we often find that our solution is constrained by the amount and configuration of available space. The constraint can be in the form of an existing building, a limitation on the size of the building site, or the availability of capital for new construction. For this reason, we must consider not only space requirements, but space availability as well.

2.8.1 Production Rate Determination

One of the major determinants of the amount of space required is the desired production rate. We have assumed the desired production rate has been previously determined by the schedule designer. Recall, we previously pointed out that schedule design decisions are closely related to layout decisions. The production rate was used in our volume–variety analysis to guide us in our choice of a product layout or a process layout.

How is the production rate determined? In brief, a marketing forecast is translated into required production quantities. As an illustration, consider an example involving seven operations and three inspections, as shown in Figure 2.53(a). We assume that bad-quality parts are rejected only at the inspections, and that the fraction defective equals 3, 2, and 4%, respectively, at the three inspection stations.

Suppose that there are 2000 operating hours in a year, and the annual demand for the product is forecast to be 180,000 units. Production planning and marketing agree that 200,000 good units should be produced to protect against forecasting errors. Consequently, 100 good units are to be produced per hour in order to maintain a balanced production line.

Since 100 good units must leave the third inspection station and, on the average, 4% of the parts entering the station are bad, then $100/0.96 \doteq 104.2$ parts should enter the third inspection station. Consequently, operations 6 and 7 must produce 104.2 parts per hour. The second inspection station rejects 2% of the incoming product. Therefore, $104.2/0.98 \doteq 106.3$ parts must be produced per hour by operations 4 and 5. Operations 1, 2, and 3 must produce $104.2/0.97 \doteq 107.4$ parts per hour.

In a process layout a given machine can be used to process a variety of different products. Typically, a given job is produced for a period of time, and then a new job is produced. The scheduling problem in batch production is a complex problem. If we are producing to inventory when equipment capacity far exceeds the demand level, then optimum batch production quantities might be computed using an appropriate inventory control model [15]. There still remains the problem of scheduling the batches on each machine. Since this is not a layout problem, the scheduling problem is not pursued further in this book. However, depending on the organization, you might not be functioning strictly as a layout analyst and might also be involved in scheduling decisions.

Process layouts are also used in job shops where "one-shot" jobs are received and processed. Rather than producing to inventory, the order is processed and shipped to the customer. In such a situation we might be tempted to determine the production quantity in much the same way as for the product layout. As an illustration, suppose that an order is received to produce 50 units of a production item having the processing sequence given in Figure 2.53(b). Suppose that, on the aver-

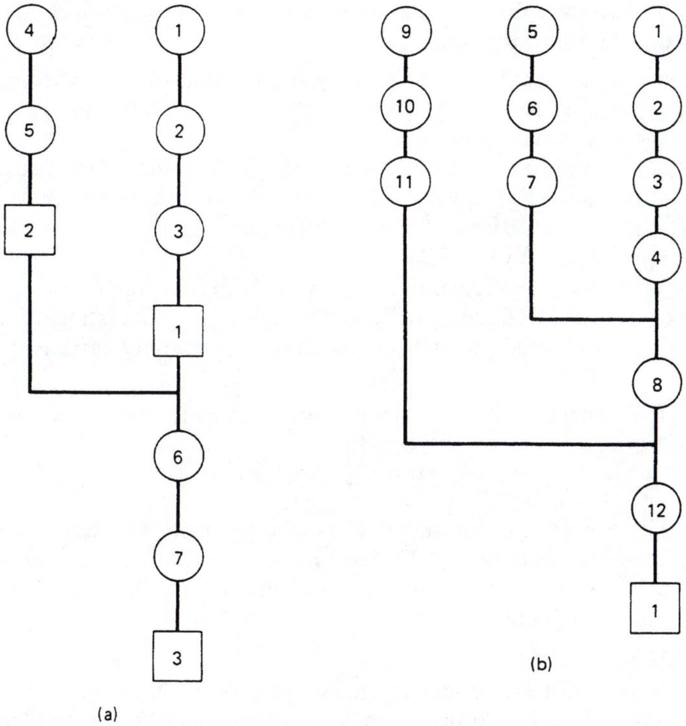

Figure 2.53 Operation process charts of manufacturing processes.

age, 10% of the units inspected are rejected. Consequently, we might schedule 50/0.90 ≐ 55.56, or 56 units to be produced. What is wrong with this approach?

In the case of the product layout, a large number of items are being produced over a long period of time. An additional 20,000 units were added to the annual production rate as a safety factor against forecasting errors. Consequently, it was reasonable to use "expected values" in the calculations. In the job shop, we are producing the batch only once. If 56 units are produced there might be only 40 good units available. In this case another setup must be made and the remainder produced. But how many should be scheduled? In fact, the production lead time might be so large that no additional units could be produced in time to meet the contracted delivery date. Suppose it is found that there are 55 good units in the 56 produced. What is done with the excess?

One approach that can be taken to answer a number of these questions is to formulate an expected profit model of the production situation and determine the optimum batch size. In such a model, the costs of producing too many units is balanced against the cost of producing too few units. The problem is referred to as the *reject allowance problem* [59].

To facilitate the formulation of one version of the reject allowance problem, consider the following notation.

X = random variable denoting the number of good castings produced
$p(x)$ = probability of producing exactly x good castings
Q = production lot size
$R(Q,X)$ = revenue resulting from producing Q castings of which exactly X are good
$C(Q,X)$ = cost of producing a lot of size Q, of which exactly X are good
$P(Q,X)$ = profit resulting from producing Q castings, of which exactly X are good
$= R(Q,X) - C(Q,X)$
$E[R(Q)]$ = expected revenue resulting from producing Q castings
$E[C(Q)]$ = expected cost resulting from producing Q castings
$E[P(Q)]$ = expected profit resulting from producing Q castings
$= E[R(Q)] - E[C(Q)]$

The expected profit model can be given as follows:

$$E[P(Q)] = \sum_{x=0}^{Q} \{R(Q,x) - C(Q,x)\}p(x) \tag{2.1}$$

The objective of the reject allowance problem is to determine the value of Q that maximizes equation (2.1). We provide the following example problem as an illustration of the reject allowance problem; it is explored further in the exercises at the end of the chapter.

Example 2.7

Suppose that a small foundry receives an order for 20 specially designed castings. The owner of the foundry quotes a price of \$4000 each for the castings. Only one production run can be made due to the customer's need for the castings "right away"! If fewer than 18 good castings are produced, the customer will not accept any of the castings; if 18, 19, or 20 good castings are produced, the customer will purchase all good castings; if more than 20 good castings are produced, the customer will only pay for 20. The remaining castings, whether good or bad, can be recycled in the foundry; hence each has a value equal to the material cost of \$500. In addition to the material cost, the cost of producing a casting is estimated to be \$2250.

The revenue and cost functions can be expressed as follows:

$$R(Q,x) = \begin{cases} \$500Q, & x = 0,\ldots,17 \\ \$4000x + \$500(Q - x), & x = 18, 19, 20 \\ \$4000(20) + \$500(Q - 20), & x = 21,\ldots,Q \end{cases}$$

and

$$C(Q,x) = \$2750Q$$

As an exercise at the end of the chapter, you are asked to determine the conditions for which equation (2.1) is a convex function. Given that such conditions exist, any of a number of unimodal search techniques can be used to determine the lot size that maximizes expected profit; otherwise, enumeration is required.

For the example, since $C(Q,x)$ is not a function of the number of good units produced, the expected profit can be computed as follows:

$$E[P(Q)] = \sum_{x=0}^{Q} \{R(Q,x)p(x)\} - C(Q)$$

The expected revenue reduces to

$$E[R(Q)] = \$500Q + \$3500 \sum_{x=18}^{20} xp(x) + \$60{,}000 \sum_{x=21}^{Q} p(x)$$

Therefore, the expected profit can be computed as follows:

$$E[P(Q)] = \$3500 \sum_{x=18}^{20} xp(x) + \$60{,}000 \sum_{x=21}^{Q} p(x) - \$2250Q$$

To complete the example, suppose that the probability mass function for the number of good castings is as given in Table 2.4. The probabilities are based on past work performed of similar complexity. The computation of the expected profit is given in Table 2.5. From the results, a lot of size 25 should be scheduled if expected profit is to be maximized.

An alternative objective might be to choose the smallest lot size having a "zero probability" of shortage. In such a case, 29 units would be scheduled for production, with an expected loss of \$5250.

TABLE 2.4 PROBABILITY MASS FUNCTIONS FOR EXAMPLE 2.7

Lot	Number of good units produced																
size	14	15	16	17	18	19	20	21	22	23	24	25	26	27	28	29	30
20	0.1	0.1	0.1	0.2	0.2	0.2	0.1	—	—	—	—	—	—	—	—	—	—
21	0.1	0.1	0.1	0.1	0.2	0.2	0.1	0.1	—	—	—	—	—	—	—	—	—
22	—	0.1	0.1	0.1	0.1	0.2	0.2	0.1	0.1	—	—	—	—	—	—	—	—
23	—	—	0.1	0.1	0.1	0.1	0.2	0.2	0.1	0.1	—	—	—	—	—	—	—
24	—	—	0.1	0.1	0.1	0.1	0.1	0.1	0.2	0.1	0.1	—	—	—	—	—	—
25	—	—	—	—	—	0.1	0.1	0.1	0.2	0.2	0.2	0.1	—	—	—	—	—
26	—	—	—	—	—	0.1	0.1	0.1	0.1	0.2	0.2	0.1	0.1	—	—	—	—
27	—	—	—	—	—	—	0.1	0.1	0.1	0.1	0.2	0.2	0.1	0.1	—	—	—
28	—	—	—	—	—	—	—	0.1	0.1	0.1	0.1	0.2	0.2	0.1	0.1	—	—
29	—	—	—	—	—	—	—	0.1	0.1	0.1	0.1	0.1	0.1	0.2	0.1	0.1	—
30	—	—	—	—	—	—	—	—	—	—	0.1	0.1	0.1	0.2	0.2	0.2	0.1

TABLE 2.5 DETERMINING THE OPTIMUM PRODUCTION LOT SIZE

Lot size	$3500 \sum_{x=18}^{20} xp(x)$	$60{,}000 \sum_{x=21}^{Q} p(x)$	Expected profit
20	\$26,600	—	(\$18,400)
21	26,600	6,000	(15,150)
22	33,600	12,000	(3,900)
23	26,950	24,000	800
24	19,950	30,000	4,050
25	13,650	48,000	5,400
26	13,650	48,000	3,150
27	7,000	54,000	250
28	—	60,000	(3,000)
29	—	60,000	(5,250)
30	—	60,000	(7,500)

2.8.2 Equipment Requirements

Given the desired production rate at each processing stage, we can determine the number of machines required. To carry out the calculation, we need to know the operation efficiencies for the equipment and the standard production times for each operation. To make the process explicit, let

P_{ij} = desired production rate for product i on machine j, measured in pieces per production period
T_{ij} = production time for product i on machine j, measured in hours per piece
H_{ij} = number of hours in the production period available for the production of product i on machine j
M_j = number of machines of type j required per production period
n = number of products

Therefore, M_j can be expressed as

$$M_j = \sum_{i=1}^{n} \frac{P_{ij} T_{ij}}{H_{ij}} \tag{2.2}$$

As an illustration of the use of equation (2.2), consider the data displayed in Table 2.6. The same type of machine is used in the processing of six different products or parts. The standard number of pieces to be produced per hour, $1/T_{ij}$, is given for each product, along with the desired production rate, P_{ij}. The number of hours in a month is 150. Therefore, for this example M_j equals 2.643. In this case, we will probably use three machines for the production of the six products. If an increase in demand is anticipated or if protection is to be provided against loss of production due to machine breakdown, four machines might be purchased.

All the calculations were based on deterministic values for P_{ij} and T_{ij}. Realistically, there will be some variation in the number of units to be produced and the processing times. If such variation is believed to be significant, one might develop

TABLE 2.6 SAMPLE DATA FOR MACHINE REQUIREMENT CALCULATION

Product number, i	Required production rate, P_{ij}	Standard production rate, $1/T_{ij}$	Hours/month, H_{ij}	$P_{ij}T_{ij}/H_{ij}$
2501	6,000	120	150	0.333
2502	9,000	150	150	0.400
3104	15,000	100	150	1.000
3206	2,000	100	150	0.133
3617	8,000	120	150	0.444
3618	4,000	80	150	0.333
				2.643

the probability distribution for M_j and determine the optimum number of machines to purchase, based on the costs involved. However, such determinations seldom are made by the layout analyst.

2.8.3 Employee Requirements

Previously, we considered ways of determining the desired production rate and the number of machines required to satisfy this rate. In this section we discuss the determination of the number of employees required.

In the case of manual assembly operations, we can determine the number of employees required in the same way we calculated machine requirements. That is,

$$A_j = \sum_{i=1}^{n} \frac{P_{ij} T_{ij}}{H_{ij}}$$

where A_j = number of operators required for assembly operation j
P_{ij} = desired production rate for product i and assembly operation j, pieces per day
T_{ij} = standard time to perform operation j on product i, minutes per piece
H_{ij} = number of hours available per day for assembly operation j on product i
n = number of products

The number of machine operators required is dependent on the number of machines tended by one or more operators. In many cases this number is determined by the existing labor contract and the requirements of the job. However, if highly automatic equipment is used, there is the possibility of one operator tending a number of machines. Again, depending on the particular organization, the determination of the number of machines supervised by one operator will be the function of the scheduling analyst or the methods and standards analyst, not the layout analyst. If such is not the case, the following discussion might be beneficial to you in this determination.

A determination of the number of machines to be supervised by one operator can take two approaches. First, you can assume all time values are deterministic. As an alternative, you can treat the activity times as random variables, and perform a probabilistic analysis.

One deterministic approach used to determine the assignment of operators to machines is to employ the multiple activity chart. The multiple activity chart is a descriptive, analog model showing the multiple activity relationships graphically against a time scale. An example of the use of the multiple activity chart is given in Figure 2.54. When the multiple activity chart is used, there still remains the choice of the criterion to be used in determining the number of operators to assign.

The multiple activity chart is especially useful in analyzing the multiple activity relationships when nonidentical machines are being supervised by one operator. Also, the multiple activity chart has been used in selling all concerned on the feasibility of a particular assignment.

As an alternative to the multiple activity chart when identical machines are involved, we present a prescriptive, symbolic model that can be used to determine the number of machines to assign an operator. Let

a = concurrent activity time (e.g., loading, unloading)
b = independent operator activity time (e.g., walking, inspecting, packaging)
t = independent machine activity time (e.g., automatic run time)
n' = number of machines to assign an operator for neither machine nor operator idle time
m = number of machines assigned an operator
T_c = repeating cycle time
I_o = idle operator time during a repeating cycle
I_m = idle time per machine during a repeating cycle
$TC(m)$ = cost per unit produced, based on an assignment of m machines per operator
C_1 = cost per operator-hour
C_2 = cost per machine-hour

Notice that it takes $a + b$ time units for an operator to perform the work content required on a single machine during one complete production cycle. Also, it takes the machine $a + t$ time units to complete a production cycle. Consequently, the ideal assignment, n', would be

$$n' = \frac{a + t}{a + b} \tag{2.3}$$

For the example illustrated in Figure 2.54, a equals 4 minutes, t equals 10 minutes, and b equals 1 minute if we include both inspection and travel between machines. Consequently, n' equals 2.8 machines.

Since a fractional number of machines cannot be assigned to an operator, consider the effects of assigning m machines to an operator, where m is integer valued and $m < n'$. In this case, the operator will be idle and the machine will be kept busy. Consequently, the repeating cycle will equal $a + t$. On the other hand, if $m > n'$, then the operator will be kept busy, and it will require $m(a + b)$ time units to complete a cycle. Therefore,

$$T_c = \begin{cases} a + t, & m \leq n' \\ m(a + b), & m > n' \end{cases} \tag{2.4}$$

Furthermore, it follows that

$$I_o = \begin{cases} (a + t) - m(a + b), & m \leq n' \\ 0, & m > n' \end{cases} \tag{2.5}$$

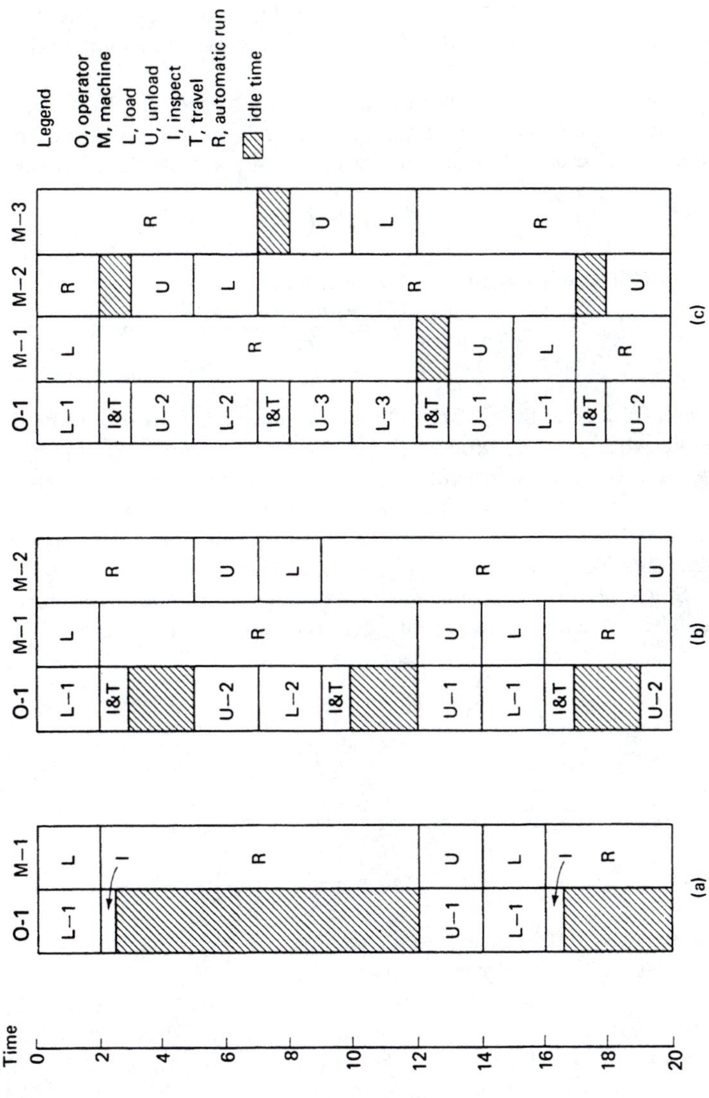

Figure 2.54 Multiple activity chart analysis of multimachine assignment.

and

$$I_m = \begin{cases} m(a + b) - (a + t), & m > n' \\ 0, & m \leq n' \end{cases} \tag{2.6}$$

For our example problem, if m equals two machines, the repeating cycle is 14 minutes, and the operator is idle 4 minutes during a repeating cycle. If m equals three machines, the repeating cycle is 15 minutes, and each machine is idle 1 minute during a repeating cycle.

If we wish to determine the cost per unit produced by an m machine assignment, notice that the cost of such an assignment per unit time equals $C_1 + mC_2$. Furthermore, during a repeating cycle m units are produced. Consequently,

$$TC(m) = (C_1 + mC_2)\frac{T_c}{m} \tag{2.7}$$

Substituting (2.4) in (2.7), we see that

$$TC(m) = \begin{cases} \dfrac{(C_1 + mC_2)(a + t)}{m} & m \leq n' \\ (C_1 + mC_2)(a + b), & m > n' \end{cases} \tag{2.8}$$

From (2.8), we see that if $TC(m)$ is to be minimized, m should be made as large as possible when $m \leq n'$, and m should be made as small as possible when $m > n'$. Consequently, if n' is integer valued, n' machines should be assigned to minimize (2.8). However, if n' is not integer valued, such that $n < n' < n + 1$, where n is the integer portion of n', then either n or $n + 1$ machines should be assigned, depending on whether $TC(n) \leq TC(n + 1)$, or vice versa.

Since we must determine which is more economic, an n machine assignment or an $n + 1$ machine assignment, form the ratio of $TC(n)$ and $TC(n + 1)$ such that

$$\Phi = \frac{TC(n)}{TC(n + 1)}$$

$$= \frac{(C_1 + nC_2)(a + t)}{[C_1 + (n + 1)C_2]n(a + b)} \tag{2.9}$$

Letting $\epsilon = C_1/C_2$ and substituting (2.3) in (2.9) gives

$$\Phi = \frac{\epsilon + n}{\epsilon + n + 1}\frac{n'}{n} \tag{2.10}$$

Consequently, if $\Phi < 1$, assign n machines; $\Phi > 1$, assign $n + 1$ machines; and if $\Phi = 1$, assign either n or $n + 1$ machines.

For the example given in Figure 2.54, suppose that C_1 equals \$3 per hour and C_2 equals \$10 per hour for each machine. From (2.3), n' equals 2.8. Therefore, n equals 2.0 and $n + 1$ equals 3.0. From (2.10), Φ equals 6.44/6.60. Since $\Phi < 1$, we would assign two machines per operator.

The deterministic machine-assignment model is considered further in the problems. The prescriptive, symbolic model is appropriate when identical machines are to be assigned, times are deterministic, and cost per unit produced is to be minimized. If we wish to consider nonidentical machine assignment and/or two or more operators jointly operating two or more machines, then the man-machine chart, a deterministic, descriptive, analog model, can be used.

If we wish to perform a probabilistic analysis, we might use appropriate queueing models or Monte Carlo simulation to determine the number of machines to assign an operator. In fact, when random variation is present, we might find that it is more economical to establish a pool of operators who service a group of machines, rather than have each operator assigned to specific machines.

Queueing theory and simulation are two subjects that can be extremely valuable to the layout analyst. Not only can they be used in determining the number of operators required to tend the production equipment, but they are also useful in balancing assembly lines, designing conveyor systems, and determining maintenance crew sizes, receiving and shipping crew sizes, toolroom attendant requirements, material handling crew sizes, material handling equipment pool sizes, receiving and shipping dock sizes, and rest room facility requirements, to list a few applications that quickly come to mind.

2.8.4 Space Determination

Once we know how many machines will be tended by a single operator, we can determine space requirements for equipment. Also, the space requirements for the assembly department can be determined, since we know the number of assembly operators required.

Some commonly used methods of determining space requirements are introduced and discussed below.

1. *Production-center method.* The production center consists of a single machine plus all the associated equipment and space required for its operation. Work space (front, rear, left side, right side), additional maintenance space, and storage space are added to the space requirements for the machine. As illustrated in Figure 2.55, all equipment and storage locations are arranged in the production center, and the floor space is determined. The space requirement is multiplied by the number of similar pieces of equipment to determine the total space requirement for, say, vertical milling machines.

2. *Converting method.* Using this method, the present space requirements are converted to those required for the proposed layout. If this method is used, be very careful about the assumptions. Remember that total space required is not a linear function of the production quantity. Therefore, just because production doubles, twice as much space is not necessarily needed. Furthermore, you may not be utilizing the present space most efficiently. This method is commonly used to determine space requirements for supporting service and storage areas, whereas the production-center method is used to determine the space requirements for manufacturing areas.

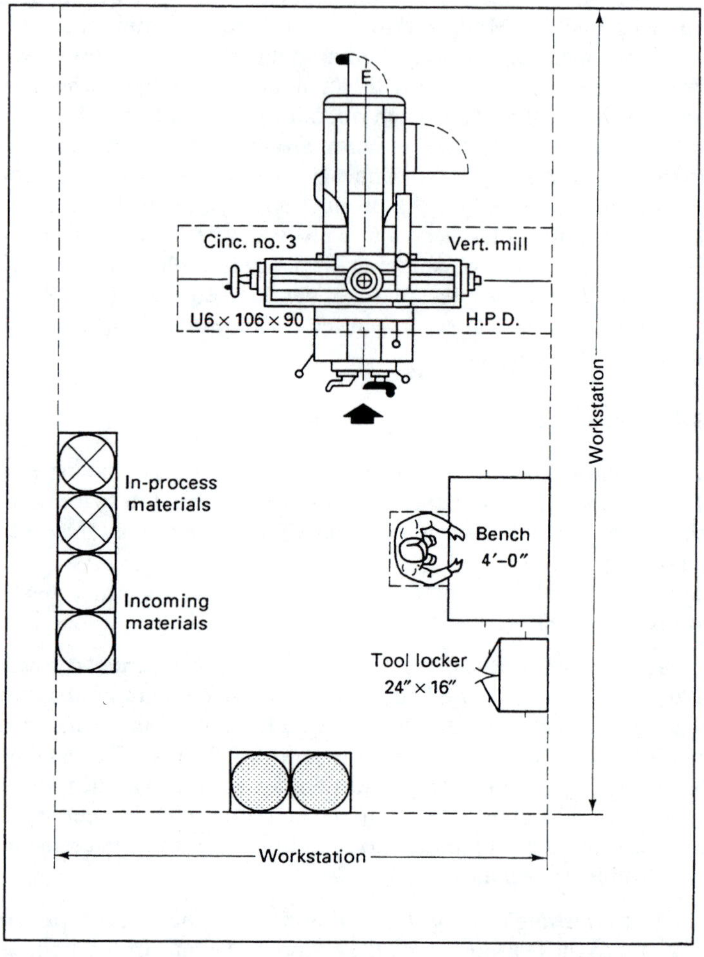

Figure 2.55 Space required for a production center. (Reprinted with permission from *Facilities Planning* by J.A. Tompkins and J.A. White, Copyright © 1984 by John Wiley & Sons, Inc.; reprinted by permission of John Wiley & Sons, Inc.)

3. *Roughed-out layout method.* Templates or models are placed on the layout to obtain an estimate of the general configuration and space requirements.

4. *Space-standards method.* In certain cases industry standards can be used to determine space requirements. Additionally, standards may be established based on past successful applications. The use of such standards without an understanding of their underlying assumptions is dangerous. Such standards adopted by others should be closely scrutinized and compared with the present layout.

5. *Ratio trend and projection method.* This method is limited to general space requirements. It is probably the least accurate of the methods presented. To use this method, one establishes a ratio of square feet to some other factor that can be measured and predicted for the proposed layout. Examples are square feet per direct labor hour, square feet per unit produced, and square feet per supervisor. In making the space determination, space must be included for the following:

1. Raw material storage
2. In-process inventory storage
3. Finished-goods storage
4. Aisles, cross aisles, and main aisles
5. Receiving and shipping
6. Material handling equipment storage
7. Toolrooms and tool cribs
8. Maintenance
9. Packaging
10. Supervision
11. Quality control and inspection
12. Health and medical facilities
13. Food service
14. Lavatories, washrooms, etc.
15. Offices
16. Employee and visitor parking
17. Receiving and shipping parking
18. Other storage

As an illustration of the calculation of space requirements, consider the determination of production space requirements given in Table 2.7. To the 5503 ft^2 of floor space for production, we must add the 4600 ft^2 shown in Table 2.8 to give an estimate of 10,103 ft^2 of total floor space required.

TABLE 2.7 PRODUCTION SPACE REQUIREMENTS

Process	Equipment	No.	Machine center dimensions per machine (ft) Depth Width	Machine center area per machine (ft²)	Total process area (ft²)
Saw	Armstrong hacksaw	3	10×19	190	570
Mill	K & T plain mill	5	$13\frac{1}{2} \times 10\frac{1}{2}$	142	710
	Vertical mill	7	$11 \times 10\frac{1}{4}$	113	791
	Hand mill	4	$7\frac{1}{4} \times 9\frac{3}{4}$	71	284
Drill	2-Spindle Avey	2	$8\frac{1}{4} \times 6\frac{1}{2}$	54	108
	1-Spindle Delta	2	$7\frac{1}{2} \times 4\frac{1}{2}$	34	68
	6-Spindle Delta	1	$7\frac{3}{4} \times 10\frac{1}{2}$	82	82
Turn	Gisholt	1	$9\frac{1}{4} \times 17\frac{3}{4}$	164	164
	Monarch	2	$14 \times 6\frac{1}{4}$	88	176
	Hardinge	1	$9\frac{1}{2} \times 5$	48	48
	W & S turret	1	$8\frac{1}{2} \times 20\frac{1}{4}$	173	173
	B & S automatic	1	$7\frac{1}{2} \times 15\frac{1}{2}$	116	116
Form	Gas furnace	1	8×7	56	56
	Arbor press X				
Paint	Dip tank	2	7×12	84	168
	Spray booth	1	9×11	99	99
Clean	Tumble	1	7×6	42	42
Assemble	Bench	1	8×7	56	56
	Bench	1	8×7	56	56
	Avey drill	2	$8\frac{1}{4} \times 6\frac{1}{2}$	54	108
Packaging	Bench	1	8×7	56	56
			Total square feet required		3931
			40% aisle space		1572
			Production space required		5503

TABLE 2.8 NONPRODUCTION ACTIVITY SPACE REQUIREMENTS

Activity	Area (ft²)
Storage	
Warehousing	540
Other	540
Office	
Main office	1500
Hallway	360
Rest rooms	225
Locker rooms	
Men	250
Women	275
Foreman	
Desk	25
Maintenance	
Desk	25
Parts	160
Tool crib	200
Receiving and shipping	500
Total space required	4600

2.9 DESIGNING THE LAYOUT

Having analyzed the flow of materials and the relationship of activities, determined the space requirements, and made space allocations to activities, alternative layouts can be designed. In terms of the phases of the design process, we have completed the analysis phase and are entering the search phase. A number of alternative layouts are to be designed, based on the analysis of flow, activity relationships, and space requirements.

Our consideration of the layout design process includes the design of the space relationship diagram, block plan, and detailed layout; the design of flexible layouts; the design of the material handling system; and the presentation of the layout design.

2.9.1 Designing the Space Relationship Diagram, the Block Plan, and Detailed Layout

In brief, the overall layout is designed by first combining space considerations with the REL diagram. Following the SLP approach, the impact of the space requirements is manifested in the design of the space relationship diagram. Maintaining the same spatial relationships as in the REL diagram, the space relationship diagram is constructed by replacing the unit squares with space templates.

For each activity, a space template is constructed, to scale, to represent the size and shape of the activity. Since different shapes can have the same area, it is possible to construct different space relationship diagrams from the same REL diagram.

Example 2.8

Recall the REL diagram given in Figure 2.25 and consider the space requirements given in Table 2.9. Figure 2.56 gives one resulting space relationship diagram.

After constructing the space relationship diagram, the space templates are modified and adjusted and the relative locations of the activities are shifted as

TABLE 2.9 SPACE REQUIREMENTS FOR EXAMPLE 2.8[a]

Number	Department name	Area (ft^2)	Number of unit squares	Template dimensions
1	Offices	1200	48	6 × 8
2	Foreman	150	6	2 × 3
3	Conference room	300	12	3 × 4
4	Parcel post	400	16	4 × 4
5	Parts shipment	600	24	4 × 6
6	Repair and service parts	300	12	3 × 4
7	Service area	900	36	6 × 6
8	Receiving	600	24	4 × 6
9	Testing	1000	40	5 × 8
10	General storage	3000	120	10 × 12
		8450 ft^2	338	

[a] Scale: 1 unit square = 5 ft × 5 ft = 25 ft^2.

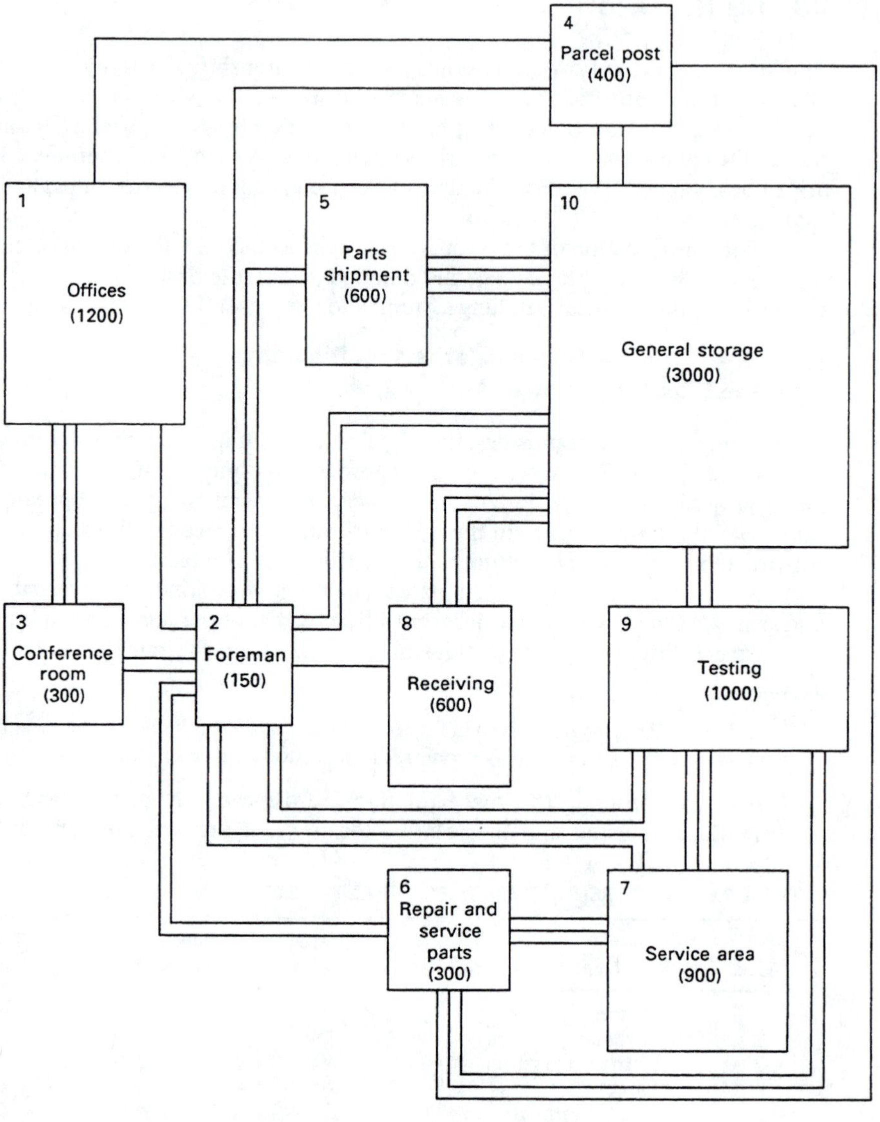

Figure 2.56 Space relationship diagram.

necessary to accommodate practical limitations and other considerations. The resulting block plan is then constructed.

The block plan is a scaled diagrammatic representation of the building and normally shows the locations of internal partitions and columns. Detailed locations of machinery and facility components are usually included in the block plan. Examples of alternative block plans based on the space relationship diagram in Figure 2.56 are given in Figure 2.57.

Once a number of block plans has been generated, the designer has the choice of seeking approval of the preferred block plan before proceeding with detailed design or performing detailed design before the selection step is performed. Again, the choice depends on the situation and the culture of the organization. If we are free to choose, we prefer to narrow the alternative block plans to the preferred one before performing detailed design. There will be at least as many alternative ways of designing the details of each department as there are alternative block plans. If sequential screening is not performed, the number of choices will become unwieldy.

2.9.2 Designing Flexible Layouts

In designing both the overall layout and the detailed layout, we must remember to consider the possibility of future expansion, future compression, and other types of changes. It is extremely important that the layout be flexible enough to accommodate, say, changes in product design, process design, and schedule design. In general, layout studies result from changes that occur in the requirements for space, equipment, and people. At the beginning of the chapter we enumerated a number of causes for layout problems. If we design a flexible layout initially, we can postpone the need for a major redesign until much later.

In a sense, the best way to achieve a flexible layout is to anticipate the changes that might occur. For example, if expansion is anticipated, what kind of expansion will occur? There are basically two types of expansion: expansion in the sizes of existing activities and expansion in the number of activities to be included (e.g., increasing the capacity to produce existing products versus increasing capacity to support the introduction of new products). The former occurs with increases in demand; the latter occurs with increases in the scope of activity supported by the layout.

2.9.3 Designing the Material Handling System

A detailed layout should not be designed without giving consideration to material handling requirements. The choice of handling methods and equipment is an integral part of layout design. It is extremely important to incorporate effective material handling methods in the layout. Material handling involves *moving*, *storing*, and *controlling* material. A systems view of material handling would result in a definition similar to the following: *providing the right amount of the right material at the right place, at the right time, in the right sequence, in the right position, in the right condition, in the right orientation, and at the right cost by using the right method(s).*

The design of material handling systems follows basically the same sequence of steps outlined for designing plant layouts. Many of the tools we have employed in the analysis of the layout problem are commonly used in analyzing material

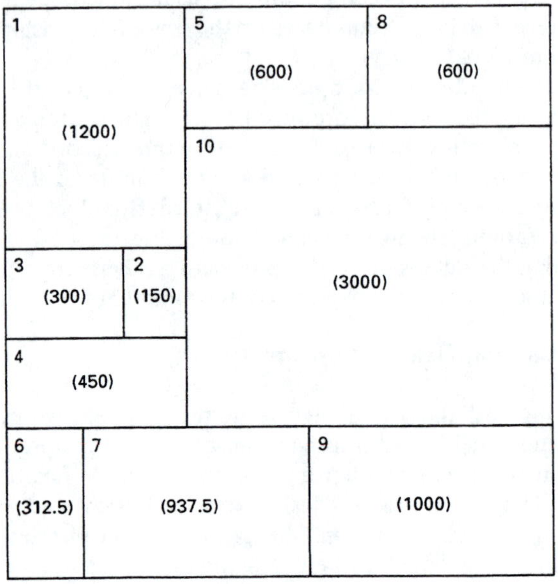

(a) Total area = 8,550 sq. ft

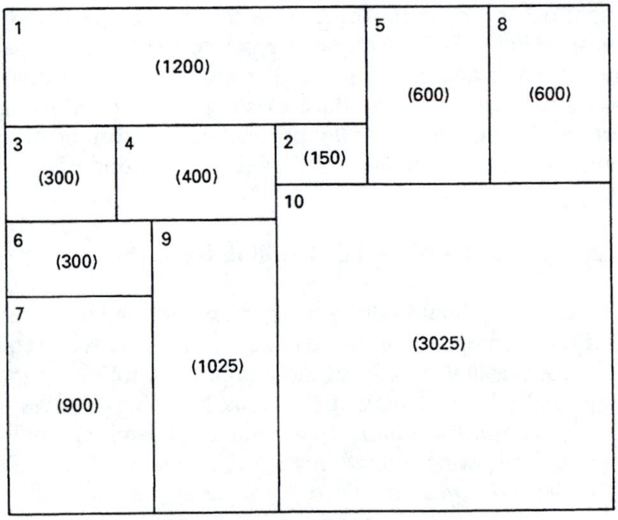

(b) Total area = 8,500 sq. ft.

Figure 2.57 Alternative block plans.

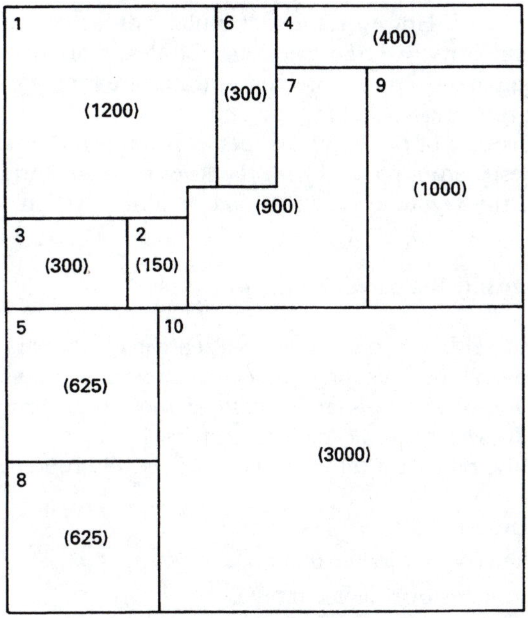

(c) Total area = 8,500 sq. ft

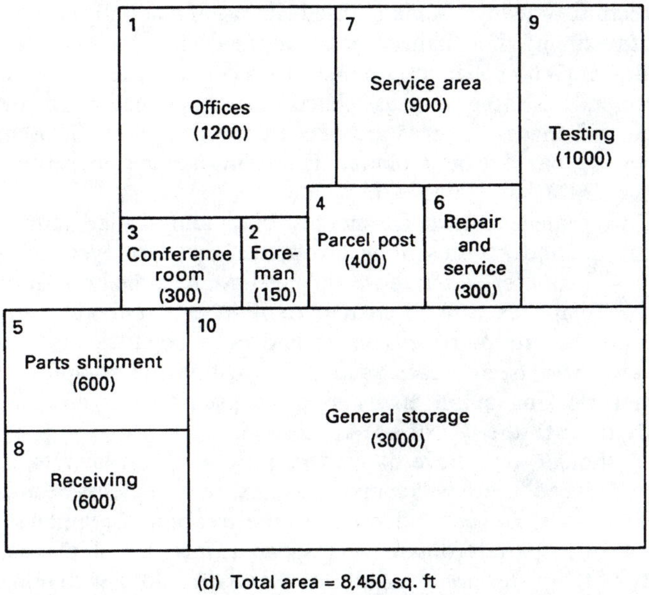

(d) Total area = 8,450 sq. ft

Figure 2.57 *(Continued)*

handling problems. However, the search phase of the design process requires a high degree of familiarity with the types, capabilities, limitations, and costs of material handling equipment. For a more concentrated treatment of material handling systems design, see Refs. 1, 2, 55, and 56.

For purposes of designing layouts, it is important that the material handling system be designed in parallel with the layout. In answering the question "Which comes first, the layout or the material handling system?," the answer must be "both!" [63].

2.9.4 Presenting the Layout Design

At this point, either a number of alternative component designs, a number of alternative overall designs, or a number of alternative detailed designs should have been generated from the search phase of the design process. But how are they presented? In what form do the designs exist?

Basically, there are three methods of visually representing layouts:

1. Drawings or sketches
2. Two-dimensional iconic models
3. Three-dimensional iconic models

Drawings and sketches have the advantage of being easy to make. This, of course, is not necessarily the case if we are referring to manually prepared engineering or architectural drawings. A major disadvantage of manually prepared drawings is their lack of flexibility; if a change is made in the design, the entire layout might need to be redrawn. When changes are likely to occur, the use of computer-aided design is recommended. Rather than use pencil and pen to create the drawings, the computer is used. Whenever changes are needed the computer file is updated and a new drawing is created using a plotter. For additional consideration of computer-aided design (CAD), see Hales [17].

Two-dimensional iconic models, commonly called templates, are the most popular method of presenting layout designs. However, as with drawings and sketches, computer-generated templates have become important aspects of a CAD system. Templates can be created to represent individual machines, individual workstations, groups of machines, and even departments. When the necessary templates have been created and stored within the computer system, the layout designer working at an engineering workstation can create alternative layouts quickly through the use of a CAD system.

Although we believe the vast majority of plant layouts in the future will be CAD produced, there will continue to exist requirements best satisfied using traditional methods. Hence, in addition to the creation of templates in a CAD system, they can be prepared from cardboard, paper, sheet metal, plastic, and wood. A wide variety of templates are available commercially, along with adhesive tapes that can be used to denote walls, aisles, columns, pallets, tables, benches, and so on. Figure 2.58 provides some sample templates.

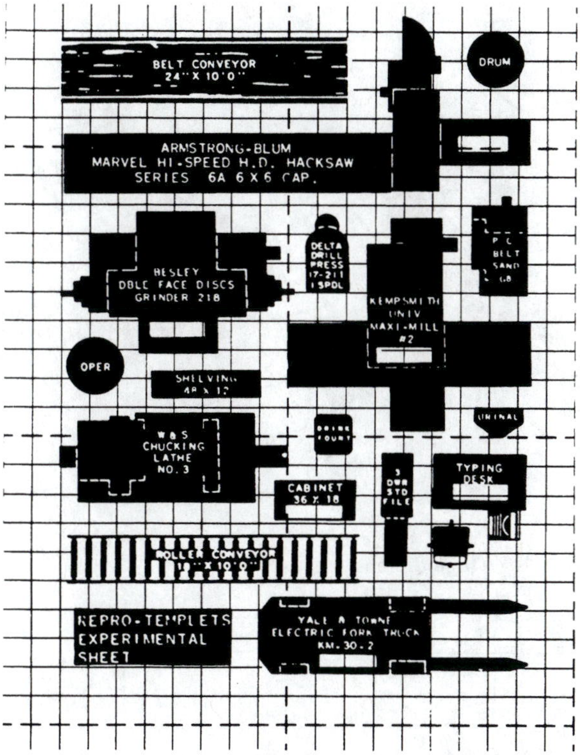

Figure 2.58 Sample templates.

Whether computer generated or prepared manually, it is important for the layout to have a professional appearance. The use of color, tapes, grid sheets, transfer lettering, templates, and overlays can supply the finishing touch to the layout.

Three-dimensional scale models can also be created physically or within the computer. Samples of physical models are given in Figure 2.59, with a CAD drawing of a three-dimensional representation given in Figure 2.60. Many of the available CAD systems have the capability of creating and storing three-dimensional representations. Although the designer only sees a two-dimensional representation of the CAD representation of the layout, the representation can be rotated and viewed from a variety of angles.

Even though there have been dramatic improvements in the field of computer graphics in recent years, there do not yet exist CAD systems that truly provide the depth perception available through the use of physical models. However, our experience has not provided a situation in which two-dimensional views of three-dimensional layouts was not sufficient for our purposes. Furthermore, recent improvements in holography encourage us that high-quality computer-generated three-dimensional layouts will become available through the use of holographic techniques.

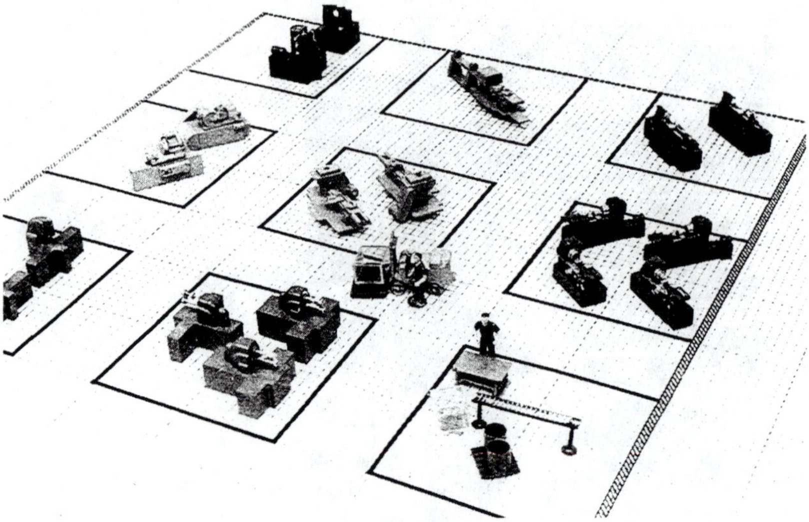

Figure 2.59 Three-dimensional model.

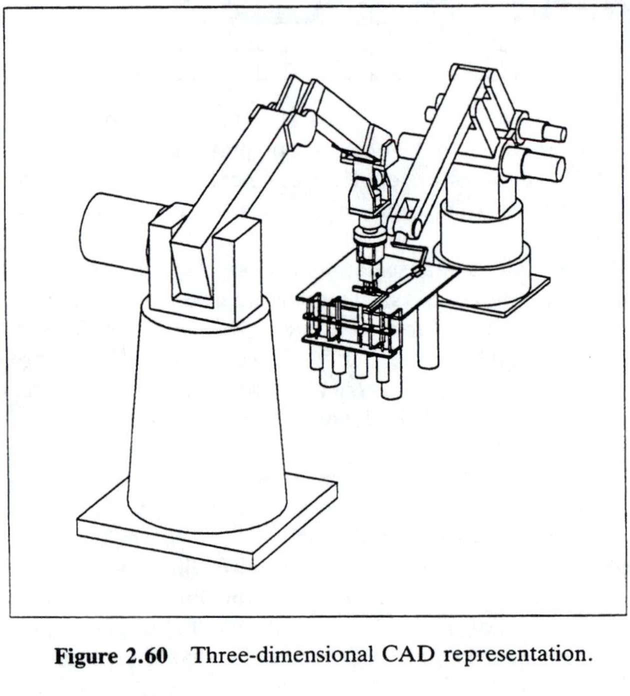

Figure 2.60 Three-dimensional CAD representation.

Some aspects of computer-generated iconic models might be inferior to the use of physical iconic models. However, there is one important feature available with the computer that is not readily available in other ways. Specifically, colorgraphic, animated computer simulations can be performed to test the layout and demonstrate vividly the way in which the system will operate. In particular, animating two- and three-dimensional representations of machines and people can provide a degree of realism not available in other ways. The use of simulation transforms the static representation of the layout into a dynamic representation of the physical system.

The ultimate decision as to the method of presenting the layout might not be yours to make. Precedent might dictate the approach you will use. The ultimate choice should be influenced by the scope of the problem. If you are designing a new plant, three-dimensional models are probably justified. If you are designing the layout for, say, a shipping area, two-dimensional models should suffice.

Although the use of iconic models can enhance the generation of layout alternatives, their strongest contribution is in selling the design. However, they also can be quite valuable in evaluating the practicality of alternative layout designs.

2.10 SELECTION, SPECIFICATION, IMPLEMENTATION, AND FOLLOW-UP

Now that alternative layout designs have been prepared, it still remains to select that design from among the alternatives which best meets your objectives. The design selected must be specified, all concerned must be sold, and the layout must be installed, observed, and periodically evaluated in an operating environment. We shall treat these steps briefly, realizing they were considered in Chapter 1 in a discussion of the total design cycle.

The selection of the "best" layout normally means the selection of the design that results in the most favorable compromise among competing objectives. Among these objectives was the minimization of cost. Thus far, little has been said about the cost of the design. If costs are to be considered, we are interested in estimating the cost of installing the new design, as well as the long-run operating cost resulting from the new layout. If there is an existing layout design, then we shall be interested in comparing the performance of the new design with that of the old design. We might find that the cost of installing the new design offsets any of the benefits it provides. In making this evaluation, remember that the new design is intended for the future. Performance costs for the existing system probably reflect present and past operating levels. Therefore, it will be necessary to forecast future costs for both the new and the old designs.

If costs are an important consideration in the evaluation of alternative layouts, it is necessary that the relevant costs be measured. This is no easy undertaking for several reasons. First, we are interested in incremental costs rather than standard costs. The latter include overhead items, which many times will be unaffected by the layout design. Second, we are interested in future costs, rather than present or past costs. Furthermore, in the case of a new layout we have no prior experience on which we can base our estimates of future costs.

Many times costs are not the major consideration in evaluating layout designs. Typically, a number of alternative layouts will have approximately the same costs, and other considerations are used in choosing the preferred design. We enumerated a number of these factors in Chapter 1 and in Section 2.2.

Although we have already discussed ways of performing the evaluation step in the design process, one additional factor should be recognized. That is, the layout design which is chosen must be sold. Consequently, it is prudent at this point to consider the amount of resistance to change that will accompany each design. No matter how good the design might be, if certain individuals are opposed to it, once it is implemented they can make the design look so bad that a new layout study will be required. Therefore, you should strive to assess and reduce such resistance and anticipate the amount of resistance that will be encountered for each alternative.

In the final analysis, the ability to sell the layout design is strongly influenced by the ability to cope with resistance to change. The following specific causes of resistance to change on the part of persons having a veto power over the proposed layout design are given by Krick [22]:

1. Inertia
2. Uncertainty
3. Failure to see the need for the proposed change
4. Failure to understand the proposal
5. Fear of obsolescence
6. Loss of job content
7. Personality conflict with the analyst
8. Resentment of outside help, or interference
9. Resentment of criticism
10. Lack of participation in the formulation of the proposed change
11. Tactless approach on the part of the layout analyst
12. Lack of confidence in the analyst
13. Inopportune timing

Some methods of minimizing resistance to change were suggested by Krick [22]:

1. Convincingly explain the need for the change.
2. Thoroughly explain the need for the change.
3. Facilitate participation or at least the feeling of participation in formulation of the proposed method.
4. Use a tactful approach in introducing your proposal.
5. Watch your timing.
6. In the case of major changes, if possible introduce the change in stages.

7. Capitalize on the features that provide the most personal benefit to the person(s) you are trying to sell.
8. If possible, by appropriate questioning maneuver a prospective rejector into "thinking" of the (your) idea himself.
9. Show a personal interest in the welfare of the person directly affected by the change.
10. Whenever possible have changes announced and introduced by the immediate supervisor of those affected.

Assuming a layout design has been accepted by the appropriate persons within the organization, it must be installed. When installing the layout it is important to remember that a considerable amount of planning must precede the actual location and installation of equipment. Once the plans are made, all activities should be scheduled. Since the installation of the layout can involve a number of activities, a project scheduling model, such as the critical path method (CPM), can be quite useful.

Once the layout has been installed, you should follow up to see that the layout was installed as designed. If modifications were made during installation, either they should be accepted or arrangements should be made to correct the discrepancies. Periodic checks should be made to see that the layout is performing satisfactorily. Also, you should be on the lookout for the problem indicators we pointed out earlier so that a redesign of the layout is begun when justified.

2.11 WHERE TO FROM HERE?

Give some serious thought to the following questions:

1. Would it surprise you to learn that some people actually believe that all, or certainly most, layout problems are solved by moving templates around until a design is found that looks good?
2. Would it surprise you to learn that a very large number of layout problems are solved in this way?
3. Do you believe the solution obtained from the template-shuffling approach can be improved upon by applying a systematic approach?
4. Do you believe that analytical approaches can assist the layout designer in developing even better layout designs than are obtained when using the traditional approach?

The answers to the first two questions probably provide some measure of your familiarity with the real world of plant layout. However, it is your answers to the last two questions that concern us most. If your answer to the third question is No!, then

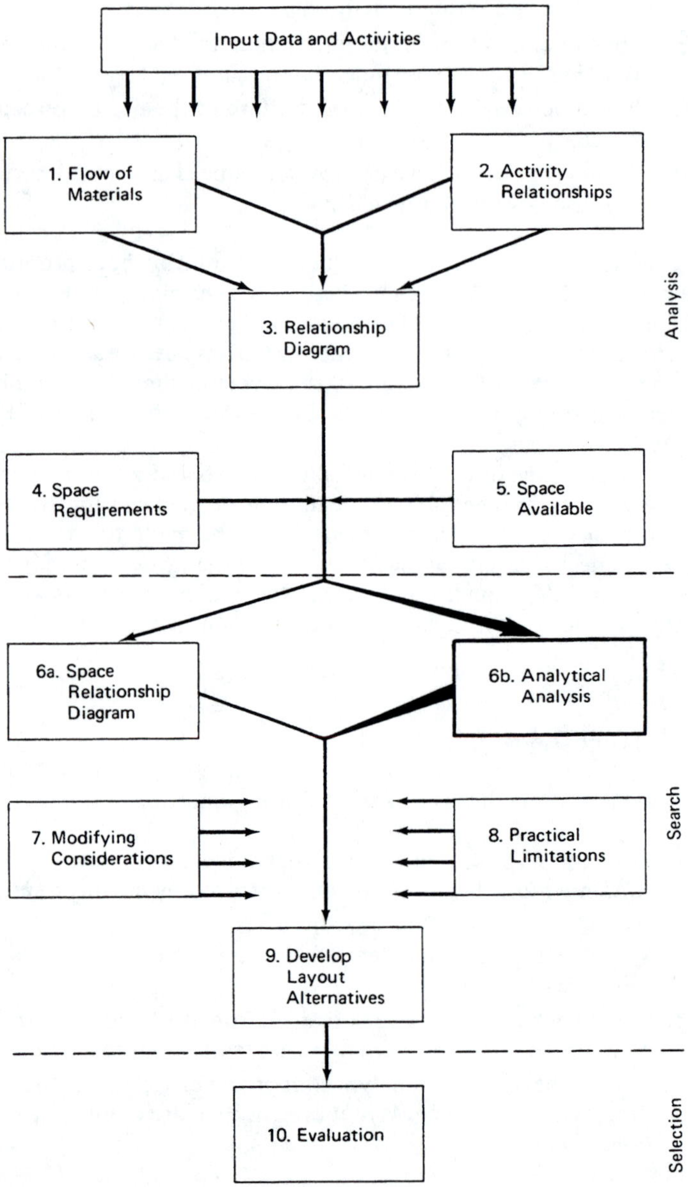

Figure 2.61 Modified SLP procedure incorporating analytic analysis.

you certainly wasted your time in reading this chapter. In fact, if your answer to the fourth question is also No!, then unless you are willing to revise your belief, you will probably waste your time by reading further.

Now, for those of you who are still with us, here is where *we* are going from here. The remainder of this book is devoted to the presentation of analytic approaches that can be used to assist the designer in developing both overall and detailed layouts. We assume in our subsequent discussions that the preliminary steps of problem formulation and analysis have been performed. It is our opinion that traditional approaches have placed too much emphasis on intuition alone. Of course, depending on the scope of the problem, commonsense solutions are sometimes preferred to a more costly analysis. However, when very complex problems are encountered, analytic approaches can serve as very helpful aids to design. Furthermore, it has been our experience that a person's intuition can be wrong. Also, what might pass for a *commonsense* solution may, in reality, be a *nonsense* solution when subjected to the test of analysis.

We view the use of analytical models in layout planning to be an activity that parallels the development of the space relationship diagram. Just as modifying considerations and practical limitations influence the development of layout alternatives in traditional SLP, they are also necessary steps when using analytical approaches. Consequently, we recommend that the steps shown in Figure 2.61 be followed in *analyzing* the problem, *searching* for alternative layout designs, and *selecting* the preferred design. Furthermore, it is recommended that all the steps in the design cycle be followed.

2.12 FURTHER READING

As noted in Chapter 1 and in our earlier discussion of SLP in this chapter, the plant layout problem embodies quantitative and qualitative considerations. Among the qualitative considerations we treated quantitatively was that of activity relationships. The aggregate treatment of material flows (quantitative) and closeness relationships (qualitative) is not necessarily a straightforward exercise. Rosenblatt [45] and Rosenblatt and Sinuany-Stern [46] address this issue and employ an efficient frontier approach to the problem. Drawing on the use of efficiency frontiers in capital budgeting and decision theory [42], Rosenblatt and Sinuany-Stern examine convex combinations of material flow values and activity relationship measures.

Our treatment of graph-based approaches to designing facility layouts was quite brief. For the reader who is interested in pursuing the subject further, we should note that a number of different graph-based approaches have appeared in the research literature. A number of papers that relate closely to graph-based layout models are included in the References. For the reader who desires a recent assessment of the literature, see Refs. 6, 10, 30, and 31.

Montreuil and Ratliff [29] unified a vast amount of the plant layout algorithmic literature through the use of what they called *design skeletons*. Examples of design

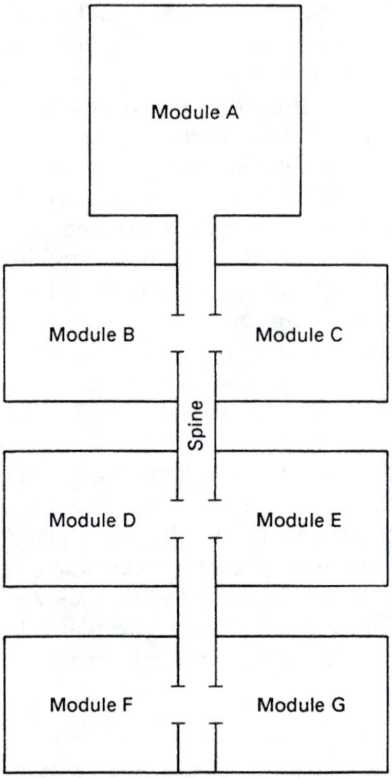

Figure 2.62 Example of the use of a spine in a facility layout.

skeletons that we have used thus far are the REL diagram, primal and dual graphs, space relationship diagram, assembly chart, precedence diagram, and operations process chart. Montreuil et al. [30] note that design skeletons have been used for more than 25 years in generating layouts; examples include flow graphs used by Reed [43], bubble diagrams used by Muther [36], planar adjacency graphs used by Carrie et al. [6], matching-based adjacency graphs used by Montreuil et al. [27], centroid locations used by Drezner [9], cut trees used by Montreuil and Ratliff [29], and spines described by Collins [7, 8], Tompkins [54], Tompkins and White [55], and White [62]. A comparison of several graph-based approaches is provided by Montreuil et al. [30], including an efficient linear programming model they developed for generating layouts from a design skeleton.

A related body of research literature is that which deals with queueing networks. Specifically, Smith [50], Smith and Bouanaka [51], Smith and Kerbache [52], and Yuhaski and Smith [66] have examined applications of queueing networks to modeling facilities planning issues. The approach taken by Smith et al. is to develop a dual graph from the existing block layout; next, Steiner–Star duals are developed from the dual graphs. The Steiner–Star topology is the fundamental unit used by Smith et al. to construct queueing networks of facilities. Depending on the flow requirements in the facility, Smith et al. use the Steiner–Star duals to develop open networks, egress graphs, closed networks, or mixed networks.

The importance of modular and flexible material handling systems and facilities was presented by Collins [7, 8] and White [62, 64, 65] based on their work with Texas Instruments. Practically all of Texas Instruments' manufacturing sites have been designed on the basis of the spine concept. The spine concept is illustrated in Figure 2.62; notice that it is an extension of the approach used to design shopping malls. Subsequently, Gupta [13], Gupta and Tompkins [14], Shore and Tompkins [48], and Tompkins [54] considered the issues of flexible facilities and application of the spine concept in facilities design.

Undoubtedly, the use of the computer in facility layout will increase in importance. Further, it is expected that expert systems will be developed through the use of artificial intelligence [18]. However, we believe that the human designer will continue to be needed. In fact, we believe the direction of the future will be toward human–computer interaction. The work of Montreuil et al. [27], for example, was the result of an interactive approach to facility layout [26, 41]. Among the earlier efforts to link the human designer with the computer was that of Moore [33]. Because of the role of the computer in facility layout, we devote the next chapter to the subject of computerized layout planning.

REFERENCES

1. Apple, J. M., *Plant Layout and Material Handling,* 3rd ed., John Wiley & Sons, Inc., New York, 1977.
2. Apple, J. M., *Material Handling Systems Design,* John Wiley & Sons, Inc., New York, 1972.
3. Blair, E. L., and S. Miller, "Interactive Approach to Facilities Design Using Microcomputers," *Computers and Industrial Engineering,* Vol. 9, No. 1, 1st Quarter 1985, pp. 91–102.
4. Buffa, E. S., G. C. Armour, and T. E. Vollman, "Allocating Facilities with CRAFT," *Harvard Business Review,* Vol. 42, No. 2, March 1964, pp. 136–159.
5. Carrie, A. S., "Numerical Taxonomy Applied to Group Technology and Plant Layout," *International Journal of Production Research,* Vol. 11, No. 4, 4th Quarter 1973, pp. 399–416.
6. Carrie, A. S., J. M. Moore, R. Roczniak, and J. J. Seppanen, "Graph Theory and Computer Aided Facilities Design," *OMEGA, The International Journal of Management Science,* Vol. 6, No. 4, July/August 1978, pp. 353–361.
7. Collins, J. D., "Material Handling System Concept," presentation at the *26th Annual Material Handling Management Course,* American Institute of Industrial Engineers, Mt. Pocono, PA, June 1979.
8. Collins, J. D., "Strategic Planning for Material Handling and Storage," presentation at the *Computer Integrated Manufacturing: Productivity for the 1980's, A Joint Industry–DoD Manufacturing Technology Workshop,* Society of Manufacturing Engineers, Detroit, MI, September 1979.

9. Drezner, Z., "DISCON: A New Method for the Layout Problem," *Operations Research,* Vol. 28, No. 6, November 1980, pp. 1375–1384.

10. Foulds, L. R., "Techniques for Facilities Layout: Deciding Which Pairs of Activities Should Be Adjacent," *Management Science,* Vol. 29, No. 12, December 1983, pp. 1414–1426.

11. Francis, R. L., and J. A. White, *Facility Layout and Location: An Analytical Approach,* Prentice Hall, Englewood Cliffs, NJ, 1974.

12. Gilbreath, S. G., III, "Facility Layout," presentation to the *28th Annual Material Handling Management Course,* Institute of Industrial Engineers, Airlie, VA, June 1981.

13. Gupta, R. M., "Flexibility in Layouts: A Simulation Approach," *Material Flow,* Vol. 3, No. 4, June 1986, pp. 243–250.

14. Gupta, R. M., and J. A. Tompkins, "An Examination of the Dynamic Behavior of Part Families in Group Technology," *International Journal of Production Research,* Vol. 20, No. 1, June 1982, pp. 73–86.

15. Hadley, G., and T. M. Whitin, *Analysis of Inventory Systems,* Prentice Hall, Englewood Cliffs, NJ, 1967.

16. Hales, H. L., and H. C. Jones, Jr., "Facilities Decision Support," masters thesis, Sloan School of Management, Massachusetts Institute of Technology, Boston, 1980.

17. Hales, H. L., *Computer-Aided Facilities Planning,* Marcel Dekker, Inc., New York, 1984.

18. Heragu, S. S., and A. Kusiak, "Machine Layout: An Optimization and Knowledge Based Approach," *International Journal of Production Research,* Vol. 28, No. 4, April 1990, pp. 615–635.

19. Hopcroft, J., and R. Tarjan, "Efficient Planarity Testing," *Journal of the Association for Computing Machinery,* Vol. 21, No. 4, October 1974, pp. 549–568.

20. Immer, J. R., Layout Planning Techniques, McGraw-Hill Book Company, New York, 1950.

21. Konz, S., *Facility Design,* John Wiley & Sons, Inc., New York, 1985.

22. Krick, E. V., *Methods Engineering,* John Wiley & Sons, Inc., New York, 1962.

23. Kusiak, A., and S. S. Heragu, "The Facility Layout Problem," *European Journal of Operational Research,* Vol. 29, No. 3, June 1987, pp. 229–251.

24. Lee, R. C., and J. M. Moore, "CORELAP—Computerized Relationship Layout Planning," *Journal of Industrial Engineering,* Vol. 18, No. 3, 1967, pp. 195–200.

25. Mallick, R. W., and J. H. Sansonetti, "Adopt the Best in Layout: Case of Westinghouse Electric Corporation," *Factory Management and Maintenance,* Vol. 103, No. 8, August 1945, p. 102–109.

26. Montreuil, B., "Interactive Optimization Based Facilities Layout," doctoral dissertation, Georgia Institute of Technology, Atlanta, GA, 1982.

27. Montreuil, B., H. D. Ratliff, and M. Goetschalckx, "Matching Based Interactive Facility Layout," *IIE Transactions,* Vol. 19, No. 3, September 1987, pp. 271–279.

28. Montreuil, B., and H. D. Ratliff, "Optimizing the Location of Input/Output Stations within Facilities Layout," *Engineering Costs and Production Economics,* Vol. 14, No. 3, September 1988, pp. 177–187.

29. Montreuil, B., and H. D. Ratliff, "Utilizing Cut Trees as Design Skeletons for Facility Layout," *IIE Transactions,* Vol. 21, No. 2, June 1989, pp. 136–143.

30. Montreuil, B., U. Venkatadri, and H. D. Ratliff, "Generating a Layout from a Design Skeleton," *IIE Transactions,* to appear.

31. Montreuil, B., "A Modelling Framework for Integrating Layout Design and Flow Network Design," *Progress in Material Handling and Logistics: Material Handling '90,* Vol. 2 (J. A. White, I. W. Pence, Jr., R. J. Graves, L. F. McGinnis, M. R. Wilhelm, and R. E. Ward, editors), Springer-Verlag, New York, 1991, pp. 96–115.

32. Moore, J. M., *Plant Layout and Design,* The Macmillan Company, New York, 1962.

33. Moore, J. M., "Computer Program Evaluates Plant Layout Alternatives," *Industrial Engineering,* Vol. 3, No. 8, August 1971, pp. 19–25.

34. Moore, J. M., and A. S. Carrie, "Impact of List Processors and Graph Theory on Use of Computers for Solving Facilities Design Problems," *Proceedings of the 3rd International Conference on Production Research,* Taylor & Francis Ltd., London, 1975.

35. Muther, R., *Practical Plant Layout,* McGraw-Hill Book Company, New York, 1955.

36. Muther, R., *Systematic Layout Planning,* Industrial Education Institute, Boston, 1961.

37. Muther, R., and H. L. Hales, *Systematic Planning of Industrial Facilities,* Vol. I, Management and Industrial Research Publications, Kansas City, MO, 1979.

38. Muther, R., and H. L. Hales, *Systematic Planning of Industrial Facilities,* Vol. II, Management and Industrial Research Publications, Kansas City, MO, 1980.

39. Nadler, G., *Work Design: A Systems Concept,* Richard D. Irwin, Inc., Homewood, IL, 1961.

40. Nozari, A., and E. E. Enscore, Jr., "Computerized Facility Layout with Graph Theory," *Computers and Industrial Engineering,* Vol. 5, No. 3, 3rd Quarter 1981, pp. 183–193.

41. O'Brien, C., and S. E. Z. Abdel Barr, "An Interactive Approach to Computer Aided Facility Layout," *International Journal of Production Research,* Vol. 18, No. 2, March/April 1980, pp. 201–211.

42. Park, C. S., and G. P. Sharp-Bette, *Advanced Engineering Economics,* John Wiley & Sons, Inc., New York, 1990.

43. Reed, R., *Plant Layout: Factors, Principles, and Techniques,* Richard D. Irwin, Inc., Homewood, IL, 1961.

44. Reed, R., *Plant Location, Layout, and Maintenance,* Richard D. Irwin, Inc., Homewood, IL, 1967.

45. Rosenblatt, M. J., "The Facilities Layout Problem: A Multigoal Approach," *International Journal of Production Research,* Vol. 17, No. 4, July/August 1979, pp. 323–332.

46. Rosenblatt, M. J., and Z. Sinuany-Stern, "A Discrete Efficient Frontier Approach to the Plant Layout Problem," *Material Flow,* Vol. 3, No. 4, June 1988, pp. 277–281.

47. Seppanen, J., and J. M. Moore, "Facilities Planning with Graph Theory," *Management Science,* Vol. 17, No. 4, December 1970, pp. 242–253.

48. Shore, R. H., and J. A.Tompkins, "Flexible Facilities Design," *AIIE Transactions,* Vol. 12, No. 2, June 1980, pp. 200–205.

49. Simon, H. A., "Style in Design," in C. M. Eastman, ed., *Spatial Synthesis in Computer-Aided Building Design,* John Wiley & Sons, Inc., New York, 1975.

50. Smith, J. M., "Queueing Networks and Facility Planning," *Building and Environment,* Vol. 17, No. 1, 1st Quarter 1982, pp. 33–45.

51. Smith, J. M., and B. Bouanaka, "Queueing Network Decomposition in Facilities Plan-

ning," *Computers and Operations Research,* Vol. 12, No. 1, January/February 1985, pp. 1–16.

52. Smith, J. M., and L. Kerbache, "Design of Large Scale Facilities with Open Finite Queueing Networks," *Proceedings 26th IEE Conf. on Design and Control,* Los Angeles, CA, Dec, 1987, pp. 1088–1091.
53. Sule, D. R., *Location, Planning, and Design,* PWS-Kent Publishing Company, Boston, 1988.
54. Tompkins, J. A., "Modularity and Flexibility: Dealing with Future Shock in Facilities Design," *Industrial Engineering,* Vol. 12, No. 3, March 1980, pp. 78–81.
55. Tompkins, J. A., and J. A. White, *Facilities Planning,* John Wiley & Sons, Inc., New York, 1984.
56. U.S. Department of the Navy, *Warehouse Modernization and Layout Planning Guide,* 0530-LP-529-0000, NAVSUP Publication 529, Naval Supply Systems Command, Department of the Navy, Washington, DC, March 1985.
57. Warnecke, H. J., W. Dangelmeier, and H. Kuhnle, "Computer-Aided Layout Planning," *Material Flow,* Vol. 1, No. 1, May 1982, pp. 35–48.
58. Warnecke, H. J., and W. Dangelmeier, *Progress in Computer Aided Plant Layout,* Technical Report, Institute of Manufacturing Engineering and Automation, Fraunhofer, Germany, 1982.
59. White, J. A., "On Absorbing Markov Chains and Optimum Batch Production Quantities," *AIIE Transactions,* Vol. 2, No. 1, March 1970, pp. 82–88.
60. White, J. A., J. W. Schmidt, and G. K. Bennett, *Analysis of Queueing Systems,* Academic Press, New York, 1975.
61. White, J. A., *Yale Management Guide to Productivity,* Industrial Truck Division, Eaton Corporation, Philadelphia, 1979.
62. White, J. A., "Modular Material Handling Systems," *AIIE-MHI Seminar,* Long Beach, CA, March 1979.
63. White, J. A., "Layout: The Chicken or the Egg?" *Modern Materials Handling,* Vol. 35, No. 9, September 1980, p. 39.
64. White, J. A., "Flexible Manufacturing Facilities," *Modern Materials Handling,* Vol. 38, No. 4, March 1983, p. 13.
65. White, J. A., "Flexible Warehouse Facilities," *Modern Materials Handling,* Vol. 38, No. 10, July 1983, p. 39.
66. Yuhaski, S., and J. M. Smith, "Modeling Circulation Systems in Buildings Using State Dependent Queueing Models," *Queueing Systems,* Vol. 4, No. 4, September 1989, pp. 319–338.

PROBLEMS*

2.1. Briefly describe each of the following, giving their application to layout planning.

- **(a)** Assembly chart
- **(b)** Operation process chart
- **(c)** Route sheet
- **(d)** Flow process chart
- **(e)** From–to chart
- **(f)** Multiproduct process chart
- **(g)** Activity relationship chart
- **(h)** Flow diagram

* Additional problems requiring SLP solutions are given at the end of Chapter 3.

2.2. Discuss the advantages and disadvantages of using standard times, rather than actual times, in making calculations in layout planning.

2.3. Discuss the use of analytical approaches in designing plant layouts. Contrast the use of analytical approaches with traditional approaches. Relate both the traditional approach and the analytical approach to the design process. Compare the use of iconic models when traditional approaches and analytical approaches are used.

2.4. Prepare assembly charts for the chair on which you are sitting and the lamp on your desk. (If you are standing and do not have either a desk or a lamp on your desk, you are exempt from this question.)

2.5. Prepare a precedence diagram and an assembly chart for one of the following products.

(a) Charcoal grill
(b) Bicycle
(c) Toy wagon
(d) Lamp
(e) Gas grill
(f) Toaster
(g) Sprinkler
(h) Thermostat
(i) Faucet

2.6. Compare the primary layout objectives for the following situations.

(a) Grocery store
(b) Drugstore
(c) Doctor's office
(d) Bank
(e) Sheriff's office
(f) Department store
(g) Warehouse
(h) Parking lot
(i) Post office
(j) Elementary school
(k) Movie theater
(l) Laundromat
(m) Beauty shop
(n) Automotive repair shop
(o) Restaurant
(p) Carry-out sandwich shop

2.7. Develop alternative layout designs for one or more of the situations listed in Problem 2.6.

2.8. Using the black-box approach, identify states A and B for a layout design problem encountered in one of the following situations.

(a) Airport
(b) Hospital
(c) Library
(d) Grocery store
(e) Bank
(f) Fast-food shop
(g) Warehouse
(h) Grocery store
(i) Meat-processing plant
(j) Printing shop

2.9. An assembly chart for a simple gate valve is given in Figure P2.9(a). Processing information for the gate valve is given in Figure P2.9(b). The following scheduling information is available.

(1) All designing and planning for the production of the gate valves has been completed.
(2) Patterns for the various cast parts can be obtained in 4 working days.
(3) Two days are required for the production of castings. Castings cannot be produced in the foundry until after receipt of the patterns.
(4) Various sizes of bar stock can be obtained from a local metal supply house 3 days after orders for stock have been placed.
(5) Nuts, screws, packing, etc., can be obtained in 3 days.
(6) The design engineers have specified that the body, bushing, and handle be made of cast bronze, and that the cap and stem of the valve be machined from standard brass bar stock.
(7) The bill of material indicates that the fiber packing and the nut are purchased parts.

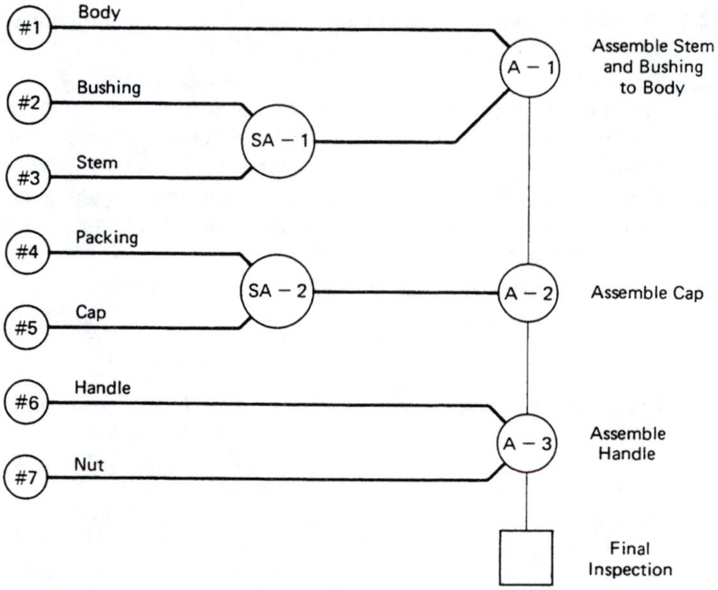

Figure P2.9(a)

Parts List

Parts	Part No.	Material
Body	001	Cast bronze
Bushing	002	Cast bronze
Stem	003	3/8 in. bar stock
Fiber packing	004	(Purchased part)
Cap	005	3/4 in. hex bar stock
Handle	006	Cast bronze
Nut	007	(Purchased part)

Assembly Operations

No.	Operation	Machine	Standard time (min/part)
1	Final assembly	Bench	2.00
2	Clean	Solvent tank	0.60
3	Inspect (pressure test)	Water test stand	1.20
4	Pack in boxes	Bench	0.12

Figure P2.9(b) (Based on a problem given in A. L. Roberts, *Production Management Workbook*, John Wiley & Sons, Inc., New York, 1962, p. 22.)

Fabrication Operations

Part No.	Operation No.			Standard time (min/part)
001	001–1	Cast	Bench mold	1.50
	001–2	Clean	Tumble barrel	0.40
	001–3	Machine-thread and face three surfaces	Turret lathe	2.40
002	002–1	Cast	Bench mold	0.75
	002–2	Clean	Tumble barrel	0.20
	002–3	Machine all i.d. and o.d.	Turret lathe	1.00
003	003–1	Machine all surfaces and cut off	Automatic screw machine	0.30
005	005–1	Machine all surfaces	Automatic screw machine	0.12
006	006–1	Cast	Bench mold	1.20
	006–2	Clean	Tumble barrel	0.30
	006–3	Machine two surfaces	Turret lathe	0.75
	006–4	Broach square hole	Broach	0.75

Note: 100% inspection is required for all fabricated parts prior to delivery to final assembly.

Figure P2.9(b) *(Continued)*

(8) The machine shop estimates the following fabrication times for processing batches of 500 parts:

Body: 3 days
Bushing: 3 days
Stem: 1 day
Cap: 1 day
Handle: 2 days

Construct an operation process chart, precedence diagram, route sheets, and a Gantt chart for the production of the gate valve. Develop a summary of equipment requirements for the production of 500 gate valves per hour.

2.10. Eight parts (*A* to *H*) are to be manufactured and assembled to produce two products (*X* and *Y*). The processing sequences for the parts and assembly sequences for the products are given below. The daily production rate for product *X* is 400 and for product *Y* is 600.

(a) Determine the daily production rates for each part.
(b) Design a product layout for manufacturing and assembling each product.
(c) Design a process layout for manufacturing both products.
(d) Design a group layout for manufacturing the two products.
(e) Construct a from–to chart for the daily flow of parts between processes.
(f) Construct an activity relationship chart for each product and a combined REL chart for the two products.
(g) Using the traditional SLP approach, design a REL diagram for each REL chart constructed in part **(f)**.

Part	Processing sequence (grouped by machine)
A	(cast), (tumble), (mill), (turn, thread, drill, tap)
B	(cast), (tumble), (mill), (turn, drill, tap)
C	(turn, thread)
D	(punch), (bend), (drill)
E	(punch), (bend), (drill)
F	(punch), (bend), (drill)
G	(turn)
H	(turn, thread)

DAILY MACHINE REQUIREMENTS

	Parts							
Machine	*A*	*B*	*C*	*D*	*E*	*F*	*G*	*H*
Bench mold	3.0	4.2	—	—	—	—	—	—
Tumble mill	1.1	0.7	—	—	—	—	—	—
Milling machine	2.4	3.4	—	—	—	—	—	—
Turret lathe	3.5	1.7	—	—	—	—	—	—
Automatic screw machine	—	—	1.3	—	—	—	0.6	0.9
Drill press	—	—	—	1.5	2.2	1.1	—	—
Punch press	—	—	—	0.3	0.5	0.4	—	—
Press brake	—	—	—	0.8	1.2	1.1	—	—

2.11. For the two graphs shown in Figure P2.11, construct the dual graphs.

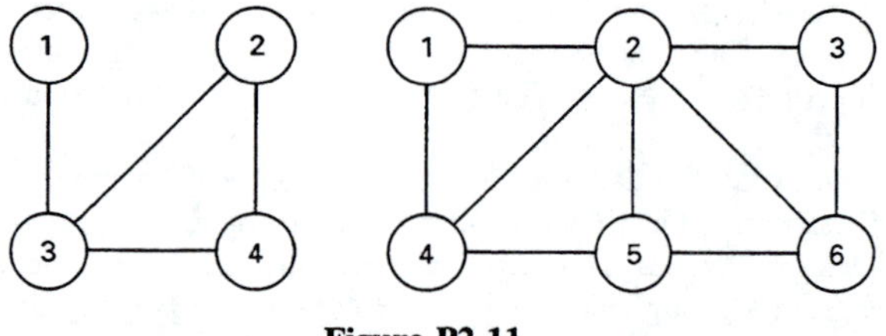

Figure P2.11

2.12. Are the two graphs shown in Figure P2.12 equivalent? Justify your answer.

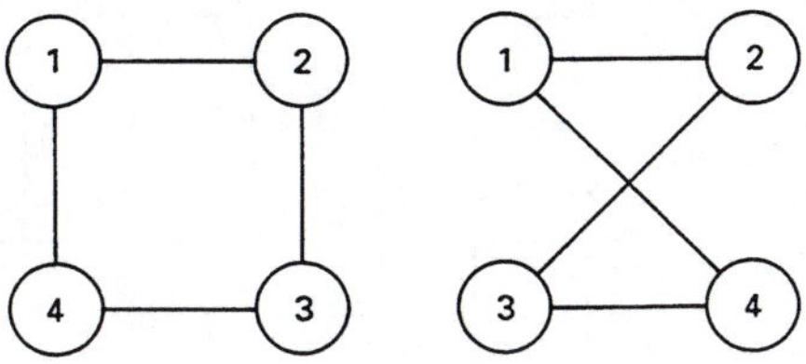

Figure P2.12

2.13. Are the two graphs shown in Figure P2.13 planar? Why or why not?

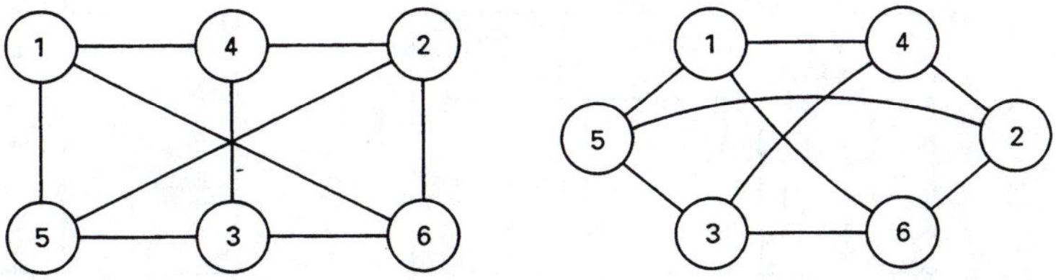

Figure P2.13

2.14. The graph-based approach for constructing REL diagrams depends on having some department adjacent to the exterior. Suppose that it is not necessary for any department to be adjacent to the exterior. Propose some reasonable approaches that will allow use of the graph-based approach.

2.15. Is the graph shown in Figure P2.15 planar? If so, construct the dual graph and the corresponding block layout.

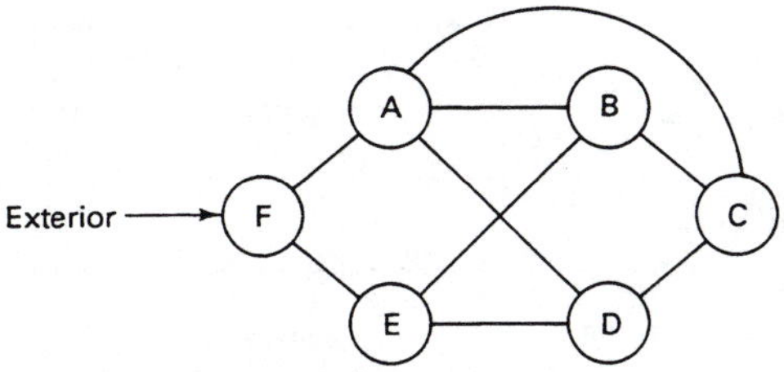

Figure P2.15

2.16. For each of the block layouts shown in Figure P2.16, construct the graph and the corresponding dual graph to obtain the REL diagram.

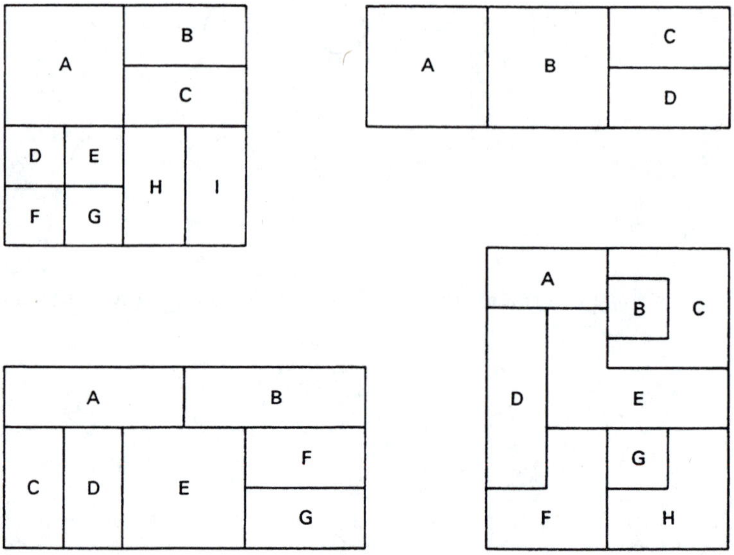

Figure P2.16

2.17. Construct the graph for the block layout shown in Figure P2.17; construct the dual graph and show that it is the same as the REL diagram for the layout.

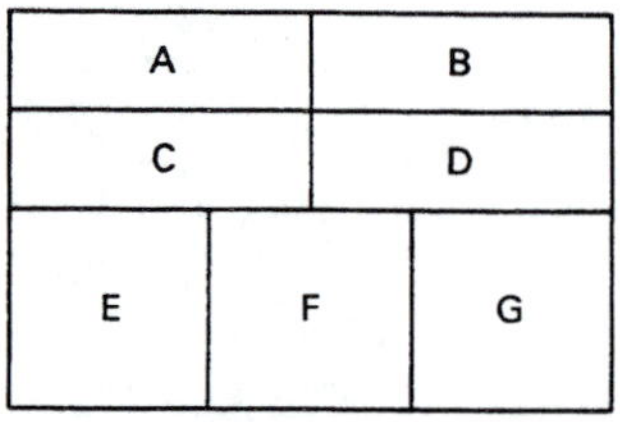

Figure P2.17

2.18. Three adjacency matrices are shown in Figure P2.18. In an adjacency matrix, 1 depicts a requirement for adjacency and 0 indicates adjacency is not required. For each, does planarity exist? Justify your answer.

	A	B	C	D	E
A	–	1	1	1	1
B	1	–	1	1	1
C	1	1	–	1	1
D	1	1	1	–	1
E	1	1	1	1	–

(a)

	A	B	C	D	E	F	G
A	–	1	1	1	0	1	1
B	1	–	0	1	1	1	1
C	1	0	–	0	1	0	0
D	1	1	0	–	0	1	1
E	0	1	1	0	–	0	1
F	1	1	0	1	0	–	1
G	1	1	0	1	1	1	–

(b)

	A	B	C	D	E	F	G	H
A	–	1	1	1	1	0	0	0
B	1	–	1	0	1	0	0	0
C	1	1	–	0	1	0	0	0
D	1	0	0	–	1	1	0	0
E	1	1	1	1	–	1	1	1
F	0	0	0	1	1	–	1	0
G	0	0	0	0	1	1	–	1
H	0	0	0	0	1	0	1	–

(c)

Figure P2.18

2.19. A foundry receives an order for specialty castings. The cost of producing each casting is estimated to be \$15,000. The customer requires that exactly four good castings be supplied. The customer will pay \$20,000 for each good casting. If fewer than four good castings are available, none will be purchased. Each casting is produced independently and the probability of an individual casting being acceptable to the customer is 0.80.
(a) Develop the expected profit function to be maximized.
(b) Determine the production lot size that maximizes expected profit.
(c) Determine the lot size that will yield a probability of 0.90 of producing at least four good castings. What is the expected profit for the lot size obtained?

2.20. A job shop has received an order for high-precision formed parts. The cost of producing each part is estimated to be \$20,000. The customer requires that either four or five good parts be supplied. Each good part sold will yield a revenue of \$30,000. However, if fewer than four good parts are produced, none will be purchased; if more than five good parts are produced, the excess will not be purchased. The probability of an individual part being acceptable to the customer is 0.80. Determine the optimum batch production quantity given that the job shop has agreed to supply the formed parts. Is the situation a profitable one for the job shop? If not, how must the selling price be changed for the expected profit to be positive?

2.21. Precision Castings Incorporated has received an order for precision castings. The cost of producing each casting is estimated to be \$2000. The customer agrees to pay a total of \$15,000 for three good castings, a total of \$18,000 for four good castings, and a total of \$20,000 for five good castings. If fewer than three good castings are produced, none will be purchased; good castings in excess of five will not be purchased. If fewer than three good castings are available, a shortage cost of \$10,000 in excess of the cost of production will be incurred regardless of the number short. Each casting is produced independently; the probability of an individual casting being acceptable is believed to be 0.70.

(a) Develop the expected profit equation to be maximized.

(b) Determine the number of units to schedule to maximize expected profit. (*Hint*: The number is greater than 4.)

(c) If five castings are produced, what is the probability that the firm will lose money?

2.22. A firm has received a production order for 10 injection-molded parts. The parts will sell for \$4000 each; it will cost \$2500 to produce an individual part. The probability of a part meeting the final inspection standards is 0.80. Any good parts in excess of 10 can be salvaged for \$2000; any bad parts produced can be salvaged for \$1000. If an insufficient number of good parts is produced, each shortage must be overcome at a cost of \$3000 in excess of the cost of production. Determine the number of parts to be scheduled for production to maximize expected profit.

2.23. A customer places an order for 50 units of a product, which your company can manufacture on a single machine. The product is custom-made and you doubt that the order will be repeated. The customer will pay you \$1000 per unit of product produced within stated quality specifications, subject to the following conditions: no more than 55 units will be purchased and the entire shipment will be refused if it contains less than 45 good units.

The cost of manufacturing and delivering the order is estimated to be \$800 per unit. The manufacturing process is such that the probability of any unit of product being "out of specification" is 0.15. Because of a lack of instrumentation, you cannot inspect the product; inspection will be done by the customer after the shipment is delivered. It is desired to determine the number of units of product to manufacture to maximize expected profits. Develop a formulation of expected profit as a function of Q, the production lot size.

2.24. Two products (X and Y) are produced on two machines (A and B). It is required that 10 units/day of product X be produced on machine A and 12 units/day of product X be produced on machine B; it is required that 20 units/day of product Y be produced on machine A. Production times per unit are $T_{AX} = 1$ hour, $T_{BX} = 2$ hours, and $T_{AY} = 1.6$ hours. During a day, 8 hours are available for production. Machine A is efficiently utilized 80% of the time; machine B is efficiently utilized 70% of the time.

(a) Determine the number of machines of type A and B that should be provided each day.

(b) If $SD[P_{ij}]$ denotes the standard deviation of daily production for product i on machine j and $SD[P_{XA}] = 2$, $SD[P_{YA}] = 3.4$, $SD[P_{XB}] = 2.2$, determine the standard deviation for the number of machines of each type required.

(c) If $SD[T_{ij}]$ denotes the standard deviation of production time/unit for product i on machine j and $SD[T_{XA}] = 0.1$, $SD[T_{YA}] = 0.2$, $SD[T_{XB}] = 0.15$, determine the standard deviation for the number of machines of each type required.

(d) Combine the standard deviation data for parts (b) and (c) to determine the standard deviation for the number of machines of each type required.

2.25. Semiautomatic mixers are used to mix a particular chemical. It requires 5 minutes to load a mixer; the mixer runs automatically and produces a batch of 1000 gallons of chemical before automatically unloading. The automatic mixing time equals 60 minutes. The automatic unloading time equals 4 minutes. Travel between machines requires 15 seconds. While mixers are automatically mixing chemicals the operator performs a quality control inspection on the previous batch, which requires 5 minutes per batch. An operator costs \$15 per hour; each mixer costs \$80 per hour.

(a) Determine the number of mixers to be assigned per operator to minimize the cost per batch produced.

(b) Suppose that there are 15 mixers to be assigned. How should they be assigned to minimize the long-run cost per batch produced?

(c) For what range of inspection time will the optimum number of mixers per operator be 7?

2.26. A highway construction firm is working on a cut-and-fill stretch of roadway for which dump trucks and power shovels are the primary equipment being used. Time studies reveal the following time values (minutes):

Time to load a truck	10
Travel time to dumping point	15
Dumping time	2
Return time	12

(a) What is the minimum number of trucks that will prevent idle time on the part of the power shovel?

(b) Suppose that it costs \$125 per hour to operate the power shovel and \$75 per hour to operate each truck. How many trucks should be assigned to the job to minimize the cost per truckload hauled?

(c) Construct a multiple activity chart for the most economical arrangement between dump trucks and the power shovel.

2.27. Consider a toaster that toasts one side of each of two pieces of bread at the same time. It takes two hands to insert or remove each slice. To turn the slice over, it is necessary to push the toaster door all the way down and allow the spring to bring it back. Thus both slices can be turned at the same time, but only one slice can be inserted or removed at one time. The time required to toast one side of a slice of bread is 0.50 minute. The time required to turn a slice over is 0.02 minute. It takes 0.05 minute to remove a toasted slice and place it on a plate. The time required to secure a piece of bread and place it in the toaster is 0.05 minute. Determine the minimum amount of time required to toast three slices of bread on both sides. Begin with three untoasted slices of bread on a plate and end with all three slices of bread toasted and placed on a plate. Illustrate your solution with a multiple activity chart.

2.28. Presently, one operator is tending five identical machines. Each machine is used to produce similar products. With five machines it has been observed that during a cycle a total of 20 machine-minutes is consumed in machines waiting for service to begin. In the past, it was observed that the operator was idle for 10 minutes each cycle when assigned to tend three machines. If machining time is 25 minutes and independent operator time is 1 minute per machine cycle, what is the value for concurrent activity? If C_1 equals \$8 per hour and C_2 equals \$25 per hour, what is the economic assignment? What is the value of the minimum cost per unit produced?

2.29. An operator is currently operating two machines. Because of an increase in sales the company wishes to have the operator run the maximum number of machines without having machine idle time. The union opposes such a move, arguing that the operator should have idle time during a cycle. After considerable controversy, the company and union agree the operator will handle the maximum number of machines that will allow the operator to be idle at least 3 minutes per cycle. A time study is made, and the following standard times are determined for a cycle under the existing assignment of two machines:

Activity	Standard time (min)
Unload, load, and start machine	2.00
Inspect parts and package	0.80
Walk to machine 2	0.20
Wait for machine 2 to stop	3.50
Unload, load, and start machine 2	2.00
Inspect part and package	0.80
Walk to machine 1	0.20
Wait for machine 1 to stop	3.50

(a) Based on the time-study data and the agreement between the company and the union, how many machines should be assigned the operator?

(b) What is the added cost per unit produced due to the union demand? Assume that the operator costs \$15 per hour and each machine costs \$30 per hour.

2.30. With a multiple activity chart show how one operator can handle two machines of type *A* and one machine of type *B* during a repeating cycle using the following data.

Machine *A*		Machine *B*	
Activity	Time (min)	Activity	Time (min)
Load	1	Load	$1\frac{1}{2}$
Inspect	$\frac{1}{2}$	Inspect	0
Travel	$\frac{1}{2}$	Travel	$\frac{1}{2}$
Machining	7	Machining	$6\frac{1}{2}$
Unload	1	Unload	1

2.31. The Wonderful Widgit Manufacturing Company has an operator who currently operates four machines: A, B, C, and D. A time study has been made of the present assignment with the following results:

Operator's activities	Standard time (min)
Load A and start	2.00
Inspect part from A	0.20
Package part	0.20
Travel to B	0.10
Unload B	1.50
Load B and start	2.00
Inspect part from B	0.20
Package part	0.20
Travel to C	0.10
Unload C	1.50
Load C and start	2.00
Inspect part from C	0.20
Package part	0.20
Travel to D	0.10
Unload D	1.50
Load D and start	2.00
Inspect part from D	0.20
Package part	0.20
Travel to A	0.10
Unload A	1.50

It has been observed that total machine idle time per cycle is 10 minutes. An attempt to secure cost data was not as successful as hoped. However, the following has been determined:

$$\$8 \le \text{cost/hour for operator} \le \$20$$
$$\$10 \le \text{cost/hour for machine} \le \$50$$

(a) What is the most economic number of machines based on total cost per unit produced?

(b) What would be the savings (cost) over the present assignment if the operator were assigned two machines with $C_1 = \$15$ per hour and $C_2 = \$30$ per hour?

2.32. The Lotta-Nuthin Construction Company is building a shopping center near White Pine, Tennessee. Concrete is being poured at the job site using wheelbarrows. The company wishes to have a mixing truck with a supply of concrete at the site at all times. Furthermore, they want to have sufficient laborers available with wheelbarrows to eliminate the possibility of the truck standing idle once the unloading begins. The following data are available:

Truck	Time (min)
Travel from supply to job site	30
Travel from job site to supply	30
Load truck at supply	10
Unload truck at job site	20

Laborer	Time (min)
Load wheelbarrow	1
Unload wheelbarrow	$\frac{1}{2}$
Travel from truck to pour	3
Travel from pour to truck	1

(a) Determine the minimum number of trucks required and the minimum number of laborers with wheelbarrows to meet the company objectives.

(b) How many minutes will a truck have to wait at the job site before the unloading begins?

(c) How many times, on the average, will a given laborer obtain concrete from any given truck?

2.33. Presently, one operator is tending five identical machines. Each machine is used to produce similar products. With five machines it has been observed that during a cycle a total of 20 machine-minutes are consumed by machines waiting for service to begin. In the past, it was observed that the operator was idle for 10 minutes each cycle when assigned to tend three machines. Machining time is 25 minutes and independent operator time is 1 minute per machine cycle. The operator costs $4 per hour and each machine costs $6 per hour.

(a) Determine the optimum number of machines to be assigned the operator.

(b) Compute the cost per unit produced based on a three-machine assignment.

(c) If there are nine machines available for assignment and an operator is assigned sole responsibility for a specified number of machines, which assignment combination is preferred: $(3, 3, 3)$ or $(4, 5)$? Base the choice on the average cost per unit produced.

2.34. The Acme Chemical Company has a drag-line conveyor system that operates between two manufacturing buildings. Large hoppers are attached to the conveyor to transport chemical mixes from building A to building B for further batch processing. Empty hoppers are returned to building A on the same loop conveyor. Empty hoppers are cleaned automatically on the return trip to A. Activities and times for this process are:

Activities	Average time (min)
Unload empty hoppers from conveyor at A, fill and place on conveyor	40
Travel from A to B	10
Unload full hopper at B, empty, and place on conveyor	50
Travel from B to A (including cleaning)	15

To ensure that chemical mixes are always available at building B, what is the minimum number of hoppers required? If the chemical mix cannot stay in the hopper longer than 20 minutes from the time loading is completed and unloading begins, what is the maximum number of hoppers that could be used? Demonstrate graphically that your solution is feasible.

2.35. It takes 2 minutes to load and 2 minutes to unload a machine. Inspection, packaging, and travel between machines equal a total of 1 minute. The machine runs automatically 8 minutes. Operators cost \$6 per hour, each. Each machine costs \$12 per hour.

(a) What is the maximum number of machines one operator can handle and not have machine idle time during a repeating cycle?

(b) What is the assignment that will minimize cost per unit produced?

(c) What is the cost per unit produced, based on a four-machine assignment?

(d) If the unload and load time had not been known, for what range of values for part (a) would the economic assignment be 2? 3?

(e) If operator cost is normally distributed with a mean of 4 and a variance of 2, and machine cost is normally distributed with a mean of 6 and a variance of 2, what is the probability that the economic assignment equals 2?

(f) If the company has seven identical machines, how should they be distributed among operators to minimize cost per unit produced?

2.36. Semiautomatic machines are used to produce a particular machined part. It requires 4 minutes to load and 4 minutes to unload a machine. A machine runs automatically for 20 minutes in producing 1 unit of product. Travel time between machines is 15 seconds. While machines are automatically running, the operator inspects the part previously produced; inspection requires 75 seconds to perform. Each operator costs \$15 per hour and each machine costs \$50 per hour.

(a) How many machines should be assigned to an operator to minimize the cost per unit produced?

(b) Suppose that there are 11 machines to be assigned. How should they be assigned to minimize the long-run cost per unit produced?

(c) Suppose that machine costs are not known. For what range of values of machine cost will the optimum assignment be three machines per operator?

2.37. Automatic palletizers are used to palletize cases of finished goods at a production plant. The palletizer places a given number of layers of cartons on each pallet, with each layer having a specified number of cartons. The time required for the palletizer to perform a palletizing cycle is 2 minutes per pallet.

An operator is responsible for programming and setting up the palletizer for each "palletizing run" based on the dimensions of the case. A palletizing run consists of a batch of 100 pallets. Hence the palletizer can operate without interruption for 200 minutes before programming and setup are required. It takes the operator 5 minutes to program and set up the palletizer.

Additionally, the operator must supply empty pallets to the palletizer. The empty pallets are placed in a magazine with a capacity to hold 25 pallets. It requires 6 minutes for the operator to travel to the empty pallet storage area, return with 25 empty pallets, and place the 25 empty pallets in the magazine.

Determine the number of palletizers to be assigned to an operator without creating idle time for the palletizer. Assume negligible travel time between palletizers.

2.38. A firm has a number of semiautomatic machines that require manual loading and unloading. The automatic machine run time is 12.5 minutes per cycle; it requires 1.25 minutes to load and 1.25 minutes to unload the machine. The time required to perform the independent operator activities is not known. However, it is known that an operator assigned four machines will be utilized (busy) 80% of the time. The cost of an operator is \$10 per hour and the cost of a machine is \$40 per hour.

(a) How many machines should be assigned to an operator to minimize the cost per unit produced?

(b) For what range of values of independent operator activity will the answer to part (a) be unchanged?

(c) If the firm has 18 machines available for assignment, how should they be assigned to minimize the long-run cost per unit produced?

2.39. An operator is currently operating three identical machines. The operator is utilized 78% of the repeating cycle. Concurrent activity equals 8 minutes and independent operator activity equals 5 minutes. Each operator costs $10 per hour and each machine costs $12 per hour.

(a) What is the cost per unit produced based on the minimum-cost assignment?

(b) For what range of values for concurrent activity will the economic assignment equal three machines?

2.40. The KLM Job Shop has requested that a new layout be designed for their operation in Alpharetta, Georgia. There are 12 departments involved. The department areas (in square feet) and activity relationships for the job shop are summarized in Figure P2.40. Design a block layout using the SLP approach.

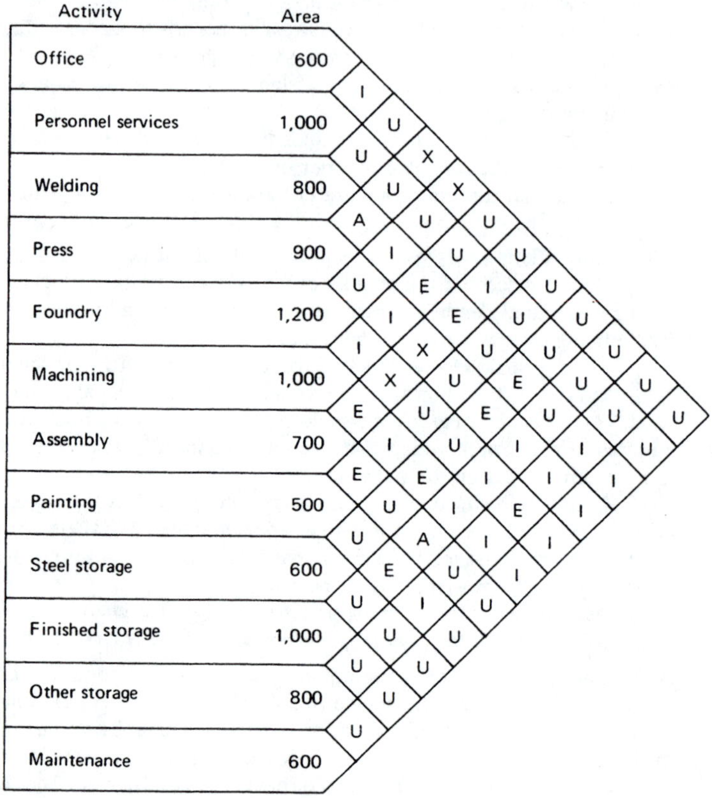

Figure P2.40

2.41. The Multiple Products Company manufactures multiple products. Six departments are involved in the processing required for the products. A summary of the processing sequences required for the 10 major products and the monthly production volumes for the products is shown in Figure P2.41 along with the departmental area.

(**a**) Develop the from–to chart giving number of units per month moving between combination of departments.

(**b**) Develop a layout design using SLP.

(**c**) Compute the total distance traveled per month based on the design obtained in part (b). Assume rectilinear travel between departmental centroids and that moves between departments are made on the basis of production lots of 100 units per move.

Product	Processing sequence	Mo. prodn.
1	A B C D E F	800
2	A B C B E D C F	1,000
3	A B E F	600
4	A B C E B C F	2,000
5	A C E F	1,500
6	A B C D E F	400
7	A B D E C B F	2,000
8	A B C B D B E B F	2,500
9	A B C D F	800
10	A B D E F	1,000

Dept.	Area (ft^2)
A	1,000
B	1,200
C	800
D	1,500
E	2,500
F	1,500

Figure P2.41

2.42. Shown in Figure P2.42 is the REL chart for a small job shop, along with the space requirements for each department. Construct REL diagrams using the traditional approach, the graph-based approach, and the relationship diagramming approach described in the text. For each, construct the space relationship diagram and design the corresponding block layout.

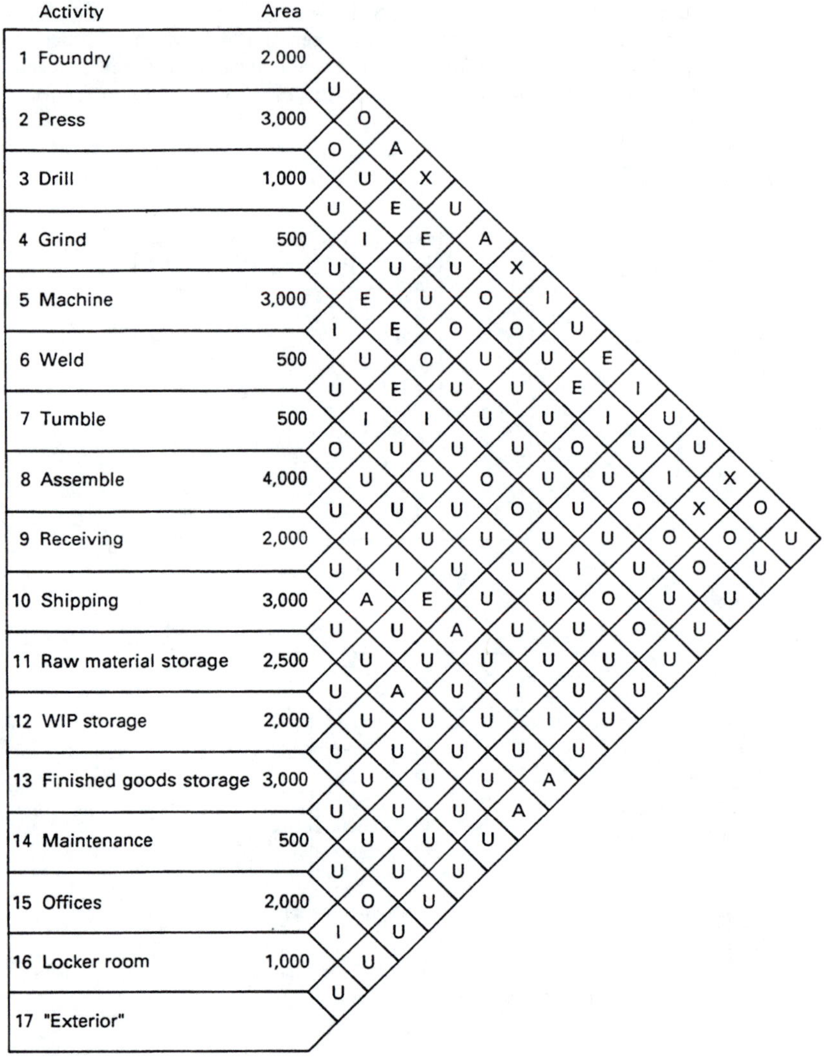
Activity
Area
1 Foundry 2,000
2 Press 3,000
3 Drill 1,000
4 Grind 500
5 Machine 3,000
6 Weld 500
7 Tumble 500
8 Assemble 4,000
9 Receiving 2,000
10 Shipping 3,000
11 Raw material storage 2,500
12 WIP storage 2,000
13 Finished goods storage 3,000
14 Maintenance 500
15 Offices 2,000
16 Locker room 1,000
17 "Exterior"

Figure P2.42

2.43. The activity relationship chart for Larry's Machine Shop is shown in Figure P2.43. Construct REL diagrams for the machine shop using the three approaches given in the text. Given the space requirements shown, construct space relationship diagrams for each REL diagram and design three block layouts for the machine shop, one for each space relationship diagram.

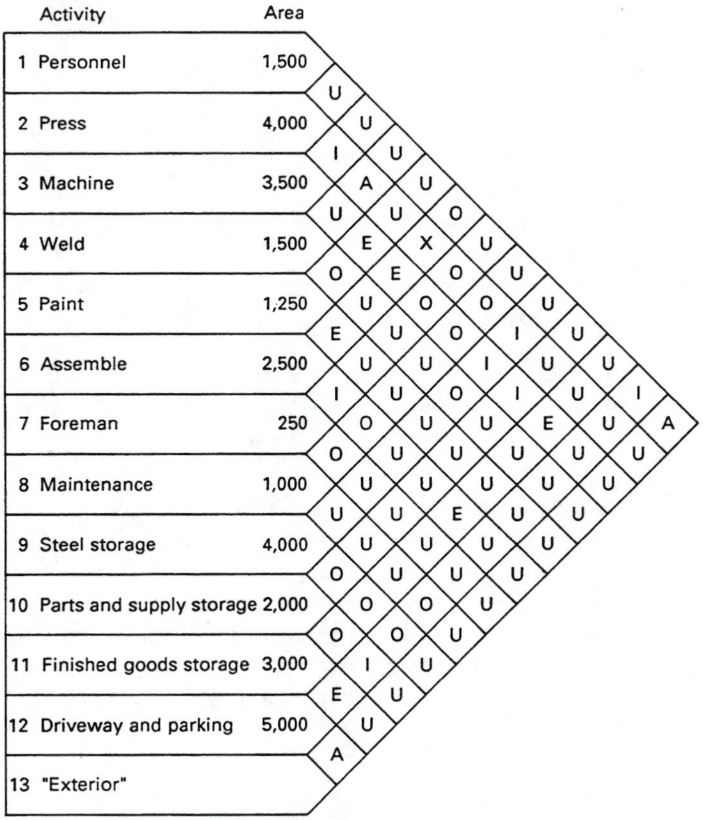

Figure P2.43

2.44. An auto-parts warehouse in Flyspeck, Texas, has requested that a new layout be designed for their main warehouse located in metropolitan Flyspeck. The warehouse has 10 major activity "centers." The current building has the dimensions of 150 ft × 225 ft. Other pertinent data are summarized in Figure P2.44. Using SLP, design a layout to be contained in the current building.

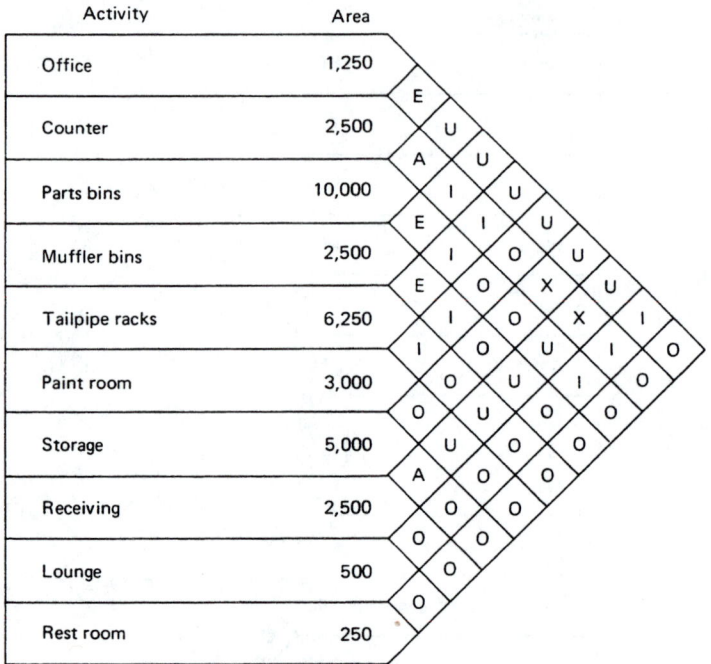

Activity	Area	Counter	Parts bins	Muffler bins	Tailpipe racks	Paint room	Storage	Receiving	Lounge	Rest room
Office	1,250	E	U	U	U	U	U	U	I	O
Counter	2,500		A	I	I	O	X	X	I	O
Parts bins	10,000			E	I	O	O	U	I	O
Muffler bins	2,500				E	I	O	U	O	O
Tailpipe racks	6,250					I	O	U	O	O
Paint room	3,000						O	U	O	O
Storage	5,000							A	O	O
Receiving	2,500								O	O
Lounge	500									O
Rest room	250									

Figure P2.44

2.45. The Original Architectural Designs (TOAD) Company specializes in designing houses to meet a client's desires. A client is asked to specify the number, type, and size of rooms required and to assign closeness ratings to pairwise combinations of the rooms. Mr. and Mrs. Snob have supplied the data shown in Figure P2.45. Using SLP, obtain a relationship diagram, space relationship diagram, and a block plan based on the REL chart. Hallways, closets, and baths will be added after meeting with the client a second time. Design both a single-story and a multistory layout.

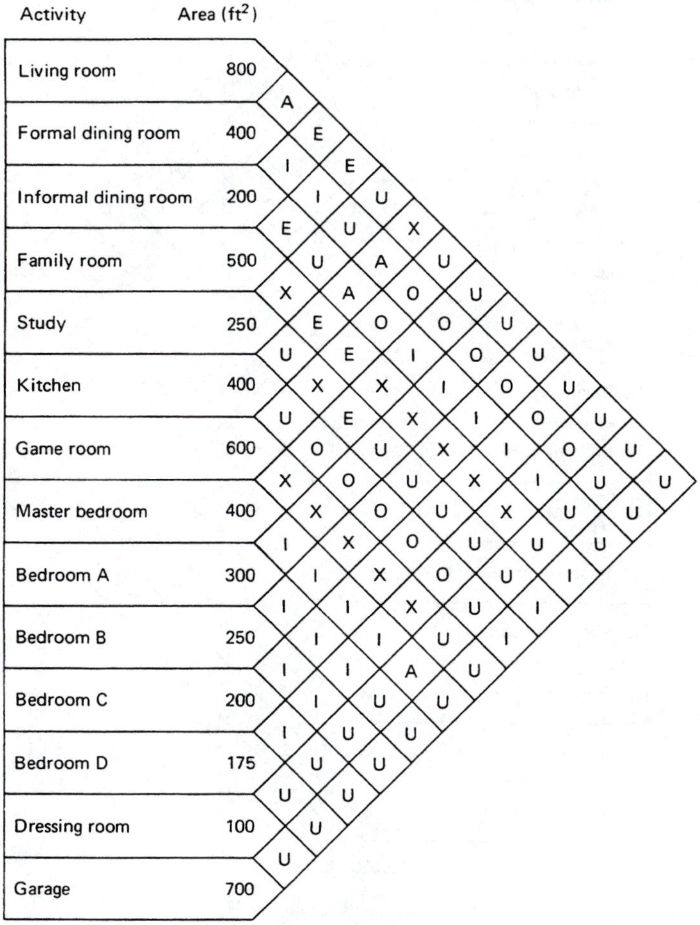
Activity
Area (ft²)
Living room 800
Formal dining room 400
Informal dining room 200
Family room 500
Study 250
Kitchen 400
Game room 600
Master bedroom 400
Bedroom A 300
Bedroom B 250
Bedroom C 200
Bedroom D 175
Dressing room 100
Garage 700

Figure P2.45

2.46. A soft-drink bottling company wishes to design a new bottling plant for their new product, Old Favorite. The activity relationship chart, including space requirements, is shown in Figure P2.46. Design a block layout for the plant using the relationship diagramming process.

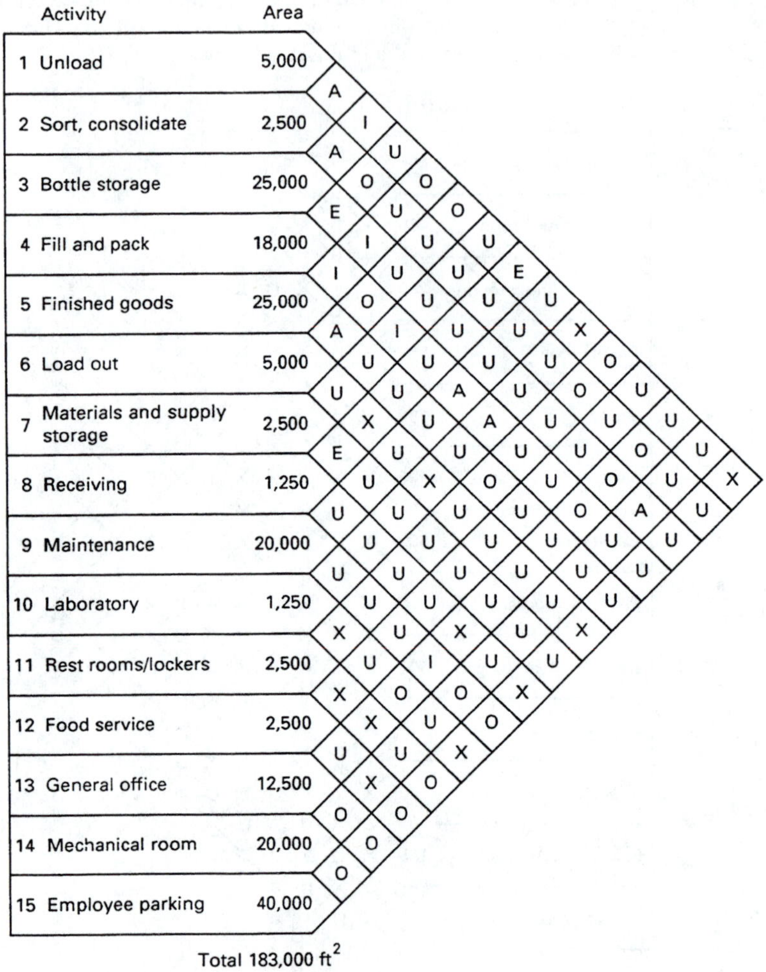

Figure P2.46

2.47. An activity relationship chart is shown in Figure P2.47 for the Rickety Furniture Company. Construct a REL diagram for the plant using the graph-based process. Based on the space requirements given, design a block layout.

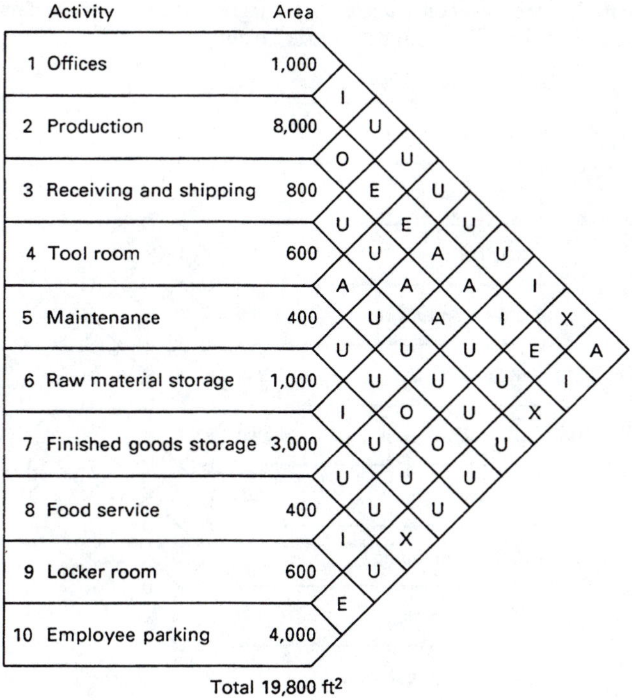

Figure P2.47

2.48. An electronics company has acquired 5000 ft^2 of space in a shopping center for its regional service facility. The space is configured 50 ft $\times$ 100 ft, with common carrier access along one of its long walls. Customers enter the facility along one of the short walls. Based on the activity relationship chart shown in Figure P2.48, use the graph-based process to construct a REL diagram for the service facility. Using the space requirements given, design a block layout.

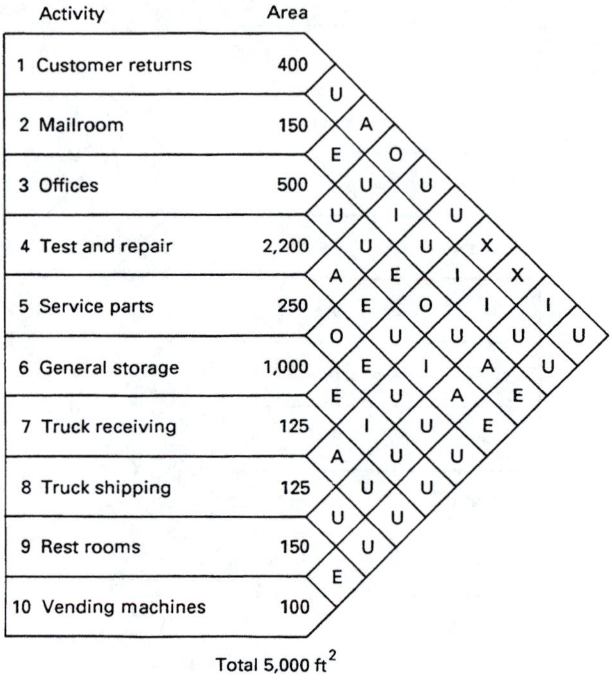
Activity
Area
1 Customer returns
400
2 Mailroom
150
3 Offices
500
4 Test and repair
2,200
5 Service parts
250
6 General storage
1,000
7 Truck receiving
125
8 Truck shipping
125
9 Rest rooms
150
10 Vending machines
100
Total 5,000 ft^2

Figure P2.48

2.49. Two attorneys are planning on opening a new office in a suburb of a major city. It will provide space for four persons: the two attorneys, a legal secretary, and a clerk-typist-receptionist. Space is also needed for storing files, meetings with clients, and copying. The space requirements are as follows:

Activity	Space (ft^2)
Attorney *A*	250
Attorney *B*	250
Secretary	175
Reception area	750
Conference room	1350
Storage	125
Copying room	100

The space available is 50 ft × 60 ft, with windows along one of the short walls and access to the office along the other short wall.

(a) Construct an activity relationship chart for the law firm.

(b) Based on the REL chart constructed, develop a REL diagram using the graph-based process.

(c) Using the space requirements for each activity, the REL diagram, and the space constraints, design a block layout for the law office.

2.50. The Unsavory Chemical Company manufactures industrial pollutants, both liquid and dry. They have purchased a new site on which a new pollution plant is to be constructed. Due to your expertise in designing layouts for pollution plants, you have been retained to design the layout. Although the activities to be performed are known, their relationships are not known. The configuration of the site and the location of the rail siding and interstate highway are such that transportation modes can be available on two sides of the new plant. Space requirements for the various activities are shown below. Management expects 50% expansion of the plant within 5 years of startup, and by 10 years it is expected to double its initial size. Two production lines are required, one for dry chemical production and the other for liquid chemical production. Liquid chemicals are packaged in disposable bottles of two sizes, 1-liter and 2-liter; dry chemicals are packaged in 1-lb and 2-lb boxes. Both are packed into cases on-line; package conveyors are used to convey the cases to automatic palletizers, where the cases are palletized and moved to storage by industrial lift truck. Pallet loads of product are block stacked in the warehouse and shipped to customers by truck; all customer orders are for full pallet loads of each product. The finished goods warehouse segregates dry chemicals from liquid chemicals, to minimize the possibility of explosions. Although it is not typical, in some instances product is shipped immediately following palletization. A work sampling study of operations at another plant making similar products yielded data on the amount of lift truck travel between activities, in pallet load equivalents per month, as shown below.

Bulk liquid and dry chemicals are received via rail and stored in tanks and silos outside the building. Screw conveyors transport the dry chemicals directly from rail cars to the bulk storage facilities; liquid chemicals are pumped to storage tanks. Approximately 25,000 ft^2 of space is required adjacent to the building for bulk storage. The space figures shown for raw material and supplies are for the palletized packing materials, labels, and miscellaneous supplies associated with each product. Dry chemical storage shown is for special chemicals that are stored inside the main plant.

Vendors supply disposable bottles on a fixed delivery schedule. Bottles are offloaded from vendor delivery trucks onto a conveyor that transports bottles directly to production. Hence no empty bottle storage occurs inside the building. Two receiving doors must be provided adjacent to the liquid production area for empty bottle receiving. The space for the liquid production area includes the space required for receiving empty bottles.

(a) Develop a from–to chart for material delivery via lift truck.

(b) Construct an activity relationship chart for the new plant.

(c) Design a block layout for the new plant using the traditional SLP process, including 5- and 10-year expansion plans.

SPACE REQUIREMENTS

Activity	Space (ft^2)
1. Receiving	1,250
2. Liquid product, raw material (R.M.), and supplies storage	7,500
3. Dry chemical supplies storage	5,000
4. Dry product, raw material, and supplies storage	10,000
5. Liquid production	10,000
6. Dry production	7,500
7. Liquid palletizer	1,250
8. Dry palletizer	1,250
9. Liquid product, finished-goods (F.G.) storage	50,000
10. Dry product, finished-goods storage	35,000
11. Shipping	1,250
12. Maintenance	2,500
13. Quality control laboratory	2,500
14. Offices	10,000
15. Food services	2,500
16. Rest rooms and locker rooms	2,500
	150,000

MATERIAL FLOW

From	To	Pallets/month
1. Receiving	2. Liquid R.M. storage	200
1. Receiving	3. Dry chemical storage	250
1. Receiving	4. Dry R.M. storage	400
2. Liquid R.M. storage	5. Liquid production	200
3. Dry chemical storage	6. Dry production	250
4. Dry R.M. storage	6. Dry production	400
5. Liquid production	7. Liquid palletizer	6,000
5. Liquid production	9. Liquid F.G. storage	100
6. Dry production	8. Dry palletizer	2,000
6. Dry production	10. Dry F.G. storage	50
7. Liquid palletizer	9. Liquid F.G. storage	5,000
7. Liquid palletizer	11. Shipping	1,000
8. Dry palletizer	11. Shipping	400
9. Liquid F.G. storage	11. Shipping	5,100
10. Dry F.G. storage	11. Shipping	1,650

Chapter 3

Computerized Layout Planning

3.1 INTRODUCTION

As we saw in Chapter 2, layout planning is a complex problem, requiring simultaneous consideration of a large number of complex, intricately related issues. Figure 3.1 illustrates some of the decisions involved in the design of a manufacturing facility, and shows that layout decisions can both depend on, and affect, other decisions. The design of other types of facilities, such as warehouses or office buildings, is equally complex.

Because of the complexity and scope of the layout planning problem, the layout planner will work together with other planners and designers, as indicated in Figure 3.1. In fact, it is not unusual to find a layout planning team rather than a single layout planner. Communication and documentation become critical in such an environment.

Documentation requirements for layout planning are substantial. In Chapter 2 we presented a summary of the sources of information required in layout planning, which include many types of charts, lists, and numerical analyses. The output of the layout planning process also can be substantial, including engineering drawings, and specifications for equipment, procedures, and staffing.

Layout planning is heavily dependent on data, documentation, and communication, and is therefore a perfect candidate for computerization. Since the mid-1970s

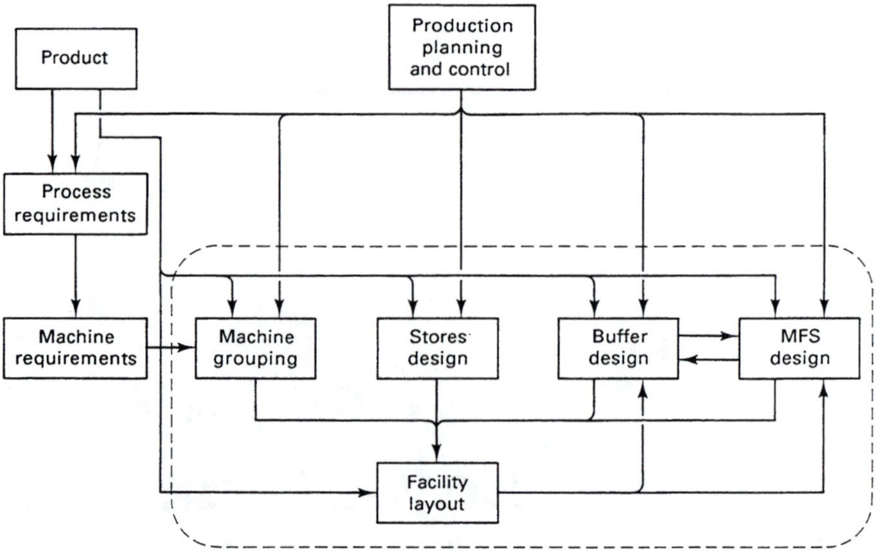

Figure 3.1 Facility planning decision interactions.

there have been astounding advancements in both the hardware and the software available for these kinds of tasks, and the trend should continue for the foreseeable future. Thus the modern facilities planner faces an enormous array of computerization opportunities in layout planning.

In this chapter we focus on the computerization opportunities that are unique to layout planning, as opposed, say, to personnel, accounting, product design, or process engineering. It is not our intent to cover the basics of database management systems (DBMS), computer-aided drafting and design (CADD) systems, or computer communication systems. Rather, our presentation is motivated by a consideration of how computerization can affect the facility layout *process*.

The facility layout process can be conceptualized as a process of design database development, where the final content of the database is the layout specification. The layout is created and modified by the layout planner(s), represented by the dashed-line box in Figure 3.2. The design database may contain data in a variety of formats, such as drawings, numerical specifications, and text. The layout data will change over time as different aspects of the layout are elaborated and coordinated with other facility design decisions. At some point in time, the layout is *frozen* and becomes the reference data for implementation.

In the layout process, the layout planner must be able to comprehend the current state of the layout in order to make additional decisions or to revise previous decisions. In making these layout decisions, the layout planner must integrate the current state of the layout database, as he or she understands it, the layout require-

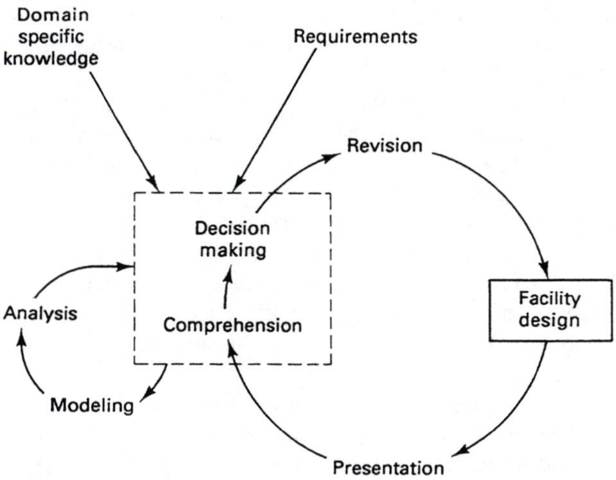

Figure 3.2 Facility layout process.

ments and constraints, as he or she understands them, and any necessary domain-specific knowledge, such as standards and local regulations. Often the layout planner will need to employ *models* in the search for good layout decisions, or to evaluate tentative layouts.

The opportunities for computerization are illustrated in Figure 3.3. A primary opportunity is the computerization of the central layout database. In reality, this may be a collection of databases: a CADD database, a product database, a process database, and so on. The layout planner will interact with these databases using

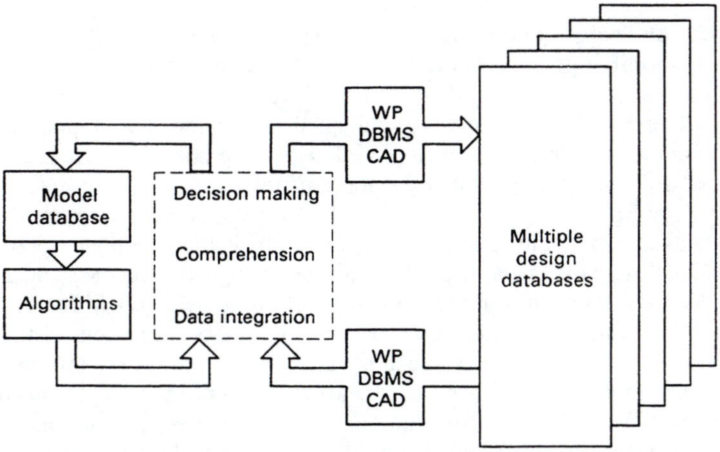

Figure 3.3 Computerized layout planning process.

CADD systems, DBMSs, and word processing (WP) systems. Modern computing technology provides the layout planner with a wide assortment of methods for displaying layouts, from graphics workstations to high-speed plotters.

Dealing with multiple databases is not the problem today that it was even a few years ago. Advances in computing capability and the corresponding advances in software provide the layout planner with access to sophisticated windowing software that will permit access to multiple programs or files simultaneously. However, as Figure 3.3 indicates, today's layout planner must still perform the data integration. That is, when data from different databases in different formats represent the same design feature, it is up to the layout planner to translate and ensure consistency. For example, when aisles are defined in a layout drawing, and lift truck specifications are in a word processing document, the layout planner must make sure that they are consistent (i.e., the lift trucks will fit in the aisles).

In Chapter 2 we summarized a number of layout *procedures* and discussed systematic layout planning (SLP) in some detail. You may wish to review that material before proceeding. A layout procedure such as SLP is simply an orderly way to execute the layout process. SLP, for example, calls for the design database to be developed in six steps:

1. Developing the basic data
2. Identifying material flows and activity relationships
3. Developing the activity relationship diagram
4. Determining space requirements and availability
5. Developing the space relationship diagram
6. Using the space relationship diagram, practical limitations, and other considerations to develop and evaluate alternative layouts

Each step involves understanding the current state of the design database and adding to or modifying its content. Each step requires decision making by the layout planner.

Generic DBMS software can be used by the layout planner for creating, storing, manipulating, retrieving, and displaying both numeric and text data. Such systems are available for computing platforms ranging from mainframes down to workstations or personal computers. Even on personal computers, quite powerful systems are available. The problem, however, is that the layout planner must build the application: that is, identify the files, design the records, design the specific DBMS functions, and implement them in the generic DBMS.

For example, in Chapter 2 we discussed the development of the from–to chart or travel chart. To construct such a chart, we need to know the production rate of each product and the flow process chart for each product. With this information we can compute the from–to flow rates between processes by summing over all products. If material is moved in unit loads, we must know the unit load quantity for each product in order to compute the from–to flows of unit loads.

In a DBMS, it would be straightforward to provide one function to create and maintain the production rate data, another to create and maintain the flow process chart data, and a third to compute the from–to matrix. Thus the layout planner could see easily the impact on from–to flows of changes in projected production rates or mixes, or variations in process plans.

Similarly, general-purpose CADD systems can be used to create, modify, store, retrieve, and display layout drawings, in effect, as an electronic drafting table and drawing file. Standard blocks, or icons, in the CADD system can automate the drawing of standard layout elements, such as building structures (doors, windows, walls, posts, etc.), furniture, or machines.

Returning to the example of the from–to chart, suppose that you want to convert from unit load flow rates to unit-load travel distances. Now you must extract from the CADD drawing the travel distances between the processes listed in the from–to chart. Currently, this would have to be done manually, because there are no generic systems that support this degree of integration between CADD and DBMS for layout planning.

While CADD and DBMS have the potential to automate a great deal of the manual work in layout planning, they do not provide much help in the area of decision making. Often, the layout planner will need some form of *decision aid* to deal with the quantity and complexity of layout planning data.

For our purposes, a decision aid has three specific elements. A *model* is a mathematical abstraction that provides an approximate representation of some aspect of the layout planning problem. For example, the from–to chart is a model, albeit a very simple one. A *model database* contains the data required for the model to be used for a specific problem. Quite often, the model data will be different from the layout planning data in the design database. Finally, a decision aid requires a computational implementation of an *algorithm* that is based on the model and operates on the model database. For example, the model could be a simple queueing model, and the algorithm simply a set of queueing formulas for computing the mean queue length and mean waiting time.

Decision aids for layout planning can take many forms. Consider, for example, the problem of developing a layout for a new warehouse. The layout planner may have access to a large quantity of historical data on customer orders. Tools for statistical analysis would be essential for building models of the customer ordering process, which in turn are necessary in layout evaluation.

As another example, in calculating area requirements, the layout planner would need a spreadsheet program, preferably one that could access the DBMS files. In this way the planner could examine alternative assumptions about the size of equipment, floor storage for materials, aisle space ratios, and so on.

These both are examples of generic decision aids that may be applied in the context of layout planning. Ultimately, however, the layout planner must create and evaluate block plans and convert them to detailed layouts. Decision aids that are specific to layout planning are almost always related to block plans.

There are a number of "classical" decision aids for block layout planning, which are more often referred to as *layout programs*. The classical layout programs (CRAFT [1,2], CORELAP [18], ALDEP [35], and PLANET [3]) all originated in the early and mid-1960s, and many variations have appeared over the years. In the remainder of the chapter we focus on the theoretical and methodological basis for the classical layout programs.

Our reasons for focusing on block layout are twofold. First, the application of computerized block plan generation and evaluation can improve the search phase of SLP by allowing the designer to generate quickly a large number of alternative solutions to a layout problem. The solutions obtained from computerized algorithms often represent radical departures from convention. Consequently, computer solutions force the designer to consider new and provocative designs. Second, the focus on the algorithmic aspects of block planning provides a logical transition from the traditional plant layout approach to the analytic approaches in subsequent chapters.

The remainder of the chapter is organized as follows. In Section 3.2 we present a discussion of some important issues in developing decision aids for the layout planning process. In Section 3.3 we address the problems that arise in automating layout evaluation. Traditional algorithms and computer programs for generating block layouts are discussed in Section 3.4, and several newer approaches are discussed in Section 3.5. Important trends are identified in Section 3.6.

3.2 INFORMATION IN LAYOUT PLANNING

The search phase of SLP involves three types of information. *Numeric information* includes such items as the space required for an activity or the total flow between two activities. *Logic information* describes the preferences of the designer regarding certain features of the layout. The closeness rating scheme is an example of logic information used to describe preferences for adjacency. Finally, the resulting block plan is presented as *graphic information*, in the form of sketches or drawings. A key problem in developing a decision aid for layout planning involves the digital representations of these three different types of information, and how the information can be manipulated.

Computers *represent* data of all types as binary numbers. Therefore, representation and manipulation of numeric data for layout planning is straightforward. Similarly, logic data are easily represented as binary numbers and easily manipulated in that form. The difficulty with logic data is not in the representation or manipulation, but in *obtaining the data in the first place*. As we discuss later, modeling and analyzing the preferences of layout designers are not a simple matter!

Our greatest difficulty is with representing and manipulating the block plans themselves. If we start from a completed block plan, it is a simple matter to represent the graphic data as points connected by lines. It is relatively simple to convert this representation to numbers, and later, with appropriate algorithms, to reconstruct the graphic data on a variety of display devices. It is much harder to manipulate subsets of points and lines that might be associated with a particular activity.

Example 3.1

Consider the simple block plan in Figure 3.4. Imagine that you are in Chicago and need to transmit the block plan to a colleague in Australia but that you may send only numeric data. Your problem is to devise an algorithm for reducing the block plan to a set of numbers. How many different ways of doing this can you develop?

In the search phase, the graphic information undergoes at least two types of manipulation. In constructing a block plan, areas representing a particular activity are assigned within a partially completed block plan. In this operation, some portion of the currently uncommitted area is committed to a particular activity. The designer may also consider a complete block plan and rearrange it to form a modified layout. This rearrangement may be as simple as interchanging the area assignments for two activities, or it may involve moving and changing the shape of several activities. Computerization involves the development of algorithms to automate some of these operations.

The problem of developing a layout planning decision aid appears to be this: A representation that is convenient for display and for mechanizing the drafting process is not well suited for the designer's purposes or for design algorithms. Conversely, a representation that is convenient for algorithmic manipulation is not well suited to display and drafting operations.

Example 3.2

In Example 3.1 you were asked to develop ways to represent the block plan as a set of numbers. Using any of the representations you devised, develop an algorithm for generating an adjacency table, that is, a from–to chart with an entry of "1" if the corresponding departments are adjacent and "0" otherwise.

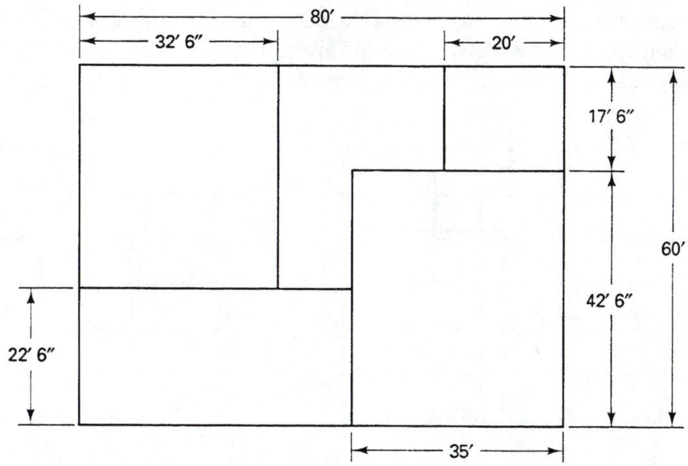

Figure 3.4 Example of block layout.

As you should have realized by now, the "points and lines" representation of a block layout is not very convenient for the types of analysis required in layout planning. Thus it is not surprising that all decision aids for block layout planning (at least all that we know of) employ a different representation, based on the concept of a "unit area square." It is assumed that the space available and the space requirement of each activity can be expressed as an integer multiple of the unit area.

Because each unit area square must be explicitly represented, and the unit area squares are the basis for all layout manipulations, it is desirable to limit their total number. Therefore, this approach to representing the block layout is an *approximation* of the real problem. The approximation is illustrated in Figure 3.5(a), where the block plan from Figure 3.4 is superimposed on a grid of unit area squares. Figure 3.5(b) illustrates one approximate representation of the activities.

This approximate representation of the block plan has some undesirable characteristics. It obviously distorts the block plan, resulting in a need to adjust the plan manually for the "true" areas and regular shapes for the activities. On the other hand, it is a very convenient representation for analyzing and manipulating the block plan.

Observe that the block plan can be represented by a two-dimensional array (or matrix) of numbers, where each cell in the array corresponds to a unit area square and the number in the cell represents the activity that occupies the corresponding unit area square in the block plan. The matrix representation for the layout of Figure 3.5(b) is shown in Figure 3.6. The adjacency table required in Example 3.2 can be constructed using an algorithm that "sweeps" through the matrix twice, once horizontally and once vertically. An activity adjacency corresponds to encountering adjacent cells in the matrix containing different numbers.

As you read the following sections on layout evaluation and layout generation, keep in mind the difference between the "real" block plan and the approximate representation of the plan used in a decision aid. The representation of the block plan is important because of the different ways we want to use the block plan—for display versus for computing an adjacency. Another problem is to ensure consistency

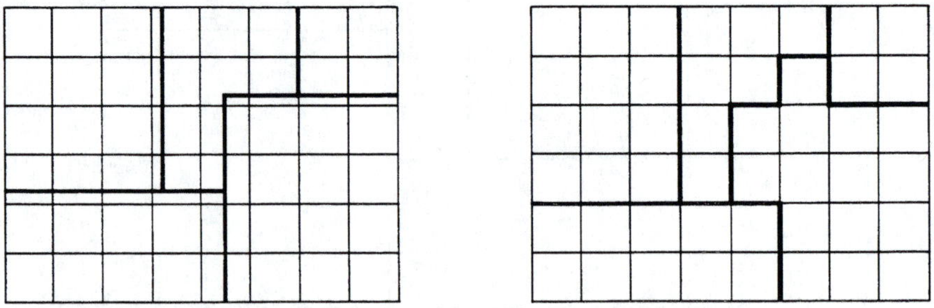

(a) (b)

Figure 3.5 Approximating the block layout with unit area squares.

1	1	1	2	2	2	3	3
1	1	1	2	2	4	3	3
1	1	1	2	4	4	4	4
1	1	1	2	4	4	4	4
5	5	5	5	5	4	4	4
5	5	5	5	5	4	4	4

Figure 3.6 Matrix representation of block layout.

between the block plan and other layout planning data. For example, each process used in the flow process chart must be assigned to exactly one block. The REL chart and the from–to chart must be coordinated with the block plan. Today, this type of *integration* is performed by the layout planner rather than being engineered into a computer-integrated layout planning procedure.

3.3 COMPUTERIZED LAYOUT EVALUATION

A number of block layout programs are capable of generating several (or many) alternative layouts but reporting only the best or the best few. This is a desirable and often helpful capability, which requires the layout program to distinguish "good" layouts from "bad" ones. Since the evaluation is to be performed by an algorithm, the method must be essentially numerical. In addition, the relative ranking of two layouts by the layout program should agree with the ranking that would have been given by the layout planner.

Before we can understand how a layout program might distinguish good layouts from bad ones, we need to understand how layout planners rank alternative layouts. A quick review of Chapters 1 and 2 will confirm that layout planners use a number of criteria, both qualitative and quantitative, in evaluating alternative layouts. They use criteria that can be evaluated directly from the layout, such as adjacency and distance between activities, and criteria that require additional computation, such as total material flow or average production time.

How does a layout planner use these criteria? Utility theory, a vast topic in itself, suggests that the planner has an implicit mental function that computes a score, or utility, for each layout. Suppose that $\mathbf{X}_i$ is a vector of attribute values for layout L_i. Then the planner implicitly computes a utility, $u_i = f(\mathbf{X}_i)$. In comparing layouts L_i and L_j, if $u_i > u_j$, layout L_i is preferred to layout L_j. So much for the concept. In practice, it is very difficult to construct an approximation for $f(\mathbf{X})$. First, we do not know all the criteria the planner uses or how to measure some of them (e.g., how do you measure "flexibility"?). Even if we did know all the criteria, we do not know the mathematical form of the planner's utility function; we can only guess at it. Finally, once we have chosen a functional form for the approximate utility function, we still face the problem of estimating its parameters.

An additional complication is that each layout planner will have a unique utility function. Therefore, even if we could construct an approximate utility function for one planner, it probably would not be accurate for another planner. In fact, it might not even be accurate for the same planner and a different application!

Nevertheless, a layout program must employ some type of approximate utility function, or scoring model. As a potential user of a layout program, you need to know how the scoring model is constructed and how well it approximates your layout utility function.

The general form of a layout scoring model is

$$s = g(\mathbf{X})$$

where $\mathbf{X}$ is a vector of attribute measures for a particular layout, $g(\cdot)$ is the scoring function, and s is the layout's score. The attributes included in $\mathbf{X}$ may be quantitative or qualitative. If a qualitative attribute is included, some arbitrary scheme has been used to assign numerical values to it. The scoring function could have any form, although it is most often a linear function.

3.3.1 Adjacency-Based Scoring

One popular scoring model is based on the adjacency rating scheme of the REL chart, which uses six classes of adjacency. (A general form of the model would utilize n classes of adjacency.) For a given layout, X_i is the number of adjacencies in class i, and w_i is the weighting factor for class i. The layout score is computed as

$$s = \sum_{i=1}^{6} w_i X_i$$

and a larger score indicates a better layout.

One implementation of the adjacency-based scoring model is used in the ALDEP package [35], with six REL chart adjacency classes, having weights of 64 (A), 16 (E), 4 (I), 1 (O), 0 (U), and −1024 (X). To illustrate the scoring, consider the 10-activity layout in Figure 3.7 and associated REL chart in Figure 3.8. The REL chart is given in the form of a triangular matrix, and the S entries along the diagonal simply indicate "same department."

The ALDEP score of 414 is obtained by noting the following combinations of adjacent departments: (1–2), (1–6), (1–10), (2–4), (2–5), (2–6), (3–4), (3–8), (3–9), (4–5), (4–8), (5–6), (5–8), (6–8), (6–9), (6–10), (7–9), and (8–9). From the REL chart in Figure 3.8, the closeness ratings associated with the combinations of adjacent departments are U, U, E, O, O, U, O, I, U, A, I, I, E, I, E, U, U, A, and E, respectively. Summing the numerical values of the closeness ratings gives a score of 211. The ALDEP program treats (10–3) and (3–10) as different adjacent combinations, so its score is 2 × 211, or 422.

Although there is some intuitive appeal in this scoring model, it has a serious flaw. The ranking of two layouts may be altered by changing the values of the weights. Thus, depending on the weight values, we may prefer layout A to layout B, or vice versa. In some cases, even relatively small changes in the weights may lead to changes in layout ranking. Therefore, it is clear that correct specification of the

TRIAL LAYOUT 15C SCORE = 422

0	0	0	0	0	0	0	0	0	0	0	0	0	0	0	0	0	0
0	7	7	9	9	9	9	6	6	6	6	10	10	10	10	0	0	0
0	7	7	9	9	9	9	6	6	6	6	10	10	10	10	0	0	0
0	7	7	9	9	9	9	6	6	6	6	10	10	10	10	0	0	0
0	7	7	9	9	9	9	6	6	6	6	10	10	10	10	0	0	0
0	7	7	9	9	9	9	6	6	6	6	10	10	10	10	0	0	0
0	7	7	9	9	9	9	6	6	6	6	10	10	10	10	0	0	0
0	7	7	9	9	9	9	6	8	6	6	10	10	10	10	0	0	0
0	7	7	9	9	9	9	8	8	6	6	10	10	10	10	0	0	0
0	7	9	9	9	9	9	8	8	6	6	10	10	10	10	0	0	0
0	9	9	9	9	9	9	8	8	6	6	1	10	10	10	0	0	0
0	9	9	9	9	9	9	8	8	6	6	1	1	10	10	0	0	0
0	9	9	9	9	9	9	8	8	6	6	1	1	10	10	10	10	0
0	9	9	9	9	9	9	8	8	6	6	1	1	10	10	10	10	0
0	9	9	9	9	9	9	8	8	6	6	1	1	10	10	10	10	0
0	9	9	9	9	9	3	8	8	6	6	1	1	10	10	10	10	0
0	9	9	9	9	3	3	8	8	6	6	2	1	10	10	10	10	0
0	9	9	9	9	3	3	8	8	6	6	2	2	10	10	10	10	0
0	9	9	9	9	3	3	8	8	6	6	2	2	10	10	10	10	0
0	9	9	9	9	3	3	8	8	6	6	2	2	10	10	10	10	0
0	9	9	9	9	3	3	8	8	6	6	2	2	10	10	10	10	0
0	9	9	9	9	3	3	8	8	6	6	2	2	10	10	10	10	0
0	9	9	9	9	3	3	8	8	6	6	2	2	10	10	10	10	0
0	9	9	9	9	3	3	8	8	5	5	2	2	10	10	10	10	0
0	9	9	9	9	3	3	8	8	5	5	2	2	10	10	10	10	0
0	9	9	9	9	3	3	8	8	5	5	2	2	10	10	10	10	0
0	9	9	9	9	3	3	8	8	4	4	2	2	10	10	10	10	0
0	9	9	9	9	3	3	8	8	4	4	2	2	10	10	10	10	0
0	9	9	9	9	3	3	8	8	4	4	2	2	10	10	10	10	0
0	9	9	9	9	3	3	8	8	4	4	2	2	10	10	10	10	0
0	9	9	9	9	3	3	3	3	4	4	2	2	10	10	10	10	0
0	9	9	9	9	3	3	3	3	4	4	2	2	10	10	10	10	0
0	9	9	9	9	3	3	3	3	4	2	2	2	10	10	10	10	0
0	0	0	0	0	0	0	0	0	0	0	0	0	0	0	0	0	0

GROUND FLOOR

Figure 3.7 Sample ALDEP layout.

111	S									
112	U	S								
113	U	U	S							
114	I	O	I	S						
115	U	O	U	I	S					
116	U	U	U	I	E	S				
117	U	U	U	O	U	U	S			
118	U	U	U	I	I	E	U	S		
119	U	I	A	I	U	U	A	E	S	
120	E	O	U	O	U	U	U	U	U	S

Figure 3.8 Relationship chart for example problem, shown as triangular matrix.

weights should be very important to the user of this scoring model. Interestingly, we are not aware of *any* published study on theoretically justified schemes for specifying the weights, even for a specific application. Therefore, you should exercise great caution when using an adjacency-based scoring model.

3.3.2 Distance-Based Scoring

Another popular approach to scoring models attempts to approximate the "cost" of flow between activities. For a given layout, let D_{ij} be a measured distance between activities i and j. The distance could be rectilinear, if flow is confined to an aisle structure, or straight line, if there are no such restrictions. In the simplest case, the rectilinear distance is measured between activity centroids. The scoring model for m activities then becomes

$$s = \sum_{i=1}^{m-1} \sum_{j=i+1}^{m} c_{ij} D_{ij}$$

where c_{ij} is the cost per unit distance for flow between activities i and j. Note that c_{ij} covers both the i-to-j and the j-to-i material flows. Part of the appeal of this scoring model is that it avoids examination of each pair of activities to specify a closeness rating. On the other hand, it does require explicit evaluation of the flow volumes and costs.

To illustrate this scoring model, consider the example shown in Figure 3.9, and assume that the rectilinear distance between activity centroids is used. To compute the area centroids, the layout is placed in a coordinate system, as shown in Figure 3.10. The centroids of the activities are marked (x) and have the following coordinate locations:

$$(x_A, y_A) = (25, 30), \qquad (x_C, y_C) = (20, 10)$$
$$(x_B, y_B) = (65, 30), \qquad (x_D, y_D) = (60, 10)$$

The from–to flow chart for the flow data, the rectilinear distance chart, and the total cost chart are given in Figure 3.9. For simplicity, we have assumed that all cost elements are equal to unity in the from–to chart for the cost data.

This particular form of the scoring model implies that the material flow system has already been specified, and that the only remaining decision is the distance between activities. The assumption that the variable flow cost will be proportional to distance between centroids could be difficult to justify in many practical situations. Consider, for example, Figure 3.11, which illustrates four departments surrounded by a contiguous aisle, each department having a designated material drop zone. The rectilinear distances between department centroids are identical for all pairs of departments, yet the travel distances clearly are not. In-floor and floor-level material handling systems almost always involve movement around the perimeter of the departments rather than centroid to centroid.

You might consider using the distance between department boundaries. It would be cumbersome, although not too difficult, to determine for each pair of

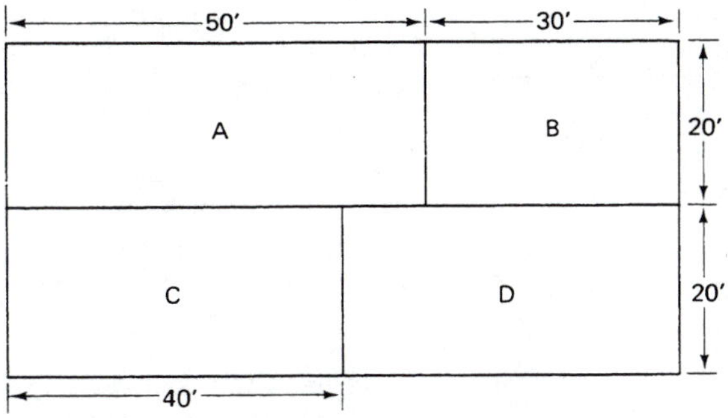

(a)

From \ To	A	B	C	D
A		2	4	4
B	1		1	3
C	2	1		2
D	4	1	0	

(b)

From \ To	A	B	C	D
A		40	25	55
B	40		65	25
C	25	65		40
D	55	25	40	

(c)

From \ To	A	B	C	D	Total
A		80	100	220	400
B	40		65	75	180
C	50	65		80	195
D	220	25	0		245
Total	310	170	165	375	1,020

(d)

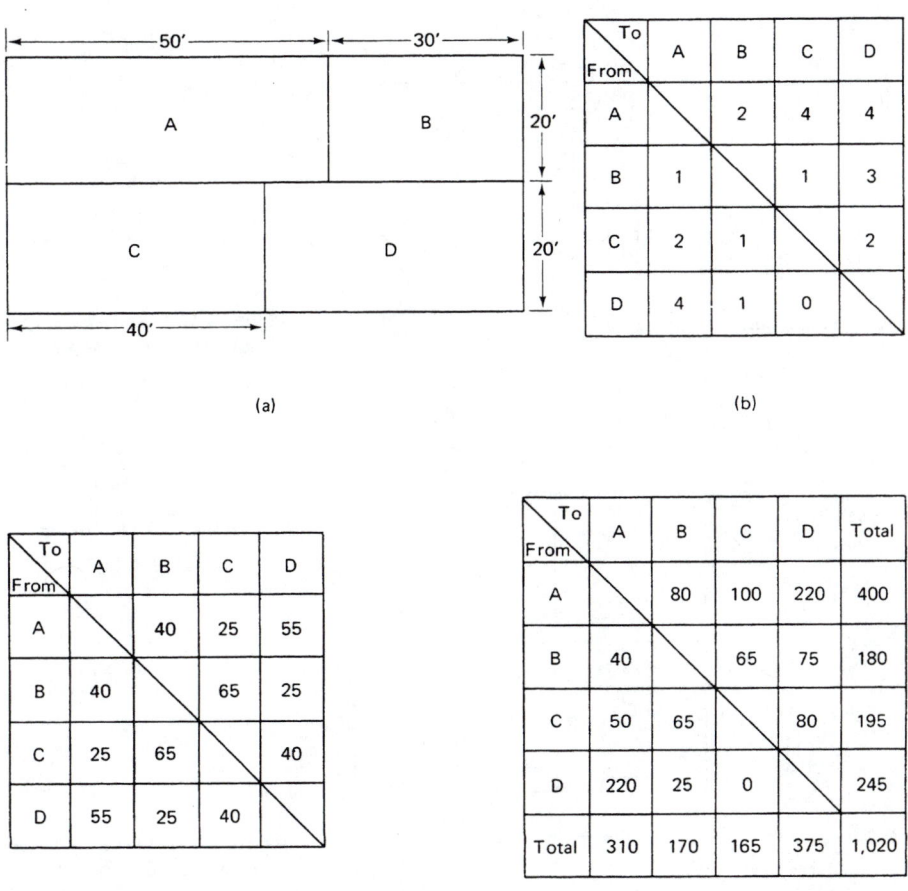

Figure 3.9 (a) Initial layout, (b) flow data, (c) distance data, and (d) total cost for distance-based scoring example.

departments the distance between closest points on their respective boundaries. As Figure 3.11 illustrates, this also may not work, since in the figure all four departments share a common point, but the material handling distances between them are not zero and not equal. Even knowing the aisle layout does not necessarily solve the problem, since some material handling systems, such as conveyors and many automated guided vehicle systems, have an essentially "one-way" movement. Thus the direction of movement may have a major impact on travel distance. In Figure 3.11, for example, if the departments are served by an automated guided vehicle that travels clockwise around the loop, the distance from B to A is much greater than the distance from A to B.

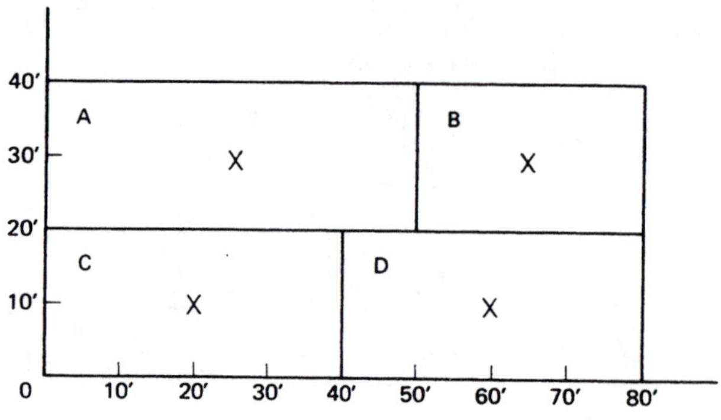

Figure 3.10 Activity centroids for distance-based scoring example.

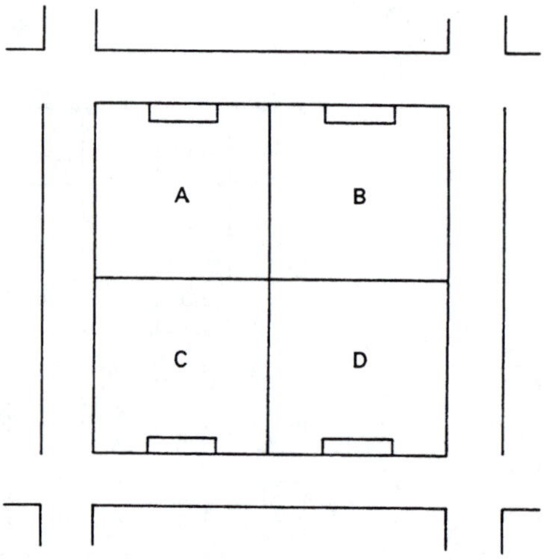

Figure 3.11 Distance-based scoring with aisle travel.

You should recognize that any layout program that computes a distance-based score is likely to provide only an approximation of the real travel distances in the final layout. The potential approximation error depends on the type of material handling system used and the nature of the layout itself.

3.3.3 Distance-Weighted Adjacency-Based Scoring

At least one computerized layout package uses a scoring model that combines distance and adjacency in a unique way. In the scoring model for interactive CORELAP [29], X_{ij} is the shortest rectilinear distance between activities i and j. There are six classes of adjacency, corresponding to the REL chart classes. The scoring model is

$$s = \sum_{i=1}^{6} \sum_{j=1}^{6} w_{ij} X_{ij}$$

where w_{ij} is the weight for the adjacency class assigned to the pair of activities (i,j). Our comments on adjacency-based scoring models apply also to this model.

3.3.4 More Complex Scoring Models

The scoring models described so far have required only the basic SLP data, a block layout, and simple arithmetic operations. This means that a score could be computed very quickly—an important point if many scores need to be calculated. There is no conceptual reason why the scoring model should not incorporate more complex computations. For example, a scoring model based on material handling system cost might employ a simulation model to determine equipment utilization, or the number of vehicles required.

More complex forms of the scoring model might more adequately reflect the true preferences of the layout planner. However, they would be more difficult to implement, probably would require more parameters, and certainly would require more effort to compute. The latter aspect might limit the number of alternatives that could be evaluated.

3.4 COMPUTERIZED LAYOUT GENERATION

Computerized layout generation algorithms fall into one of two distinct categories. *Construction* algorithms start with the basic SLP data, or some subset of it, and "build" a block layout by iteratively adding one more activity to a partial layout until all activities have been placed. The algorithm may, in fact, build a large number of layouts and report only the best of them. *Improvement* algorithms, as the name suggests, require, in addition, an initial block layout which they then attempt to improve. Without the scoring model, construction algorithms would not know which were the best layouts, nor would improvement algorithms know when an improvement had been made.

3.4.1 Construction Algorithms

Suppose that there are n activities to be arranged in a block layout, and the standard SLP data are available. A generic construction algorithm has the following form:

```
Procedure CONSTRUCT
   FOR i = 1 to n
          SELECT an activity not yet placed
          PLACE the selected activity in the layout
   END FOR
END CONSTRUCT
```

The basic differences between construction algorithms are in the implementation of the SELECT and PLACE steps. If either the SELECT or the PLACE step has a random (or variable) outcome, a number of different layouts can be obtained simply by embedding CONSTRUCT in a suitable loop.

Selection rules. If we observe the operation of a construction algorithm, we can record a *sequence* of activity placement, which is the result of the SELECT step. There are many different ways to generate this sequence.

In the original ALDEP package [35], for example, the SELECT procedure is based on the REL chart. A "cutoff" for importance is specified (e.g., using *E* as a cutoff means that *A* and *E* relationships are to be used). The first activity is randomly selected. After that, the next activity added to the partial layout is randomly selected from among all those with *A* or *E* relationships to activities already in the layout. If there are no such activities to choose from, the next activity to be added is randomly selected from among all the remaining activities.

In the context of the selection rule, "randomly" has the following meaning. There is a candidate list of activities, any one of which may be selected for placement. All activities are placed in a lottery, and one is selected by a process that is intended to be unbiased. For example, if there are 10 candidates, a computer-generated pseudorandom number between 1 and 10 might be used as the selection mechanism.

One of the first construction algorithms, CORELAP [18,36], uses the closeness ratings, the activity sizes, and a *total closeness rating* in the SELECT step. The total closeness rating (TCR) for an activity is simply the sum of the weights for relationships between that activity and all other activities, and can easily be calculated from the REL chart. The activity with the largest TCR is selected first. After that, activities are selected on the basis of the highest-rated relationship to activities already selected, with ties broken first by largest TCR and then largest size.

Neither ALDEP nor CORELAP considers the flows between activities in selecting activities for placement. The PLANET algorithm [3] provides a choice of three different selection rules, all based on the flow cost chart and a set of "placement priorities" specified by the user. All three rules use placement priorities to determine which subset of activities to consider, and then, within that subset, use the flow cost to make the selection.

The SELECT rules described above are similar in that the sequence of activity placement could be determined before actual construction of the layout. As far as we know, there are no construction algorithms with a SELECT rule that considers the partial layout as well as the REL and flow cost data to sequence the activity placements.

Placement rules. If all activities had the same area and a square shape, activity placement would be a conceptually straightforward, though very difficult problem. For example, suppose that we have 10 activities to place in a layout. For the first activity, there are exactly 10 possible placements, corresponding to the 10 empty unit area squares in the layout. For the *k*th activity placed in the layout, there are exactly $10 - k + 1$ possible locations, $1 \leq k \leq 10$. Thus there are 10!, or 3,628,800, possible layouts (there will be many that are rotations or mirror images of others).

Even though there is an enormous number of possible layouts, we can imagine a straightforward algorithm for constructing them all. We would start by enumerating all the possible placements for the first activity, which would give us 10 partial layouts. Then, for each of the 10 partial layouts, we would enumerate all possible placements for the second activity, which would yield $10 \times 9 = 90$ partial layouts. Continuing in this manner, we would finish with all 3,628,800 complete layouts.

Why is it so easy to construct the layouts for equal-sized, square activities? The reason is that we can predefine the activity boundaries in the final layout before we begin constructing it. This is also why we can consider a partial layout like the one shown in Figure 3.12. Even though the first two activities placed in this partial layout are not adjacent, we know that all the other activities still "will fit" without overlapping (i.e., that this partial layout leads to a feasible complete layout). (In fact, *every* completion is a feasible layout from the perspective of fitting the activities into the layout.)

The situation changes completely when the activities are represented by a collection of unit area squares. To begin with, we *must* now address the question: What is a feasible shape for this activity in the final layout? To be sure that the shapes are "sensible," we need the following:

Contiguity Rule. If an activity is represented by more than one unit area square, every unit area square representing the activity must share at least one edge with at least one other unit area square representing the activity.

The contiguity rule says that the unit area squares representing an activity must share more than just a corner point. For example, the shape shown in Figure 3.13(a) is not a feasible shape.

Unfortunately, the contiguity rule does not prohibit the shape illustrated in Figure 3.13(b). Thus we also need the following:

Connectedness Rule. The perimeter of an activity must be a single closed loop that is always in contact with some edge of some unit area square representing the activity.

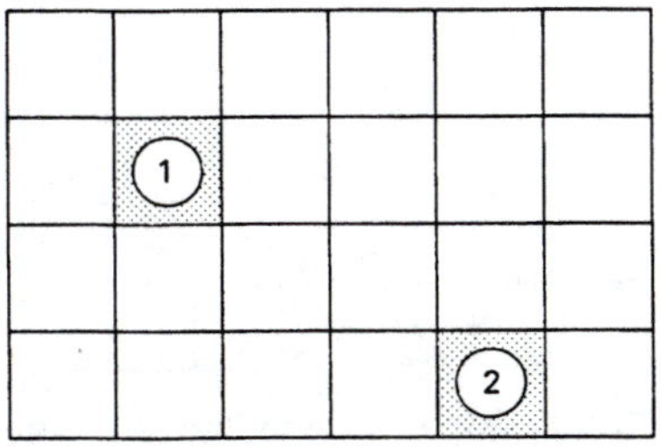

Figure 3.12 Feasible placement for equal-sized square activities.

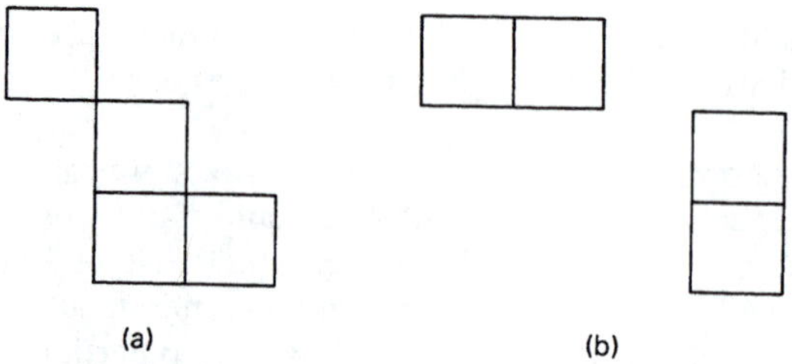

Figure 3.13 Contiguity and connectedness rules.

Both the contiguity rule and the connectedness rule are easy to understand and easy for you to check visually. As an exercise, see if you can develop an algorithm for checking these two rules, using a matrix representation of the block layout.

Consider an activity represented by 4 unit area squares. How many possible shapes are there for this activity in the layout? Figure 3.14 shows the five *basic* shapes for the activity, but if you consider all the possible mirror images and rotations, there actually are 19 different ways that this activity may appear in a layout. You should make sure that you can identify the 14 mirror images and rotations. Now determine how many different shapes there are for an activity represented by 6 unit area squares.

As the preceding exercise shows, simply identifying all the possible shapes becomes a nontrivial problem when the number of unit area squares is greater than 4 or 5. If an activity has a large number of unit area squares, among the possible shapes would be some bizarre configurations, as illustrated in Figure 3.15. Thus layout generation programs usually employ rules in addition to the contiguity and connectedness rules to ensure "reasonable" shapes. Some examples are:

Enclosed Voids Rule. No activity shape shall contain an enclosed void.

Shape Ratio Rule. The ratio of a feasible shape's greatest length to its greatest width shall be constrained to lie between specified limits.

Corner Count Rule. The number of corners for a feasible shape may not exceed a specified maximum.

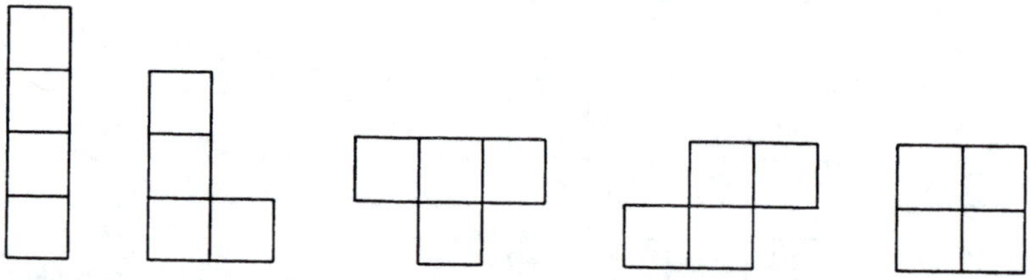

Figure 3.14 Basic shapes for 4 unit area squares.

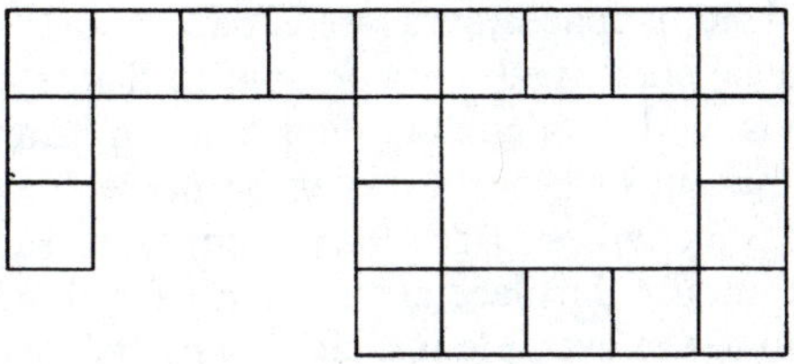

Figure 3.15 Possible shape for an activity with 20 unit area squares.

Now consider the number of ways that a shape may be placed in the layout. Suppose that you want to place the shape from Figure 3.14(b) into a layout that is 6 unit area squares high by 10 unit area squares long. There are 36 possible placements of the basic shape, but 144 possible placements including just its mirror images. There are another 160 possible placements if you consider its rotations and their mirror images, or a total of 304 possible placements for this basic shape. Now, if you are trying to decide how to place an activity with 4 unit area squares, how many alternatives must you consider?

Obviously, when an activity is larger than 1 unit area square, the number of possible placements for it becomes very large. There is an additional complication to the placement process. Consider the partial layout shown in Figure 3.16(a). Suppose that you decide to place the next activity as shown in Figure 3.16(b). If you

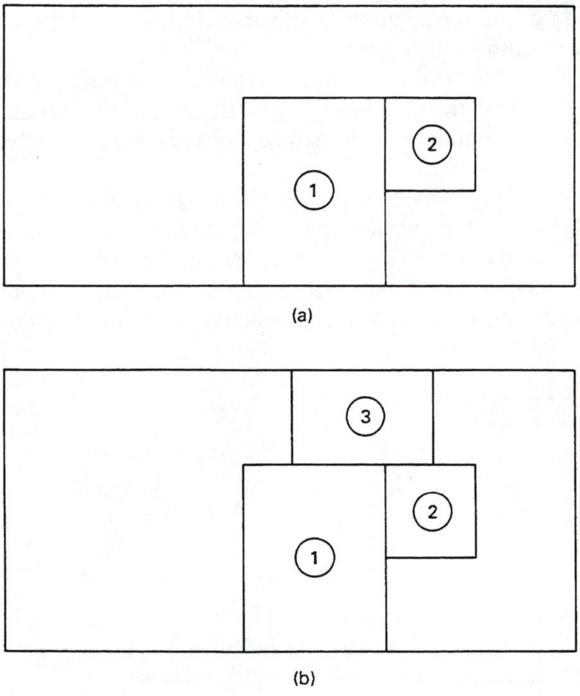

Figure 3.16 Partial block layouts.

do so, you have created two disjoint spaces, with areas of 32 and 18, respectively. If the contiguity rule is to be satisfied, you must be able to divide the remaining unplaced activities into two sets, with total area requirements of 32 and 18, respectively. It is entirely possible that no such partition can be made.

The problem with placing activities larger than 1 unit area square is that a partial layout generates constraints on the placements that are feasible for the as-yet-unplaced activities. In addition to the example just cited, a partial layout may force an infeasible shape ratio or corner count.

This difficulty with constructing the layout has been resolved by developing two classes of placement rules. *Bounded placement* rules accept a specified facility configuration and use simple but robust rules to fit the activities into the facility. As you might expect, a bounded placement rule is not able to enforce some of the activity shape rules. *Free placement* rules, on the other hand, create a layout without regard to the resulting facility configuration. As a result, they are likely to produce layouts that require considerable manual adjustment to conform to conventional building configurations.

A bounded placement rule is used in the ALDEP algorithm [35] and is based on a "sweep width" set by the planner. The sweep width divides the layout into a series of "lanes" as illustrated in Figure 3.17. The placement rule starts in one corner of the layout and follows a serpentine path, as shown in Figure 3.17, through the lanes. Activities are placed along this path simply by filling the lane. Thus a particular sweep width may lead to some undesirable shapes, as illustrated in Figure 3.18, so that a variety of sweep widths should be tried.

The free placement rule used in CORELAP [18,36] evaluates a number of different locations for the activity to be placed, as well as several different rectangular shapes for the activity. The placement decision is made using a *placing rating* and a *boundary length* evaluation.

The *placing rating* is the sum of the *weighted ratings* between the new activity to be placed in the layout and its neighbors in the layout. A neighbor is defined to be any activity already in the layout that has a common border with the new activity. The weighted rating values are specified by the user of the program. *Boundary length* refers to the length of the boundaries common to the new activity and all activities already in the layout.

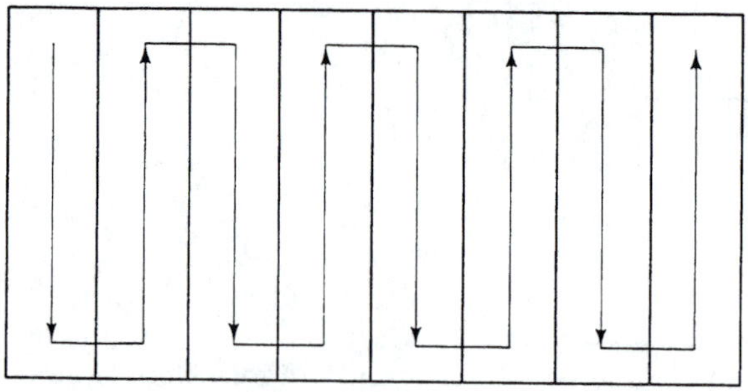

Figure 3.17 Serpentine scanning pattern used by ALDEP.

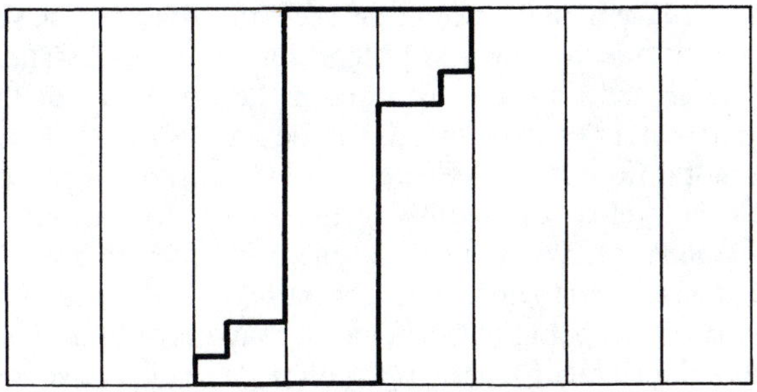

Figure 3.18 Possible ALDEP placement for an activity.

As an illustration of the calculation of the placing rating and the boundary length, consider Figure 3.19. Activities 1, 2, and 3 are already in the layout, and activity 4 has been selected for placement in the layout. Assume that weighted rating values have been assigned as follows: $A = 600$, $E = 200$, $I = 50$, $O = 10$, $U = 0$, and $X = -200$. Furthermore, suppose that the closeness ratings between activity 4 and activities 1, 2, and 3 are A, E, and I, respectively. Using the placement given in Figure 3.19(a), the placing rating is 800. The placement given in Figure 3.19(b) has a placing rating of 650. Were it not for the fact that activity 3 is already located in the layout, we could obtain a placing rating of 850 with the placement given in Figure 3.19(c). The boundary lengths are 4, 3, and 6 for placements (a), (b), and (c), respectively, in Figure 3.19.

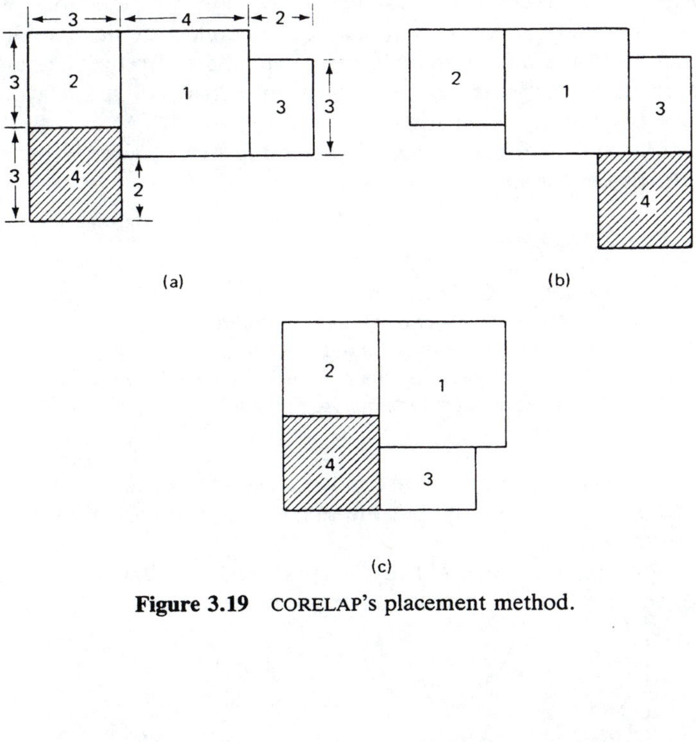

Figure 3.19 CORELAP's placement method.

The evaluation of activity placement in CORELAP uses an adjacency-based scoring model. A similar placement rule in the PLANET algorithm [3] uses a distance-based scoring model. Each unassigned unit area square on the boundary of the current partial layout is evaluated as if it were the centroid of the activity to be placed. The location with the smallest total flow cost is selected, and the activity is placed in such a way that the centroid is as close as possible to the assumed location.

The result of a free placement rule will depend on the way in which specific unit area squares in the layout are assigned to an activity, because this will determine its shape in the layout. As the layout is being constructed, it may become difficult or impossible to preserve rectangular shapes for activities without leaving unassigned unit area squares completely enclosed by activities. Thus very irregular activity shapes may be generated, resulting in a need for manual adjustment of the computer-generated layout.

3.4.2 Improvement Algorithms

Suppose that we are given an initial block layout, along with the flow cost chart and a rule for calculating the distances between activities. Our task is to modify this initial layout so that the resulting layout has a smaller total flow cost. Obviously, what we want to do is to "move" activities around in the block plan. What we need is a systematic method for doing so.

The problem we face is much like a jigsaw puzzle, except that the shapes of the pieces are not fixed. In this form, the problem simply has too many degrees of freedom for us to devise a good method for modifying the block plan. Thus all the improvement algorithms with which we are familiar are based on limiting the kinds of changes that are permitted. In particular, these improvement algorithms permit two, or sometimes three, activities to be "interchanged" in the layout, provided that certain conditions are satisfied.

The general form of an algorithm for improving a given layout plan by exchanging the places for two activities is:

```
Procedure TWOWAYX
   REPEAT
      CHOOSE a pair of activities
      ESTIMATE the effect of exchanging them
      EXCHANGE if the effect is to reduce total cost
      CHECK to be sure that the new layout is better
   UNTIL no more improvements are possible
END TWOWAYX
```

In implementing TWOWAYX, the contiguity, connectedness, and enclosed voids rules can be incorporated easily, but the shape ratio and corner count rules usually are not enforced.

To CHOOSE a pair of activities that can be exchanged without disturbing the other activities, one of two conditions must be satisfied: (1) the activities have the same area; or (2) the activities share a common boundary. When adjacent activities with unequal areas are exchanged, the resulting activity shapes depend on the method used to assign the unit area squares to the activities. In Figure 3.20, activities *A* and *B* are to be exchanged, and *A* is the smaller.

Figure 3.20 also illustrates the results from two different methods for performing EXCHANGE. Both initially "erase" all the unit area square assignments and then reassign, starting with the smaller activity and the unit area square that is farthest to the right and lowest in the layout. The first method assigns all the unit area squares in a column before moving to the adjacent column [Figure 3.20(b)], while the second method assigns all the unit area squares in a row before moving to an adjacent row [Figure 3.20(c)].

As the figure illustrates, the result of EXCHANGE depends on the method used to fill in the activities. There are many possible methods in addition to the ones we have described. Note that EXCHANGE may lead to the violation of the shape ratio or corner count rules. Such violations could be prohibited, but this would severely limit the opportunities for improving flow cost. Therefore, improvement algorithms generally do not enforce these two rules, but depend on manual adjustment of the final layout to satisfy them.

From our small example it should be clear that actually performing EXCHANGE can be a time-consuming process. Thus it is desirable to know that exchanging a pair of activities will, in fact, reduce the flow cost. The CRAFT algorithm, probably the earliest widely known improvement algorithm, uses an estimate of the flow cost change that is based on the activity centroids. The flow cost for the original layout is computed from the original activity centroids using rectilinear distances and the given flow cost chart. If activities *A* and *B* are being considered, the new flow cost is estimated by assuming that the new centroid for activity *A* will be the old centroid for activity *B* and that the new centroid for activity *B* will be the old centroid for activity *A*.

This method of estimating the flow cost for the new layout is exact if activities *A* and *B* have the same area, but can be in error if the areas are different. As an illustration, consider the layout shown in Figure 3.9. The rectilinear distance between, say, the current centroids for activities *A* and *B* is

$$|x_A - x_B| + |y_A - y_B| = |25 - 65| + |30 - 30| = 40$$

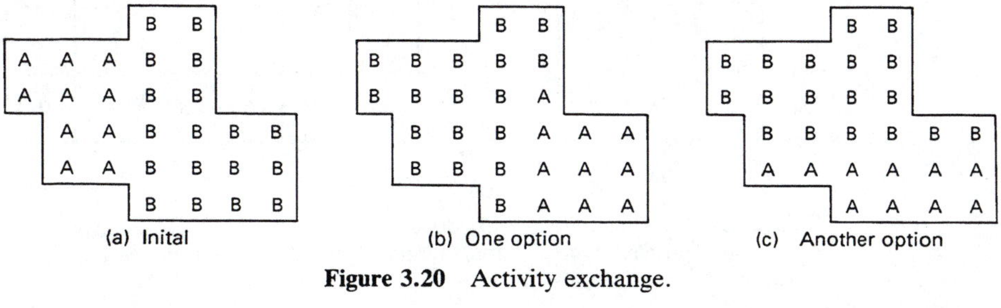

(a) Inital (b) One option (c) Another option

Figure 3.20 Activity exchange.

as shown in Figure 3.21. Notice that Figure 3.21(b) shows the distances that would be realized if *A* and *B* were interchanged. The total cost is estimated to be 1060, as shown in Figure 3.21(c). Obviously, we will not want to interchange these activities, due to an anticipated increase of 40 in total cost.

Suppose that *A* and *C* are interchanged. If we develop the resulting distance chart and compute the total cost, we obtain a cost of 955. Interchanging *A* and *D* produces a total cost of 1095. Notice that *B* and *C* are not the same area and do not have a common border. Thus an interchange of *B* and *C* is not considered. Interchanging *B* and *D* results in a total cost of 945. Interchanging *C* and *D* produces a total cost of 1040. Consequently, the best interchange is *B* and *D*.

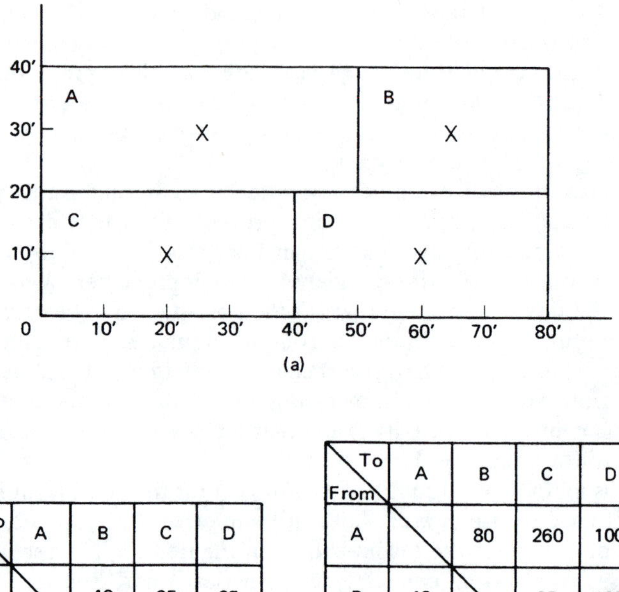

(a)

To / From	A	B	C	D
A		40	65	25
B	40		25	55
C	65	25		40
D	25	55	40	

(b)

To / From	A	B	C	D	Total
A		80	260	100	440
B	40		25	165	230
C	130	25		80	235
D	100	55	0		155
Total	270	160	285	345	1,060

(c)

Figure 3.21 Initial layout with department centroids shown. Distance chart and total cost chart if departments *A* and *B* are interchanged in the initial layout.

On interchanging B and D, we obtain the layout given in Figure 3.22(a). Since B and D are not the same size, we notice that D is no longer rectangularly shaped and that the activity centroids were not exchanged exactly. Rather, a calculation shows that

$$(x_A, y_A) = (25, 30), \qquad (x_C, y_C) = (20, 10)$$
$$(x_B, y_B) = (55, 10), \qquad (x_D, y_D) = (67.5, 25)$$

Since the centroid locations have changed, our estimate of total cost might be incorrect. Therefore, we develop a distance chart for the new layout, as given in Figure 3.22(b). As before, these distances are the rectilinear distances between activity centroids. On multiplying the flow values in Figure 3.9 and the distance values in Figure 3.22, we obtain a total cost of 985 rather than the estimated cost of 945.

As this example shows, the estimates upon which the activities to exchange are chosen can be in error. Thus the "best" exchange could be passed over, due to

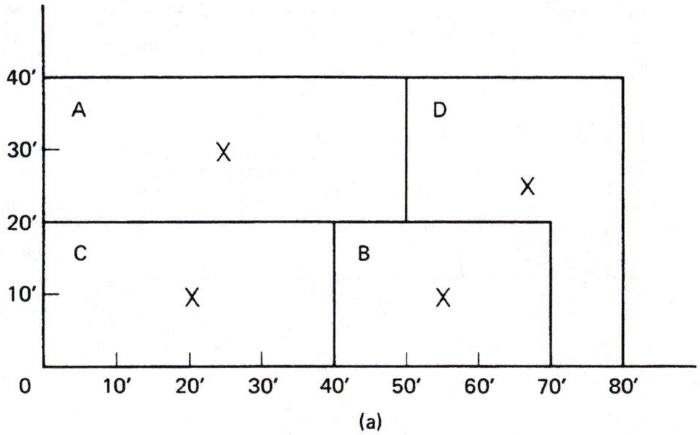

(a)

From \ To	A	B	C	D
A		50	25	47.5
B	50		35	27.5
C	25	35		62.5
D	47.5	27.5	62.5	

(b)

Figure 3.22 First improved layout and its associated distance chart.

LOCATION PATTERN

	1	2	3	4	5	6	7	8	9	10	11	12	13	14	15	16
1	A	A	A	A	A	A	A	C	C	C	I	I	I	I	I	I
2	A					A	A	C		C	I					I
3	A					A	C	C	C	C	I	I	I	I	I	I
4	A					A	C	C	C	H	H	H	H	J	J	J
5	A					A	G	G	G	H			H	J		J
6	A	A	A	A	A	A	G	G	G	H			H	J	J	J
7	B	B	B	B	B	B	G	G	H	H	H	H	H	K	K	K
8	B				B	B	F	F	L	L	L	L	L	L	K	K
9	B				B	F	F	F	L					L	L	L
10	B				B	F		F	L							L
11	B	B	B	B	B	F	F	F	L							L
12	D	D	D	D	E	L	L	L	L	L	L	L	L	L	L	L
13	D		D	E	E	E	L	L	L	L	M	M	M	M	M	M
14	D		D	E		E	M	M	M	M	M					M
15	D		D	E		E	M									M
16	D	D	D	E	E	E	M	M	M	M	M	M	M	M	M	M

(a)

LOCATION PATTERN

	1	2	3	4	5	6	7	8	9	10	11	12	13	14	15	16
1	A	A	A	A	A	A	A	L	L	L	J	J	J	I	I	I
2	A					A	A	L		L	J		J	I		I
3	A					A	L	L		L	J	J	J	I		I
4	A					A	L	L	L	L	L	L	L	I		I
5	A					A	G	G	G	L			L	I		I
6	A	A	A	A	A	A	G	G	G	L			L	I	I	I
7	B	B	B	B	B	B	G	G	L	L			L	K	K	K
8	B				B	B	C	C	L			L	L	L	K	K
9	B				B	C	C	C	L			L	H	H	H	H
10	B				B	C		C	L	L	L	L	H			H
11	B	B	B	B	B	C	C	C	L	F	F	F	H			H
12	D	D	D	D	E	C	C	F	F	F	F	H	H	H	H	H
13	D		D	E	E	E	F	F	F	F	M	M	M	M	M	M
14	D		D	E		E	M	M	M	M	M					M
15	D		D	E		E	M									M
16	D	D	D	E	E	E	M	M	M	M	M	M	M	M	M	M

TOTAL COST 87963.81 EST. COST REDUCTION 1410.00
MOVEA J MOVEB I MOVEC INTERATION 4

(b)

Figure 3.23 Initial (a) and final (b) layouts from CRAFT.

estimation errors. This generally will be the case for any improvement algorithm that does not actually perform every exchange in order to evaluate the cost.

Another characteristic of improvement algorithms is that a long sequence of activity exchanges may result in some very odd activity shapes. Figure 3.23 shows an initial layout and the final layout produced by the CRAFT program. For activity *L*, the final layout comes very close to violating the contiguity assumption. Clearly, a manual adjustment to the final CRAFT layout will be required.

We have discussed only two-way exchanges. Conceptually, we could perform exchanges of three or more activities at the same time. However, as the number of activities being exchanged increases, the difficulty of constructing the modified layout increases. Also, the number of possible exchanges to be evaluated increases tremendously. For example, for 10 activities, there are 45 distinct pairs of activities but 120 distinct triples. Not only would there be nearly three times as many three-way exchanges to evaluate, but each would require more work to evaluate than a two-way exchange.

Before leaving the topic of improvement algorithms, we should note that both the ESTIMATE and the CHECK functions require a scoring model. Our discussion has dealt only with scoring models based on distance and flow cost. More complex scoring models could be used. For example, the COFAD algorithm [38,39] allows a more detailed description of the various material handling alternatives and the costs associated with flows between activities. Other types of scoring models might also be considered.

3.5 TRENDS IN COMPUTERIZED LAYOUT

The first widely known and widely available computerized layout algorithm, CRAFT [1,2], appeared in the early 1960s. CRAFT and its many successors essentially approached the layout problem as a *quadratic assignment problem* (QAP)—a notoriously difficult problem to solve. We discuss the QAP in Chapter 9. Furthermore, these layout programs used the computer hardware and software technology of that era, which is to say, they were batch oriented, not user friendly, and did not make use of graphics technology as we know it today. On the other hand, they were reasonably portable and could be obtained at almost no cost. For roughly 20 years there was no significant change to either the algorithmic structures or the technology for computerized layout, although there were efforts to add interactive "front ends" for data input to the batch-oriented programs.

In the decade of the 1980s, the developers of CADD systems began to market applications software for computerized layout as an integrated component of their CADD offerings. Examples include the ability to generate stack plans (assignment of activities to floors in a multifloor building) automatically, interactive color graphic systems for block plan generation and manipulation, and integrated planning, engineering, and financial systems. Interestingly, most of these developments have come from architectural and engineering firms rather than from the engineering research community. Also, these systems were available primarily on mainframe computers and were quite expensive.

The decade of the 1980s also brought an explosion of computing technology: high-powered desktop computers, new programming languages, desktop CADD systems, desktop database management systems, artificial intelligence and expert systems, and so on. Along with the basic advances in computing, there were also significant advances in algorithms and methods for simulation, optimization, queueing analysis, and statistics. Despite these advances, the capabilities of mainframe CADD systems for facility design still are not available on the typical engineer's desktop.

However, there are some promising trends. The decade of the 1980s also brought us new approaches to the layout problem, based on models other than the QAP. For example, other discrete models, including matching [23] and cut trees [22], were proposed, and the importance of combining optimization models with interactive graphics [22] was recognized. The research community recognized that people have some unique pattern recognition capabilities that should be exploited in layout planning [32,34]. Thus a strong trend has developed toward *interactive* approaches that provide a range of analytic models as decision aids.

Although these trends have yet to bring highly functional computerized layout programs to the desktop of the typical facilities engineer, there have been some promising developments. For example, the integration of analytical models with commercial CADD software has been demonstrated for the special case of laying out an automated guided vehicle system [11]. This type of integration is at least a step in the right direction.

The future promises to bring us greater computing power, better software tools, such as CADD and DBMS, and more useful and powerful analytic models. What is not yet clear is whether or not these developments will be integrated in a widely available system for facility layout planning.

3.6 SUMMARY

There are two categories of computerized layout programs: (1) those that simply automate some of the drafting or calculation tasks, leaving all the decisions to the layout planner, and (2) those that generate alternative layouts. This chapter has focused on the general principles underlying the latter.

Probably the single most important issue for the potential user of computerized layout programs is the layout scoring model. A variety of models are used, and all are based on very strong assumptions. The potential user must exercise great caution in using and interpreting the results of the scoring models.

We have not given a "how to" description for any computerized layout software. For an excellent discussion of "public domain" layout packages, see Tompkins and Moore [37]. The available CADD-based computerized layout packages still are largely proprietary, so you will need to consult the vendor's literature for descriptions.

Our emphasis on the shortcomings and weaknesses of computerized layout algorithms should not be interpreted as a negative recommendation. Rather, it is our way of trying to ensure that you, as a potential user of computerized layout, have enough information to make you successful.

In closing, we feel that layout planning offers great opportunities for computerization, especially in the use of computerized layout models. Computer methods can be powerful tools for significantly improving the search phase of the layout design process. As with any powerful tool, the user must read the instruction manual carefully, understand the principles of operation, and exercise caution in using the tool on specific problems.

3.7 FURTHER READING

The operation of traditional layout programs is covered very nicely in Francis and White [8] and in Tompkins and Moore [37]. Filley [5–7] surveys commercial CADD based layout software and provides a good discussion of capabilities. For an excellent general treatment of the computerization opportunities, see Hales [12,13] and McCormick and Wrennal [20].

Modeling to support layout planning remains an area rife with opportunity for the researcher and practitioner. A number of researchers have addressed this issue, including Kusiak and Heragu [16] and Montreuil [26]. The latter proposes a general framework of models to integrate the layout design decisions and the corresponding design of the material flow network. A number of issues related to integration of models within the layout process are discussed in McGinnis [21].

REFERENCES

1. Armour, F. C., and E. S. Buffa, "A Heuristic Algorithm and Simulation Approach to Relative Location of Facilities," *Management Science,* Vol. 10, 1963, pp. 294–309.
2. Buffa, E. S., Armour, G. C., and T. E. Vollmann, "Allocating Facilities with CRAFT," *Harvard Business Review,* Vol. 42, 1964, pp. 136–159.
3. Deisenroth, M. P., and J. M. Apple, "A Computerized Plant Layout Analysis and Evaluation Technique," *Technical Papers 1972,* AIIE Twenty-fifth Anniversary Conference and Convention, Norcross, GA, 1972, pp. 75–87.

4. Edwards, H. K., Gillet, W. E., and M. E. Hale, "Modular Allocation Technique (MAT)," *Management Science,* Vol. 17, 1970, pp. 161–169.
5. Filley, R. D., "CAD Systems and Software for Facilities Planning: A Product Survey," *Industrial Engineering,* Vol. 15, No. 3, 1983, pp. 66–80.
6. Filley, R. D., "Integrated Facilities Planning and Design: A Survey," *Industrial Engineering,* Vol. 16, No. 5, 1984, pp. 71–80.
7. Filley, R. D., "Emerging Computer Technologies Boost Value of and Respect for Facilities Function," *Industrial Engineering,* Vol. 17, No. 5, 1985, pp. 27–39.
8. Francis, R. L., and J. A. White, *Facility Layout and Location: An Analytical Approach,* Prentice Hall, Englewood Cliffs, NJ, 1974.
9. Graves, G. W., and A. B. Whinston, "Optimal and Suboptimal Algorithms for the Quadratic Assignment Problem," *Management Science,* Vol. 7, No. 7, 1970, pp. 453–471.
10. Goetschalckx, M., Montreuil, B., and H. D. Ratliff, "Matching Based Interactive Layout," *IIE Transactions,* Vol. 19, No. 3, 1987, pp. 271–297.
11. Goetschalckx, M., and L. F. McGinnis, "Engineering Work Station Is Design Tool for Computer-Aided Engineering of Material Flow Systems," *Industrial Engineering,* Vol. 21, No. 6, 1989, pp. 34–38.
12. Hales, H. L., *Computer-Aided Facilities Planning,* Marcel Dekker, Inc., New York, 1984.
13. Hales, H. L., "Using Computers in State-of-the-Art Facilities Planning," *Industrial Engineering,* Vol. 16, No. 5, 1984, pp. 60–70.
14. Hitchings, G. G., and M. Cottam, "An Efficient Heuristic Procedure for Solving the Layout Design Problem," *Omega,* Vol. 4, No. 2, 1976, pp. 205–214.
15. Konz, S., *Facility Design,* John Wiley & Sons, Inc., New York, 1985.
16. Kusiak, A., and S. S. Heragu, "The Facility Layout Problem," *European Journal of Operational Research,* Vol. 29, No. 3, 1987, pp. 229–251.
17. Lee, R. C., "Computerized Relationship Layout Planning," unpublished M.S. thesis, Northeastern University, Boston, 1966.
18. Lee, R. C., and J. M. Moore, CORELAP—COmputerized RElationship LAyout Planning," *Journal of Industrial Engineering,* Vol. 18, No. xx, 1967, pp. 195–200.
19. Lewis, W. P., and T. E. Block, "On the Application of Computer Aids to Plant Layout," *International Journal of Production Research,* Vol. 18, No. 1, 1980, pp. 11–20.
20. McCormick, M. J., and W. Wrennal, "A Step Beyond Computer-Aided Layout," *Industrial Engineering,* Vol. 17, No. 5, 1985, pp. 41–50.
21. McGinnis, L. F., "Computer-Aided Facility Design Revisited: A Prototype Design Workstation for AGV Systems," *Progress in Materials Handling and Logistics: Material Handling '90,* (J. A. White, I. W. Pence, Jr., R. J. Graves, L. F. McGinnis, M. R. Wilhelm, and R. A. Ward, editors), Springer-Verlag, New York, 1991, pp. 67–94.
22. Montreuil, B., "Interactive Optimization Based Facilities Layout," Ph.D. dissertation, Georgia Institute of Technology, Atlanta, GA, 1982.
23. Montreuil, B., Ratliff, H. D., and M. Goetschalckx, "Matching Based Interactive Facility Layout," *IIE Transactions,* Vol. 19, No. 3, 1987, pp. 271–279.
24. Montreuil, B., and U. Venkatadri, "From Gross to Net Layout: An Efficient Design Model," *IIE Transactions,* to appear.

25. Montreuil, B., Venkatadri, U., and H. D. Ratliff, "Generating a Layout from a Design Skeleton," *IIE Transactions,* to appear.
26. Montreuil, B., "A Modelling Framework for Integrating Layout Design and Flow Network Design," *Progress in Materials Handling and Logistics: Material Handling '90,* (J. A. White, I. W. Pence, Jr., R. J. Graves, L. F. McGinnis, M. R. Wilhelm, and R. A. Ward, editors), Springer-Verlag, New York, 1991, pp. 95–116.
27. Moore, J. M., *Plant Layout and Design,* The Macmillan Company, New York, 1962.
28. Moore, J. M., "Computer Aided Facilities Design: An International Survey," *International Journal of Production Research,* Vol. 12, No. 1, 1974, pp. 21–44.
29. Moore, J. M., "Computer Program Evaluates Plant Layout Alternatives," *Industrial Engineering,* Vol. 3, No. 8, 1971, pp. 19–25.
30. Muther, R., *Systematic Layout Planning,* Industrial Education Institute, Boston, 1961.
31. Nugent, C. E., Vollmann, T. E., and J. Ruml, "An Experimental Comparison of Techniques for the Assignment of Facilities to Locations," *Operations Research,* Vol. 16, No. 1, 1968, pp. 150–173.
32. O'Brien, C., and S. E. Z. Abdel Barr, "An Interactive Approach to Computer Aided Facility Layout," *International Journal of Production Research,* Vol. 18, No. 2, 1980, pp. 201–211.
33. Ritzman, L. P., "The Efficiency of Computer Algorithms for Plant Location," *Management Science,* Vol. 18, No. 5, 1972, pp. 240–248.
34. Scriabin, M., and R. C. Vergin, "Comparisons of Computer Algorithms and Visual Based Methods for Plant Layout," *Management Science,* Vol. 22, No. 2, 1975, pp. 172–181.
35. Seehof, J. M., and W. O. Evans, "Automated Layout Design Program," *Journal of Industrial Engineering,* Vol. 18, No. 12, 1967, pp. 690–695.
36. Sepponen, R., *CORELAP8 Users Manual,* Department of Industrial Engineering, Northeastern University, Boston, 1969.
37. Tompkins, J. A., and J. M. Moore, *Computer Aided Layout: A User's Guide,* FP&D Monograph Series No. 1, American Institute of Industrial Engineers, Norcross, GA, 1978.
38. Tompkins, J. A., and R. Reed, Jr., "COFAD—A New Approach to Computerized Layout," *Modern Materials Handling,* Vol. 30, No. 5, 1975, pp. 40–43.
39. Tompkins, J. A., and R. Reed, Jr., "An Applied Model for the Facilities Design Problem," *International Journal of Production Research,* Vol. 14, No. 5, 1976, pp. 583–595.
40. Warnecke, H. H., and W. Dangelmaier, *Progress in Computer Aided Plant Layout,* Technical Report, Institute of Manufacturing Engineering and Automation, Fraunhofer, Germany, 1982.

PROBLEMS

3.1. Given the initial layout, flow matrix, and cost matrix shown in Figure P3.1, use the CRAFT scoring method and procedure TWOWAYX to obtain a final layout.

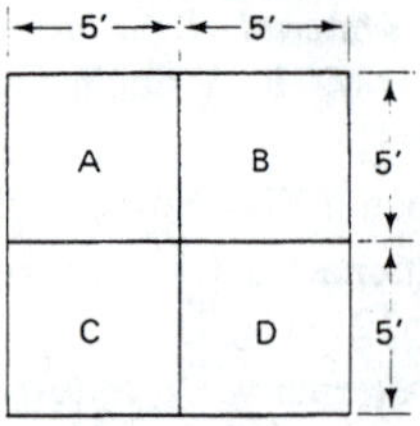

Initial Layout

From \ To	A	B	C	D
A		1	1	2
B	1		2	1
C	1	2		1
D	2	1	1	

Cost Matrix

From \ To	A	B	C	D
A		1	1	3
B	2		2	1
C	0	2		2
D	1	0	0	

Flow Matrix

Figure P3.1

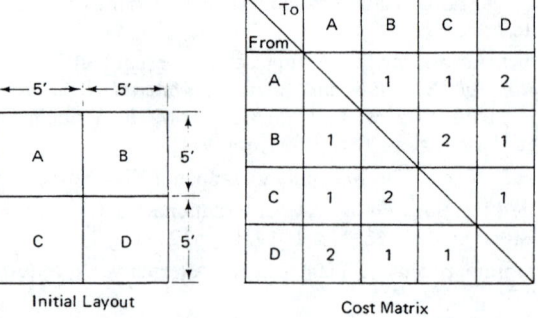

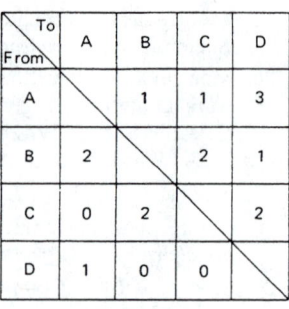

3.2. For the initial layout, flow matrix, and cost matrix shown in Figure P3.2, use the CRAFT scoring method and procedure TWOWAYX to obtain the optimum layout (i.e., consider all possible exchanges).

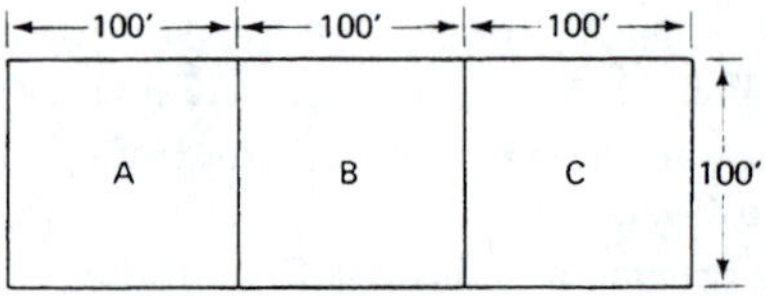

Initial Layout

From \ To	A	B	C
A		1	3
B	1		3
C	4	2	

Flow Matrix

From \ To	A	B	C
A		1	1
B	1		1
C	1	1	

Cost Matrix

Figure P3.2

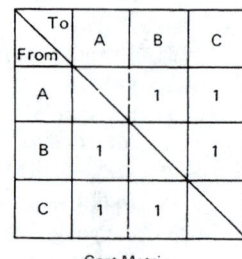

3.3. Given the initial layout, flow matrix, and cost matrix shown in Figure P3.3, use the CRAFT scoring method and procedure TWOWAYX to obtain a final layout. If there are several equally good solutions, list them all.

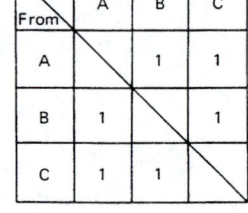

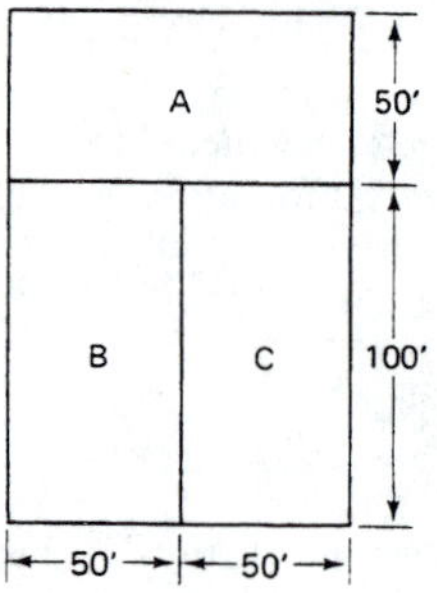

Initial Layout

From \ To	A	B	C
A		4	1
B	2		2
C	2	1	

Flow Matrix

From \ To	A	B	C
A		1	1
B	1		1
C	1	1	

Cost Matrix

Figure P3.3

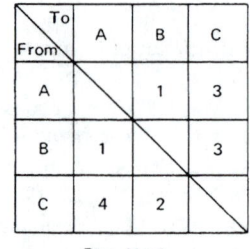

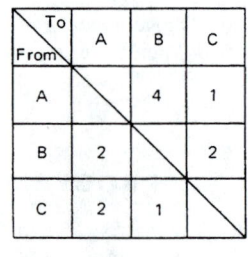

3.4. Consider the example problem associated with Figure 3.19. Design a layout that would have a zero score using the CRAFT scoring method. (*Hint*: The layout probably will violate one or more of the placement rules.)

3.5. Suppose that instead of using distances between centroids in the scoring model, you use the shortest distance between activities. Solve Problem 3.2 using this revised scoring model.

3.6. Suppose that instead of using total cost, you use the maximum cost of flow between any two activities as the score. Solve Problem 3.1 using this revised scoring model.

3.7. Manually adjust the layout given in Figure 3.21 so that departments F and L take on more realistic shapes (e.g., limit the number of corners, or specify a limit for the shape ratios). Use the resulting modified layout as the initial layout for procedure TWOWAYX. Compare the "new" final layout with the one given in Figure 3.21. What is the difference in cost?

3.8. A manufacturing concern has four departments located in two buildings, as shown in Figure P3.8. The firm is contemplating a rearrangement of departments to minimize the total flow cost between department centroids. Use procedure TWOWAYX to find the optimum layout for the data given.

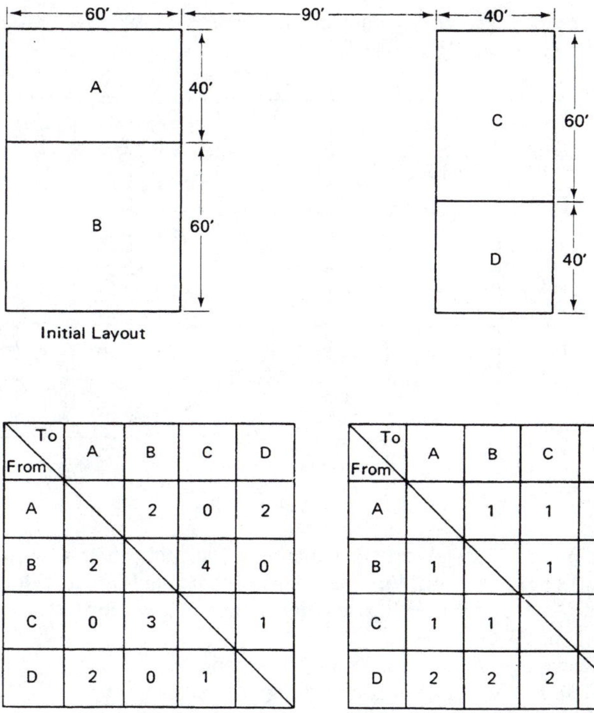

Initial Layout

To / From	A	B	C	D
A		2	0	2
B	2		4	0
C	0	3		1
D	2	0	1	

Flow Data

To / From	A	B	C	D
A		1	1	2
B	1		1	2
C	1	1		2
D	2	2	2	

Cost Data

Figure P3.8

3.9. For the data of Problem 3.8, determine the layout that minimizes the maximum cost of flow.

3.10. Use the CRAFT scoring method and procedure TWOWAYX to develop a layout for the data shown in Figure P3.10.

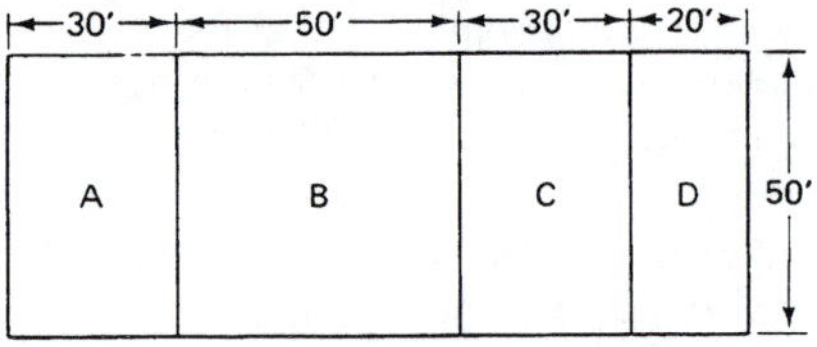

Initial Layout

From \ To	A	B	C	D
A		3	4	0
B	2		0	4
C	2	0		0
D	0	2	0	

Cost Data

From \ To	A	B	C	D
A		1	2	1
B	1		1	2
C	2	1		1
D	1	2	1	

Flow Data

Figure P3.10

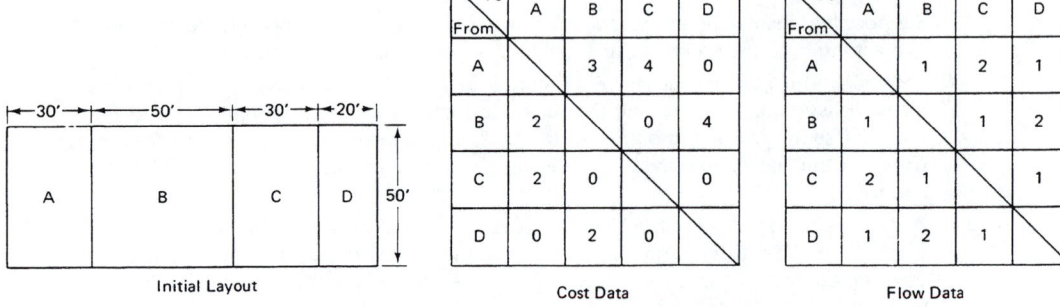

3.11. Use the CRAFT scoring method and procedure TWOWAYX to develop a layout for the data shown in Figure P3.11.

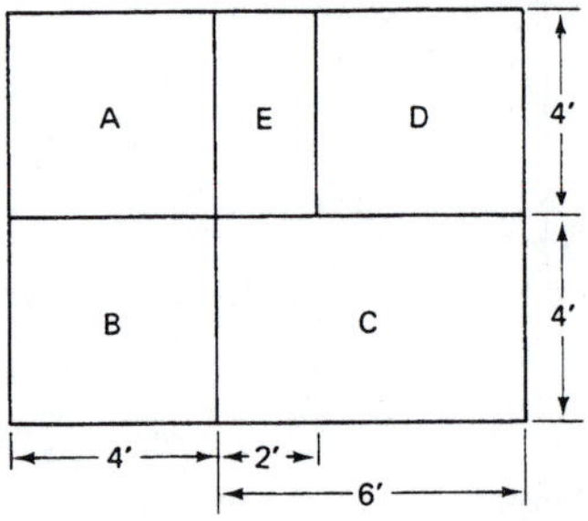

Initial Layout

From \ To	A	B	C	D	E
A		1	2	2	0
B	0		2	0	0
C	2	0		0	0
D	3	0	1		0
E	0	0	0	0	

Flow Data

From \ To	A	B	C	D	E
A		1	1	1	1
B	1		1	1	1
C	1	1		1	1
D	1	1	1		1
E	1	1	1	1	

Cost Data

Figure P3.11

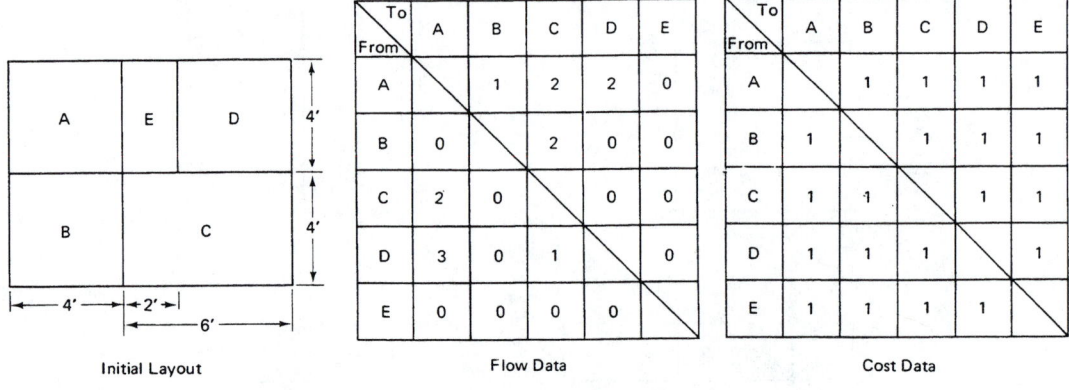

3.12. The Weird Widget Works is planning on locating their new production plant in the suburbs of Prices Fork, Virginia. Eleven departments are to be located in the production plant. Activity relationships and departmental areas for the new plant are summarized in Figure P3.12.

(a) Develop a layout using the SLP approach.

(b) Use the ALDEP heuristic to design the layout.

(c) Use the CORELAP heuristic to design the layout.

(d) Assign numerical values to the closeness ratings, and use the CRAFT heuristic to design the layout.

(e) What assumptions are inherent in the CRAFT solution?

Dept. No.	Department	Area (ft^2)										
11	Lobby	600										
12	General manager's office	400	A									
13	Bookkeeping	350	E	E								
14	Supervisor's office	150	U	E	U							
15	Snack bar	350	O	U	O	O						
16	Maintenance	150	U	U	U	E	O					
17	Receiving and shipping	200	X	U	O	O	U	U				
18	Storage	800	U	U	O	O	U	O	A			
19	First aid	200	U	U	O	O	U	I	O	O		
20	Rest rooms	350	U	U	O	O	I	O	O	O	O	
21	Production	5,000	U	O	U	A	I	E	A	A	E	I

Relationship Chart

Figure P3.12

3.13. Carry-On Baggage, Inc. manufactures carry-on luggage for airline travel. A new production plant is being planned and a layout is to be designed. Fourteen activities are to be provided for in the new plant. The relationships for the activities are given in Figure P3.13, along with the area requirements.*

(a) Design a layout using SLP.

(b) Design a layout using the ALDEP heuristic.

(c) Design a layout using the CORELAP heuristic.

(d) Assign appropriate numerical values to the closeness ratings, and design a layout using the CRAFT heuristic.

(e) Modify the layout designs obtained (based on practical limitations and modifying considerations) and recommend your preferred design.

Activity	Area (ft^2)													
Cutting	1,500													
Art area	500	U												
Dark room	500	U	E											
Silk screen	2,000	E	O	E										
Inspection	1,500	U	U	U	U									
Subassembly	1,000	I	U	U	U	U								
Final assembly	6,000	U	U	U	E	E	E							
Receiving and shipping	2,500	U	U	X	O	I	U	U						
Material storage	1,500	A	U	U	U	I	I	U	A					
Finished stores	2,500	U	U	U	U	E	U	U	A	I				
Maintenance	500	U	U	U	U	U	O	I	U	U	U			
Office	2,500	I	O	O	I	O	U	O	I	O	O	U		
Rest rooms	500	U	U	U	O	U	O	O	O	U	U	U	O	
Lunch room	1,500	U	U	U	O	U	O	O	O	U	U	U	O	O

Figure P3.13

* Based on a problem entitled "Novelty Luggage" developed by Richard Muther, Richard Muther and Associates, Kansas City, MO.

3.14. The town council of the Burg of Black has decided to construct a new municipal building to replace the old building, which was recently condemned. A study has been made and data were collected. The number of daily interdepartmental personnel contacts, departmental area requirements, closeness ratings between departments, and relative ratings of the importance of each department have been obtained. The importance rating is a composite evaluation based on the number of people in the department and the average hourly wage of the departmental personnel.

(a) Using the REL chart data, Figure P3.14(a), design a layout for the new municipal building using (1) the ALDEP heuristic and (2) the CORELAP heuristic.

Dept.	Area (ft^2)	Ratings	M	L	K	J	I	H	G	F	E	D	C	B	A
A	1,500	12	U	I	U	U	U	U	U	U	O	O	I	A	
B	650	5	U	U	U	U	U	U	U	U	U	U	I		
C	3,800	5	I	U	U	U	U	U	U	U	U	A			
D	300	11	U	U	U	U	U	U	U	U	U				
E	900	8	U	U	U	U	U	U	U	E					
F	900	8	U	U	U	U	I	U	E						
G	900	8	U	U	I	E	U	E							
H	900	8	U	U	I	U	E								
I	900	10	U	U	I	A									
J	2,300	13	U	U	I										
K	900	8	U	U											
L	300	9	A												
M	750	6													

Relationship Chart

Figure P3.14(a)

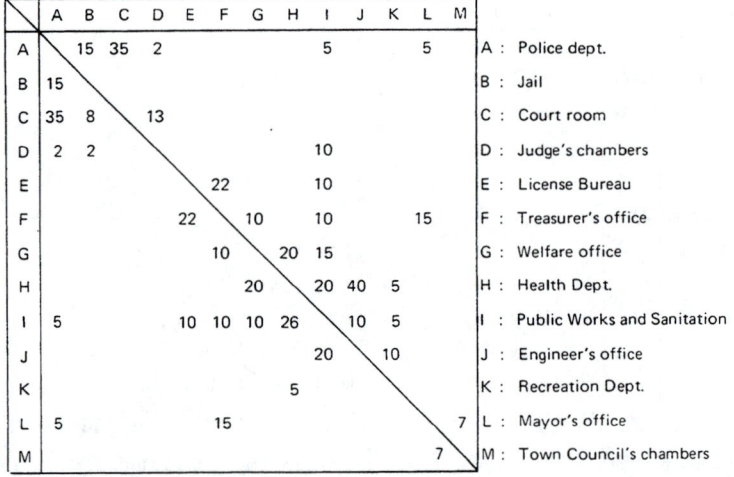

	A	B	C	D	E	F	G	H	I	J	K	L	M
A		15	35	2					5			5	
B	15												
C	35	8		13									
D	2	2							10				
E						22			10				
F					22		10		10			15	
G						10		20	15				
H							20		20	40	5		
I	5				10	10	10	26		10	5		
J									20		10		
K								5					
L	5					15							7
M												7	

A : Police dept.
B : Jail
C : Court room
D : Judge's chambers
E : License Bureau
F : Treasurer's office
G : Welfare office
H : Health Dept.
I : Public Works and Sanitation
J : Engineer's office
K : Recreation Dept.
L : Mayor's office
M : Town Council's chambers

Interdepartmental Personnel Flow Data

Figure P3.14(b)

(b) Using the interdepartmental personnel flow data, the departmental areas, and the departmental ratings, Figures P3.14(a) and (b), design a layout for the new municipal building using the CRAFT heuristic.

3.15. A layout is to be designed for the engineering department of the G. W. Quarles, Corp. A study of the face-to-face contacts per month between the members of the 10 groups within the department yielded the data shown below in Figure P3.15(a). Also shown are the area requirements and average wage rates for the groups. The current layout is also given, Figure 3.15(c). Obtain the final CRAFT layout using (a) two-way exchanges, (b) three-way exchanges, (c) best of two- and three-way exchanges.*

Area Requirements and Average Wage Rate Data

No.	Group	Area (ft^2)	Average Wage ($hr)
01	Filing	400	3.00
02	Supervision	600	10.50
03	Blueprint	750	3.50
04	Product support	750	4.00
05	Structural design	1,500	8.50
06	Electrical design	750	8.50
07	Hydraulic design	1,250	8.50
08	Mechanical design	1,000	8.50
09	Systems design	750	8.50
10	Safety design	750	8.50
11	Production liaison	1,250	4.00
12	Detailing and checking	1,250	6.50
13	Secretarial pool	1,000	3.50

Figure P3.15(a)

From \ To	01	02	03	04	05	06	07	08	09	10	11	12	13
01	–	15	0	0	0	5	5	10	15	10	10	0	15
02	40	–	25	40	100	90	80	70	40	50	160	85	60
03	0	20	–	0	0	0	0	0	0	0	0	0	0
04	10	60	0	–	0	0	20	30	210	100	280	0	10
05	50	120	600	0	–	40	0	60	0	50	0	340	50
06	5	90	350	0	40	–	0	20	60	40	160	270	60
07	10	80	400	40	20	0	–	30	0	10	140	320	45
08	10	70	350	30	10	10	60	–	10	20	150	310	40
09	15	110	215	60	10	15	10	20	–	60	210	180	110
10	30	85	75	85	120	115	105	135	60	–	180	150	105
11	30	160	0	450	160	350	380	300	400	100	–	0	680
12	0	300	700	40	550	510	400	410	250	370	0	–	20
13	20	90	0	10	50	60	50	30	410	320	680	30	–

Number of Face-to-Face Contacts per Month

Figure P3.15(b)

* Based on a problem given by E. S. Buffa, *Modern Production Management,* 3rd ed. (New York: John Wiley & Sons, Inc., 1969).

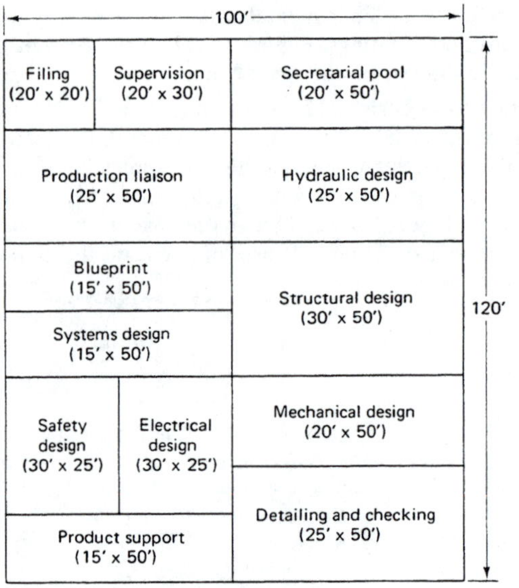

Initial Layout

Figure P3.15(c)

3.16. Manufactured Products, Inc., has requested a layout study of their White Pine, Tennessee, plant. The plant consists of 17 departments. Forklift trucks are used for all material handling. A summary of the number of truck loads delivered per day between combinations of departments is given in Figure P3.16. Obtain the final CRAFT layout using pairwise exchanges.

No.	Department	Area (ft^2)	01	02	03	04	05	06	07	08	09	10	11	12	13	14	15	16	17
01	Receiving	500		100															
02	Raw material storage	1,000			25	22	30	62		15							18		
03	Shearing	200											20	5					
04	Sawing	200						10	5				5						
05	Automatic screw machine	4,000						5	10		7	5					3		
06	Turret lathe	2,000							9		5	3	4	2			31		
07	Engine lathe	500						1			3		10	8			20		
08	Punch press	1,000											15	10	5	7	8		
09	Hobbing	400							2			2	5		8				

Figure P3.16

No.	Department	Area (ft^2)	01	02	03	04	05	06	07	08	09	10	11	12	13	14	15	16	17
10	Broaching	300						2	3					2	3				
11	Milling	3,000						2						30	25	7	1		
12	Drilling	1,200						24	15			2	5		5	2	5		
13	Heat treating	500						2								15	30		
14	Plating	300							1								32		
15	Assembly	700																175	
16	Finished goods storage	1,000																	175
17	Shipping	500																	

Figure P3.16 *(Continued)*

3.17. The Agee Mechanical Works includes 10 departments with areas shown on the REL chart given in Figure P3.17.*

(a) Develop a layout for the plant using SLP.

(b) Solve part (a) using the ALDEP heuristic.

(c) Solve part (a) using the CORELAP heuristic.

(d) Solve part (a) using the CRAFT heuristic by assigning numerical values to the closeness ratings and by treating the converted REL chart as flow data.

(e) Consider the designs obtained in parts (a) through (d) and make appropriate modifications to obtain a preferred layout.

Dept. No.	Department	Area (ft^2)									
301	Lobby	600									
302	Plant manager's office	400	A								
303	Accounting	330	E	E							
304	Supervisor's office	180	U	O	U						
305	Snack bar	340	O	U	O	O					
306	Maintenance	140	U	U	U	E	O				
307	Receiving and storage	690	X	U	O	O	U	U			
308	First aid	210	U	U	O	O	U	O	O		
309	Rest rooms	350	U	U	O	O	I	O	O	O	
310	Production	4,800	U	O	U	A	I	E	A	E	I

Relationship Chart

Figure P3.17

* Based on a problem entitled "ABC Mechanical Works," developed by Richard Muther, Richard Muther and Associates, Kansas City, MO.

3.18. The Flimsy Furniture Company consists of 10 production departments. The existing layout is shown in Figure P3.18(a), and a REL chart, flow data, and cost data are given in Figure P3.18(b).

(a) Use SLP to develop a layout.

(b) Apply the ALDEP placement method to develop a layout.

(c) Use the CORELAP placement method to develop a layout.

(d) Use the CRAFT scoring method and procedure TWOWAYX to develop a layout.

(e) For each of the four layouts, perform any adjustments that seem reasonable. Which of the resulting layouts would you recommend? Why?

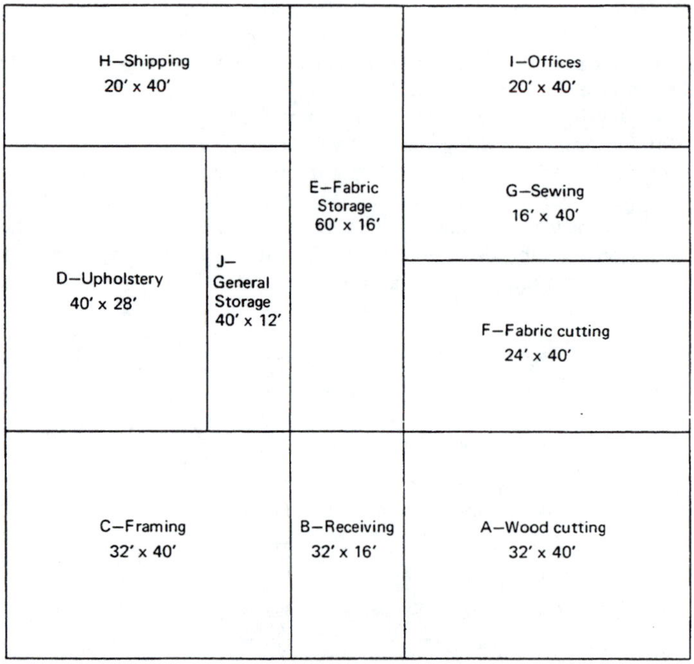

Current Layout

Figure P3.18(a)

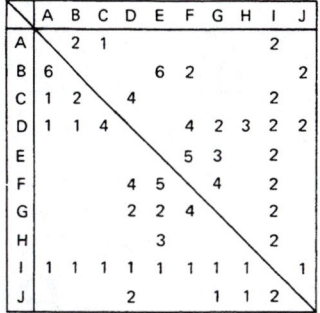

	A	B	C	D	E	F	G	H	I	J
A		2	1						2	
B	6				6	2				2
C	1	2		4					2	
D	1	1	4			4	2	3	2	2
E						5	3		2	
F				4	5		4		2	
G				2	2	4			2	
H					3				2	
I	1	1	1	1	1	1	1	1		1
J				2			1	1	2	

Flow Data

	A	B	C	D	E	F	G	H	I	J
A		2	2						1	
B	3				2	2				2
C	1	1		5					1	
D	1	1	2			2	2	3	1	1
E						4	3		1	
F				1	3		2		1	
G				4	2	2			1	
H					2				1	
I	1	1	1	1	1	1	1	1		1
J				1			1	2	1	

Cost Data

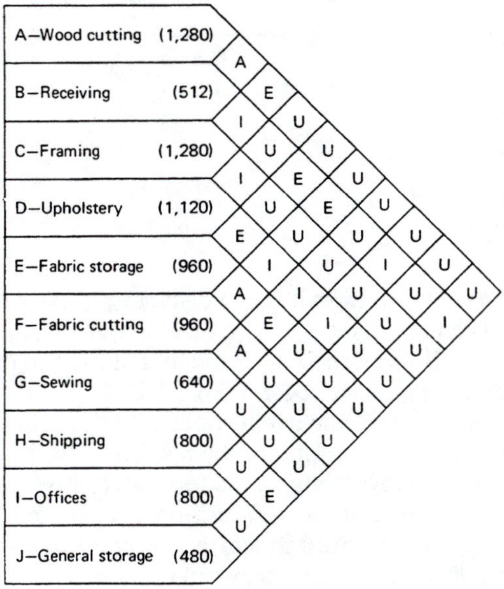

Relationship Chart

Figure P3.18(b)

Chapter 4

Planar Single-Facility Location Problems

4.1 INTRODUCTION

In earlier chapters we have seen that travel distance plays a very important role in evaluating facility layouts, since the cost of material handling is often directly proportional to travel distance; the longer a material handling transport, the more it can cost. Thus in this chapter we focus on travel distance. Travel distances depend on locations, of course: starting and ending points of trips. Thus it is natural that we focus on the simplest types of location problems, single-facility location problems. We shall develop some relatively simple models that can be useful in studying and solving such problems. These problems occur on a regular basis when working on layout problems (e.g., we may need to locate a machine in a shop, or items inside a warehouse). Also, on a larger scale, they can occur in, say, choosing the location of a warehouse to serve customers to whom goods must be delivered. We give an extensive list of problem contexts later.

We can view the models we shall study as being "quick and dirty." They are "quick" in the sense that they can be used quickly and easily, and "dirty" in the sense that they are approximate. The use of these models should be considered particularly when some location decision must be made quickly and with limited resources available for decision analysis. At the end of the chapter we discuss some of the limitations of these models. For the time being we simply claim that you will be in a better position to solve single-facility location problems if you have these models available than if you do not. By the time you reach the end of the chapter, you should be able to evaluate our claim for yourself.

When we wish to locate a single new facility in the plane, often we would like to minimize an objective function involving Euclidean or rectilinear distances between the new facility and a collection of existing facilities having known planar locations. The first objective function we consider is that of total travel distance, or total travel cost. We believe that this objective function is the one of most general interest, and hence Section 4.2, which treats the total travel cost model, is perhaps the most important section of the chapter.

In Section 4.3 we consider two multiobjective function location problems, one involving Euclidean distance and the other again involving rectilinear distance. Such problems are more specialized than those of Section 4.2 and tend to occur in public service contexts, where multiple objectives are of interest. However, Section 4.3.2, which deals with rectilinear distances, can be viewed as a sensitivity analysis of the rectilinear distance problem of Section 4.2.1 and hence provides useful supplemental information.

In Section 4.4 we consider minimax problems; Section 4.4.2 treats a minimax problem involving Euclidean distances called the *circle covering problem*, which can be interpreted as the problem of covering all the existing facility locations with a circle of minimum radius. Section 4.4.3 treats analogous problems with rectilinear distance. Minimax problems are more specialized than minisum problems and seem to be of interest principally in cases where a worst-case analysis is quite important.

4.1.1 Minisum Examples

To begin our study of the problems of this chapter, we consider a specific example for which we construct a model representing a total cost function to be minimized. Subsequently, we shall generalize the model we construct.

The model we consider first involves locating a machine in a shop. However, the model need not be restricted to this particular application. We can think of a number of other location problems for which such a model might be useful. Examples of such problems include the following:

- Locate a tool crib in a manufacturing facility.
- Locate a new warehouse relative to production facilities and customers.
- Locate a hospital, fire station, police station, or library in a metropolitan area.
- Locate a new classroom building on a college campus.
- Locate a new airfield to be used to provide supplies for a number of military bases or to provide service to a number of communities.
- Locate a new pump in a chemical operation to minimize the total cost of the pipe lines running to and from the pump.

- Locate a component in an electrical network to minimize the total cost of wire connected to the component.
- Locate a wrecker along a section of highway to minimize its average time to respond to an accident.
- Locate a new component on a control panel to minimize total eye movement between the new component and other components already located on the panel.
- Locate a dock in a warehouse for purposes of loading and unloading goods in the warehouse.
- Locate a new appliance in a kitchen.
- Locate a water fountain in an office building.
- Locate a new power generating plant.
- Locate a copying machine in a library.

Our development of the model now begins. Consider a new machine to be located in a manufacturing shop. The new machine receives parts from existing machines and supplies parts to existing machines. Thus there is a total cost of moving parts to and from the new machine. This total cost depends on the location of the new machine, and we wish to locate the new machine so as to *minimize* the total cost. Let m denote the number of existing machines, and for each existing machine i, denote its location by $\mathbf{P}_i$. We assume that we know the number of trips per month, say t_i, between $\mathbf{P}_i$ and $\mathbf{X}$, where $\mathbf{X}$ is the location of the new machine. Thus if $d(\mathbf{X}, \mathbf{P}_i)$ denotes the distance between $\mathbf{X}$ and $\mathbf{P}_i$, $t_i d(\mathbf{X}, \mathbf{P}_i)$ is the total distance items travel per month to and from the new machine and the existing machine. Suppose that we also know the average velocity, v_i, of items traveling to and from the two machines. Then $(t_i/v_i)d(\mathbf{X}, \mathbf{P}_i)$ is the total travel time per month between the two machines. Hence if we know c_i, the cost per hour of travel between the two machines, then $(c_i t_i/v_i)d(\mathbf{X}, \mathbf{P}_i)$ is the cost per month involving the two machines. To simplify the notation, let

$$w_i = c_i \frac{t_i}{v_i}$$

We conclude now that $w_i d(\mathbf{X}, \mathbf{P}_i)$ is the cost per month of item movement between the two machines. Because there are m existing machines, to obtain the total cost of movement between the new machine and all the existing machines, which we denote by $f(\mathbf{X})$, we simply add the costs involving the new machine and each existing machine. We obtain

$$f(\mathbf{X}) = w_1 d(\mathbf{X}, \mathbf{P}_1) + \cdots + w_m d(\mathbf{X}, \mathbf{P}_m) = \sum_{i=1}^{m} w_i d(\mathbf{X}, \mathbf{P}_i)$$

A choice of $\mathbf{X}$ that minimizes $f(\mathbf{X})$ will thus be a location of the new machine which minimizes the total cost of movement between it and the m existing machines. We remark that it is convenient, and customary, to call the w_i terms *weights*, and we shall do so from now on.

Now consider again the list of other applications of the model and ask yourself for each of the examples what is the "cost" to be minimized and what the weights are. You will discover that the units of the weights may change from one example to the next, but in every case the weights will be positive numbers. Thus we shall assume from this point on that *the weights are some given positive constants*. It will be up to you to determine what the weights actually are for any specific problem. Also, in the interest of having a more general model to consider, we call the facility to be located a *new facility* and call the facilities with locations $\mathbf{P}_1, \ldots, \mathbf{P}_m$ *existing facilities*. We call the problem of finding a new facility location that minimizes the function $f(\mathbf{X})$ a *minisum* location problem, because a sum of weighted distances is being minimized. We call a location that solves the minisum location problem an *optimal, or best, solution or location*. Usually, we shall think of $f(\mathbf{X})$ as having units of cost, so that an optimal location is one that minimizes a total cost.

4.1.2 Insights for the Minisum Problem

With reference to Figure 4.1, a useful construct for gaining insight into our location problem is to imagine a horizontal pegboard with holes chosen to represent the locations of the existing facilities. Strings, one per existing facility, are all tied together at one end into a single knot, which is above the board. The knot represents the location of the new facility, and the strings radiate out from the knot to the chosen holes, with the lengths of the strings above the board representing the travel distances between the new and existing facilities. The string for each existing facility passes through the corresponding chosen hole and has attached to its end below the board a weight, which is directly proportional to the weight that multiplies the distance. Imagine that there is no fraction and that the knot is big enough not to pass through any of the holes. If we lift up the knot and then release it, letting the string and weight system go into equilibrium, the resulting location of the knot represents an optimal location of the new facility. For the case where each weight is unity, note that what the model does is to minimize the total length of all the strings lying above the pegboard.

A principal insight we can gain from this string and weight model occurs when one weight is at least as large as the sum of the other weights; that is, it constitutes a *majority* of the total weight. In this case the optimal location of the knot coincides with the hole from which this "majority weight" is suspended. In other words, *when one weight constitutes a majority of the total weight, an optimal new facility location coincides with the existing facility which has the majority weight*. (This "majority theorem," whose analytical proof is due to Witzgall, turns out to be true not only when distances are Euclidean, but for a more general class of distances, referred to as l_p distances, including rectilinear distances. We discuss distances other than Euclidean later in the chapter.)

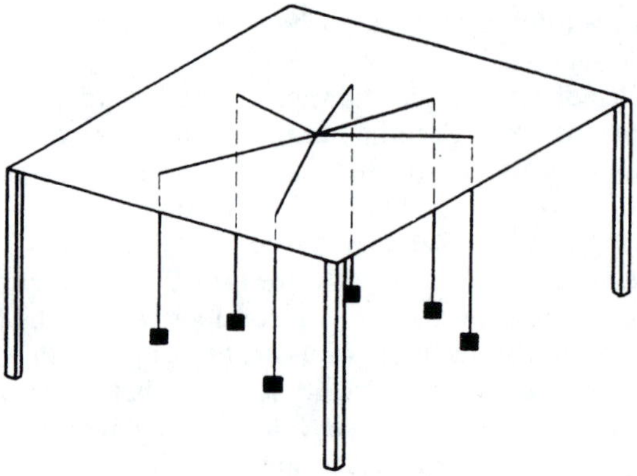

Figure 4.1 Mechanical analog model of the Euclidean problem.

For the case where distances are Euclidean, the string and weight model can also be used to gain a "convex hull" insight. Imagine placing a peg in each hole representing an existing facility location, stretching a giant rubber band around all the pegs, and releasing the rubber band. The region that the rubber band encloses is known as the *convex hull* of the existing facility locations. Any optimal solution to the location problem will be inside the convex hull, as the knot representing the optimal new facility location would never go into equilibrium outside the convex hull.

The usefulness of this convex hull insight is that *to find a best location, we need only consider points in the convex hull*. For example, if the existing facility locations should all be collinear, that is, all lie on a single line, the convex hull is just the shortest line segment containing all the existing facility locations, in which case the location problem reduces to a one-dimensional problem. If you have had a course in linear programming, you may know that each extreme point of the convex hull is the location of an existing facility. Thus, another way, in principle, of obtaining the convex hull is to compute all possible weighted averages of the existing facility locations; the set of all such weighted averages constitutes the convex hull. We shall use this alternative characterization of the convex hull later in the chapter.

Unless you were very alert while reading the last three paragraphs, you may believe that the only kind of distance that exists is the Euclidean distance, the distance we have been considering to this point in the chapter (excluding our parenthetical remarks above). In fact, not only are there many other kinds of distances than Euclidean, but *in many cases it is not appropriate to assume that distances are Euclidean*. If you doubt our assertion, pause and ask yourself the following question. Is it reasonable to assume that travel between any two points in the plane can be in a straight line? In fact, particularly in industrial contexts where there is often a rectilinear aisle network, the only time that movement between any two points can be in a straight line may be when the points lie on the same aisle. Alternatively, imagine yourself in the corner of a classroom filled with desks, and consider walking to the opposite corner. Would you be able to walk from one corner to the other in a straight line?

If, then, Euclidean distances can seldom be realized, it is reasonable to ask why we should consider Euclidean distances. One reason to consider Euclidean distances is that they may be the distances people are most familiar with, which should result in your having some helpful geometrical insights for such problems. A second and better reason to consider Euclidean distances is that they may be adequate approximations to actual distances, such as when a street network is not just rectilinear but has many "diagonal" streets as well. Yet a third reason to consider Euclidean distances is the well-known result that "the shortest distance between two points in the plane is a straight line." Hence *Euclidean distances give lower bounds on actual distances*.

To amplify our remark that Euclidean distances give lower bounds, let us call our location problem of actual interest our *original problem*. We construct a *replacement problem* by replacing the distances in our original problem by Euclidean distances. If we solve our replacement problem with Euclidean distances and compute the minimum total cost, this cost will be a lower bound on the cost we would obtain by solving our original problem. Suppose that we can find a new facility location for our original problem whose cost is only a little bit larger than the minimum cost for our replacement problem. We may well be satisfied with the location we have found for the original problem, since its cost cannot be less than the minimum cost for our replacement problem.

Later in the chapter we shall return to our location problem for the case where distances are Euclidean. Before doing so, however, we consider some simpler kinds of distances, such as rectilinear distances. We shall find that the resulting problems are easy to solve and that by solving such problems we will develop insights that will help us when we again consider the Euclidean distance problem.

4.2 MINISUM LOCATION PROBLEM

4.2.1 Minisum Location Problem with Rectilinear Distances

As we have seen, we may want to consider using rectilinear distances when we must work with a rectilinear street network or aisle network. Indeed, another name for *rectilinear distance* is *Manhattan distance*, due to the recognition that the street network on the island of Manhattan in New York City is a rectilinear network. Other names you may encounter for rectilinear distance include *right-angle distance* and *rectangular distance*. There is little agreement as to which name is best. Figure 4.2 illustrates several different paths between a new facility location **X** and an existing facility location $\mathbf{P}_i$ for which the rectilinear distances are the same. You should be able to construct many other paths which have the same total length as the two

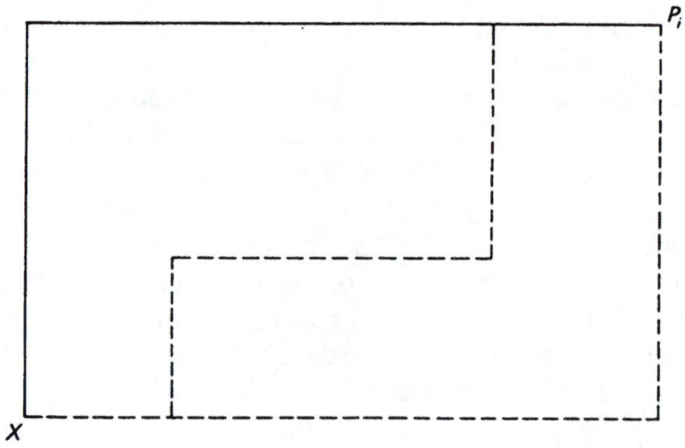

Figure 4.2 Different rectilinear paths between X and P_i having the same rectilinear distances.

illustrated in Figure 4.2. Figure 4.7(a) illustrates rectilinear "circles," sets of constant rectilinear distance from the origin which we shall call *diamonds*. You should take the time to construct a few diamonds in order to improve your insight about rectilinear distances.

If we replace the Euclidean distances of our location model in the preceding section by rectilinear distances, with the understanding that $\mathbf{X} = (x, y)$ and $\mathbf{P}_i = (a_i, b_i)$, we obtain the following equation:

$$f(x,y) = \sum_{i=1}^{m} w_i(|x - a_i| + |y - b_i|)$$

Let us define $f_1(x)$ and $f_2(y)$ as follows:

$$f_1(x) = \sum_{i=1}^{m} w_i|x - a_i|, \qquad f_2(y) = \sum_{i=1}^{m} w_i|y - b_i|$$

We now draw the very useful conclusion that

$$f(x,y) = f_1(x) + f_2(y)$$

That is, the total cost of movement is the sum of the cost of movement in the x-direction and the cost of movement in the y-direction. Thus an implication of the equation is that *we can minimize the total cost of movement by solving the two smaller—and independent—problems of minimizing the cost of movement in the x direction and minimizing the cost of movement in the y-direction*. The best choice of x has absolutely no effect on the best choice of y, and vice versa! Also, we observe that *the two cost-of-movement functions have exactly the same form, so that every conclusion we draw about one cost function applies to the other*. In particular, if we have a way to minimize one cost function, we can also apply it to minimize the other cost function. Note that *we can interpret the problem of minimizing either cost function as a location problem on the line*. The point x and the points $a_1, \ldots, a_m$ are all points on the line, and $|x - a_i|$ is the distance on the line between x and a_i. You should draw some points on the line and compute distances between them to verify that we have a location problem on the line.

Let us now develop a procedure for solving a minisum location problem on the line. Once we have such a procedure, we can apply it twice to minimize the cost functions $f_1(x)$ and $f_2(y)$ independently. To develop insight into the location problem on the line, we shall consider a specific location problem and later use the insight we have gained to draw some general conclusions.

Example 4.1

Consider Figure 4.3. Suppose that a conveyor line is being planned to run into a warehouse. The line will begin at the point $(0, 5)$ (all coordinates are given in units of tens of feet) and run parallel to the x-axis into the warehouse. Items entering the warehouse on the line are picked up at the end of the line and transported directly to one of the truck docks at the points $\mathbf{P}_1 = (7, 10)$, $\mathbf{P}_2 = (15, 7)$, $\mathbf{P}_3 = (15, 3)$ and $\mathbf{P}_4 = (12, 0)$. There is a trade-off between the cost of the conveyor and the transport cost to the truck docks. We can have a short conveyor and high transport cost to the docks, or a long, expensive conveyor and low transport cost to the docks. Thus what we seek is a best trade-off.

The distance items travel between the conveyor line end and any one of the docks may be accurately approximated by the rectilinear distance between the line end and the dock. From data on expected usage rates of docks, labor costs, and material transport times, the total annual cost per 10 ft for transporting items between the end

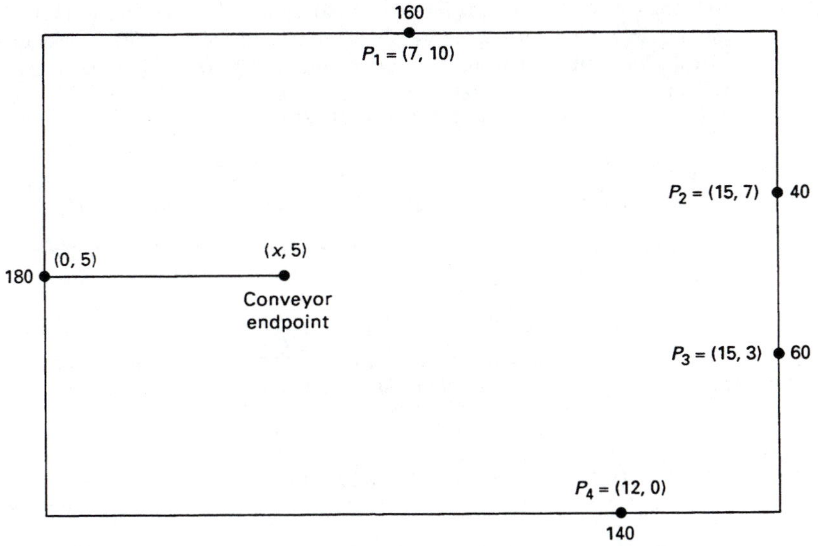

Figure 4.3 Warehouse with docks at points P_1, P_2, P_3, P_4 and conveyor joining (0,5) and $(x,5)$.

of the line and points $\mathbf{P}_1$ through $\mathbf{P}_4$ will be \$160, \$40, \$60, and \$140, respectively. The equivalent annual cost per 10 ft of conveyor is \$180.

We wish to find the length of the conveyor so as to minimize the total cost of the conveyor plus the costs of moving items between the end of the conveyor and the docks. Let us take the weights for points 1 through 4 to be given by 160, 40, 60, and 140, respectively. Further, suppose that the entering point of the conveyor into the warehouse, (0, 5), is a fifth existing facility, with a weight of 180.

If we take $m = 5$, our total cost model $f(x, y)$ represents the sum of the costs of the conveyor and of movement from the conveyor to the docks. Further,

$$f_1(x) = 180|x - 0| + 160|x - 7| + 140|x - 12| + 100|x - 15|$$

Because $y = 5$, the cost of movement in the y-direction is given by a constant, namely

$$f_2(5) = 140(5) + 60(2) + 180(0) + 40(2) + 160(5) = 1700$$

Note that we have sorted the coordinates of the "existing facilities" so that they appear in increasing order in the two equations above. Also, we have combined the two terms involving $|x - 15|$ into a single term. Hence to find the best location for the conveyor end we must minimize the cost function $f_1(x)$. Table 4.1 summarizes the data for this example.

Figure 4.4 gives a graph of the function $f_1(x)$. Note that the conveyor length that minimizes the total cost is 70 ft. To simplify the discussion of the graph, we shall simply call 0, 7, 12, and 15 "points" rather than "x-coordinates of docks and the conveyor input point." We write below each point its associated weight (e.g., 160 is the weight associated with the point 7). The function $f_1(x)$ is linear between adjacent points, and its slope increases from one interval of adjacent points to the next interval of adjacent points. To see how we compute the slope in any such interval, consider the interval [7,12]. Recall that the absolute value of a number is the number itself when the number is nonnegative, but the negative of the number when the number is negative. For example, $|3| = 3$, $|-3| = 3$, $|0| = 0$. Therefore, for $7 \le x \le 12$, we have

$$|x - 0| = x - 0, \quad |x - 7| = x - 7, \quad |x - 12| = 12 - x, \quad |x - 15| = 15 - x$$

for all values of x in the interval [7,12]. Thus it follows that

$$\begin{aligned} f_1(x) &= (180 + 160 - 140 - 100)x - 180(0) - 160(7) + 140(12) + 100(15) \\ &= (100)x + 2060 \quad \text{for } x \text{ in } [7,12] \end{aligned}$$

In general, we can conclude that *the coefficient of x in any interval defined by adjacent points is the sum of the weights of the points to the left of x minus the sum of the weights of the points to the right of x*. The slope of the function for each interval

TABLE 4.1 DATA FOR EXAMPLE 4.1: RECTILINEAR DISTANCES, $y = 5$

i	1	2	3	4	5
w_i	160	40	60	140	180
$\mathbf{P}_i$	(7,10)	(15,7)	(15,3)	(12,0)	(0,5)

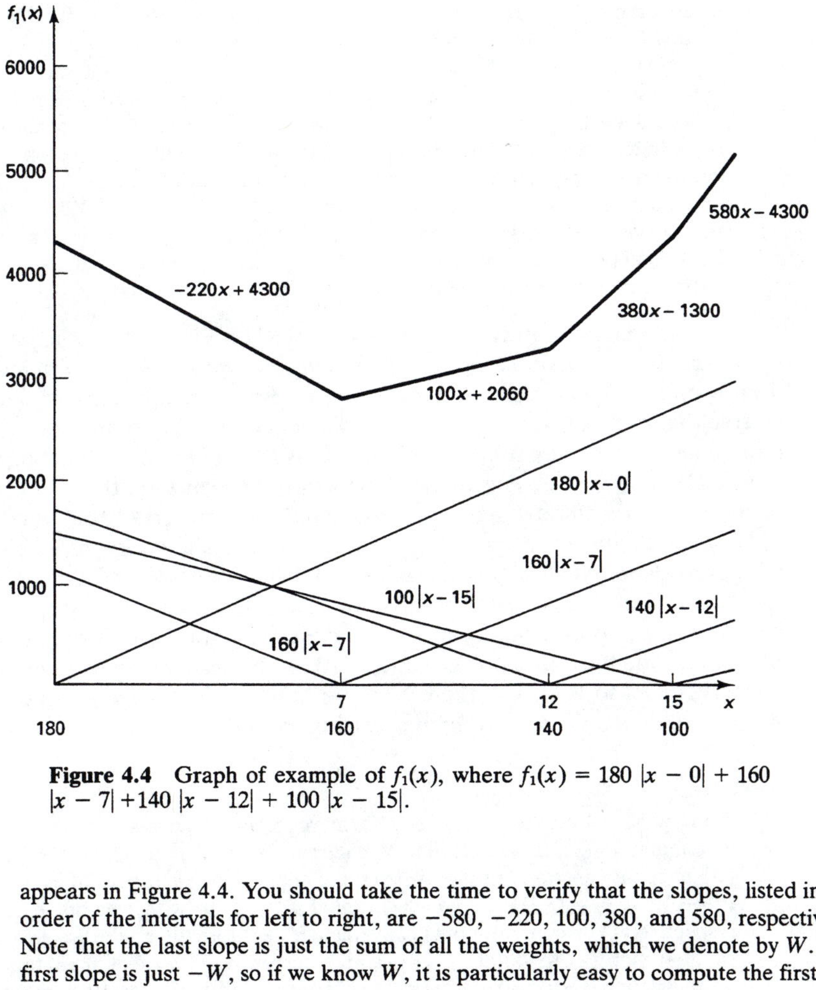

Figure 4.4 Graph of example of $f_1(x)$, where $f_1(x) = 180\,|x - 0| + 160\,|x - 7| + 140\,|x - 12| + 100\,|x - 15|$.

appears in Figure 4.4. You should take the time to verify that the slopes, listed in the order of the intervals for left to right, are −580, −220, 100, 380, and 580, respectively. Note that the last slope is just the sum of all the weights, which we denote by W. The first slope is just $-W$, so if we know W, it is particularly easy to compute the first and last slopes.

Another observation which is generally true is that *the slope for any interval but the first is the sum of the slope of the previous interval* (proceeding from left to right) *and twice the weight of the point common to both intervals*: for example, $-220 = -580 + 2(180)$, $100 = -220 + 2(160)$, and so on. The latter observation provides a simple, systematic way to compute the slopes. Start with a slope of $-W$ for the leftmost interval, then use the observation to compute the slope in the next interval to the right, use the slope for the second interval to compute the slope for the third, and so on.

Note that Figure 4.4 shows *that the function* we graphed *is minimized at the point where the slope changes from nonpositive to nonnegative*. (For Figure 4.4, the minimizing point is 7, giving $f_1(7) = 2{,}760$ as the minimum value of the function.) From this observation we can draw the general conclusion that *at least one value of x minimizing the cost function $f_1(x)$ will be an x-coordinate of an existing facility*.

Still considering Figure 4.4, an alternative and equivalent way of finding the minimizing point is to compute partial sums of the weights, starting from the left. Thus we would use the weight (180) for the leftmost point as the first partial sum, the sum of the weights for the two leftmost points ($180 + 160 = 340$) as the second partial sum, and so on. The point at which a partial sum first becomes at least $W/2$ (290 in this case) is a minimizing point. Since the first partial sum, 180, is less than $W/2 = 290$, while the second partial sum, 340, exceeds $W/2 = 290$, we conclude that point 7 associated with the second partial sum is a best location. Note in case $W = 1$ that we are comparing partial sums with $\frac{1}{2}$, which is the reason for the use of the term *median conditions*.

We can make two final observations based on Figure 4.4. *The function $f_1(x)$ is convex, since the line joining any two points on the graph of the function lies on or above the function*. An alternative way to draw the convexity conclusion is to observe that each function $w_i|x - a_i|$ is a V-shaped graph with its vertex at the point a_i and hence is a convex function. Thus $f_1(x)$ is a sum of convex functions. Because a sum of convex functions is a convex function, we conclude that $f_1(x)$ is a convex function. Why is it important to know that a function is convex? *For convex functions a "local minimum" is a "global minimum," so because the point where the slope changes from nonpositive to nonnegative is a local minimum, it follows that it is also a global minimum*. The latter observation also justifies the fact that the use of the median conditions determines a point minimizing $f_1(x)$. To end this paragraph we suggest that you think about what conclusion can be drawn if a partial sum of weights is exactly equal to $W/2$; will there be more than one minimizing value of x in this case? If so, which values will be minimizing values?

Example 4.2

We now have a way to minimize $f(\mathbf{X})$, by repeating the procedure twice above, once to minimize $f_1(x)$, a second time to minimize $f_2(y)$. Let us apply the procedure to minimize the function $f(\mathbf{X})$ for which there are $m = 4$ existing facilities, with locations $\mathbf{P}_1 = (4,2)$, $\mathbf{P}_2 = (8,5)$, $\mathbf{P}_3 = (11,8)$, and $\mathbf{P}_4 = (13,2)$, with weights 1, 2, 2, and 1, respectively. For computational convenience we "normalize" the weights by dividing by their sum (6) to obtain $\frac{1}{6}$, $\frac{1}{3}$, $\frac{1}{3}$, and $\frac{1}{6}$, respectively, so that their total is 1. The normalization is equivalent to a change of scale and will have no effect on the discussion to follow other than to change the units in which we measure $f(\mathbf{X})$. (You can always return to the original scale by multiplying by 6.) Table 4.2 summarizes the data for this example. You should verify the fact that a best location is given by $\mathbf{X} = (8,5)$. Note that $(11,5)$ is also an optimal location. Indeed, every point on the line segment joining $(8,5)$ and $(11,5)$ is an optimal location.

In Example 4.2, because one least cost location coincides with the existing facility point $\mathbf{P}_2$, in fact the location would probably not be usable. Instead, *what we would do is to try to find a nearby location that is usable*. Thus it would be useful to have some general procedure to allow us to evaluate the costs of locations other than those which have the least cost. To develop this procedure we introduce the

TABLE 4.2 DATA FOR EXAMPLE 4.2: RECTILINEAR DISTANCES

i	1	2	3	4
w_i	$\frac{1}{6}$	$\frac{1}{3}$	$\frac{1}{3}$	$\frac{1}{6}$
$\mathbf{P}_i$	(4,2)	(8,5)	(11,8)	(13,2)

concept of *contour sets*, also called *level sets*, and the associated notion of *contour lines* or *level lines*. Refer to Figure 4.5, which shows contour sets graphed for Example 4.2. The characterizing property of a contour line is that *every point on the contour line has the same value* of the function f. If you are familiar with topographical maps, you should observe that a contour line is exactly analogous to a line of constant elevation in a topographical map.

We use the contour lines to help us obtain insight into the shape of the graph of the function $f(\mathbf{X})$—which is a three-dimensional object—by using a two-dimensional analysis. Figure 4.5 shows three such contour lines, with the innermost line having the smallest value, the middle line having an intermediate value, and the outer line having the largest value. *Each contour set, whose boundary is a contour line, is the set of all points having values of $f(\mathbf{X})$ no larger than those of the points on the contour line*. Hence if we want to evaluate other possible locations for the new facility, then for the example we would first consider locations in the innermost contour set. If none of these locations is usable, we would consider locations inside the middle and outside the innermost contour set, and so on. *Contour lines and sets*

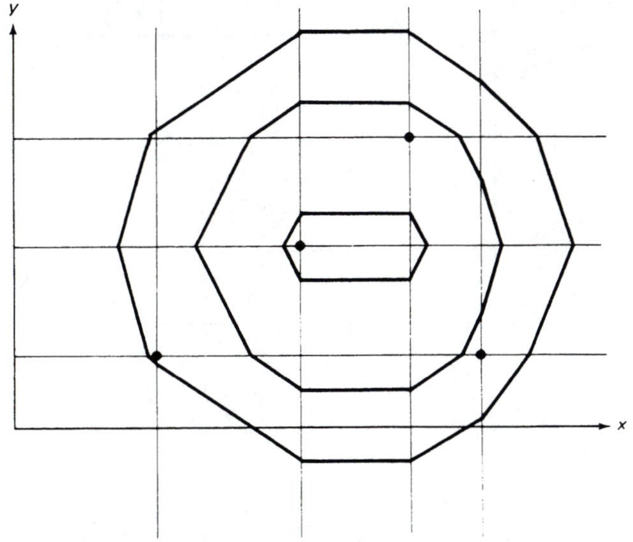

Figure 4.5 Sample contour line for Example 4.2.

can be very useful in giving us geometrical insight into the problem, and allow us to get an idea of the values of $f(\mathbf{X})$ for many different values of $\mathbf{X}$ by using a simple graphical procedure.

Because of the usefulness of contour sets, we shall take the time to develop a procedure for constructing such sets. We shall construct the contour sets for Example 4.2 and then generalize from the example. For the example of interest, we observe that

$$f_1(x) = \tfrac{1}{6}|x - 4| + \tfrac{1}{3}|x - 8| + \tfrac{1}{3}|x - 11| + \tfrac{1}{6}|x - 13|$$

$$f_2(y) = \tfrac{1}{3}|y - 2| + \tfrac{1}{3}|y - 5| + \tfrac{1}{3}|y - 8|$$

Refer now to Figure 4.6. We have plotted the existing facility locations and for each location written the associated weight adjacent to the location. Also we have drawn both a vertical and a horizontal line through each location. In addition, we have totaled the weights of the points lying on each line and written these totals at the bottom ends and left ends of the lines. Also, we have written the slopes of $f_1(x)$ and $f_2(y)$ in the bottom and left margins, respectively. We computed these slopes using the procedure we developed above in conjunction with the conveyor example. Note that the lines we have constructed partition the plane into "boxes" (some such boxes have only three sides, but it is convenient to call them boxes anyway). Let us

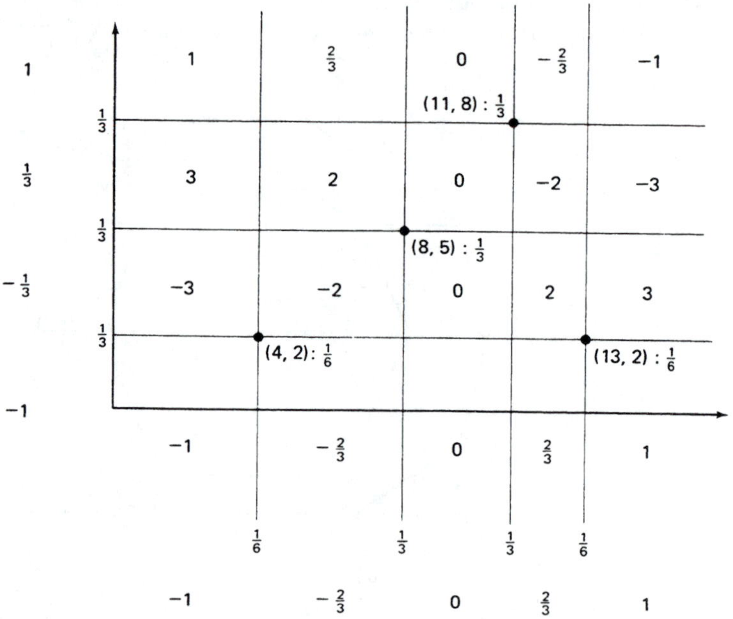

Figure 4.6 Data required for the construction of contour lines for Example 4.2.

consider the box B for which x lies in the interval [4,8] and y lies in the interval [2,5]. For any point X in the box of interest, you should be able to use the analysis developed for the conveyor example to verify that

$$\begin{aligned} f(x,y) &= f_1(x) + f_2(y) \\ &= (-\tfrac{2}{3}x + 10) + (-\tfrac{1}{3}y + \tfrac{11}{3}) \\ &= -\tfrac{2}{3}x + -\tfrac{1}{3}y + 13\tfrac{2}{3} \end{aligned}$$

Thus we can conclude that *within any given box the function f is a linear function in x and y*. Because of this conclusion, the part of any contour lines inside any box of interest will be a line segment, as you can see by examining Figure 4.5. It is easy to compute the slope of any line segment within a box. For example, for any point (x,y) in the box B we are considering, if we let k denote the value of the function at the point, then from the equation $f(x,y) = k$, which defines a contour line, we conclude that

$$-\tfrac{2}{3}x + -\tfrac{1}{3}y + 13\tfrac{2}{3} = k$$

so that solving for y gives

$$\begin{aligned} y &= \frac{2/3}{-1/3}x + \frac{k - 13\tfrac{2}{3}}{-1/3} \\ &= (-2)x + k' \end{aligned}$$

where $k' = (k - 13\tfrac{2}{3})/-\tfrac{1}{3}$ is a new constant. Hence the slope of any contour line passing through the box B is -2, which we note is the negative ratio of the bottom margin x coefficient to the left margin y coefficient.

The observation we have made about the slope of any contour line passing through the box B is in fact generally true. *The slope of every contour line passing through a box B is the negative ratio of the bottom margin x coefficient to the left margin y coefficient*. Since the slope of every contour line in a box is the same, we can associate a unique number, the slope, with each box. Once we know the slope associated with each box we can easily construct a single contour line.

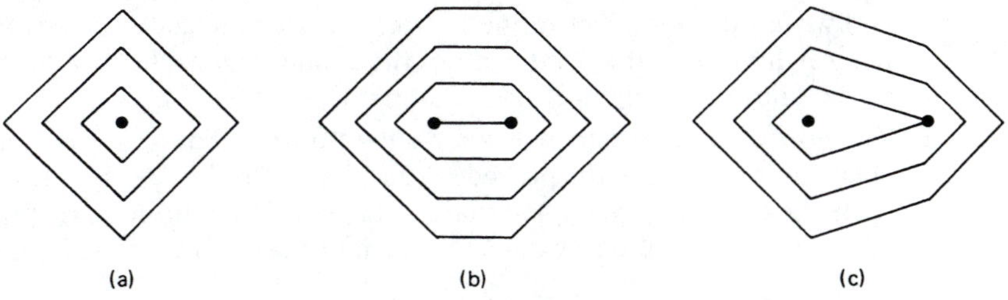

Figure 4.7 Contour lines for three simple rectilinear location problems.

Single-contour-line construction method. To construct a single contour line, start with any point in the interior of any box other than a point that minimizes $f(\mathbf{X})$. Pass a line through the point that has the slope computed for the box, and extend the line until it intersects the boundary of the box. Choose either of the points intersecting the boundary. Such a point will be in another box, so you can use the same procedure as for the first box to construct another line segment through the second box. Continue in this fashion until you have constructed a complete contour line. As a computational check, you should end at the same point you start from. We emphasize the fact that this contour line construction procedure requires that the starting point of the procedure is not a point minimizing $f(\mathbf{X})$, since no contour line can pass through such a minimizing point (except a trivial one).

Two special cases can occur in computing the slope for a box if the y coefficient in the left margin is zero, for then you may attempt to divide by 0. Provided that the x coefficient is not also zero, you should conclude that each contour line passing through the box is a vertical line and, in effect, has a slope of infinity. *In case both the x and y coefficient for the box are zero, then in fact every point in the box is a minimizing point and no contour line should pass through the box*. To test your understanding, you should explain why the assertion of the last sentence is correct.

We now summarize the procedure we have developed for constructing contour lines. The procedure we state can be implemented using computer graphics, by the way; computer graphics makes the procedure particularly easy to use.

Contour Line Construction Procedure

1. Pass a horizontal line and a vertical line through each existing facility location. Each horizontal line should extend beyond the leftmost and beyond the rightmost existing facility location. Similarly, each vertical line should extend beyond the bottommost and beyond the topmost existing facility location.
2. For each vertical line, total the weights of the existing facilities lying on the line and write the total at the bottom of the line.
3. For each horizontal line, total the weights of the existing facilities lying on the line and write the total at the left of the line.
4. The vertical lines partition the plane into columns. For each column compute the coefficient of x (using the procedure developed for the conveyor example) and write the coefficient in the bottom margin (the coefficient is the sum of the weights of the lines to the left of the column minus the sum of the weights of the lines to the right of the column).
5. The horizontal lines partition the plane into rows. For each row compute the coefficient of y (using the procedure developed for the conveyor example) and write the coefficient in the left margin (the coefficient is the sum of the weights of the lines below the row minus the sum of the weights of the lines above the row).

6. The slope for every contour line passing through a given box is the negative ratio of the number in the bottom margin to the number in the left margin.
7. Use the single contour line construction method to construct a contour line through each point of interest.
8. To determine those points that minimize $f(\mathbf{X})$, either identify the points where the margin numbers change from negative to nonnegative or else, equivalently, use the median conditions.

Next, we summarize some of the insights we have gained about the rectilinear distance location problem of this section.

If we imagine drawing a horizontal and a vertical line through each existing facility location and call the points where the lines intersect *intersection points*, it is known that *some intersection point is an optimal location*. Also, the minimization problem decomposes into two *independent* problems of the same form, one for each new facility coordinate, which makes the problem easier to solve.

We can find an optimal location by using the *median conditions*. For example, to find the optimal x-coordinate, we proceed as follows. For each vertical line we compute the sum of the weights of the existing facilities lying on the line; for convenience, call each such sum a *line weight*. Let W denote the sum of all the line weights. We next begin to compute partial sums of the line weights, comparing with $W/2$ the line weight for the leftmost line, the partial sum for the two leftmost lines, the partial sum for the three leftmost lines, and so on. An optimal x coordinate is the coordinate of the *first* line we encounter for which the corresponding partial sum is at least $W/2$. To find an optimal y coordinate we proceed in an exactly analogous fashion, now working with horizontal lines and considering the line sum for the bottommost line, the partial sum for the two bottommost lines, and so on. To evaluate locations other than those that are optimal, we can use the contour line construction procedure given above.

4.2.2 Minisum Location Problem with Euclidean Distances: Introduction

We introduce the minisum location problem with Euclidean distances by means of an example problem. We shall solve the problem using a now famous algorithm due to Weiszfeld called *Weiszfeld's algorithm*, which is quite simple and easy to implement on a computer. The following section then gives the theory needed to justify the conclusions of this section, as well as additional related theory.

Example 4.3

Let us return to our warehouse conveyor example illustrated in Figure 4.3 and consider a related problem. Suppose now that we are considering installing an overhead conveyor system, consisting of straight-line segments. Items will enter the warehouse on the conveyor system at the point now denoted by $\mathbf{P}_1 = (0, 5)$, travel along a conveyor segment to a "diverge" point $\mathbf{X} = (x, y)$, from where they will go on a conveyor segment either to the point $\mathbf{P}_2 = (15, 7)$, or on another conveyor segment to the point $\mathbf{P}_3 = (15, 2)$. Let us suppose that the cost of the first conveyor segment is \$30 per foot, while the cost of the second and third segments is \$20 per foot. Recalling that coordinates have units of tens of feet, the total cost of the conveyor is then given by 10 times the following quantity:

$$f(x,y) = 30[(x-0)^2 + (y-5)^2]^{1/2} + 20[(x-15)^2 + (y-7)^2]^{1/2}$$
$$+ 20[(x-15)^2 + (y-2)^2]^{1/2}$$

We now make the important observation that *the location of the diverge point affects the conveyor cost*. If we use Weiszfeld's algorithm to minimize the cost, we get a diverge point location of $(x^*, y^*) = (12.17, 4.64)$, with a corresponding total conveyor cost of \$5163.62. What do you think would happen to the location of the diverge point if the first segment costs \$20 per foot instead of \$30 per foot? The result is to make the first segment longer and the other two segments shorter; if we use Weiszfeld's algorithm, we now get a diverge point location of $(x^*, y^*) = (13.56, 4.65)$, with a corresponding total conveyor cost of \$3867.32. On the other hand, what do you think would happen to the location of the diverge point if the first segment costs \$40 per foot instead of \$30 per foot? The result is to make the first segment shorter and the other two segments longer; if we use Weiszfeld's algorithm, we now get a diverge point location of $(x^*, y^*) = (0.00, 5.00)$, with a corresponding total conveyor cost of \$6085.20. This third case illustrates the majority theorem, when it is cheaper to have no first segment at all.

We can see that even though this problem of choosing a conveyor system of minimal total length is an easy one to describe, it is not trivial to analyze. Would you know how to find the best diverge point location if we had not told you we were using Weiszfeld's algorithm?

Therefore, let us consider Weiszfeld's algorithm. Weiszfeld's algorithm is based on the idea of a transformation, say $\mathrm{WF}(x, y)$, that transforms a point (x, y) in the plane into another point in the plane. We choose an initial point, say (x^1, y^1), use the transformation to compute $(x^2, y^2) = \mathrm{WF}(x^1, y^1)$, then compute $(x^3, y^3) = \mathrm{WF}(x^2, y^2)$, and, in general, compute $(x^j, y^j) = \mathrm{WF}(x^{j-1}, y^{j-1}), j = 2, 3, 4, \ldots$. Thus we obtain a sequence of points, the limit of which is the optimal solution to our problem. Instead of actually computing the limit, in practice we stop when the distance between two successive computed points is sufficiently small. Computational experience shows that the choice of an initial point is not critical to the success of the algorithm.

To generalize from our example, we suppose now that our conveyor total cost expression is given by

$$f(x,y) = \sum_{i=1}^{m} w_i[(x-a_i)^2 + (y-b_i)^2]^{1/2}$$

The transformation WF comes from computing the partial derivatives of the total cost expression, say $\partial f(x, y)/\partial x$, $\partial f(x, y)/\partial y$, and setting them to zero. We find that

$$\frac{\partial f(x,y)}{\partial x} = \sum_{i=1}^{m} \gamma_i(x,y)(x - a_i)$$

$$\frac{\partial f(x,y)}{\partial y} = \sum_{i=1}^{m} \gamma_i(x,y)(y - b_i)$$

where

$$\gamma_i(x,y) = \frac{w_i}{[(x - a_i)^2 + (y - b_i)^2]^{1/2}}, \qquad i = 1,\ldots,m$$

For convenience, define

$$\Gamma(x,y) = \sum_{i=1}^{m} \gamma_i(x,y), \quad \lambda_i(x,y) = \frac{\gamma_i(x,y)}{\Gamma(x,y)}, \qquad i = 1,\ldots,m$$

Then the transformation WF is given by

$$\text{WF}(x,y) = \sum_{i=1}^{m} \lambda_i(x,y)(a_i, b_i)$$

Setting the partial derivatives to zero and solving for x and y gives $(x,y) = \text{WF}(x,y)$.

Let us summarize how to compute $\text{WF}(x,y)$, given (x,y). For $i = 1$ to m, first compute $\gamma_i(x,y)$. Next, compute $\Gamma(x,y)$. Next, for $i = 1$ to m, compute $\lambda_i(x,y)$; the sum of products of the latter terms with the (a_i, b_i) gives $\text{WF}(x,y)$. For an example of the use of the algorithm giving intermediate calculations, refer to Table 4.6 in the next section (ignoring the last two columns).

4.2.3 Minisum Location Problem with Euclidean Distances: Theory

We now develop some theory for the minisum location problem with Euclidean distances. The reader willing to accept the results of the preceding section may wish to skip this section.

With (x,y) and (a_i, b_i) denoting the locations of the new facility and of existing facility i, respectively, we obtain the following equation for the total cost:

$$f(x,y) = \sum_{i=1}^{m} w_i[(x - a_i)^2 + (y - b_i)^2]^{1/2}$$

What we want to do of course, is to find a new facility location, say $X^* = (x^*, y^*)$, which will minimize this total cost function. To simplify matters, we call such a point a *best location*.

Consider first a particularly simple case for which the existing facility locations all lie on the same line, that is, are *collinear*. In this case the convex hull of the existing facility locations, which we denote by CH, is the smallest line segment containing all the existing facility locations. Let us choose the axes so that all the existing facilities lie on the x-axis, with $\mathbf{P}_i = (a_i, 0)$ for $i = 1, \ldots, m$. Recall that we only need to consider points in CH in order to find X^*. We note that it follows for $y = 0$ that

$$f(x,0) = \sum_{i=1}^{m} w_i[(x - a_i)^2]^{1/2} = \sum_{i=1}^{m} w_i|x - a_i| = f_1(x)$$

Recall that we studied the function $f_1(x)$ in Section 4.2.1. Thus we can use the procedures of that section to find $\mathbf{X}^*$ in this case. Further, it follows, in this case, that such a best location coincides with an existing facility location. In conclusion, *when the existing facility locations are collinear, CH is a line segment.* We need only consider existing facility locations in CH and can find a best location in CH by using either the graphical solution procedure or the median conditions of the preceding section.

There is another relationship between the Euclidean and rectilinear distance problems which is not as obvious as that of the preceding paragraph. If $\mathbf{X}^*$ denotes a best location for the Euclidean problem, f_1^* and f_2^* denote the minimum values of f_1 and f_2, respectively, for the rectilinear problem, and $\mathbf{X}$ is any point in the plane, it is known that

$$[(f_1^*)^2 + (f_2^*)^2]^{1/2} \leq f(\mathbf{X}^*) \leq f(\mathbf{X}) \tag{4.1}$$

The value of expression (4.1) is that it can sometimes allow us to obtain an acceptable solution to the Euclidean problem by solving the simpler rectilinear problem. Note that the second inequality in (4.1) is obviously true, for by definition the cost of a best solution to the Euclidean problem is a lower bound on the cost of any other location, and $\mathbf{X}^*$ is one such location. The left inequality in (4.1) is not particularly obvious, however. The left inequality enables us to obtain a lower bound on the minimum cost for the Euclidean problem by solving the rectilinear problem and then computing the leftmost term in (4.1).

Example 4.4

Let us apply (4.1) to the data given in Table 4.3. We know from our earlier consideration of Figure 4.6 that $\mathbf{X}^{**} = (8,5)$ is a best solution to the rectilinear problem. You should verify that $f_1(8) = 2.5$ and $f_2(5) = 2.0$. Thus substitution into the leftmost term of (4.1) gives

$$[(2.5)^2 + (2.0)^2]^{1/2} = 3.201562119$$

Similarly, substitution of $\mathbf{X}^{**} = (8,5)$ into the rightmost term of (4.1) gives 3.219372212. Thus we know that the minimum cost for the Euclidean distance problem is between 3.2 and 3.22 *without ever solving the Euclidean distance problem.* In particular, (8, 5) may well be an acceptable solution to the Euclidean distance problem, since its cost is so close to the minimum cost. Indeed, we shall see later, when we develop a general procedure to solve the Euclidean distance problem, that (8, 5) *is* a best solution to the problem. We warn you, however, that there is no guarantee that the best solution to the rectilinear problem is also a best solution to the Euclidean problem. Further, it is usually the case that the difference between the rightmost and leftmost terms in (4.1) is smaller for small values of m than for large values of m. Thus (4.1) becomes less useful as the number of existing facilities increases.

TABLE 4.3 DATA FOR EXAMPLE 4.4: EUCLIDEAN DISTANCES

i	1	2	3	4
w_i	$\frac{1}{6}$	$\frac{1}{3}$	$\frac{1}{3}$	$\frac{1}{6}$
$\mathbf{P}_i$	(4,2)	(8,5)	(11,8)	(13,2)

Now let us return to the problem of minimizing $f(\mathbf{X}) = f(x, y)$ directly. Since we have already covered the case where CH is a line segment, let us now consider the more interesting case where CH is not a line segment. To gain some insight, we first solve a simpler location problem, obtained from the basic problem of interest by *squaring* each Euclidean distance. If we square each distance in the above equation, we get the following function:

$$f^2(x,y) = \sum_{i=1}^{m} w_i[(x - a_i)^2 + (y - b_i)^2]$$

Note that the superscript 2 on the left side of the equation is *not* an exponent but is used simply to remind us we are working with a function involving squares of Euclidean distances. For a reason we shall see shortly, finding a new facility location to minimize $f^2(x, y)$ is called the *centroid problem*.

The centroid problem is of some interest by itself, provided that we have a problem for which each transport cost between the new facility and existing facility increases with the square of the Euclidean distance. Basically, however, our reason for considering the centroid problem is to provide a transition to the problem with regular Euclidean distances.

Since we want to solve the centroid problem, we want to minimize the function $f^2(\mathbf{X})$. If you graph the function given by

$$w_i[(x - a_i)^2 + (y - b_i)^2]$$

you will see that it is a parabaloid, with its base at the point (a_i, b_i). A line segment joining any two points on the surface of the parabaloid lies strictly above the surface except at its endpoints, and thus a parabaloid is a strictly convex function. Because a sum of strictly convex functions is a strictly convex function, it follows that $f^2(x, y)$ is a strictly convex function. Hence a point in the plane for which each partial derivative of $f^2(x, y)$ is zero is a unique minimizing point of the function. Thus the natural thing to do is to compute partial derivatives of $f^2(\mathbf{X})$, set the partial derivatives to zero, and solve for $\mathbf{X}$. You should verify that the partial derivative of $f^2(\mathbf{X})$ with respect to x is given by

$$\frac{\partial f^2(\mathbf{X})}{\partial x} = 2 \sum_{i=1}^{m} w_i(x - a_i)$$

Similarly, the partial derivative of $f^2(X)$ with respect to y is given by

$$\frac{\partial f^2(\mathbf{X})}{\partial y} = 2 \sum_{i=1}^{m} w_i(y - b_i)$$

Let us put the two partial derivatives into a two-tuple, a vector with two entries. Such a vector is called the *gradient*. We denote the gradient of $f^2(\mathbf{X})$ by

$$\nabla f^2(\mathbf{X})$$

It follows from above that

$$\nabla f^2(\mathbf{X}) = 2 \sum_{i=1}^{m} w_i(\mathbf{X} - \mathbf{P}_i)$$

Setting each partial derivative to zero is equivalent to setting the gradient equal to the zero vector, that is,

$$\nabla f^2(\mathbf{X}) = \mathbf{0}$$

If we let W denote the total of all the weights, the latter equation is equivalent to

$$W\mathbf{X} - \sum_{i=1}^{m} w_i \mathbf{P}_i = \mathbf{0}$$

which has a unique solution as follows:

$$\mathbf{X}^* = \sum_{i=1}^{m} \frac{w_i}{W} \mathbf{P}_i$$

We observe now that $\mathbf{X}^*$ is a weighted average of the existing facility locations, since each of the terms w_i/W is positive and the total of the terms is 1. This particular solution is usually called the *centroid*. The centroid is the unique new facility location that minimizes the function $f^2(\mathbf{X})$. It is interesting to note that contour sets of the function are quite simple; in fact, it is known that *contour sets are disks*, each with $\mathbf{X}^*$ as a center. Thus the contour sets are particularly simple to construct, and just as with the rectilinear distance problem, are useful in evaluating other possible locations for the new facility. The closer the new facility can be to the centroid $\mathbf{X}^*$, the better the solution will be.

Example 4.5

To illustrate the centroid discussion, consider Table 4.4. The centroid for this example is the point (4.25, 4.00). Thus contour sets for this example are just disks with centers at the point (4.25, 4.00).

We return now to a consideration of the Euclidean distance problem. Consider the graph of the function

$$w_i[(x - a_i)^2 + (y - b_i)^2]^{1/2}$$

TABLE 4.4 DATA FOR EXAMPLE 4.5: SQUARED EUCLIDEAN DISTANCES

i	1	2	3	4
w_i	1	1	1	1
$\mathbf{P}_i$	(0,0)	(0,10)	(5,0)	(12,6)

You should recall from geometry that *the graph is a cone*, with its tip at the point (a_i, b_i). A line segment joining any two points on the surface of the cone lies on or above the surface. (The only time such a line segment coincides with the surface is when one endpoint lies on the line joining the other endpoint and the tip of the cone. Thus it follows that the weighted Euclidean distance is a convex function. Thus $f(x, y)$ is a convex function, since it is a sum of convex functions. In fact, $f(x, y)$ is known to be a strictly convex function except when all the existing facility locations lie on the same line, that is, are *collinear*, a case we have already treated. *If we can find a point where the partial derivatives of $f(x, y)$ are both zero, then because $f(x, y)$ is a convex or strictly convex function it follows that the point is a best location.*

Let us look for a point for which the partial derivatives of $f(x, y)$ are both zero. We shall follow the same approach we used for the centroid problem. For each existing facility location $\mathbf{P}_i$, recall the definition of $\gamma_i(\mathbf{X})$:

$$\gamma_i(\mathbf{X}) = \frac{w_i}{[(x - a_i)^2 + (y - b_i)^2]^{1/2}}$$

Note we must assume that $\mathbf{X} \neq \mathbf{P}_i$, to avoid division by zero. In other words, the new facility location must be distinct from each existing facility location. We can now compute partial derivatives. We know that the partial derivative of f with respect to $\mathbf{x}$ is given by

$$\frac{\partial f(\mathbf{X})}{\partial x} = \sum_{i=1}^{m} \gamma_i(\mathbf{X})(x - a_i)$$

Similarly, the partial derivative of f with respect to y is

$$\frac{\partial f(\mathbf{X})}{\partial y} = \sum_{i=1}^{m} \gamma_i(\mathbf{X})(y - b_i)$$

If we put the two partial derivatives into a two-tuple, we get the gradient of f evaluated at $\mathbf{X}$, namely,

$$\nabla f(\mathbf{X}) = \sum_{i=1}^{m} \gamma_i(\mathbf{X})(\mathbf{X} - \mathbf{P}_i)$$

Naturally, the next thing to try is to set the gradient equal to the zero vector and attempt to solve for $\mathbf{X}$. Recall that $\Gamma(\mathbf{X})$ is the sum of all the terms $\gamma_i(\mathbf{X})$. Then setting the gradient of f to the zero vector gives

$$\Gamma(\mathbf{X})\mathbf{X} - \sum_{i=1}^{m} \gamma_i(\mathbf{X})\mathbf{P}_i = \mathbf{0}$$

It now follows, on denoting the solution we obtain by $\mathbf{X}$, that

$$\mathbf{X} = \sum_{i=1}^{m} \frac{\gamma_i(\mathbf{X})}{\Gamma(\mathbf{X})}\mathbf{P}_i$$

To simplify the notation, define

$$\lambda_i(\mathbf{X}) = \frac{\gamma_i(\mathbf{X})}{\Gamma(\mathbf{X})} \qquad \text{for } i = 1, \ldots, m$$

We now have

$$\mathbf{X} = \sum_{i=1}^{m} \lambda_i(\mathbf{X})\mathbf{P}_i \tag{4.2}$$

You should be able to verify that each term $\gamma_i(\mathbf{X})$ is positive and that the total of these terms is 1. Thus it follows that $\mathbf{X}$ is a weighted average of the existing facility locations, which means that $\mathbf{X}$ *is in the convex hull, CH, of the existing facility locations*.

Example 4.6

Consider Figure 4.8 and Table 4.5. Because of the special structure of this example, we might guess that (4, 2) is a best location. You should verify that the gradient evaluated at the point (4, 2) is the zero vector. In other words, each partial derivative evaluated at the point (4, 2) is zero. Thus we conclude that (4, 2) is a best location. An equivalent way to verify that (4, 2) is a best location is to substitute (4, 2) into the right side of equation (4.2); we find that the left side also turns out to be (4, 2), which is another means of concluding that the gradient is zero at the point (4, 2).

Hence if we have a specific point, say $\mathbf{Y}$, which is distinct from each $\mathbf{P}_i$ and we want to check for optimality, we can simply check to see if $\mathbf{Y}$ satisfies equation (4.2). Any such point that satisfies (4.2) is a best location. Note that we assumed that $\mathbf{Y}$ coincides with no $\mathbf{P}_i$, as otherwise (4.2) is not defined due to division by zero. Hence if we want to check to see if any point $\mathbf{P}_k$ is a best location, we cannot use the gradient test implied by (4.2). Fortunately, however, we can use a modification of the gradient test. Define ρ_k to be the Euclidean distance between the origin and the following vector:

$$\sum_{\substack{i=1 \\ i \neq k}}^{m} \gamma_i(\mathbf{P}_k)(\mathbf{P}_k - \mathbf{P}_i)$$

If you compare the vector above with the gradient of f, you will see that it is similar; the point $\mathbf{X}$ has been replaced by $\mathbf{P}_k$, and the term in the sum for $i = k$ has been omitted. It is known that an equivalent condition for $\mathbf{P}_k$ to be a best location is for ρ_k to satisfy

$$\rho_k \leq w_k \tag{4.3}$$

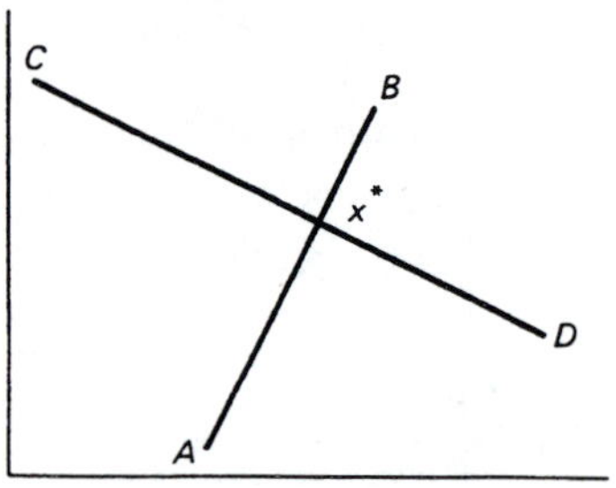

Figure 4.8 Graphical solution procedure for the Euclidean problem of Example 4.6.

TABLE 4.5 DATA FOR EXAMPLE 4.6: EUCLIDEAN DISTANCES

i	1	2	3	4
w_i	1	1	1	1
$\mathbf{P}_i$	(0,0)	(0,10)	(5,0)	(12,6)

Hence to check to see if any P_k is a best location, simply check to see if (4.3) *holds.* To illustrate the use of (4.3), consider the data of Example 4.4. If you compute ρ_1, the ρ_k term for the point (4, 2), you will find that it is 0.4441465833, which is greater than $\frac{1}{6}$, the weight of the point. Thus (4, 2) is not a best solution. However, if you compute ρ_2, the ρ_k term for the point (8, 5), you will find that (8, 5) is a best solution, since $\rho_2 = 0.109092477 \leq \frac{1}{3} = w_2$.

Using either (4.2) or (4.3) as appropriate, we now have a way to check any point of interest to see if it is a best location. However, it may be the case that we do not have a specific point we wish to check for optimality. Nevertheless, we can still use (4.2) to help us develop *Weiszfeld's algorithm*, to find a best location. The algorithm works in the following manner. Choose a first trial new facility location and substitute it into the right side of (4.2) to obtain a second trial value, substitute the second trial value into the right side to obtain a third trial value, and so on, stopping when the distance between two successive trial values is "almost" zero.

Computational experience indicates that the algorithm performs well, although it can sometimes run into difficulties when a best location coincides with an existing facility location, in which case an attempt to divide by zero can occur.

A simple way to avoid the problem of division by zero when using the algorithm is to "perturb" the problem by replacing each Euclidean distance between $\mathbf{X} = (x, y)$ and $\mathbf{P}_i = (a_i, b_i)$ by the following quantity:

$$[(x - a_i)^2 + (y - b_i)^2 + \epsilon]^{1/2}$$

By choosing ϵ positive but sufficiently close to zero, the distances are not changed enough to have much effect on the accuracy of the model, while division by zero is avoided. In the remainder of this section we assume that $\epsilon > 0$. All of the analysis above for the partial derivatives of $f(\mathbf{X})$ goes through as before when we use $\epsilon > 0$, with the exception that we will now have partial derivatives always defined. Thus all the gradients we compute are guaranteed to exist, and it is no longer necessary to compute the ρ_k numbers as defined above.

Table 4.6 illustrates the use of the Weiszfeld algorithm to solve Example 4.6 for the case where $\epsilon = 0.000001$. You can see that the initial location is chosen to be the centroid location, (4.25, 4.00), which has a cost (f value) of 25.2626. Ignoring the last two columns of the table for now, note that successive rows of the table display successive trial locations obtained by means of the algorithm, together with their f values. At the last iteration, the distance between the current trial location and the previous trial location is less than the stopping criterion, so that the algorithm terminates. Note that the steps between successive trial locations decrease each time,

TABLE 4.6 ILLUSTRATION OF THE WEISZFELD ALGORITHM, WITH LOWER BOUND, FOR EXAMPLE 4.6

Iteration	x value	y value	f value	Lower bound	Gap
0	4.2500	4.0000	25.2626	22.2003	3.0624
1	4.0234	3.1160	24.8083	23.4582	1.3501
2	3.9488	2.6268	24.6653	24.0595	0.6058
3	3.9352	2.3584	24.6196	24.3173	0.3023
4	3.9440	2.2086	24.6047	24.4314	0.1734
5	3.9580	2.1237	24.5996	24.4971	0.1025
6	3.9708	2.0745	24.5978	24.5360	0.0619
7	3.9805	2.0454	24.5972	24.5593	0.0378
8	3.9873	2.0279	24.5969	24.5736	0.0233
9	3.9919	2.0173	24.5968	24.5824	0.0144
10	3.9948	2.0107	24.5968	24.5878	0.0089
11	3.9968	2.0067	24.5968	24.5917	0.0056
12	3.9979	2.0042	24.5968	24.5933	0.0034
13	3.9987	2.0026	24.5967	24.5946	0.0021
14	3.9992	2.0016	24.5967	24.5954	0.0013

a frequent occurrence with the algorithm. The first few steps are usually "large," after which the steps quickly become quite small, and little further reduction in the total cost occurs.

Example 4.7

A case when the algorithm does not perform well—in a sense—is when the function $f(\mathbf{X})$ is almost "flat" over an extended region. To be specific, consider Table 4.7. If an initial trial location of $(1, 1)$ is chosen, the algorithm requires 320,000 iterations to reach the point $x = y = 97.46398$, and requires 640,000 iterations to reach the point $x = y = 99.83249$. The majority theorem implies that $(100, 100)$ is a best location. Because the existing facility locations are not collinear, the function f in this case is strictly convex, so $(100, 100)$ is the unique best location. Hence in one sense the algorithm performs poorly and would typically terminate prior to reaching the unique optimum solution. In another sense the algorithm does not do badly, however, since $f(1, 1)/f(100, 100) = 1.001376915$, so that even the initial trial location has a cost that is very close to the minimum cost.

It is known that the Weiszfeld algorithm is a steepest descent algorithm which chooses the step lengths automatically; what can happen in unusual cases is that the algorithm will choose much smaller step lengths than are best. If such behavior is

TABLE 4.7 DATA FOR EXAMPLE 4.7: EUCLIDEAN DISTANCES

i	1	2	3	4
w_i	1	1	1	3
$\mathbf{P}_i$	(0,0)	(1,0)	(1,0)	(100,100)

suspected, it may be useful to compute a lower bound on $f(\mathbf{X}^*)$, the minimum value of f. The lower bound is given by the right side of inequality (4.4), where $\mathbf{X}_k$ is the trial location the algorithm provides at iteration k and the row vector $\nabla f(\mathbf{X}_k)$ denotes the gradient of f evaluated at $\mathbf{X}_k$:

$$f(\mathbf{X}^*) \geq f(\mathbf{X}_k) - \nabla f(\mathbf{X}_k)\mathbf{X}_k^T + \min\{\nabla f(\mathbf{X}_k)\mathbf{P}_i^T: i = 1, \ldots, m\} \tag{4.4}$$

When a value of the function f is found that is sufficiently close to the lower bound, we may wish to terminate the algorithm. Table 4.6 illustrates the use of (4.4) for Example 4.6. The column titled "lower bound" shows the right side of (4.4) at each iteration. You can see that the bound values increase and, by iteration 14, agree with the f value to two significant digits to the right of the decimal point. The last column of the table, titled "gap," gives the difference between the f value and the bound value at each iteration (discrepancies are due to round-off error). The gap is zero, of course, if the algorithm gives an exactly optimal solution, and you can see how the gap decreases at each iteration. The gap can be used as part of an alternative stopping criterion, called the *relative error*. The relative error is the ratio of the gap to the lower bound, and it might be useful if we encounter a pathological problem such as the one we give above.

You may have noted we have not discussed the construction of contour lines for $f(\mathbf{X})$. There is a very good reason for our omission. We know of no really elementary means to construct contour lines when distances are Euclidean. However, you should be happy to learn that your friendly local computer center is very likely to have programs and the necessary auxiliary graphics equipment for constructing contour lines. We illustrate several such contour lines for Example 4.4 in Figure 4.9. We suggest that you compare Figures 4.5 and 4.9 to observe the way the contour lines differ depending on whether or not the distances are rectilinear or Euclidean. Needless to say, contour lines for the Euclidean distance problem have exactly the same uses as those for the rectilinear distance problem; a few contour lines allow us to evaluate easily the values of many possible new facility locations.

4.3 SOME MULTIOBJECTIVE LOCATION PROBLEMS

4.3.1 Multiobjective Euclidean Distance Location Problem

The location problems we considered previously all have a common aspect, a single objective function. Either an individual, a corporation, a governmental body, or a single entity of some sort has decided on a *single* objective function, often total cost, which must be optimized. There are, however, instances in which there are *multiple* objectives. Hence the purpose of this section is to introduce you to a multiobjective location problem.

As an introduction to multiobjective location problems, consider two nearby towns, connected by a highway, which will share an airport to be constructed. Each

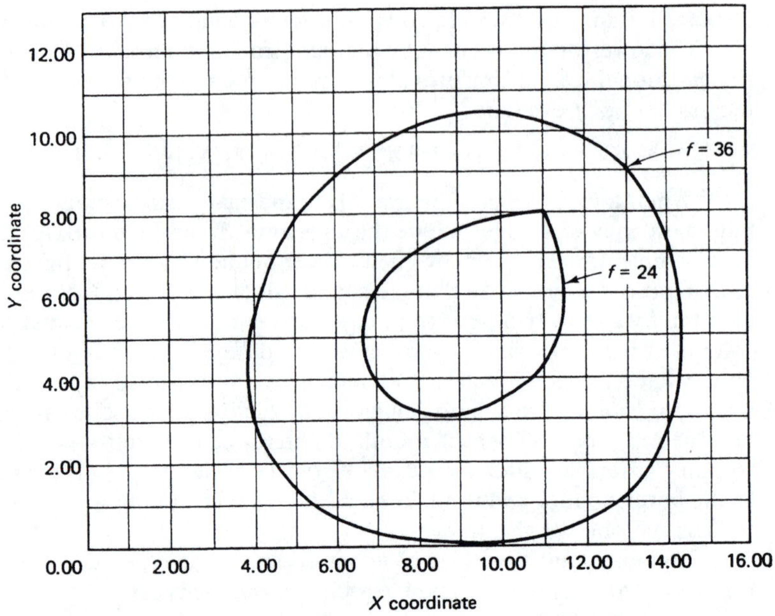

Figure 4.9 Contour lines for Euclidean example problem.

town would like the airport to be readily accessible. The terrain is such that many locations are feasible for the airport. Also, since the airport will be built in a rural area, the possible airport sites all have about the same cost per acre. We might then address the following question. Where should the airport *not* be built? In other words, what possible sites can be eliminated from consideration? You might think that answering the question is a somewhat indirect way of choosing a site, and in fact you would be right to think so. However, it may well be easier for the two towns to choose which sites to eliminate than to finalize a choice of a single site. Hence *a reasonable first step in the process of choosing a site might be to determine which sites can be eliminated from consideration*.

Hopefully, you will have decided by now that there is no point in considering airport sites that are not close to the highway connecting the towns. We can say that any site **X** that is not close to the highway is *dominated*, in the sense that there will be another site **X**′, close to the highway, which will be at least as convenient as **X** for both towns and more convenient than **X** for at least one of the towns. On the other hand, if we choose any site **X** that is on the highway, it will be impossible to find another site that dominates **X**; in order for any site **X**′ to be closer to one town than **X** will be, **X**′ must be farther from the other town than **X** will be. We shall call a site that cannot be dominated an *efficient* site. Those sites that are efficient will be the ones on the highway joining the two towns. For any two efficient sites, say **X** and **Y**, one site would be more convenient than the other for one town but less convenient than the other site for the second town. A final choice from among the efficient sites may well be a political decision, but we can at least use quantitative criteria to eliminate the dominated sites from consideration. In particular, we have used *two* criteria for eliminating dominated sites and determining efficient sites: convenience of access for town 1 and convenience of access for town 2.

Keeping in mind this initial example of a multiobjective location problem, let us try now to quantify a bit more what we mean by "convenience of access." For each town i, which has a location $\mathbf{P}_i$, let $d(\mathbf{X}, \mathbf{P}_i)$ denote the Euclidean distance between the town and a site at $\mathbf{X}$. We shall measure "convenience of access" by $d(\mathbf{X}, \mathbf{P}_i)$, the Euclidean distance between town i and the airport site. The closer $\mathbf{X}$ is to town i, the more convenient will be the access of town i to $\mathbf{X}$. Because there are two towns, we must consider convenience of access for each town, so we put the Euclidean distances into a two-tuple, a vector with two entries, defined as follows:

$$\mathbf{D}(\mathbf{X}) \equiv (d(\mathbf{X}, \mathbf{P}_1), d(\mathbf{X}, \mathbf{P}_2))$$

Similarly, we can define $\mathbf{D}(\mathbf{Y})$ for any other site $\mathbf{Y}$ of interest simply by replacing $\mathbf{X}$ by $\mathbf{Y}$. We can then be more precise in our definition of a dominated site. A site $\mathbf{X}$ will *dominate* a site $\mathbf{Y}$ if each entry in $\mathbf{D}(\mathbf{X})$ is no more than the corresponding entry in $\mathbf{D}(\mathbf{Y})$, and at least one entry in $\mathbf{D}(\mathbf{X})$ is strictly less than the corresponding entry in $\mathbf{D}(\mathbf{Y})$. Those sites that are not dominated we call the *efficient* sites.

With our more precise definitions of dominance and efficiency, let us consider the airport site problem again. Suppose that we choose the axis and scale of measurement so that town 1 is at the origin, $(0, 0)$, while town 2 is at the point $(10, 0)$. Which sites will be dominated, and which sites will be efficient? Let us denote by H the line segment joining the two town locations, so that H is the line segment joining $(0, 0)$ and $(10, 0)$. We see that no point on H is dominated. However, for any point $\mathbf{X}$ not in H, if we choose the closest point in H to $\mathbf{X}$, say $\mathbf{X}'$, then $\mathbf{X}'$ dominates $\mathbf{X}$. Hence we conclude that the set of efficient sites is the line segment H, while the set of dominated sites consists of those points not on the line segment H. H, of course, corresponds to the highway joining the two towns. Hence we obtain reasonably close agreement between our initial qualitative discussion of airport sites and our second, more quantitative discussion of dominated and efficient sites.

From our discussion of the convex hull in Section 4.1, you should recognize that H is just the convex hull of the two points $(0, 0)$ and $(10, 0)$. This fact provides a clue for finding dominated and efficient points when there are more than two towns. For example, suppose that there are now three towns, at locations $(0, 0)$, $(10, 0)$, and $(5, 10)$. Thus for any site $\mathbf{X}$, the vector $\mathbf{D}(\mathbf{X})$ now has three entries. Ask yourself which will be the efficient sites for this example. Hopefully, you have concluded that the efficient sites will be those points in or on the boundary of the triangle, say T, whose vertices are the locations of the three towns. Further, you should see that T, *the set of efficient points, is the convex hull of the locations of the points* $\mathbf{P}_i$, the towns. For any site $\mathbf{X}$ not in T, if $\mathbf{X}'$ is the closest point in T to $\mathbf{X}$, then $\mathbf{X}'$ will dominate $\mathbf{X}$. On the other hand, if $\mathbf{X}$ is any point in T, there is no other point that will dominate $\mathbf{X}$.

It is now easy to consider the multiobjective location problem, with m locations in the plane, $\mathbf{P}_1, \ldots, \mathbf{P}_m$, and hence m entries in the vector $\mathbf{D}(\mathbf{X})$, with entry i being the Euclidean distance between the planar points $\mathbf{X}$ and $\mathbf{P}_i$. Let CH denote the convex hull of the points $\mathbf{P}_1, \ldots, \mathbf{P}_m$. As you might guess, *CH consists of the set of efficient points, while the points not in CH are the dominated points*. It should be clear that points $\mathbf{P}_1, \ldots, \mathbf{P}_m$ need not be towns, and the location problem to be analyzed need not be for an airport. What is essential is that there be m entities of some sort for which there is no agreement upon a single objective function, and that convenience of access is measured using Euclidean distance. The location of any activity that provides or receives service from m distinct entities of some sort is a candidate for multiobjective analysis. Of course, it may be the case that some other sort of distance than Euclidean distance may be most appropriate. Therefore, in the next section we consider a multiobjective location problem using rectilinear distances.

4.3.2 Multiobjective Location Problem with Rectilinear Distances

We have seen many instances previously in which the use of rectilinear distance is more appropriate than the use of Euclidean distance. Hence in this section we consider a multiobjective location problem with rectilinear distances. We obtain the problem we consider by making the following modification to the problem of the preceding section; we use rectilinear distance instead of Euclidean distance in the definition of the vector $\mathbf{D}(\mathbf{X})$. Thus the entries in the vector are the rectilinear distances between $\mathbf{X}$ and the points $\mathbf{P}_1, \ldots, \mathbf{P}_m$. We call these m points existing facility locations. We shall denote the set of all efficient locations by E. We shall see that E is *not* the convex hull of the given m existing facility locations; this is one instance in which the rectilinear distance problem is more complicated than the corresponding Euclidean distance problem. Efficiency is defined using the vector $\mathbf{D}(\mathbf{X})$ as in the preceding section, with the exception that distances are now rectilinear.

In addition to applications similar to those we discussed in the preceding section, the problem of finding E is related to doing a sensitivity analysis of the minisum rectilinear distance location problem we studied in Section 4.2. It is known that every optimum solution to this minisum location problem is an efficient point. Further, it is possible to choose the weights of the minisum problem so that each efficient point will be an optimum solution to some minisum problem. Therefore, *finding the efficient set E is equivalent to doing a complete sensitivity analysis of the weights of the rectilinear distance minisum problem*. Since we may be uncertain of the values of the weights, or since the weights may possibly change over time, it is clearly useful to know the set E of efficient points; *no matter what the weights are, the set E will contain all the optimum solutions to the minisum problem*.

In what follows in this section we develop an algorithm, which we call the *arrow algorithm*, for finding the efficient set E. We shall always assume that the existing facility locations do not all lie on the same horizontal or the same vertical line, for in this case the set E is just the line segment joining the two existing facility locations that are farthest apart, so that in this special case it is a trivial matter to construct E.

To develop insight we first consider some miniature efficient point location problems, shown in Figure 4.10. Once we have the insight to solve these miniature problems, we will have most of the insight to solve the general problem. Consider first Figure 4.10(a), with four existing facility locations $\mathbf{P}_1, \ldots, \mathbf{P}_4$ as shown. Note that a horizontal and a vertical line through each of the existing facility locations partition the plane into a "box" B as well as eight other regions. If $\mathbf{X}$ is any point not in the box B, we see that the closest point in B to $\mathbf{X}$, say $\mathbf{X}'$, dominates $\mathbf{X}$. Thus no points outside B can be efficient. Further, all the points inside or on the boundary of B are efficient points; given any point $\mathbf{X}$ in the box, if $\mathbf{Y}$ is another point that is closer to one or more of the corners of the box than $\mathbf{X}$ is, $\mathbf{Y}$ must be farther from at least one other corner of the box.

As Figure 4.10(a) illustrates, it is convenient to give names to some of the planar regions obtained when the plane is partitioned. Those points that lie on or above the top line, and on or to the left of the leftmost line, we call the *northwest direction of B* and denote by NW(B). Similarly, we define three other directions, denoted NE(B), SE(B), and SW(B). We say that *a direction is occupied* if it contains at least one of the existing facility locations $\mathbf{P}_i$. In Figure 4.10(a) all four directions are occupied. *We call a box with all four of its directions occupied a null box*; the reason for the name is that the number of unoccupied directions of the box is 0. Figure 4.11 illustrates the directions of an arbitrary box.

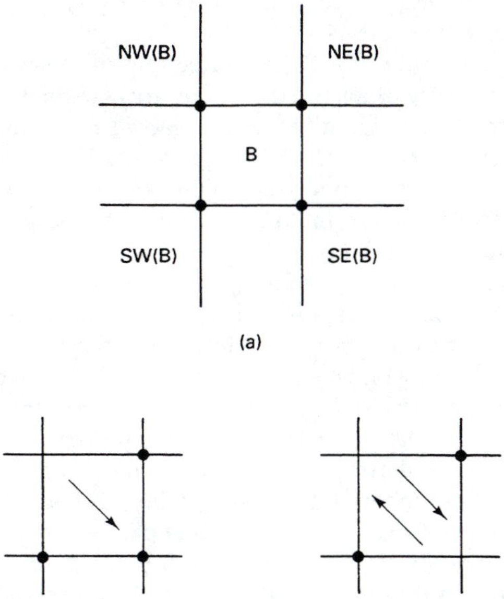

Figure 4.10 Example boxes: (a) null box B; (b) 1-box; (c) 2-box.

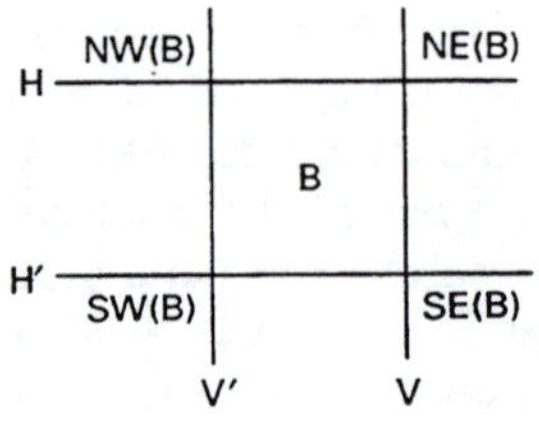

Figure 4.11 Directions of B.

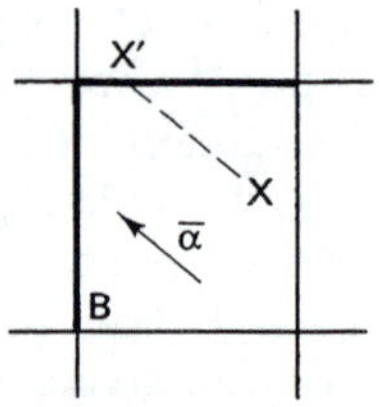

Figure 4.12 Leading edges of SE 1-box.

Now consider Figure 4.10(b), with $m = 3$ existing facility locations. Only one direction, NW(B), is unoccupied, while the other three directions of the box are occupied. The arrow in the box, called a NW arrow, points away from the unoccupied direction, thus indicating which direction is unoccupied. The arrow also points toward the efficient set E, which consists in this case of the right and bottom edges of the box, which are called the *leading edges* of the box. Because only one direction of the box is unoccupied, there is only one arrow in the box; we call such a box a *1-box*.

Figure 4.12 illustrates a SE 1-box. Note that the point **X** can be moved along the dashed line to the point **X**′ lying on a leading edge of the box while remaining the same distance from points in the SW and NE directions of the box and becoming closer to the points in the NW direction of the box, which explains why the points in a 1-box that are not on the leading edge of the box are dominated. Figure 4.12 also illustrates the general definition of the leading edges of a 1-box; the two edges of the box pointed to by the arrow in the box.

Next consider Figure 4.10(c), with $m = 2$. Two directions of the box, NW(B) and SE(B), are unoccupied, while the other two directions are occupied. Hence we draw NW and SE arrows in the box to identify the unoccupied directions. A NW arrow always points away from an unoccupied NW direction, a NE arrow always points away from an unoccupied NE direction, and so on. We call a box with two arrows a *2-box*. You can see for this figure that the set E consists of all the points lying inside and on the boundaries of the box.

We now have almost all of the ideas we need to state the arrow algorithm. Refer to Figure 4.13(a), which shows a sample problem with $m = 5$ and illustrates the *line construction procedure*, with a vertical and horizontal line drawn through each of the five existing facility locations. Note, with the line construction procedure, that all vertical (horizontal) lines extend above (to the left of) the topmost (leftmost) point, and below (to the right of) the bottommost (rightmost) point. We denote by BB the region lying on or to the right of the leftmost vertical line, on or to the left of the rightmost vertical line, on or below the topmost horizontal line, and on or above the bottommost horizontal line, and call BB the *big box*. You can see that it is never possible to have efficient points outside the big box; each point **X** outside BB is dominated by the point **X**′, which is the closest point in BB to **X**. You should be able to see from this example how to define the big box, as well as to draw the necessary

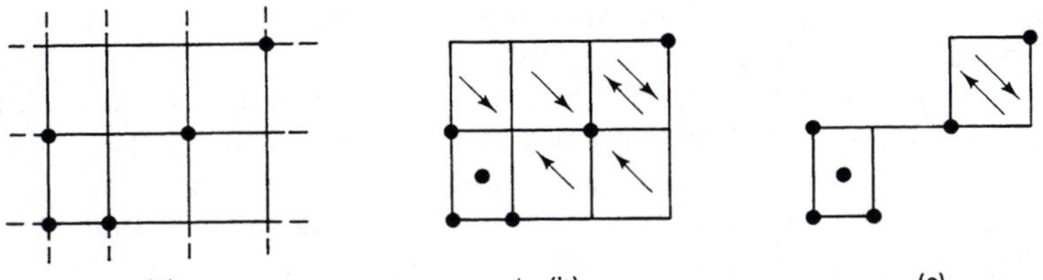

Figure 4.13 (a) Line construction; (b) arrow drawing; (c) $S^* = B^*$.

horizontal and vertical lines through each $\mathbf{P}_i$ for any value of m; we call the construction of these horizontal and vertical lines the *line construction procedure*.

We can now state the *arrow algorithm*. To construct the set E of all efficient locations, carry out the line construction procedure. Then, for each box B inside BB, draw an arrow corresponding to each unoccupied direction of B. Classify each resultant box inside BB as a null-box, 1-box, or 2-box, according to the number of arrows the box contains. If there are no 1-boxes, take $E = BB$. Otherwise, choose a 1-box B not yet chosen and delete from BB all points in B except those on the leading edges of B; repeat this deletion procedure for every 1-box. Take E to be the subset of BB remaining after the completion of the deletion procedure.

Figure 4.13(a) and (b) illustrates the line construction and the arrow drawing step, respectively. Figure 4.13(c) shows the results of the deletion process. E consists of a null-box, a 2-box, and an edge joining the two boxes, called a *connecting edge*. By definition, a *connecting edge is an edge e of a box B for which every arrow in B points toward e and e is contained in the boundary of BB, or else there is another box, say B', such that e is the common edge of B and B', and every arrow in each box points toward e*.

We can see from Figure 4.13 that the ideas of the arrow algorithm are really rather simple, particularly the steps of drawing the lines and arrows. If we consider the deletion process proceeding from left to right in the top row, and then in the bottom row, the first box B we encounter is a NW 1-box, so we delete from BB all points in B except the leading edges of B. The next box we encounter is also a NW 1-box, so we delete all of it except for its leading edges, consequently deleting one leading edge of the previous box we considered. The third box in the first row is a 2-box, so we leave it unchanged. In row 2, the first box we encounter is a null-box, so we leave it unchanged as well. The remaining two boxes are SE 1-boxes, which have all of their points deleted except for their leading edges. Note that all or part of a leading edge of a 1-box may subsequently be deleted due to deleting dominated points in other 1-boxes.

This example illustrates a result that is generally true: *The set E of efficient points is the union of the following: the null-boxes, the 2-boxes, the connecting edges.* In essence, what the arrow algorithm does is identify and delete all the dominated points; the points remaining then constitute the set E. The arrows that are drawn in 1-boxes point away from dominated points and indicate directions of boxes that can be deleted. It is known that the only other possible boxes besides 1-boxes are null-boxes and 2-boxes. Null-boxes and 2-boxes, together with connecting edges, are not deleted from BB, and when taken together constitute the efficient set E.

Figure 4.14 shows a second example, based on the conveyor analysis example of Section 4.2. We shall modify the conveyor example very slightly, not restricting the conveyor to be a horizontal line segment, as previously, but instead allowing the conveyor to consist of at most two rectilinear line segments; that is, the length of the conveyor is now the rectilinear distance between the unloading end of the conveyor and the point $(0, 5)$ where the conveyor enters the warehouse. You can see that one of the connecting edges, the one joining the points $(0, 5)$ nd $(7, 5)$, actually coincides with the best conveyor length we determined in Section 4.2. If the weight of the dock at the location $(7, 10)$ were to become much larger, we might want to run the conveyor from $(0, 5)$ to $(7, 5)$ to $(7, 10)$, thus "using" two connecting edges and one edge of a null-box. Alternatively, if the weight at the point $(12, 0)$ were to become much larger, we might want to run the conveyor from $(0, 5)$ to $(12, 5)$ to $(12, 0)$. Similarly, if the weights for the docks at $(15, 7)$ or $(15, 3)$ became much larger, we might want to locate the unloading end of the conveyor close to these two docks. However, *in no case would we ever want to locate the unloading end in the interior of any of the 1-boxes, since all such locations are dominated*. Thus by knowing the efficient set E we can exclude from consideration more than half of the warehouse, *regardless* of the actual values of the weights associated with the conveyor and the docks.

Figure 4.15 shows a third example, with $m = 16$. To simplify the figure the arrows have been omitted. You should be able to solve this example using the arrow algorithm and obtain the efficient set E shown in the figure.

Note that the example of Figure 4.15 has 120 boxes. If you had to draw the appropriate number of arrows for each of the boxes, one box at a time, you might well conclude that you would rather not. Thus we mention that there is an alternative approach to identifying the efficient set E, called the *coloring algorithm*, which requires less computational effort than the arrow algorithm. The coloring algorithm also requires more analysis effort and is thus a bit more difficult to understand than the arrow algorithm. As is often the case, to solve a problem really well requires more

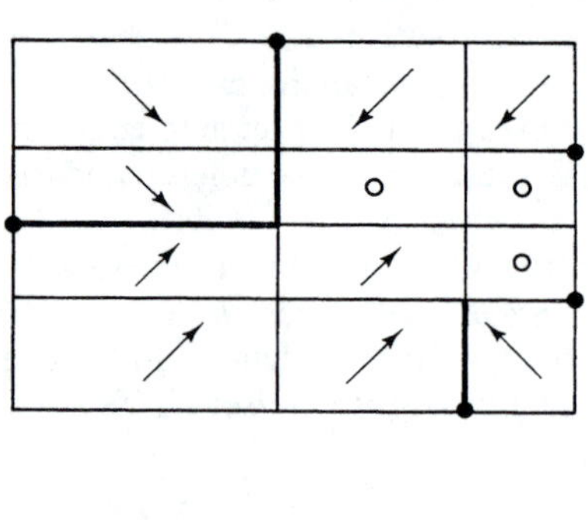

Figure 4.14 Efficient and dominated points for the conveyor example of Figure 4.3.

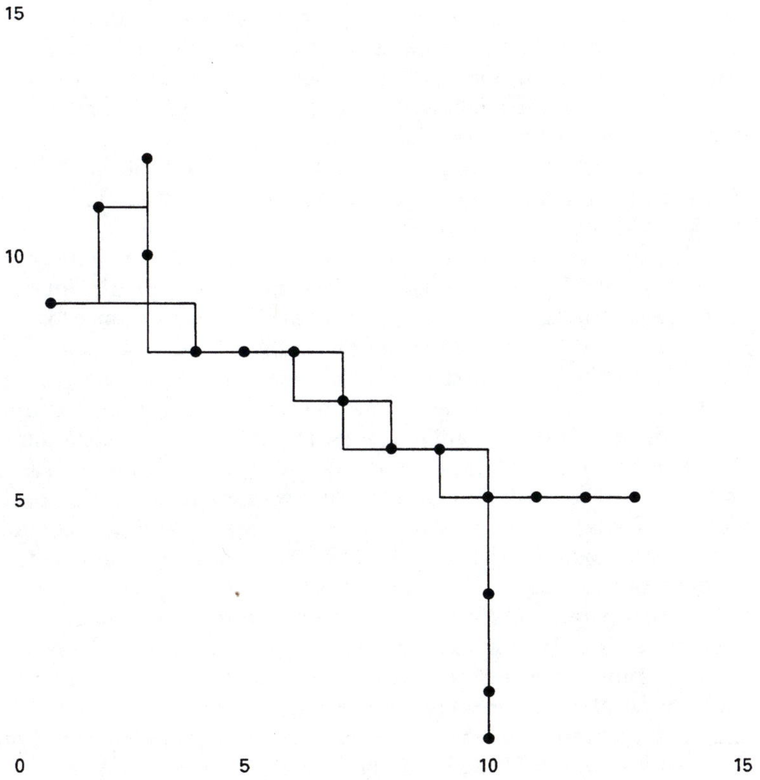

Figure 4.15 Second efficient set example.

effort than is needed to obtain an adequate solution. For details on the coloring algorithm we refer you to Chalmet et al. [7].

4.4 SINGLE-FACILITY MINIMAX LOCATION PROBLEMS

4.4.1 Introduction

There is another class of single-facility location problems that we should mention, called *minimax problems*. One of the best known such problems, called the *circle covering problem*, involves enclosing m known points in the plane within a circle of minimum radius. The circle covering problem is equivalent to the problem of locating a new facility with respect to m existing facilities so as to minimize the maximum Euclidean distance from the new facility to the existing facilities. The circle covering problem may be of interest in locating a transmitter of some kind, or a receiver, so as to "cover" m stations with as strong a signal strength as possible. Also, the problem of stationing a helicopter so as to minimize the maximum time for it respond to an emergency at any one of m sites is closely related to the circle covering problem. Contrary to what one might think, the circle covering problem cannot be solved by inspection (or at least no one has yet been able to do so). There are, however, very efficient and relatively simple algorithms for solving the problem; the paper by Hearn and Vijay [30] gives a good exposition and evaluation of such algorithms, which have quite a long history.

We note that it is easy to draw contour sets for the circle covering problem. If we draw a circle about each of the m points of radius r, the intersection of the m circles consists of all points whose maximum Euclidean distance to the m given points is r or less. If we imagine reducing r until the intersection of the circles is a single point, we obtain the solution to the circle covering problem. Such an approach is now quite feasible when one has a computer terminal with the facility for displaying circles, provided that m is not too large. The value of having contour sets is that they allow the ready evaluation of other locations than the one that is optimal to the problem. In closing the initial discussion of the circle covering problem, we shall mention an interesting insight it provides. There is some choice of either two or three of the given points so that constructing a circle through this choice of points gives a minimum covering circle; if we knew this choice of points, we could throw away all the other points and still solve the original problem. In effect, this choice of points "dominates" all the other choices, so we would need to have good physical arguments to rationalize this dominance if we are to convince anyone that the circle covering problem is a meaningful operational problem to solve.

Another problem, which we might term the *diamond covering problem*, occurs when we replace the Euclidean distances of the circle covering problem by rectilinear distances. Here by "diamond" we mean a square with each edge making an angle of ± 45 degrees with an axis; the radius of the diamond is half the length of the line segment joining opposite vertices. The diamond covering problem is easy to solve. By applying a 45-degree rotation we obtain an equivalent square covering problem, which we solve by constructing a smallest enclosing rectangle. If the rectangle is a square, its center is a minimax location. Otherwise, we extend the shorter of one pair of edges to have the same length as the longer pair, and take the center of any such square so constructed as a minimax location. Of course, we must apply a reverse 45-degree rotation to translate our answer back to a solution of the original problem. In some cases there will be more than one smallest enclosing diamond, resulting in alternative optimum locations. As with the circle covering problem, it is easy to construct contour sets. The set of all points such that the maximum rectilinear distance between the points and the m existing facilities is at most r consists of the intersection of m diamonds, with diamond i having a center at point i and a radius of r, for $i = 1, \ldots, m$. Much the same comments apply to these contour sets as those made above for the circle covering problem.

While diamond and circle covering problems have substantial geometrical interest, it is reasonable to ask why they should be of interest as location problems.

Usually, the reason has to do with a greater concern for the "worst case" than for the average case. Consider buying life insurance. If, on the average, a life insurance company pays out more than it takes in, it has serious financial difficulties. Thus, assuming that we purchase life insurance from a solvent company, on the average we can expect to pay out more than we receive. Does this mean that it is not a good idea to buy life insurance? In fact, we buy life insurance to protect us from the consequences of having a short life. There are some situations where a long-run analysis may not apply. We may be more concerned about the "worst case" than about the long-run expected net benefits. Similarly, with minimax location problems, we must be more concerned about the worst case than about the expected total cost, and thus find a minimax approach advocated for locating such things as fire extinguishers, fire stations, and various emergency service facilities, such as a helicopter service to respond to automobile accidents.

4.4.2 Circle Covering Problem

In this section we consider two approaches for solving the circle covering problem. The first approach is basically geometrical in nature and is particularly well suited for planar circle covering problems. The second approach is more general, in the sense that it can be used not only for planar problems but for analogous problems in three or more dimensions. For the first approach we present an algorithm. For the second approach we show how to convert the problem into an equivalent quadratic programming problem. Most available algorithms for solving quadratic programming can then be applied to solve the equivalent problem.

Let us first state the problem of interest precisely. We wish to minimize the function $g(x,y)$ defined by

$$g(x,y) = \max\{[(x - a_i)^2 + (y - b_i)^2]^{1/2}: 1 \le i \le m\}$$

Here, as usual, the points (a_i, b_i) are m existing facility locations, and (x,y) is a new facility to be located in such a way as to minimize $g(x,y)$.

A problem equivalent to minimizing $g(x,y)$ is as follows:

$$\text{minimize} \quad z$$

subject to

$$[(x - a_i)^2 + (y - b_i)^2]^{1/2} \le z, \qquad 1 \le i \le m$$

The equivalent problem has the following geometrical interpretation. The constraints state that each existing facility location must lie in a circle with center (x,y) and radius z, so that the geometrical problem is to find a smallest circle that encloses all the existing facility locations.

The idea of the algorithm is to construct a sequence of circles, with each current circle CC defined by a set of existing facility locations denoted by S. Each circle in the sequence is the minimum covering circle for the points in S and has a larger radius than the previous circle in the sequence. Once a circle has been constructed that contains all the existing facilities, the algorithm terminates, taking the last circle constructed to be the minimum covering circle, with the center of the last circle being the minimax location.

Some preliminaries are necessary prior to stating the algorithm. For convenience, we shall subsequently refer to the existing facility locations as *points*. Throughout the algorithm, the set S is the most recent set of points used to define the circle and will always consist of two or three points.

The algorithm is based on the ability to determine quickly the minimum covering circle for sets of two and three points. For any two points, say **A** and **B**, the minimum covering circle has the line segment **AB** as its diameter. Finding the minimum covering circle of any three points, say **A**, **B**, and **D**, involves constructing the circle that passes through the three points. You may recall from geometry that the procedure is as follows. Construct perpendicular bisectors of any two of the line segments **AB**, **AD**, and **BD**. The intersection of the two perpendicular bisectors is the center of the circle passing through **A**, **B**, and **D**. It is elementary to observe that if the triangle **ABD** has an obtuse angle, say at **A**, then a smaller circle covering the three points is the circle with the line segment **BD** as a diameter (rather than as a chord). Thus the minimum covering circle for three points is the circle passing through the three points if they form an acute (or right) triangle. Otherwise, the two-point circle defined by the points opposite the obtuse angle is the minimum covering circle.

Figure 4.16 illustrates a means of constructing the new set S when the current set S consists of two points, say **A** and **B**. This means can be used in a procedure of the covering algorithm named TwoPointsPlus. The points **A** and **B** are on opposite sides of CC, the current circle. We draw tangent lines to CC at the points **A** and **B**. The lines, together with the circle CC, partition the part of the plane outside CC into three distinct regions. Supposing **D** to represent a point chosen that is outside CC, the location of the point **D** in one of the three regions determines the new definition of the set S. The set S can consist of either {**B, D**}, {**A, D**}, or {**A, B, D**}. In

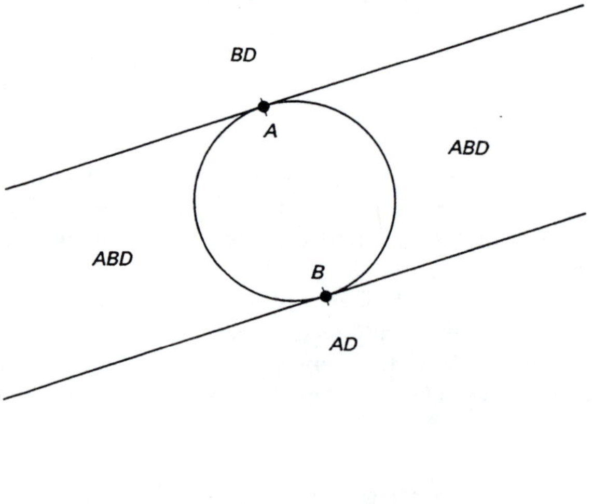

Figure 4.16 Constructing the new defining set when $S = \{A, B\}$.

case the point **D** lies "outside" the two tangent lines, the points **A**, **B**, and **D** form an obtuse triangle and a two-point circle (based on {**A**, **D**} or {**B**, **D**}) is the minimum covering circle for the points **A**, **B**, and **D**. Otherwise, the points **A**, **B**, and **D** form an acute triangle and the minimum covering circle is based on all three points.

Figure 4.17 shows the means of constructing the new defining set S that is used in the procedure named FourPoints. Currently, we have points **A**, **B**, and **C** constituting the set S, and a circle CC passing through the three points of S. For each of the three points **A**, **B**, and **C** we construct corresponding points **A**′, **B**′, and **C**′, with each on the opposite side of the circle from its corresponding point. Next we partition that part of the plane outside the circle into regions as follows. Draw a line segment with one endpoint at **C** through the point **B**′, and remove the part of the line segment joining **B**′ and **C**. Similarly, draw a line segment with one endpoint at **B** through the point **C**′ and remove the part of the line joining **B** and **C**′. Likewise, we use the points **A** and **B**′ and **B** and **A**′ to construct two line segments, and we use the points **A** and **C**′ and **C** and **A**′ to construct two line segments. The result, shown in Figure 4.17, is to partition the part of the plane outside the circle into six regions. Supposing **D** to be a point chosen that is outside the circle, depending on which of the six regions contains the point **D**, the new defining set will be, as shown, one of the six following sets: {**B**, **C**, **D**}, {**C**, **D**}, {**A**, **C**, **D**}, {**A**, **D**}, {**A**, **B**, **D**}, {**B**, **D**}. The rationale for this construction is as follows. If the point **D** is "outside" the **AC**′ line segment, the **ACD** triangle will have an obtuse angle at **A**. If at the same time it is "outside" the **BC**′

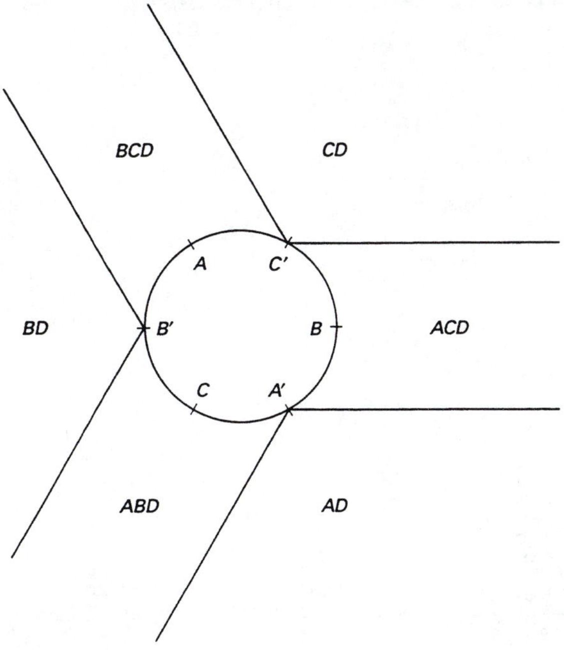

Figure 4.17 Constructing the new defining set when $S = \{A, B, C\}$.

line segment, the **BCD** triangle will have an obtuse angle at **B**, and the minimum covering circle for **A**, **B**, **C**, and **D** will be the **CD** circle, as indicated. In the region in which **B**, **C**, and **D** define the minimum covering circle, note that the **BCD** triangle is acute by construction and that the **BCD** circle contains **A**. In case the point **D** lies in more than one of the regions, any one of the regions in which it lies can be used to determine the new defining set S. Note that Figure 4.16 is the limiting case of Figure 4.17 as the points **B** and **C** become identical.

It is perhaps simplest to motivate the algorithm by considering an example. Suppose that there are nine existing facilities with locations $\mathbf{P}_1$ to $\mathbf{P}_9$ given, respectively, by the following points: (0.0, 1.75), (5.75, 3.50), (6.75, 8.00), (8.50, 1.50), (8.00, 0.75), (3.50, 9.75), (9.50, 8.50), (9.00, 0.50), and (5.25, 1.00). The nine points, as well as the minimax location, (x^*, y^*), are shown in Figure 4.18. The steps the algorithm goes through are shown in Table 4.8. The first procedure used is TwoPoints, and points 1 and 9 define a circle with a radius of 2.652. Point 2 lies outside the circle, so the algorithm applies the procedure TwoPointsPlus, obtaining different defining points, until the points 1, 3, and 4 are determined to form an acute triangle in TwoPointsPlus. A circle is constructed through these three points with a radius of 4.803. Not all the points lie within this circle, so the algorithm applies the procedure FourPoints, choosing point 6 as an outside point. The algorithm continues in this fashion until finally the points 1, 7, and 8 define a circle with radius 5.844 that encloses all the points. The center of this last circle, which is the minimax location, is the point (5.006, 4.765).

Each time the algorithm constructs a circle, the circle is defined by either two

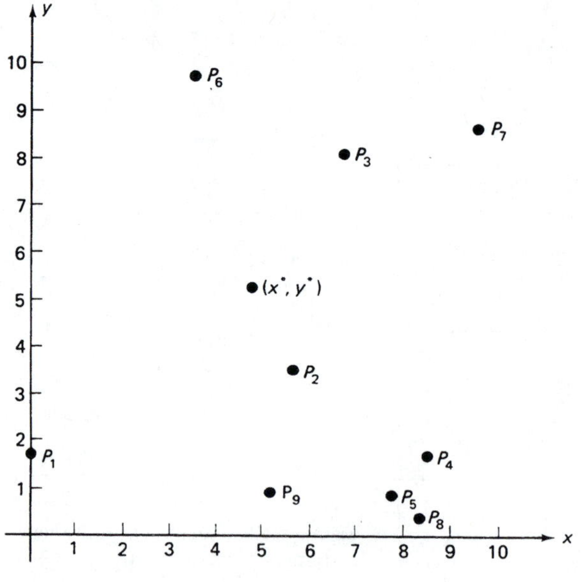

Figure 4.18 Plot of example solved using the Elzinga–Hearn algorithm.

Minimum Covering Circle Algorithm

```
BEGIN
PROCEDURE TwoPoints;
   BEGIN
      Let the two points in S define the diameter of CC
   END;

PROCEDURE TwoPointsPlus;
   {Figure 4.16 illustrates an alternative approach}
   BEGIN
      IF the three points of S define a right or obtuse triangle
         THEN
            let the two points opposite the right or obtuse triangle
            define the diameter of the circle and S
         ELSE
            construct the circle CC through all three points in S
   END;

PROCEDURE FourPoints;
   BEGIN
      Construct; CC covering the four points in S as illustrated in Figure 4.17
   END;

PROCEDURE Construct the minimum covering circle of S;
   BEGIN
      IF |S| = 2
         THEN
            TwoPoints
         ELSE IF |S| = 3
            THEN
               TwoPointsPlus
         ELSE
            FourPoints
   END;

BEGIN {main program}
   Choose any two points to constitute S;
   TwoPoints;
   WHILE not every point is in CC DO
      BEGIN
         Discard from S those points not lying on CC;
         Choose any point D outside CC, and augment the set S by D;
         Construct the minimum covering circle of S;
      END
   {the current CC is optimal and its center is the minimax location}
END.
```

TABLE 4.8 EXAMPLE OF THE CIRCLE COVERING ALGORITHM[a]

Procedure called	Defining points	Outside point	Radius
TwoPoints	1, 9		2.652
TwoPoints	1, 9	2	
TwoPointsPlus	1, 9, 2		
TwoPoints	1, 2	3	3.005
TwoPointsPlus	1, 2, 3		
TwoPoints	1, 3	4	4.600
TwoPointsPlus	1, 3, 4		4.803
FourPoints	1, 3, 4	6	
TwoPointsPlus	1, 4, 6		5.200
FourPoints	1, 4, 6	5	
TwoPointsPlus	1, 6, 5		5.247
FourPoints	1, 6, 5	7	
TwoPointsPlus	1, 5, 7		
TwoPoints	1, 7	8	5.827
TwoPointsPlus	1, 7, 8		5.844

[a] Radii rounded to three significant digits to right of decimal point.

or three points. It is known that each new circle has a radius that is strictly greater than that of the previous circle, so the algorithm never examines any combination of two points or three points more than once. If there are m points, the number of possible choices of two points is m-choose-2, while the number of possible choices of three points is m-choose-3, so that an upper bound on the number of circles constructed is m-choose-2 + m-choose-3, guaranteeing the algorithm is finite. At some iteration the constructed circle must contain all the points, in which case the algorithm terminates.

Table 4.8 illustrates all the aspects of the algorithm, and we recommend that you use the algorithm to solve completely the example given. We note that the algorithm can be considered to be a dual-based algorithm, since it computes a sequence of increasingly larger radii, with only the last one being the minimum objective function value of the original problem and with only the last constructed circle containing all of the points. Computational experience with the algorithm indicates that a random choice of any two initial points to constitute S outperforms heuristic starting procedures such as choosing two most distant points as the initial set S.

We now consider briefly an algebraic approach to the circle covering problem, which is valid in any number of dimensions. With $\mathbf{X}$ and $\mathbf{P}_i$ denoting the locations of the new facility and existing facility i, respectively, the Euclidean distance between the two locations is the square root of the following term (where the superscript T denotes the transpose operation):

$$(\mathbf{X} - \mathbf{P}_i)^T(\mathbf{X} - \mathbf{P}_i)$$

Hence an equivalent version of the circle covering problem is as follows:

$$\text{minimize} \quad u$$

subject to

$$(\mathbf{X} - \mathbf{P}_i)^T(\mathbf{X} - \mathbf{P}_i) \leq u, \qquad i = 1, \ldots, m$$

Because

$$(\mathbf{X} - \mathbf{P}_i)^T(\mathbf{X} - \mathbf{P}_i) = \mathbf{X}^T\mathbf{X} - 2\mathbf{P}_i^T\mathbf{X} + \mathbf{P}_i^T\mathbf{P}_i$$

an equivalent way to write the constraints is as follows:

$$\mathbf{X}^T\mathbf{X} - 2\mathbf{P}_i^T\mathbf{X} + \mathbf{P}_i^T\mathbf{P}_i \leq u, \qquad i = 1, \ldots, m$$

But now if we make the following change of variables,

$$v = \mathbf{X}^T\mathbf{X} - u$$

we obtain the following equivalent version of the problem:

$$\text{minimize} \quad \mathbf{X}^T\mathbf{X} - v$$

subject to

$$2\mathbf{P}_i^T\mathbf{X} - v \geq \mathbf{P}_i^T\mathbf{P}_i, \qquad i = 1, \ldots, m$$

The latter problem is a quadratic programming problem with a convex objective function and linear constraints, and thus is solvable by most quadratic programming algorithms.

4.4.3 Minimax Location Problems with Rectilinear Distances

The problem we consider now is one of finding a new facility location that will minimize the following function:

$$g(x,y) = \max\{w_i[|x - a_i| + |y - b_i|] + h_i\text{: } i = 1, \ldots, m\}$$

As a possible example of the problem, suppose that (x, y) is the location of a "convenience" center and that "users" of the center are located at the existing facility locations, the points (a_1, b_1) through (a_m, b_m). User i requires a time of h_i to prepare to go to the center and then travels to the center at a time per unit distance of w_i, so that $w_i[|x - a_i| + |y - b_i|] + h_i$ is the total time to prepare to go to the center and then go there. The center is to be located so that the maximum such time for any user will be minimized. Alternatively, consider the problem of locating an ambulance, with the existing facilities being the locations of possible accidents and h_i being the travel distance from existing facility i to the nearest hospital. In this case, with all the weights set equal to 1, $|x - a_i| + |y - b_i| + h_i$ could be the total ambulance travel distance, the travel distance for the ambulance from its permanent location (x, y) to (a_i, b_i) plus the travel distance to the nearest hospital. We want to find an ambulance location so as to minimize the maximum such distance. With these two examples you should be able to construct other examples yourself; you might consider the problem of locating a fire extinguisher within a building and ask yourself how the model above could apply.

An alternative but equivalent formulation of the problem of minimizing $g(x, y)$ is the following one:

$$\text{minimize} \quad z$$

subject to

$$w_i[|x - a_i| + |y - b_i|] + h_i \leq z, \qquad i = 1, \ldots, m$$

Because all the weights are positive, we can also write the problem as follows:

$$\text{minimize} \quad z$$

subject to

$$|x - a_i| + |y - b_i| \leq \frac{z - h_i}{w_i}, \qquad i = 1, \ldots, m$$

The constraints of the latter formulation state that the existing facility location (a_i, b_i) is to be in a diamond with center (x, y) and radius $(z - h_i)/w_i$ for $i = 1, \ldots, m$. Let us denote the problem with all unit weights and all zero addends by UP, which represents "unweighted problem." Similarly, we let WPA denote the weighted problem with some nonzero addends, and we let UPA denote the unweighted problem with some nonzero addends. For UP, the simplest problem, we conclude that the problem is one of finding a diamond of minimum radius that will contain all the existing facility locations. Hence we obtain the diamond covering problem we referred to in Section 4.4.1.

Let us use geometry to solve the diamond covering problem, UP. Figure 4.19 shows a plot of the following existing facility locations: (9, 13), (9, 18), (10, 7), (14, 15), (15, 15), and (16, 6). We solve UP by constructing a smallest diamond that contains all the locations. Equivalently, we can construct a smallest square, with each side making an angle of ±45 degrees with an axis. We can solve the latter problem by first constructing a smallest rectangle with each side making an angle of ±45 degrees with an axis and then extending the shorter edges (if necessary) to obtain a square.

Figure 4.19 shows the smallest rectangle, whose four edges pass through the points (15, 15), (16, 6), (10, 7), and (9, 18). Obviously, the rectangle is not a square, so we must extend the shorter edges to obtain a square. One such extension is illustrated by the dotted lines, which, together with the rectangle, provide a square whose center is at the point (13.5, 13). The square we thus obtain is a diamond of smallest radius containing all the points plotted. Another such extension of the rectangle is the one illustrated by the dashed lines, which, together with the rectangle, provide a square whose center is at the point (10.5, 10). Hence we get another diamond of minimum radius containing all the plotted points. Each diamond has a radius of 9.5, which is the minimum objective function value. We make one more observation about the example: Any diamond with a radius of 9.5 whose center is

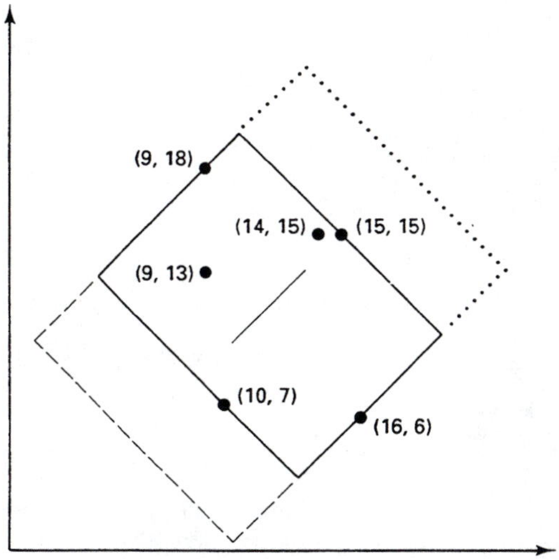

Figure 4.19 Diamond covering problem example.

on the line segment L joining (10.5, 10) and (13.5, 13) also covers all the points and hence is also optimal. Thus we can conclude that every location on the line segment L is an optimal location, and the minimum objective function value is 9.5.

We now give an approach to solve the more general problem UPA. This approach is based on the fact, illustrated in Figure 4.20, that the inequality

$$|x - a_i| + |y - b_i| \leq r_i \equiv z - h_i$$

is equivalent to the following four inequalities:

$$x - a_i + y - b_i \leq r_i$$

$$x - a_i - y + b_i \leq r_i$$

$$-x + a_i - y + b_i \leq r_i$$

$$-x + a_i + y - b_i \leq r_i$$

We can use this fact to transform UPA to a linear program which can be solved as follows. Compute the following numbers:

$$c_1 = \min\{a_i + b_i - h_i: i = 1, \ldots, m\}$$

$$c_2 = \max\{a_i + b_i + h_i: i = 1, \ldots, m\}$$

$$c_3 = \min\{-a_i + b_i - h_i: i = 1, \ldots, m\}$$

$$c_4 = \max\{-a_i + b_i + h_i: i = 1, \ldots, m\}$$

$$c_5 = \max\{c_2 - c_1, c_4 - c_3\}$$

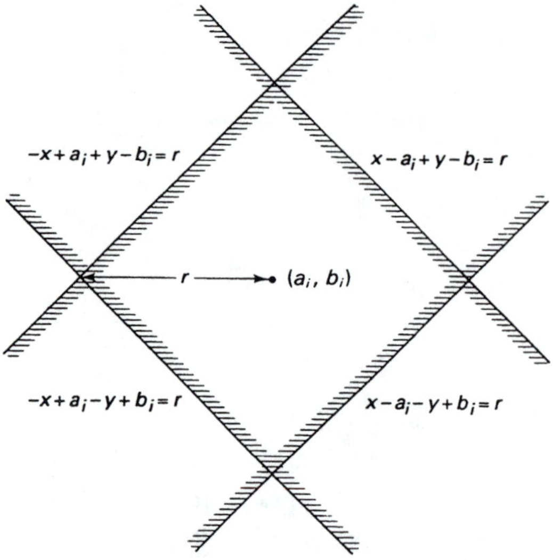

Figure 4.20 Diamond with center (a_i, b_i) and radius r as the intersection of four half-spaces.

The minimum objective function value is $c_5/2$, and the minimax locations are the locations on the line segment L joining the following two points:

$$(x_1^*, y_1^*) = \tfrac{1}{2}(c_1 - c_3, c_1 + c_3 + c_5)$$

$$(x_2^*, y_2^*) = \tfrac{1}{2}(c_2 - c_4, c_2 + c_4 - c_5)$$

If you apply this solution procedure to the example of Figure 4.19, you will find that the lines $x + y = c_1 = 17$ and $x + y = c_2 = 30$ pass through the points $(10, 7)$ and $(15, 15)$, respectively, thus defining two edges of the smallest rectangle containing all the points. Similarly, the lines $y - x = c_3 = -10$ and $y - x = c_4 = 9$ pass through the points $(16, 6)$ and $(9, 18)$, respectively, thus defining the other two edges of the smallest rectangle. Hence computing the numbers c_1 through c_4 corresponds to finding the four edges of the rectangle. For this example, $c_5/2 = 9.5$ corresponds to computing the radius of a smallest diamond containing all the points. Finally, you will find that the points (x_1^*, y_1^*) and (x_2^*, y_2^*) are given by $(13.5, 13)$ and $(10.5, 10)$, respectively.

Consider the construction of level lines for the function $g(x, y)$ of the problem WPA. To construct level lines we construct level sets; the boundaries of the level sets are the level lines. Level lines are of interest for exactly the same reasons we discussed earlier for the minisum problems; they allow us to evaluate easily locations other than the optimal locations [those that minimize the function $g(x, y)$]. Let us denote a level set of $g(x, y)$ of boundary value z by $S(z)$, so that $S(z) = \{(x, y): g(x, y) \leq z\}$. We construct $S(z)$ as follows. Given a value of z of interest, we first compute the following numbers:

$$c_1(z) = \min\left\{a_i + b_i + \frac{z - h_i}{w_i}: i = 1, \ldots, m\right\}$$

$$c_2(z) = \max\left\{a_i + b_i + \frac{-z + h_i}{w_i}: i = 1, \ldots, m\right\}$$

$$c_3(z) = \min\left\{-a_i + b_i + \frac{z - h_i}{w_i}: i = 1, \ldots, m\right\}$$

$$c_4(z) = \max\left\{-a_i + b_i + \frac{-z + h_i}{w_i}: i = 1, \ldots, m\right\}$$

The level set $S(z)$ is then as follows:

$$S(z) = \{(x, y): c_2(z) \leq x + y \leq c_1(z), c_4(z) \leq -x + y \leq c_3(z)\}$$

If at least one of the inequalities $c_2(z) \leq c_1(z), c_4(z) \leq c_3(z)$ does not hold, you have chosen a value of z that is smaller than the minimum value of $g(x, y)$ and the level set will be empty; thus you must pick another value of z which is large enough so that both of the inequalities hold. Supposing the level set to be nonempty, you should be able to see that the level set is a rectangle, with two parallel sides making a +45-degree angle with the x-axis and the other two parallel sides making a −45-degree angle with the x-axis. The vertices of the level set, starting at the top corner and proceeding clockwise, are as follows:

$$\mathbf{V}_1(z) = \tfrac{1}{2}(c_1(z) - c_3(z), c_1(z) + c_3(z))$$

$$\mathbf{V}_2(z) = \tfrac{1}{2}(c_1(z) - c_4(z), c_1(z) + c_4(z))$$

$$\mathbf{V}_3(z) = \tfrac{1}{2}(c_2(z) - c_4(z), c_2(z) + c_4(z))$$

$$\mathbf{V}_4(z) = \tfrac{1}{2}c_2(z) - c_3(z), c_2(z) + c_3(z))$$

Hence you can plot a level set of value z by first choosing z, computing $c_1(z)$ through $c_4(z)$, checking to be sure that the inequalities $c_2(z) \leq c_1(z), c_4(z) \leq c_3(z)$ are satisfied by the z you have chosen, computing $\mathbf{V}_1(z)$ through $\mathbf{V}_4(z)$ and plotting the four points, and then constructing lines joining $\mathbf{V}_1(z)$ and $\mathbf{V}_2(z)$, $\mathbf{V}_2(z)$ and $\mathbf{V}_3(z)$, $\mathbf{V}_3(z)$ and $\mathbf{V}_4(z)$, and $\mathbf{V}_4(z)$ and $\mathbf{V}_1(z)$. The rectangle the lines enclose is the level set of value z, and its boundary is the level line of value z.

We illustrate the procedure for constructing level sets using the example of Figure 4.19, for which all the weights are 1 and the addends are zero. Recall that we earlier computed a minimum objective function value of 9.5 for the example of Figure 4.19. Thus if we construct $S(9.5)$ it should agree with the line segment L joining (10.5, 10) and (13.5, 13). Using the formulas we have developed, we find $c_1(9.5) = 26.5$, $c_2(9.5) = 20.5$, $c_3(9.5) = -0.5$, and $c_4(9.5) = -0.5$. It now follows that $\mathbf{V}_1(9.5) = (13.5, 13) = \mathbf{V}_2(9.5)$ and $\mathbf{V}_3(9.5) = (10.5, 10) = \mathbf{V}_4(9.5)$, and thus the level set, which is degenerate, is in fact identical with the line L. Now suppose that we choose a value of $z = 11.5$; we find $c_1(11.5) = 28.5$, $c_2(11.5) = 18.5$,

$c_3(11.5) = 1.5$, and $c_4(11.5) = -2.5$. It then follows that the points $\mathbf{V}_1(11.5)$ through $\mathbf{V}_4(11.5)$ are given, respectively, by (13.5, 15), (15.5, 13), (10.5, 8), and (8.5, 10). We suggest that you go through the computations involved to be sure you understand what we have done. Here is a question you should be able to answer. What would the level set be if we choose a value of z strictly less than 9.5?

4.4.4 Minimax Problems with Tchebychev and Rectilinear Distances

In the preceding section we saw that the diamond covering problem could be interpreted, given a 45-degree rotation, as a square covering problem, and that this interpretation led to a simple solution procedure for the diamond covering problem. What we do now is exploit this discovery in a systematic way. As a result, we will obtain an efficient means of solving WPA, the minimax weighted rectilinear distance problem with addends. Also, we shall find that the results of this section will be useful in Chapter 5, in which we consider certain warehouse layout problems involving cranes.

Consider two points **X** and **Y** in the plane. Suppose that **X** and **Y** are the endpoints of the hypotenuse H of a right triangle, with the other two sides of the triangle, denoted by A and B, being parallel to the horizontal and vertical axes, respectively. We have seen earlier that the length of the hypotenuse H is the Euclidean distance between **X** and **Y**, while the sum of the lengths of sides A and B is the rectilinear distance between **X** and **Y**. We now introduce a new distance, called the *Tchebychev distance* between **X** and **Y**, defined to be the maximum of the lengths of sides A and B. In other words, if A is longer than B, then A is the Tchebychev distance between **X** and **Y**, while if A is not longer than B, then B is the Tchebychev distance between **X** and **Y**. We denote the Tchebychev distance between **X** and **Y** by $t(\mathbf{X}, \mathbf{Y})$. Thus if $\mathbf{X} = (x_1, x_2)$, and $\mathbf{Y} = (y_1, y_2)$, then

$$t(\mathbf{X}, \mathbf{Y}) = \max\{|x_1 - y_1|, |x_2 - y_2|\}$$

What does a contour set of Tchebychev distance look like? Consider the set of all points **X** whose Tchebychev distance from the origin is at most 1, that is, the set of all points $\mathbf{X} = (x_1, x_2)$ satisfying $t(\mathbf{X}, \mathbf{0}) \leq 1$, or, equivalently, satisfying $\max\{|x_1|, |x_2|\} \leq 1$. The latter inequality is equivalent to $|x_1| \leq 1$ and $|x_2| \leq 1$. But these last two inequalities are in turn equivalent to

$$-1 \leq x_1 \leq 1 \quad \text{and} \quad -1 \leq x_2 \leq 1$$

Hence the set of all points **X** whose Tchebychev distance from the origin is at most 1 is a square with its center at the origin and each side of length 2. This square is the Tchebychev analog of a circle with center at the origin and radius 1 and the Tchebychev analog (for rectilinear distance) of a diamond with center at the origin and radius 1. Of course, if we rotate a diamond by 45 degrees, we obtain a square, a result that should give you a good clue about the relationship between Tchebychev and rectilinear distances.

Let us now explore the relationship between Tchebychev and rectilinear distances. It is convenient to introduce the following linear transformation, which we denote by $Q(x,y)$:

$$Q(x,y) = (x,y)\begin{bmatrix} 1 & -1 \\ 1 & 1 \end{bmatrix} = (x + y, -x + y)$$

You should be able to verify that the inverse transformation, denoted by $Q^{-1}(u,v)$, is given by

$$Q^{-1}(u,v) = (u,v)\begin{bmatrix} \frac{1}{2} & \frac{1}{2} \\ -\frac{1}{2} & \frac{1}{2} \end{bmatrix} = \tfrac{1}{2}(u - v, u + v)$$

To illustrate what the transformation Q does, suppose that we have a diamond D with center the origin and radius 1, so that the vertices of D are the points $(0,1)$, $(-1,0)$, $(0,-1)$, and $(1,0)$. If we use the transformation Q to evaluate each of the latter four points, we obtain the points $(1,1)$, $(-1,1)$, $(-1,-1)$, and $(1,-1)$, respectively. But the latter four points are the vertices of a square, say S, with center at the origin, and a side length of 2 (a "radius" of 1). In effect, what the transformation Q does is to rotate the diamond D by 45 degrees to obtain a square, as well as effect an "expansion" of scale. We obtain an expansion of scale in the sense that while the Euclidean distance between each of the vertices and the origin is 1, the Euclidean distance between each of the rotated points and the origin is 2. (As an exercise, you ought to determine what we obtain if we apply the inverse transformation to the vertices of the square S.)

Another interesting aspect of this example is that the rectilinear distance between any two vertices of D, namely 2, is the same as the Tchebychev distance between the corresponding transformed points, the vertices of the square S. This result is no coincidence. Given any points $\mathbf{X}$ and $\mathbf{Y}$ in the plane, let $r(\mathbf{X},\mathbf{Y})$ denote the rectilinear distance between $\mathbf{X}$ and $\mathbf{Y}$. If we compute points $\mathbf{X}'$ and $\mathbf{Y}'$ using the equations $\mathbf{X}' = Q(\mathbf{X})$ and $\mathbf{Y}' = Q(\mathbf{Y})$, it is known that

$$r(\mathbf{X},\mathbf{Y}) = t(\mathbf{X}',\mathbf{Y}')$$

That is, the rectilinear distance between $\mathbf{X}$ and $\mathbf{Y}$ is the same as the Tchebychev distance between the transformed points $\mathbf{X}'$ and $\mathbf{Y}'$. Equivalently, given any points $\mathbf{X}'$ and $\mathbf{Y}'$ in the plane, if we compute points $\mathbf{X}$ and $\mathbf{Y}$ using the equations

$$\mathbf{X} = Q^{-1}(\mathbf{X}') \quad \text{and} \quad \mathbf{Y} = Q^{-1}(\mathbf{Y}')$$

we conclude that

$$t(\mathbf{X}',\mathbf{Y}') = r(\mathbf{X},\mathbf{Y})$$

The consequence of the two equations above is that we can transform a planar location problem involving rectilinear distances into an equivalent problem involving Tchebychev distances, and vice versa. Hence we obtain an equivalence between planar location problems involving Tchebychev and rectilinear distances. This equivalence is useful since it is often the case that one problem is easier to analyze than the other. Further, many planar location problems involving rectilinear distance are well studied, and thus we may well be able to apply existing results for rectilinear distance location problems (e.g., median conditions) to analogous results for Tchebychev distance location problems.

To make matters more concrete, consider the diamond covering problem illustrated in Figure 4.19. Figure 4.21 shows the points obtained by applying the transformation Q to the following existing facility locations: (9, 13), (9, 18), (10, 7), (14, 15), (15, 15), and (16, 6). The corresponding transformed points we obtain are (22, 4), (27, 9), (17, −3), (29, 1), (30, 0), and (22, −10), respectively. Now consider the problem of enclosing the transformed points in a smallest square that has each side parallel to an axis; this problem should be equivalent to the diamond covering problem that Figure 4.19 illustrates. To construct a smallest square we first construct a smallest covering rectangle; this rectangle has its left and right edges passing through the points $u_1 = 17$ and $u_1 = 30$, respectively. Similarly, the bottom and top edges lie on the lines $u_2 = -10$ and $u_2 = 9$, respectively. Since the horizontal edges are shorter than the vertical edges, we must expand the horizontal edges to be of length 19 in order to obtain a square. One such expansion is indicated by the dotted lines, while another such expansion is indicated by the dashed lines. You can see that the center of the first square we obtain (the one with a dotted right edge) is the point (26.5, −0.5). Similarly, the center of the second square we obtain is the point (20.5, −0.5). Each of these two squares is a smallest containing square. Further, each

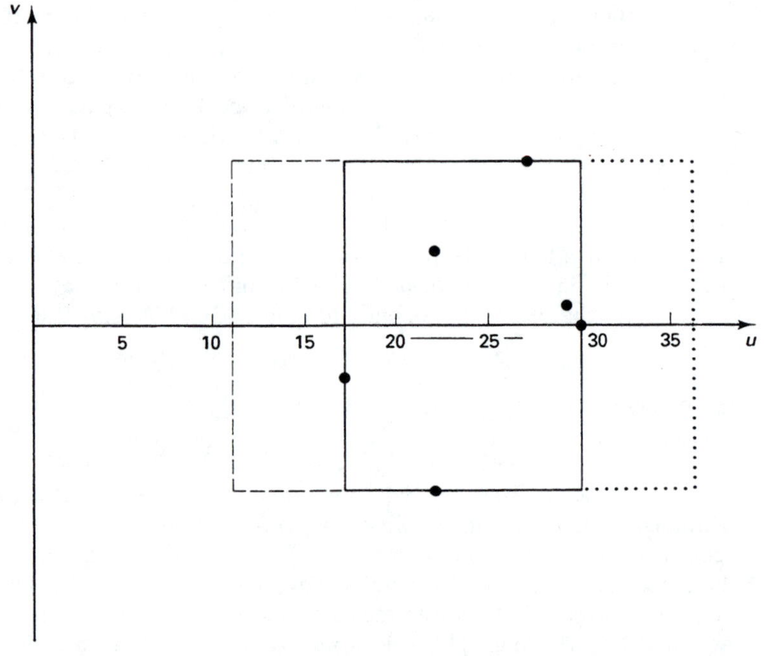

Figure 4.21 Square covering problem example: Figure 4.19 transformed.

square with side length 19 whose center is on the line segment joining (20.5, −0.5) and (26.5, −0.5) is also a smallest containing square. Thus the set of optimal locations for the square covering problem example is the line segment joining (20.5, −0.5) and (26.5, −0.5). We can obtain the set of optimal locations for the diamond covering problem by applying the inverse transformation to the line segment joining (20.5, −0.5) and (26.5, −0.5): we obtain the line segment joining the points (10.5, 10) and (13.5, 13), respectively. Hence we have verified the conclusion of the previous section that the optimal locations for the diamond covering problem example are the points on the line segment joining the points (10.5, 10) and (13.5, 13); further, the minimum objective function is the radius of a minimum covering diamond, 19/2 = 9.5. Therefore, we have illustrated how we can use the transformation Q to go from a diamond covering problem to an equivalent square covering problem, and vice versa.

Let us now generalize from our experience with the example of Figure 4.21 and consider converting the general minimax problem with rectilinear distances, denoted by WPA, into an equivalent problem with Tchebychev distances. Recall that WPA is the problem of minimizing the function $g(x, y)$, where

$$g(x,y) = \max\{w_i[|x - a_i| + |y - b_i|] + h_i: i = 1, \ldots, m\}$$

Let (u, v) be the result of applying the transformation Q to the point (x, y), and let (α_i, β_i) be the result of applying the transformation Q to the point (a_i, b_i). We know that

$$\begin{aligned} w_i[|x - a_i| + |y - b_i|] + h_i &= w_i \max[|u - \alpha_i|, |v - \beta_i|] + h_i \\ &= \max[w_i|u - \alpha_i| + h_i,\ w_i|v - \beta_i| + h_i] \end{aligned}$$

We can thus conclude that

$$g(x,y) = \max\{\max[w_i|u - \alpha_i| + h_i,\ w_i|v - \beta_i| + h_i]: i = 1, \ldots, m\}$$

Suppose that we now define the function $g_1(u)$ and $g_2(v)$ as follows:

$$\begin{aligned} g_1(u) &= \max\{w_i|u - \alpha_i| + h_i: i = 1, \ldots, m\} \\ g_2(v) &= \max\{w_i|v - \beta_i| + h_i: i = 1, \ldots, m\} \end{aligned}$$

It then follows that

$$\max\{\max[w_i|u - \alpha_i| + h_i, w_i|v - \beta_i| + h_i]: i = 1, \ldots, m\} = \max\{g_1(u), g_2(v)\}$$

The reason for the latter equality is that regardless of the order in which we compute the maximum of a collection of numbers, we obtain the same result.

If you examine the latter equation, you can see that the term on the left is the same as the term on the right in our most recent equation for $g(x, y)$; hence we obtain a very useful result as follows: Given

$(u, v) = Q(x, y)$

we have

$$g(x,y) = \max\{g_1(u),\ g_2(v)\}$$

The consequence of our result is that we can minimize $g(x, y)$ by solving two independent minimization problems as follows:

1. We minimize $g_1(u)$ and obtain a minimizing point, say u^*.
2. Next, we minimize $g_2(v)$ and obtain a minimizing point, say v^*.
3. We can then apply the inverse transformation to (u^*, v^*) to obtain a point, say (x^*, y^*), and conclude that (x^*, y^*) minimizes $g(x, y)$.
4. Further, the minimum value of $g(x, y)$ is equal to $\max\{g_1(u^*), g_2(v^*)\}$.

In effect, we can decompose our original minimization problem into two independent minimization problems, solve these two problems, put the solutions into a vector, apply the inverse transformation of Q to the vector, and obtain a solution to our original problem. Further, each of the two independent minimization problems is a location problem on the line. Also, the two problems have exactly the same structure, so any procedure we develop to solve one problem can also be used to solve the other problem.

As an example, suppose that we consider again the problem of Figure 4.19, a problem for which all the weights are 1 and the addends are zero. Recall that when we apply the transformation Q to the existing facility locations, the corresponding transformed points we obtain are (22, 4), (27, 9), (17, −3), (29, 1), (30, 0), and (22, −10), respectively. We then conclude that

$$g_1(u) = \max\{|u - \alpha_i|: i = 1, \ldots, 6\}$$

where α_1 to α_6 are given by the points 22, 27, 17, 29, 30, and 22. Similarly, we conclude that

$$g_2(v) = \max\{|v - \beta_i|: i = 1, \ldots, 6\}$$

where β_1 to β_6 are given by the points 4, 9, −3, 1, 0, and −10.

Continuing with the example, consider minimizing the functions $g_1(u)$ and $g_2(v)$. Another way to minimize the former function is as follows: Minimize z_1 subject to $|u - \alpha_i| \le z_1$ for each i. The latter constraint simply states that each point α_i must lie within an interval with center u and length $2z_1$; that is, each point α_i must lie within an interval with center u and "radius" z_1. In other words, we want to construct an interval of minimum length containing each point α_i and then choose its midpoint to be u^*; but this problem is particularly simple to solve: An interval of minimum length is [17, 30], having midpoint $u^* = 23.5$ and radius 6.5. Similarly, to minimize the function $g_2(v)$ we must construct an interval of minimum length containing each point β_i. Such an interval is [−10, 9], having midpoint $v^* = -0.5$ and radius 9.5. Thus we conclude that $g_1(u^*) = g_1(23.5) = 6.5$ and $g_2(v^*) = g_2(-0.5) = 9.5$. To obtain a point (x^*, y^*) minimizing $g(x, y)$, we compute $(x^*, y^*) = Q^{-1}(u^*, v^*) = Q^{-1}(23.5, -0.5) = (12, 11.5)$. Because $(12, 11.5) = (\frac{1}{2})(10.5, 10) + (\frac{1}{2})(13.5, 13)$, we conclude that (12, 11.5) is on the line segment joining the points (10.5, 10) and (13.5, 13), which we know from our earlier analysis is the set of all optimal locations.

To find other optimal locations using this approach, we first observe that any (u, v) for which $g_1(u) \leq 9.5$, and for which $v = -0.5$ will minimize $\max\{g_1(u), g_2(v)\}$. Hence we can let u satisfy $20.5 \leq u \leq 23.5$ and $v = -0.5$ and obtain minimizing points. Applying the inverse transformation of Q to the points $(20.5, -0.5)$ and $(23.5, -0.5)$ gives $(10.5, 10)$ and $(13.5, 13)$ as minimax locations, which agrees with our earlier analysis.

With the insight that we have obtained from the example above, let us consider the general problem of minimizing

$$g_1(u) = \max\{w_i|u - \alpha_i| + h_i: i = 1, \ldots, m\}$$

A procedure to minimize $g_1(u)$ is as follows:

1. For $1 \leq i < j \leq m$, compute γ_{ij} as follows:

$$\gamma_{ij} = \frac{w_i w_j|\alpha_i - \alpha_j| + w_j h_i + w_i h_j}{w_i + w_j}$$

2. Compute

$$\gamma_{pq} = \max\{\gamma_{ij}: 1 \leq i < j \leq m\}$$

3. Compute

$$h_r = \max\{h_i: 1 \leq i \leq m\}$$

4. If $\gamma_{pq} \leq h_r$, take $u^* = \alpha_r, g_1(u^*) = h_r$, and stop. Otherwise (when $\gamma_{pq} > h_r$), take u^* to be the unique point between α_p and α_q satisfying

$$|u^* - \alpha_p| = \frac{\gamma_{pq} - h_p}{w_p}$$

and stop.

We can use exactly the same procedure to minimize $g_2(v)$, simply by replacing each α by β and replacing u^* by v^*.

Once we have u^* and v^*, we can find (x^*, y^*) to minimize $g(x, y)$ as follows:

$$(x^*, y^*) = Q^{-1}(u^*, v^*), \qquad g(x^*, y^*) = \max\{g_1(u^*), g_2(v^*)\}$$

We now illustrate the foregoing approach with an example. There are $m = 3$ existing facilities, with facilities 1 to 3 having locations given by $(5, 0)$, $(5, 10)$, and $(10, 10)$, respectively, having weights given by 2, 1, and 2, respectively, and addends given by 0, 12, and 0, respectively. Using the transformation Q, we obtain $Q(5, 0) = (5, -5)$, $Q(5, 10) = (15, 5)$, and $Q(10, 10) = (20, 0)$. Thus

$$g_1(u) = \max\{2|u - 5| + 0, 1|u - 15| + 12, 2|u - 20| + 0\}$$

$$g_2(v) = \max\{2|v - (-5)| + 0, 1|v - 5| + 12, 2|v - 0| + 0\}$$

Figures 4.22 and 4.23 show graphs of the two functions. The figures illustrate several results that are generally true; each function is convex, piecewise linear, and has a unique minimizing point. Let us use our minimization procedure to verify that we obtain the same minimizing points as shown in the figures. When we use the procedure to minimize $g_1(u)$, we find that

$$\gamma_{12} = 14\tfrac{2}{3}, \quad \gamma_{13} = 15, \quad \gamma_{23} = 11\tfrac{1}{3}$$

giving

$$\gamma_{pq} = \gamma_{13} = 15$$

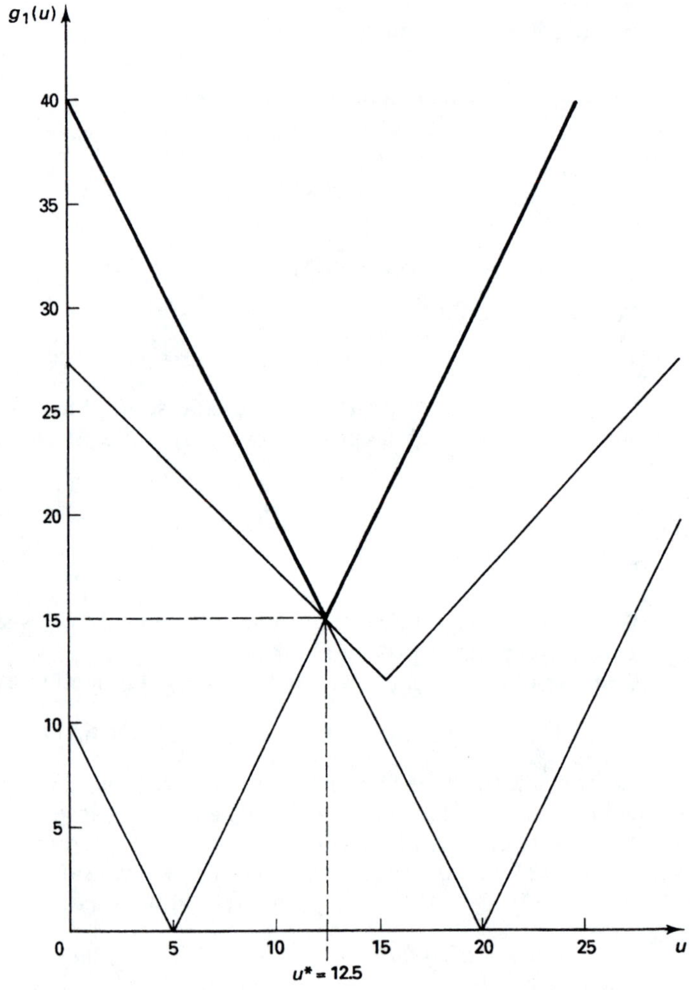

Figure 4.22 Graph of $g_1(u) = \max\{2|u-5|, 1|u-15|+12, 2|u-20|\}$.

Further,

$$h_r = \max\{0, 12, 0\} = 12 = h_2$$

Because $12 < 15$, we conclude u^* is the unique point between $\alpha_1 = 5$ and $\alpha_3 = 15$ satisfying

$$|u^* - \alpha_1| = \frac{\gamma_{13} - h_1}{w_1}$$

That is, $|u^* - 5| = (15 - 0)/2 = 7.5$; so $u^* = 5 + 7.5 = 12.5$. Also, we conclude that $g_1(u^*) = 15$.

When we use the procedure a second time to minimize $g_2(v)$, we find that

$$\gamma_{12} = 14\tfrac{2}{3}, \quad \gamma_{13} = 5, \quad \gamma_{23} = 11\tfrac{1}{3}$$

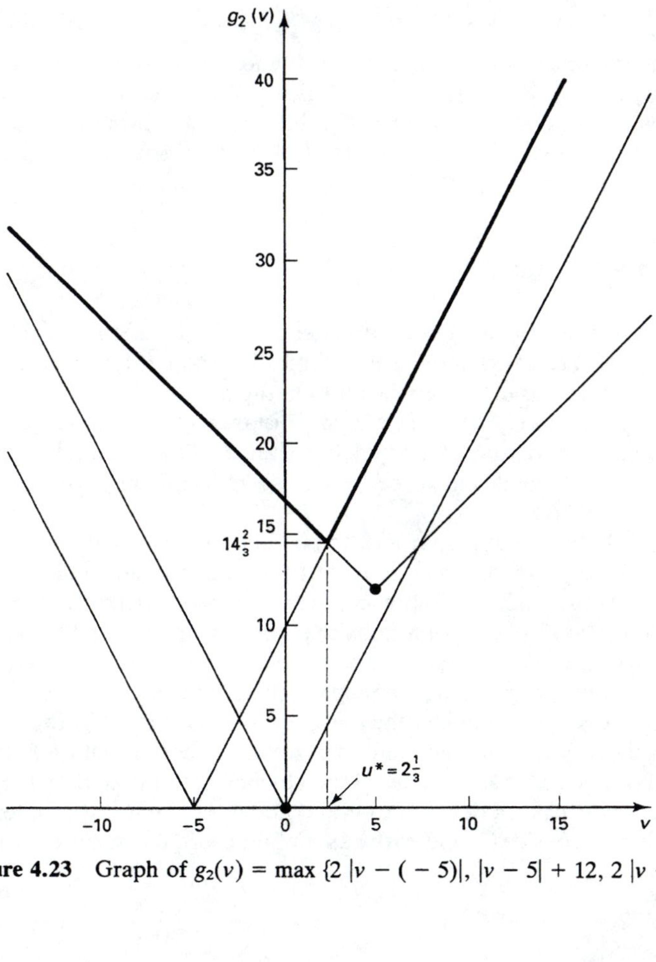

Figure 4.23 Graph of $g_2(v) = \max\{2|v - (-5)|, |v - 5| + 12, 2|v - 0|\}$.

giving

$$\gamma_{pq} = \gamma_{12} = 14\tfrac{2}{3}$$

Further,

$$h_r = \max\{0, 12, 0\} = 12 = h_2$$

Because $14\frac{2}{3} > 12 = h_2$, we conclude that v^* is the unique point between $\beta_1 = -5$ and $\beta_2 = 5$ satisfying

$$|v^* - (-5)| = \frac{14\frac{2}{3} - 0}{2} = 7\tfrac{1}{3}$$

that is, $v^* = 2\frac{1}{3}$. Also, we conclude that $g_2(v^*) = 14\frac{2}{3}$.

Thus the minimum value of $g(x, y)$ is $\max\{15, 14\frac{2}{3}\} = 15$. To find a minimax location, we compute

$$(x^*, y^*) = Q^{-1}(u^*, v^*) = Q^{-1}(25/2, 7/3) = (61/12, 89/12)$$

Other minimax locations can be found by observing that $g(x^*, y^*) = 15 = \max\{g_1(u^*), g_2(v)\}$ for $g_2(v) \leq 15$, which can be seen to be equivalent to having v satisfy $2 \leq v \leq 2.5$. Applying the inverse transformation to $(12.5, 2)$ and to $(12.5, 2.5)$ gives $(5.25, 7.25)$ and $(5, 7.5)$, respectively. Thus each point on the line segment joining $(5.25, 7.25)$ and $(5, 7.5)$ is a minimax location, with a value of 15.

4.5 CONCLUDING REMARKS

Now that we have considered a number of planar location models, let us try to make some general remarks about them. Typically, a planar location problem involves the location of one or more new facilities in the plane, with costs incurred that depend on an appropriately chosen "planar" distance (e.g., rectilinear or Euclidean distance) between the new facility and existing facilities that have known planar locations. The new facility is to be located so as to minimize an appropriately chosen objective function.

We believe that planar location models are principally of value for the insight they provide, and the simplicity of their construction and use, rather than the accuracy with which they represent the problems of interest. Planar location models involve certain basic assumptions that limit their realism. Thus we think that the construction of very detailed and involved planar location models may well be a self-defeating exercise. Consequently, the planar models we have considered are among the simpler models; they are "quick and dirty," in the sense that they are relatively easy to construct and analyze and to obtain data for. If a more detailed model is required than the ones of this chapter, we suggest that it may be appropriate to use a network model or a discrete model, as discussed in later chapters.

Let us consider some of the assumptions often associated with planar location models.

1. A plane is an adequate approximation of a sphere.
2. Any point in the plane is a valid location to consider.
3. The facilities to be located may be idealized as points (and thus have an area of zero).
4. Distances traveled between the facility to be located and the existing facilities may be adequately represented by "planar" distances.
5. Travel costs are directly proportional to the planar distance used, with the constants of proportionality being independent of the values of the distances.
6. Fixed costs can be ignored.
7. There are no associated distribution costs of interest.

Clearly, some of the assumptions above are quite stringent. A multinational corporation would probably have little or no interest in planar models due to assumption 1; rather, the problem should be "regional" or "local" in nature for assumption 1 to be valid.

Assumption 2 is similarly a strong assumption; a model that recommends a location in the middle of a lake, for example, is unlikely to meet with much enthusiasm; what we hope for is that if the location the model suggests is unacceptable, some "nearby" location will be acceptable (e.g., on the shore of the lake).

Assumption 3 may or may not be a stringent assumption, depending on the scope of the model. If the model is being used to determine the location of a machine within a factory, assumption 3 may be quite stringent, and the best we can hope for is that some location which is close to that indicated by the model, perhaps found using a contour line approach, will be acceptable. On the other hand, if the model is of a regional or national scope, assumption 3 will probably cause little concern, but in such a case assumption 1 would be more important.

Assumption 4 is more involved than it appears. Perhaps due to the influence of a first course in geometry, the distance that one immediately thinks of is the Euclidean distance. In some ways this is unfortunate, for it requires only a little reflection to conclude that it is unusual for objects (unless they fly) to travel between any two points in a straight line. Rather, objects must travel on an existing aisle, street, road, or transport network of some sort. Thus the best we can hope for is that the Euclidean distance will be an adequate approximation to the actual travel distance. In cases where the aisle or street network is rectilinear, the use of the rectilinear distance is clearly more appropriate than the Euclidean distance, and it is unfortunate that the former distance is not better known.

Concerning assumption 5, it is easy to draw upon one's own experience to conclude that the farther away a facility of interest is, the less likely one is to use it. Hence simple personal experience suggests that this assumption may be rather stringent: to use a *constant* of proportionality may be to ignore economies of scale.

Perhaps the strongest criticism of planar location models is due to assumption 6: Fixed costs, including such things as site purchase and preparation costs, are likely to be quite substantial and may be a function of the type of facility to be located as well. What we hope for in ignoring fixed costs is that they will be much the same regardless of the location of interest and thus will have no effect on the choice of an optimal location. If, in fact, fixed costs are quite important and vary depending on the location chosen and the type of facility to be located, a mixed-integer programming model deserves careful consideration; the best we could hope for in this case from a planar model would be that it would facilitate a trade-off analysis between travel distances and fixed costs.

Concerning assumption 7, it is often the case that the travel that occurs between a new and an existing facility is, to some extent, controllable, in which case the assumption may be invalid. Especially if a number of new facilities are to be located, the degree to which any new facility interacts with a particular existing facility may be a decision variable, and simple location models typically do not take this fact into account. Thus if distribution aspects are important, it is again probably best to consider a mixed-integer programming model.

In view of the above, it is tempting to conclude that planar location models are of limited usefulness. Such a conclusion, however, ignores the fact that there is almost always a trade-off between model tractability and model realism. Having the most realistic model in the world is of little value if the model is computationally intractable. Further, the more realistic a model is, the more expensive it may be to construct and to obtain data for, in which case there is surely something to be said for a "quick and dirty" model, particularly if the scope of the location project is not large and there are limits on the time and resources available. In fact, in some cases having a very realistic and detailed model can actually put its builder at a disadvantage in trying to explain the model to a client, who may have little or no knowledge of the theory the model relies on, and who may want to see the model provide qualitative insights that are consistent with the client's experience. Of course, such a situation changes over time, as analytical models become more widely used, but a model that is too sophisticated can appear to a client to be a mysterious "black box." The box somehow transforms problem data into a solution, possibly at a high computational cost. High computer costs also have a way of inhibiting sensitivity analysis, which is of major importance in gaining insight into the problem and in gaining confidence in the model. Giving qualitative insights is what planar location models are most likely to do well, and in some cases one may actually want to consider the option of having *both* a planar and a more sophisticated model of the same problem, using the planar model to explain some of the behavior of the more sophisticated model.

4.6 FURTHER READING

The literature on single-facility location problems is quite large. Our list of references, which is by no means comprehensive, lists more than 70 papers.

The earliest relevant paper we know of, by Lamé and Clapeyron, dates from 1829 and is discussed in detail by Franksen and Grattan-Guinness [27], who give a complete translation of the work from French into English. The string model we present is related to this work from 1829.

There is a long history on the median conditions, as traced by Mole [56] and Rosenhead [63], including work by Edgeworth discussed by Bowley [3]; the contour line material is due to Francis [24], generalizing work by Bindschedler and Moore [1].

Weiszfeld's algorithm [70] was an important early contribution to location theory that languished unrecognized for many years; until computers came into general use it was not a computationally practical approach. A discussion of the string model appears in the now classic book by Weber [69]. For convergence analysis for the algorithm, see work by Katz [37, 38] and Ostresh [60]. For interesting related theory, see the references by Kuhn [39–42]. The majority theorem is due to Witzgall [77].

The l_p distance, which we introduce in Chapter 6, generalizes the Euclidean and rectilinear distances we have considered; references with l_p in the title address single-facility location problems with the l_p distance. For work analyzing how well Euclidean travel distances approximate actual network travel distances, see Love and Morris [49, 50].

The efficient location material is based on work by Chalmet et al. [7]; for related work, see Wendell et al. [72] and Juel and Love [36].

There is a very extensive literature on the circle covering problem, going back at least as early as a paper by Sylvester [65], much of which is traced in the paper by Hearn and Vijay [30]; they find the Elzinga–Hearn algorithm [23] to be the fastest one for solving the problem. For the use of computer graphics for the circle covering problem, see Brady and Rosenthal [4]. The material on the related "diamond" covering problem, with rectilinear distances, is based on work by Francis [25].

The discussion of assumptions in Section 4.5 is based on work by Francis et al. [26]. For a complete chapter on the minisum single-facility location problem, see Love et al. [53]. For applications of location theory to public sector problems, see Thisse and Zollers [67].

REFERENCES

1. Bindschedler, A. E., and J. M. Moore, "Optimum Location of New Machines in Existing Plant Layouts," *Journal of Industrial Engineering*, Vol. 12, No. 1, 1961, pp. 41–48.
2. Blumenthal, L. M., and G. E. Wahlin, "On the Spherical Surface of Smallest Radius Enclosing a Bounded Subset of *n*-Dimensional Euclidean Space," *American Mathematical Society Bulletin,* Vol. 47, 1941, pp. 771–777.
3. Bowley, A. L., *F. Y. Edgeworth's Contributions to Mathematical Statistics,* Royal Statistical Society, London, 1928.
4. Brady, S. D., and R. E. Rosenthal, "Interactive Computer Graphical Solutions of Constrained Minimax Location Problems," *AIIE Transactions,* Vol. 12, 1980, pp. 241–248.
5. Burstall, R. M., and R. A. Leaver, "Evaluation of Transport Costs for Alternative Factory Sites: A Case Study," *Operational Research Quarterly,* Vol. 13, 1962, p. 4.

6. Chakraborty, R. K., and P. K. Chaudhuri, "Note on Geometrical Solutions for Some Minimax Location Problems," *Transportation Science,* Vol. 15, 1981, pp. 164–166.
7. Chalmet, L. G., R. L. Francis, and A. Kolen, "Finding Efficient Solutions for Rectilinear Distance Location Problems Efficiently," *European Journal of Operational Research,* Vol. 6, 1981, pp. 117–124.
8. Chandrasekaran, R., and M. J. A. P. Pacca, *Weighted Min-Max Location Problems: Finite and Polynomially Bounded Algorithms,* Report, University of Texas at Dallas, 1980.
9. Charalambous, C., and A. R. Conn, "An Efficient Method to Solve the Minimax Problem Directly," *SIAM Journal of Numerical Analysis,* Vol. 15, 1978, pp. 162–187.
10. Chatelon, J. A., D. W. Hearn, and T. J. Lowe, "A Subgradient Algorithm for Certain Minimax and Minisum Location Problems," *Mathematical Programming,* Vol. 15, 1978, pp. 130–145.
11. Chrystal, G., "On the Problem to Construct the Minimum Circle Enclosing n Given Points in the Plane," *Proceedings of the Edinburgh Mathematical Society,* Vol. 3, 1885, pp. 30–33.
12. Cooper, L., "Location-Allocation Problems," *Operations Research,* Vol. 11, No. 3, 1963, pp. 331–344.
13. Cooper, L., "Heuristic Methods for Location-Allocation Problems," *SIAM Review,* Vol. 6, No. 1, 1964, pp. 37–52.
14. Cooper, L., "Solutions of Generalized Locational Equilibrium Models," *Journal of Regional Science,* Vol. 7, No. 1, 1967, pp. 1–18.
15. Domschke, W., and A. Drexl, *Location and Layout Planning: An International Bibliography, Lecture Notes in Economics and Mathematical Systems,* Vol. 238, Springer-Verlag, Berlin, 1985.
16. Domschke, W., and A. Drexl, *Logistik: Standorte* (in German), R. Oldenbourg Verlag BmbH, Munich, Germany, 1985.
17. Drezner, Z., and G. O. Wesolowsky, "Single Facility l_p-Distance Minimax Location," *SIAM Journal on Algebraic and Discrete Methods,* Vol. 1, 1980, pp. 315–321.
18. Drezner, Z., and G. O. Wesolowsky, "Optimal Location of a Facility Relative to Area Demands," *Naval Research Logistics Quarterly,* Vol. 27, 1980, pp. 199–208.
19. Drezner, Z., and G. O. Wesolowsky, "Single Facility l_p Distance Minimax Location," *SIAM Journal on Algebraic and Discrete Methods,* Vol. 1, 1980, pp. 315–321.
20. Eilon, S., and D. P. Deziel, "Siting a Distribution Center," *Management Science,* Vol. 12, No. 6, 1966, pp. 245–254.
21. Eilon, S., C. D. T. Watson-Gandy, and N. Christofides, *Distribution Management: Mathematical Modelling and Practical Analysis,* Hafner Publishing Company, Inc., New York, 1971.
22. Elzinga, J., and D. W. Hearn, "The Minimum Covering Sphere Problem," *Management Science,* Vol. 19, 1972, pp. 96–104.
23. Elzinga, J., and D. W. Hearn, "Geometrical Solutions for Some Minimax Location Problems," *Transportation Science,* Vol. 6, 1972, pp. 379–394.
24. Francis, R. L., "A Note on the Optimum Location of New Machines in Existing Plant Layouts," *Journal of Industrial Engineering,* Vol. 14, 1963, pp. 57–59.

25. Francis, R. L., "A Geometrical Solution Procedure for a Rectilinear Distance Minimax Location Problem," *AIIE Transactions,* Vol. 4, No. 4, 1971, pp. 328–332.

26. Francis, R. L., L. F. McGinnis, and J. A. White, "Locational Analysis," *European Journal of Operational Research,* Vol. 12, 1983, pp. 220–252.

27. Franksen, O. I., and I. Grattan-Guinness, "The Earliest Contribution to Location Theory? Spatio-Economic Equilibrium with Lamé and Clapeyron, 1829," *Mathematics and Computers in Simulation,* Vol. 31, 1989, pp. 195–220.

28. Haley, K. B., "The Siting of Depots," *International Journal of Production Research,* Vol. 2, 1963, pp. 41–46.

29. Hansen, P., J. Perreur, and J.-F. Thisse, "Location Theory, Dominance, and Convexity: Some Further Results," *Operations Research,* Vol. 28, 1980, pp. 1241–1250.

30. Hearn, D. W., and J. Vijay, "Efficient Algorithms for the (Weighted) Minimum Circle Problem," *Operations Research,* Vol. 30, No. 4, 1982, pp. 777–795.

31. Hitchings, G. G., "Analogue Techniques for the Optimal Location of a Main Facility in Relation to Ancillary Facilities," *International Journal of Production Research,* Vol. 7, No. 3, 1967, pp. 189–197.

32. Hurter, A. P., Jr., M. K. Schaefer, and R. E. Wendell, "Solutions of Constrained Location Problems," *Management Science*, Vol. 22, 1975, pp. 51–56.

33. Jacobsen, S. K., "An Algorithm for the Minimax Weber Problem," *European Journal of Operational Research,* Vol. 6, 1981, pp. 144–148.

34. Juel, H., *Properties of Location Models,* Operations Research Technical Report 3, Graduate School of Business, University of Wisconsin, Madison, WI, 1975.

35. Juel, H., "Bounds in the Generalized Weber Problem under Locational Uncertainty," *Operations Research,* Vol. 29, No. 6, 1981, pp. 1219–1227.

36. Juel, H., and R. F. Love, "Hull Properties in Location Problems," *European Journal of Operational Research,* Vol. 12, No. 3, 1983, pp. 262–265.

37. Katz, N. I., "On the Convergence of a Numerical Scheme for Solving Some Locational Equilibrium Problems," *Journal of the Society of Industrial and Applied Mathematics,* Vol. 17, No. 6, 1969, pp. 1224–1231.

38. Katz, N. I., "Local Convergence in Fermat's Problem," *Mathematical Programming,* Vol. 6, 1974, pp. 89–104.

39. Kuhn, H. W., "Locational Problems and Mathematical Programming," *Separatum-Colloquium on the Application of Mathematics to Economics,* 1965, pp. 235–242.

40. Kuhn, H. W., "On a Pair of Dual Non-linear Problems," in J. Abadie, ed., *Non-linear Programming,* Chapter 3, John Wiley & Sons, Inc., New York, 1967.

41. Kuhn, H. W., "A Note on Fermat's Problem," *Mathematical Programming,* Vol. 4, No. 1, 1973, pp. 98–107.

42. Kuhn, H. W., "Nonlinear Programming: A Historical View," *SIAM-AMS Proceedings,* Vol. 9, 1975, pp. 1–26.

43. Kuhn, H. W., and E. Kuenne, "An Efficient Algorithm for the Numerical Solution of the Generalized Weber Problem in Spatial Economics," *Journal of Regional Science,* Vol. 4, No. 2, Winter 1962, pp. 21–33.

44. Lawson, C. L., "The Smallest Covering Cone or Sphere," *SIAM Reviews,* Vol. 7, 1965, pp. 415–417.

45. Love, R. F., "The Location of Single Facilities in Three-Dimensional Space by Nonlinear Programming," *Journal of the Canadian Operational Research Society,* Vol. 5, 1967, pp. 136–143.

46. Love, R. F., "A Note on the Convexity of Siting Depots," *International Journal of Production Research,* Vol. 6, 1967, pp. 153–154.

47. Love, R. F., "A Computational Procedure for Optimally Locating a Facility with Respect to Several Rectangular Regions," *Journal of Regional Science,* Vol. 12, 1972, pp. 233–242.

48. Love, R. F., "One-Dimensional Facility Location-Allocation Using Dynamic Programming," *Management Science,* Vol. 22, 1976, pp. 614–617.

49. Love, R. F., and J. G. Morris, "Modeling Inter-city Road Distances by Mathematical Functions," *Operational Research Quarterly,* Vol. 23, 1972, pp. 61–71.

50. Love, R. F., and J. G. Morris, "Mathematical Models of Road Travel Distances," *Management Science,* Vol. 25, 1979, pp. 130–139.

51. Love, R. F., and W. E. Yeong, "A Stopping Rule for Facilities Location Algorithms," *AIIE Transactions,* Vol. 13, 1981, pp. 357–362.

52. Love, R. F., and L. Yerex, "An Application of a Facilities Location Model in the Prestressed Concrete Industry," *Interfaces,* Vol. 6, 1976, pp. 45–49.

53. Love, R. F., J. G. Morris, and G. O. Wesolowsky, *Facilities Location: Models and Methods,* North-Holland Publishing Company, New York, 1988.

54. McHose, A. H., "A Quadratic Formulation of the Activity Location Problem," *Journal of Industrial Engineering,* Vol. 12, No. 5, 1961, pp. 334–338.

55. Miehle, W., "Link Length Minimization in Networks," *Operations Research,* Vol. 6, No. 2, 1958, pp. 232–243.

56. Mole, R. H., "Comments on 'A Note on the Location of Depots,'" *Management Science,* Vol. 19, 1973, pp. 832–833.

57. Morris, J. G., "Convergence of the Weiszfeld Algorithm for Weber Problems Using a Generalized 'Distance' Function," *Operations Research,* Vol. 29, 1981, pp. 37–48.

58. Morris, J. G., and W. A. Verdini, "Minisum l_p Distance Location Problems Solved via a Perturbed Problem and Weiszfeld's Algorithm," *Operations Research,* Vol. 27, 1979, pp. 1180–1187.

59. Nair, K. P. K., and R. Chandrasekaran, "Optimal Location of a Single Service Center of Certain Types," *Naval Research Logistics Quarterly,* Vol. 18, No. 4, 1971, pp. 503–510.

60. Ostresh, L. M., "On the Convergence of a Class of Iterative Methods for Solving the Weber Location Problem," *Operations Research,* Vol. 26, 1978, pp. 597–609.

61. Palermo, F. P., "A Network Minimization Problem," *IBM Journal of Research and Development*, Vol. 5, No. 4, 1961, pp. 335–337.

62. Rademacher, H., and O. Toeplitz, *The Enjoyment of Mathematics* (translated from *Von Zahlen und Figuren,* 2nd ed., Julius Springer, Berlin, 1933), Princeton University Press, Princeton, NJ, 1957.

63. Rosenhead, J., "Some Comments on 'A Note on the Location of Depots' by Ralph D. Snyder," *Management Science,* Vol. 19, 1973, pp. 831–832.

64. Snyder, R. D., "A Note on the Location of Depots," *Management Science,* Vol. 18, 1971, pp. 97–97.

65. Sylvester, J. J., "A Question in the Geometry of Situation," *Quarterly Journal of Pure and Applied Mathematics,* Vol. 1, 1857, pp. 79–79.

66. Sylvester, J. J., "On Poncelet's Approximate Linear Valuation of Surd Forms," *Philosophical Magazine,* Vol. 20 (Fourth Series), 1860, pp. 203–222.

67. Thisse, J.-F., and H. G. Zollers, eds., *Locational Analysis of Public Facilities,* North-Holland Publishing Company, Amsterdam, 1983.

68. Tideman, M., "Comment on 'A Network Minimization Problem,'" *IBM Journal of Research and Development,* Vol. 6, No. 2, 1962, p. 259.

69. Weber, A., *Theory of the Location of Industries,* The University of Chicago Press, Chicago, 1929 (translation of *Ueber den Standort der Industrien,* Tuebingen, Germany, 1909).

70. Weiszfeld, E., "Sur le point pout lequel la somme des distances de *n* points donnes est minimum," *Tohoku Mathematics Journal,* Vol. 43, 1937, pp. 355–386.

71. Wendell, R. E., and A. P. Hurter, Jr., "Location Theory, Dominance, and Convexity," *Operations Research,* Vol. 21, 1973, pp. 314–320.

72. Wendell, R. E., A. P. Hurter, Jr., and T. J. Lowe, "Efficient Points in Location Problems," *AIIE Transactions,* Vol. 9, 1977, pp. 338–346.

73. Wesolowsky, G. O., "Probabilistic Weights in the One-Dimensional Facility Location Problem," *Management Science,* Vol. 24, No. 2, 1977, pp. 224–229.

74. Wesolowsky, G. O., "Rectangular Distance Location under the Minimax Optimality Criterion," *Transportation Science,* Vol. 6, No. 2, 1972, pp. 103–113.

75. Wesolowsky, G. O., and R. F. Love, "A Nonlinear Approximation Method for Solving a Generalized Rectangular Distance Weber Problem," *Management Science,* Vol. 18, 1972, pp. 656–663.

76. White, J. A., "A Quadratic Facility Location Problem," *AIIE Transactions,* Vol. 3, No. 2, 1971, pp. 156–157.

77. Witzgall, C., *Optimal Location of a Central Facility: Mathematical Models and Concepts,* National Bureau of Standards Report 8388, NBS, Gaithersburg, MD, 1965.

PROBLEMS

4.1. Go through the complete procedure for constructing contour lines for the rectilinear distance problem for which the data are as follows: $\mathbf{P}_1 = (4,4)$, $\mathbf{P}_2 = (4,11)$, $\mathbf{P}_3 = (7,2)$, $\mathbf{P}_4 = (11,11)$, $\mathbf{P}_5 = (14,7)$; $w_1 = 3$, $w_2 = 2$, $w_3 = 2$, $w_4 = 4$, $w_5 = 1$. Also find all least-cost new-facility locations.

4.2. Existing facilities are located as follows: $\mathbf{P}_1 = (4,4)$, $\mathbf{P}_2 = (4,10)$, $\mathbf{P}_3 = (6,5)$, $\mathbf{P}_4 = (10,5)$, $\mathbf{P}_5 = (10,9)$, and $\mathbf{P}_6 = (12,3)$. It is desired to locate one new machine with respect to the existing facilities. Travel between facilities is along a rectilinear aisle structure. The amount of item movement between the new facility and each existing facility is given as $w_1 = 4$, $w_2 = 4$, $w_3 = 2$, $w_4 = 3$, $w_5 = 5$, and $w_6 = 6$. Where should the new facility be located to minimize distance traveled? Plot contour lines and rank the following potential locations: $A = (4,6)$, $B = (8,8)$, $C = (8,5)$, $D = (10,3)$.

4.3. Four hospitals located within a city are cooperating to establish a centralized blood-bank facility that will serve the hospitals. The new facility is to be located such that distance traveled is minimized. The hospitals are located as follows: $\mathbf{P}_1 = (5, 10)$, $\mathbf{P}_2 = (7, 6)$, $\mathbf{P}_3 = (4, 2)$, and $\mathbf{P}_4 = (16, 3)$. The number of deliveries to be made per year between the blood-bank facility and each hospital is estimated to be 450, 1200, 300, and 1500, respectively. Assuming rectilinear travel, determine the optimum location. What assumption must be made concerning the deliveries of blood? Is it reasonable to expect delivery trips will always be to only one hospital when a delivery is made?

4.4. A large corporation has employed a campus plan layout for its corporate headquarters. A district heating system is to be installed, which will heat each of four buildings. Considering the cost of installation and the heat losses, it is agreed that the cost for the system is proportional to the square of the Euclidean distance between the heating facility and each building. The constant of proportionality is the Btus per hour required at each building. The buildings to be served by the heating system are located as follows: $\mathbf{P}_1 = (18, 5)$, $\mathbf{P}_2 = (13, 9)$, $\mathbf{P}_3 = (23, 11)$, and $\mathbf{P}_4 = (7, 11)$. The Btu requirements per hour are 12,000, 5000, 4000, and 15,000, respectively. Find the least-cost location for the central heating facility.

4.5. An airport terminal is served by Pokey, Early, Delayed, and Undecided airlines. The airlines unload baggage from arriving airplanes at the points indicated in Figure P4.5. The number of arriving flights per day for each airline is 36, 22, 28, and 18, for Pokey, Early, Delayed, and Undecided, respectively. If a separate trip is made from the baggage receiving area to a passenger pickup point for each arriving flight, and rectilinear distance is assumed, where should the pickup point be located? Suppose that the pickup point must be outside and on either the south or west side of the terminal; where should the pickup point be located? Suppose that a conveyorized system is to be installed with separate conveyor belts joining each airline and the pickup point. If the conveyor layout follows a rectilinear layout pattern, what is the optimum location for the pickup point?

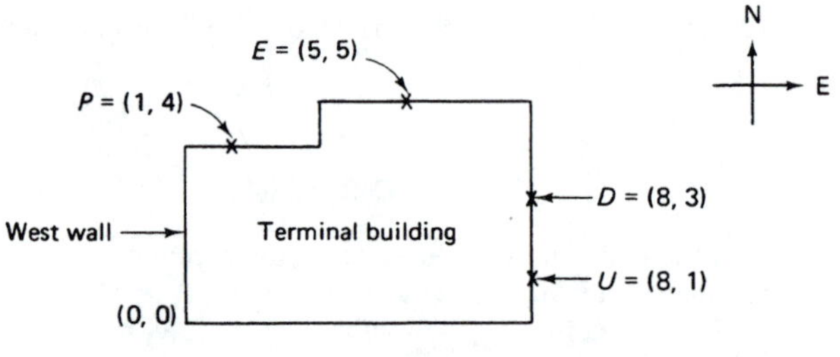

Figure P4.5

4.6. Mrs. Black has to pick up several items for her daughter's party at various locations. She wants to locate a parking space such that she can get to all the stores in a minimum amount of time without reparking. Each block is square—100 ft on a side. Streets running north to south are numbered consecutively. Those running east to west are lettered consecutively. The bakery is at 6th and E; she must walk half as fast as normal, so that she won't drop the cake or pastries. At 10th and D is the grocery store. The dress shop is at 12th and G. Mrs. Black picks up her daughter, Kathy, from the beauty parlor at 10th and G, and they walk twice as fast as normal back to the car, so that no one will see her with her hair up. It is assumed she must stay on the sidewalks that enclose each block—distance used crossing streets is considered negligible. Determine the location of the parking space that satisfied her objective.

4.7. A portable lunch wagon is to be located along the main aisle of an industrial plant. The vendor wishes to locate the wagon so that the total walking distance for his customers is minimized. On a coordinate system the main aisle runs parallel to the x-axis at $y = 10$. Workstations are located at the following coordinate points. Travel from the workstations to the main aisle is rectilinear.

Station	Location	Station	Location
1	(4, 4)	6	(12, 12)
2	(4, 12)	7	(10, 14)
3	(6, 4)	8	(12, 6)
4	(6, 12)	9	(8, 14)
5	(8, 8)	10	(10, 4)

(a) If one customer is located at each station, where should the wagon be located along the main aisle to minimize cumulative customer walking distance?

(b) If one customer is located at stations 1, 2, 3, 4, and 5, and three customers are located at stations 6, 7, 8, 9, and 10, where should the wagon be located along the main aisle to minimize cumulative customer walking distance?

4.8. Assuming Euclidean distances, what is the optimum location for a single new facility if $\mathbf{P}_1 = (0, 9)$, $\mathbf{P}_2 = (1, 2)$, $\mathbf{P}_3 = (7, 10)$, $\mathbf{P}_4 = (8, 3)$, and $w_1 = w_2 = w_3 = w_4$?

4.9. Plot the contour line passing through the point $(1, 0)$ for the problem

$$f(x, y) = 3|x - 1| + 2|y - 1| + 1|y - 2|$$

What is the value of $f(x, y)$ for this contour line?

4.10. Suggest an iterative procedure for solving the problem

$$\text{minimize} \quad \sum_{i=1}^{m} w_i[(x - a_i)^2 + (y - b_i)^2]^k, \qquad \text{where } k > 0$$

The procedure should agree with the one given in the case for $k = \frac{1}{2}$. What reasons can you suggest for considering values of k other than $\frac{1}{2}$?

4.11. Given existing facilities at points $\mathbf{P}_1 = (0, 0)$ and $\mathbf{P}_2 = (6, 0)$, with $w_1 = w_2 = 1$, plot several contour lines for the (a) rectilinear distance model, (b) squared distance model, and (c) Euclidean distance model.

4.12. For Problem 4.11, which points will minimize the total cost function for each of the three cases?

4.13. Suppose that two people go to a shopping center. They have agreed beforehand that each will visit one store alone and then return to their parked car. The two stores are at opposite ends of the shopping center, and the car can be parked at any spot along a straight sidewalk running between the two stores. Where do you think they would park their car and why? Does your answer suggest that there may be quantifiable approaches to location problems other than minimum total cost approaches; if so, what are they?

4.14. Using the contour-line approach for evaluating locations for a new machine, determine the optimal location based on the following data:

Locations of existing machines			Possible locations for new machines		
i	x_i	y_i	j	x_j	y_j
1	8	7	1	8	10
2	13	2	2	16	17
3	13	13	3	2	16
4	18	10	4	16	3
5	2	10	5	16	8
6	8	16			
7	18	20			

The pallet load movement per day between the new machine and the ith existing machine, w_i, is given to be

i	w_i
1	100
2	100
3	150
4	100
5	200
6	250
7	100

Rank the alternative locations in order of preference using contour lines. Rectilinear travel is used.

4.15. The ABC Auto Parts Company has 11 retail sales stores in the city of Greenville. The ABC Company needs a new warehouse facility to service its retail stores. The locations of the stores and the expected deliveries per week from the warehouse to each store are

Store	*Location (miles)*	*Expected deliveries*
1	(1, 0)	1
2	(1, 3)	1
3	(2, 0)	1
4	(2, −1)	1
5	(1, −2)	3
6	(−1, 0)	5
7	(−1, 1)	6
8	(−1, 3)	2
9	(−2, −2)	4
10	(−3, −1)	1
11	(−3, 1)	2

Assume that travel within the city of Greenville is rectilinear and that after each delivery the delivery truck must return to the warehouse.

(a) If there are no restrictions on the warehouse location, where should it be located?

(b) If the ABC Company is going to open a new retail outlet at the warehouse location and this outlet must be at least 1 rectilinear mile away from an existing store, where should it be located?

4.16. A plumbing contractor wishes to minimize the cost of pipe for a water waste disposal system in a building under construction. The locations of all the facilities that require water waste removal have already been determined. However, all the individual drain pipes must be connected at some point along the inside basement wall of the building (i.e., $y = 0$). From the connection one main drain pipe is used to connect with the sewer system. Assume that the (x, y, z) coordinates of the connection points within the building are $\mathbf{P}_1 = (8, 20, -8)$, $\mathbf{P}_2 = (12, 10, 2)$, $\mathbf{P}_3 = (25, 35, 2)$, $\mathbf{P}_4 = (40, 20, 2)$, $\mathbf{P}_5 = (25, 35, 12)$, $\mathbf{P}_6 = (30, 40, 12)$, $\mathbf{P}_7 = (15, 25, 22)$, and $\mathbf{P}_8 = (25, 35, 22)$. The main drain must be connected to the sewer line at $\mathbf{P}_0 = (20, -50, -10)$. The cost per foot of drain line from the connection point to each facility is $w_1 = \$4$, $w_2 = \$5$, $w_3 = \$3$, $w_4 = \$1.50$, $w_5 = \$3$, $w_6 = \$1.50$, $w_7 = \$2$, and $w_8 = \$3$, respectively. The cost per foot of pipe for the main drain is \$15. Pipe lines will be laid out rectilinearly. Determine the optimum location for the main drain connection. What is the optimum location if the main drain connection is constrained to have $x \leq 1$ and $z \leq -8$?

4.17. As Minister of Defense for the planet Xerxes in the galaxy Euclid, you are seriously concerned about your four most aggressive rivals and plan to place one outpost in space fully equipped with an army of laser ships to be dispatched, as necessary, to attack any of the four rival planets. The positions in space of the four rivals, relative to an arbitrarily selected origin, are $R_1 = (2, 2, 2)$, $R_2 = (12, 6, 8)$, $R_3 = (10, 0, -4)$, and $R_4 = (6, 20, -10)$, measured in millions of miles. The critical travel time is a function of the square of the Euclidean distance traveled (due to the repelling forces generated from the four planets). You have assessed the probability of an attack originating from each planet and assigned the following values: $p_1 = 0.4$, $p_2 = 0.6$, $p_3 = 0.4$, and $p_4 = 0.9$. (Notice that attacks are not mutually exclusive events.) Fortunately, due to the magnitude of the diameters of the orbits of all planets and the outpost (order of 10 million light-years), the entire coordinate system is considered to orbit, such that any changes in the relative positions of the planets due to orbiting can be ignored. Where should the outpost be located to minimize the expected distance traveled?

4.18. One new facility is to be located with respect to three existing facilities, and straight-line movement is used. Annual item movement between the new facility and each existing facility is given to be $w_1 = 5$, $w_2 = 2$, and $w_3 = 10$. The coordinate locations of the existing facilities are $\mathbf{P}_1 = (2, 6)$, $\mathbf{P}_2 = (4, 0)$, and $\mathbf{P}_3 = (8, 3)$.

(a) Give upper and lower bounds on the straight-line distance traveled per year. Use (4.1).

(b) Determine the optimum location for the new facility, assuming cost is proportional to the square of the straight-line distance traveled.

(c) Determine the optimum location for the Euclidean problem.

4.19. Are the following statements true or false? If false, give a reason for the statement being false.

(a) Let $\mathbf{P}_1 = (0, 0)$, $\mathbf{P}_2 = (0, 10)$, $\mathbf{P}_3 = (x, 5)$, and $w_1 = w_2 = w_3$. With rectilinear distance, the optimum location for the new facility is not dependent on the value of x.

(b) Under some conditions the rectilinear, Euclidean, and squared Euclidean problems will yield the same location for a given problem.

(c) The partial derivatives $\partial f/\partial x_j$ are always defined for both the Euclidean and squared Euclidean problem.

(d) The single-facility rectilinear problem is easiest solved using linear programming.

(e) The rectilinear, Euclidean, and squared Euclidean problems can have multiple optimum solutions.

(f) In a single-facility location problem (of the type treated in this chapter) involving existing facilities at $\mathbf{P}_1 = (10, 10)$, $\mathbf{P}_2 = (12, 15)$, and $\mathbf{P}_3 = (16, 18)$, the location $(16, 10)$ can never be an optimum location.

4.20. Let four existing facilities be located at $\mathbf{P}_1 = (0,0)$, $\mathbf{P}_2 = (5,0)$, $\mathbf{P}_3 = (5,5)$, and $\mathbf{P}_4 = (0,5)$ with $w_1 = 2$, $w_2 = 1$, $w_3 = 2$, and $w_4 = 2$. Determine the optimum location for a single new facility when cost is proportional to (a) rectilinear distance, (b) squared Euclidean distance, and (c) Euclidean distance. Compare the minimum total cost obtained in part (c) with the upper bound from (4.1), using the solution obtained in part (a).

4.21. Let four existing facilities be located at $\mathbf{P}_1 = (0,0)$, $\mathbf{P}_2 = (6,0)$, $\mathbf{P}_3 = (6,16)$, and $\mathbf{P}_4 = (12,16)$ with $w_1 = w_2 = w_3 = w_4 = 2$. Determine the optimum location for a single new facility when cost is proportional to (a) rectilinear distance, (b) squared Euclidean distance, and (c) Euclidean distance. Compare the minimum total cost obtained in part (c) with the upper bound from (4.1) using the following rectilinear solutions: $(6,0)$, $(6,8)$, and $(6,16)$.

4.22. A city wishes to locate a fire station so as to minimize the cost of fire damage. After some analysis, five major areas have been identified in the city as being primary risk areas. The feeling is that fire damage increases in proportion to the square of the time required to answer an alarm. It is also felt that time and distance are directly proportional. For purposes of the analysis straight-line distances are assumed. Weighting factors have been obtained such that the fire loss at area k is equal to $w_k d_k^2$, where w_k is the weighting factor and d_k is the Euclidean distance from the fire station to area k. The coordinate locations of the areas, along with the weight factors, are

k	x_k	y_k	w_k
1	4	12	4
2	8	2	2
3	14	10	2
4	2	3	8
5	14	4	9

What is the optimum location for the fire station?

4.23. Four machines are located at the coordinate points $(0,4)$, $(0,8)$, $(4,0)$, and $(8,0)$. Thirty percent of the item movement between the existing machines and a new machine is between the new machine and each of the existing machines at $(0,4)$ and $(4,0)$. Twenty percent is between the new machine and each of the existing machines at $(0,8)$ and $(8,0)$. Rectilinear movement is used. Construct the contour line passing through the positive quadrant and beginning at the point $(0,12)$. If the machine locations are interpreted as docks and the contour line is interpreted as the periphery of the warehouse, what warehousing interpretation might be given to the percentage weights? Explain.

4.24. Given three existing facilities located at $\mathbf{P}_1 = (0,0)$, $\mathbf{P}_2 = (0,4)$, and $\mathbf{P}_3 = (4,2)$ with $w_1 = w_2 = w_3 = 1$, determine the upper and lower bounds on the minimum value of the objective function for the single-facility Euclidean problem using (4.1).

4.25. Solve the Euclidean-distance problem given the following data:

(a) $\mathbf{P}_1 = (0,0)$, $\mathbf{P}_2 = (2,0)$, $\mathbf{P}_3 = (0,2)$, $w_1 = \frac{1}{5}$, $w_2 = \frac{3}{5}$, $w_3 = \frac{1}{5}$.

(b) $\mathbf{P}_1 = (1,0)$, $\mathbf{P}_2 = (2,1)$, $\mathbf{P}_3 = (1,2)$, $\mathbf{P}_4 = (0,1)$, $w_i = 1$ for $i = 1, \ldots, 4$.

Be sure to justify your answers.

4.26. A new cooperative power plant is to be located to serve each of 10 towns located in a large level area devoted to wheat ranching. Separate electric cables are to be run from the power plant to each town. The locations of the 10 towns are $\mathbf{P}_1 = (0,0)$, $\mathbf{P}_2 = (1,3)$, $\mathbf{P}_3 = (2,5)$, $\mathbf{P}_4 = (3,9)$, $\mathbf{P}_5 = (4,2)$, $\mathbf{P}_6 = (5,4)$, $\mathbf{P}_7 = (6,6)$, $\mathbf{P}_8 = (7,1)$, $\mathbf{P}_9 = (8,5)$, and $\mathbf{P}_{10} = (10,8)$. Determine the location for the power plant that will minimize the total required length of electrical cable.

4.27. The Wei-Louse Army has combat outposts located at $\mathbf{P}_1 = (2,7)$, $\mathbf{P}_2 = (3,8)$, $\mathbf{P}_3 = (6,9)$, and $\mathbf{P}_4 = (9,8)$. The military commander, General B. Trayed, has decided to construct a temporary base from which helicopters will be dispatched to provide air support and supplies for the outposts. The airbase will receive its supplies by air from two supply points, $\mathbf{P}_5 = (0,4)$ and $\mathbf{P}_6 = (8,1)$. The expected number of flights per day between the new base and the point $\mathbf{P}_j$ is given by w_j, where $w_1 = 10$, $w_2 = 12$, $w_3 = 14$, $w_4 = 10$, $w_5 = 26$, and $w_6 = 34$. The local air traffic controller, Private Ama Kamikazee, indicates straight-line paths are to be used for all flights in order to confuse the enemy. Determine the location for the base such that the expected distance flown per day is minimized.

4.28. The Royal Canadian Hillies plan to locate a brandy carryout in the Yukon. A number of Saint Bernard dogs, trained by Sergeant Preston, will be used to rescue lost tourists looking for the South Pole. Based on past experience, it is anticipated that emergency rescue missions will have to be dispatched to the points $Q_1 = (18,2)$, $Q_2 = (4,0)$, $Q_3 = (6,20)$, and $Q_4 = (12,18)$ with weekly frequencies $w_1 = 6$, $w_2 = 2$, $w_3 = 7$, and $w_4 = 4$. Sergeant Preston has trained his dogs to travel along a straight-line path in carrying out the rescue mission.

(a) Determine the location of the brandy carryout that minimizes the distance traveled by the dogs.

(b) Assume the local SPCA has protested that the suffering of St. Bernards in carrying out rescue missions is proportional to the square of the distance traveled per mission. What location will minimize the total amount of suffering?

4.29. Suppose that an item is to be located in three-space instead of two-space. Suggest contexts in which (a) the location models considered in this chapter would extend in a logical manner to their three-space counterpart, and (b) would not extend in a logical manner to their three-space counterpart.

4.30. Existing facilities are located as follows:

$\mathbf{P}_1 = (0,0)$,	$\mathbf{P}_6 = (7,8)$
$\mathbf{P}_2 = (2,8)$,	$\mathbf{P}_7 = (9,0)$
$\mathbf{P}_3 = (4,0)$,	$\mathbf{P}_8 = (9,4)$
$\mathbf{P}_4 = (4,4)$,	$\mathbf{P}_9 = (13,2)$
$\mathbf{P}_5 = (7,2)$,	$\mathbf{P}_{10} = (13,8)$

There is equal item movement between the single new facility and each existing facility.

(a) Assuming rectilinear travel and assuming that the new facility must be at least 3 rectilinear distance units away from an existing facility, what is the optimum location for the new facility? What is the associated value of $f(x,y)$?

(b) Assuming that cost is proportional to the square of the straight-line distance between the new facility and each existing facility, where should the new facility be located?

4.31. Define the following approximating function for the Euclidean problem:

$$\text{minimize} \quad g(x,y) = \sum_{i=1}^{m} w_i E_i(\epsilon)$$

where

$$E_i(\epsilon) = [(x - a_i)^2 + (y - b_i)^2 + (\epsilon)]^{1/2}$$

and ϵ is some arbitrarily small, positive constant. Let

$$g_x = \frac{\partial g(x,y)}{\partial x}, \qquad g_{xx} = \frac{\partial^2 g(x,y)}{\partial x^2}$$

$$g_y = \frac{\partial g(x,y)}{\partial y}, \qquad g_{yy} = \frac{\partial^2 g(x,y)}{\partial y^2}$$

$$g_{xy} = \frac{\partial^2 g(x,y)}{\partial x\, \partial y} = g_{yx}$$

Newton's method can be employed to solve the set of equations

$$g_x = 0$$

$$g_y = 0$$

Let (x_0, y_0) be an approximation of the optimum location for the new facility, (x^*, y^*). Successive approximations of (x^*, y^*) can be generated from the recursion formulas

$$x_{k+1} = x_k - \left[\frac{g_x g_{yy} - g_y g_{xy}}{g_{xx} g_{yy} - (g_{xy})^2}\right]_k$$

$$y_{k+1} = y_k - \left[\frac{g_y g_{xx} - g_x g_{xy}}{g_{xx} g_{yy} - (g_{xy})^2}\right]_k$$

where all functions involved are to be evaluated at (x_k, y_k). Use Newton's method to solve the Euclidean distance problems in Problems 4.26, 4.27, and 4.28. Compare the solution accuracies and number of iterations required with those obtained using the iterative procedure described in the chapter.

4.32. Newton's method, as described in Problem 4.27, is not guaranteed to converge to the optimum solution. As an illustration, consider the following data for a single-facility location problem:

$$\mathbf{P}_1 = (0,0), \qquad \mathbf{P}_2 = (4,0), \qquad \mathbf{P}_3 = (2,4)$$

$$w_1 = 2, \qquad w_2 = 3, \qquad w_3 = 4$$

Based on Euclidean distances, perform four iterations using Newton's method and compare the locations with those obtained after performing four iterations using Weiszfeldt's algorithm. Let $(x^0, y^0) = (3, 3)$.

4.33. Write a computer program to solve the Euclidean problem using the HAP iterative procedure.

4.34. Solve the Euclidean problem in Problem 4.32 using the HAP iterative procedure.

4.35. Plot the contour lines for $f(x,y) = 28$ given
(a) $f(x,y) = 3y + 2x + 4|y - 3| + 2|x - 2| - |x - 4|$.
(b) $f(x,y) = 3|y| + 2|x| + 4|y - 3| + 2|x - 2| - |x - 4|$.

4.36. A new facility is to be placed in an existing layout. The new facility interacts with four existing facilities located at (0, 0), (2, 4), (4, 2), and (5, 5). Rectilinear travel is used. The new facility is located at the point (2, 2). What can be said about the values of w_i in order for the location to minimize travel to and from the existing facilities?

4.37. Solve Problem 4.36 for the case of Euclidean distances and the case of the gravity problem.

4.38. Given $a_{1j}, a_{2j} \geq 0, j = 1, \ldots, n$; the following inequality is known as the triangle inequality:

$$\left[\sum_{j=1}^{n} (a_{1j} + a_{2j})^2\right]^{1/2} \leq \left[\sum_{j=1}^{n} a_{1j}^2\right]^{1/2} + \left[\sum_{j=1}^{n} a_{2j}^2\right]^{1/2}$$

By induction, given $a_{ij} \geq 0$,

$$\left[\sum_{j=1}^{n} \left(\sum_{i=1}^{m} a_{ij}\right)^2\right]^{1/2} \leq \sum_{i=1}^{m} \left[\sum_{j=1}^{n} a_{ij}^2\right]^{1/2}$$

Use the latter inequality to derive the lower bound for the optimum Euclidean solution.

4.39. Plot contour lines for the rent-a-car example problem for the cases of (a) rectilinear distance, (b) Euclidean distance, and (c) squared Euclidean distance.

4.40. A single-facility location problem is given in which the distance is *Euclidean*; the existing facility locations are given by

$$\mathbf{P}_1 = (0,2), \quad \mathbf{P}_2 = (2,7), \quad \mathbf{P}_3 = (3,5), \quad \mathbf{P}_4 = (8,9), \quad \mathbf{P}_5 = (5,9), \quad \mathbf{P}_6 = (4,4)$$

with corresponding weights

$$w_1 = 5, \quad w_2 = 4, \quad w_3 = 6, \quad w_4 = 2, \quad w_5 = 1, \quad w_6 = 3$$

Find a new-facility location for which the transportation cost will be at most 2% greater than the minimum transportation cost and, in particular, be certain to *give a rigorous justification* for your choice of the new-facility location found.

4.41. Construct the efficient set E using the existing facility location information given in Problem 4.1, assuming rectilinear distances.

4.42. Construct the efficient set E using the existing facility location information given in Problem 4.2, assuming rectilinear distances.

4.43. Construct the efficient set E using the existing facility location information given in Problem 4.3, assuming rectilinear distances.

4.44. Construct the efficient set E using the existing facility location information given in Problem 4.4, assuming rectilinear distances.

4.45. Construct the efficient set E using the existing facility location information given in Problem 4.5, assuming rectilinear distances.

4.46. Construct the efficient set E using the existing facility location information given in Problem 4.14, assuming rectilinear distances.

4.47. Construct the efficient set E using the existing facility location information given in Problem 4.15, assuming rectilinear distances.

4.48. The points $(a_1, b_1), \ldots, (a_{12}, b_{12})$ are given by (0, 3), (2, 7), (3, 7), (3, 10), (4, 7), (8, 7), (9, 6), (8, 5), (7, 6), (5, 6), (4, 3), and (2, 1), respectively. The boundary of a region **S** may be obtained by connecting points (a_i, b_i) and (a_{i+1}, b_{i+1}) with a straight line, for $i = 1, \ldots, 11$, and by connecting points (a_{12}, b_{12}) and (a_1, b_1) with a straight line. The region represents a region within a city that must be served by a fire station. Assuming that travel distance between any two points in the region is rectilinear, at what points can the fire station be located so as to minimize its maximum travel distance to any fire within the region?

4.49. For the example location problem of Figure 4.19 for $z_0 = 9.5$, compute the four corner points of the rectangle $\mathbf{S}(z_0)$ and verify that $\mathbf{S}(z_0)$ gives the same minimax locations as obtained in the section for this example. Given z_0, notice that it is almost as easy to compute the corner points of $\mathbf{S}(z_0)$, thus obtaining minimax locations, as it is to compute minimax locations. Since this is the case, what is the point in using the procedure?

4.50. If $(r, s) = T(x, y) = (x + y, -x + y)$, and $(a', b') = T(a, b) = (a + b, -a + b)$, show that

$$|x - a| + |y - b| = \max(|r - a'|, |s - b'|)$$

[*Hint*: Use the following facts: (1) If c is any number, $|c| = \max(c, -c)$. (2) If c and d are any numbers, $|c| + |d| = \max(c, -c) + \max(d, -d) = \max(c + d, c - d, -c + d, -c - d) = \max[\max(c + d, -c - d), \max(c - d, -c + d)] = \max(|c + d|, |c - d|)$.]

4.51. Set up the problem of minimizing

$$f_1(r) = \max_{1 \le i \le m} (w_i|r - a_i'| + g_i)$$

as an equivalent linear programming problem, and write the dual of the linear programming problem. Note that the dual problem has only two constraints.

4.52. Show that $\alpha_{pq} \le f_1(r)$ for any value of r, and thus α_{pq} is a lower bound on the minimum value of $f_1(r)$, where α_{pq} is the maximum of the $\alpha_{ij}, i < j$. [*Hint*: Use the triangle inequality and the definition of $f_1(r)$ to show that $(w_p w_q|a_p' - a_q'| + w_p g_q + w_q g_p)/(w_p + w_q) \le w_q(w_p|r - a_p'| + g_p)/(w_p + w_q) + w_p(w_q|r - a_q'| + g_q)/(w_p + w_q) \le w_q f_1(r)/(w_p + w_q) + w_p f_1(r)/(w_p + w_q) \le f_1(r)$. Next, show that $g_p \le f_1(r)$ and $g_q \le f_1(r)$. It will then follow (why?) that $\alpha_{pq} \le f_1(r)$.]

4.53. **(a)** Graph the function

$$f_1(r) = \max(2|r - 6| + 1, 4|r - 9|, 2|r - 15|)$$

Verify that $r^* = 10\frac{1}{4}$ minimizes $f_1(r)$, and that the minimum value of $f_1(r)$ is $9\frac{1}{2}$. [*Hint*: To graph $f_1(r)$, first graph the three functions $2|r - 6| + 1$, $4|r - 9|$, and $2|r - 15|$; the graph of $f_1(r)$ is then the maximum of these three graphs. Furthermore, r^* is obtained by computing the value of r where the graphs of $2|r - 6| + 1$ and $2|r - 15|$ intersect.]

(b) Graph the function

$$f_2(s) = \max(2|s - 0| + 1, 3|s - 3|, 4|s - (-3)|, 2|s - 1|)$$

Verify that $s^* = -\frac{3}{7}$ minimizes $f_2(s)$, and that the minimum value of $f_2(s)$ is $10\frac{2}{7}$.

4.54. Show that the inequalities $w_i|r - a_i'| + g_i \leq z_0, i = 1, \ldots, m,$ are equivalent to $r_1 \leq r \leq r_2$. [*Hint*: First show that the inequalities are equivalent to $-(z_0 - g_i)/w_i \leq r - a_i' \leq (z_0 - g_i)/w_i, i = 1, \ldots, m$.]

4.55. Show, by computing them, that the four corners of the set $S(k)$ defined in Section 4.43 are in fact as given.

4.56. Show that the vectors $\frac{1}{2}(c_1 - c_3, c_1 + c_3 + c_5, c_5)$ and $\frac{1}{2}(c_2 - c_4, c_2 + c_4 - c_5, c_5)$ are feasible solutions to the linear programming formulation of Problem UPA.

4.57. Solve, geometrically, the problem illustrating the Elzinga–Hearn algorithm.

4.58. Write a computer program for solving problems with the Elzinga–Hearn algorithm.

4.59. Solve the problem for which the existing facilities are given by (2, 10), (7, 9), (7, 12), (3, 15), and (1, 14), respectively, using the Elzinga–Hearn algorithm.

4.60. Solve Problem 4.3 using a minimax objective.

4.61. Solve Problem 4.4 using a minimax objective based on Euclidean distances and assuming the central heating facility is to be located to minimize the maximum length of pipe between the central heating facility and the buildings it services.

4.62. Consider Problem 4.5 in which a central baggage delivery point is to be determined. Based on rectilinear distances between the baggage receiving areas and the baggage pickup point, determine the location for the baggage pickup that minimizes the maximum of the weighted distances traveled.

4.63. Solve Problem 4.7 using a minimax objective.

4.64. Determine the minimax solution to Problem 4.8.

4.65. Given existing facilities at points $\mathbf{P}_1 = (0, 0)$ and $\mathbf{P}_2 = (6, 0)$, with $w_1 = w_2 = 1$, plot several contour lines for the single-facility minimax location problem based on rectilinear distances.

4.66. Show that the minimax solution to the Euclidean problem is also a minimax solution to the squared Euclidean problem.

4.67. Obtain a minimax solution to Problem 4.20 assuming (a) rectilinear distance, (b) Euclidean distance, and (c) squared Euclidean distance.

4.68. Obtain a minimax solution to Problem 4.22 assuming rectilinear distances.

4.69. Solve Problem 4.23 based on a minimax formulation.

4.70. Solve Problem 4.27 based on a minimax formulation.

4.71. Solve Problem 4.28 based on a minimax formulation.

4.72. Consider the following alternative minimax formulation of the multifacility location problem:

$$\text{minimize} \quad \phi = \sum_{j=1}^{n} \alpha_j z_j$$

where α_j is an appropriately chosen weighting factor and

$$z_j = \max[v_{jk}\, d(\mathbf{X}_j, \mathbf{X}_k), w_{ji}\, d(\mathbf{X}_j, \mathbf{P}_i): 1 \leq k \leq n, 1 \leq i \leq m]$$

with $d(\mathbf{X}_j, \mathbf{X}_k)$ and $d(\mathbf{X}_j, \mathbf{P}_i)$ the rectilinear distances between the points $\mathbf{X}_j$ and $\mathbf{X}_k$ and the points $\mathbf{X}_j$ and $\mathbf{P}_i$, respectively. Formulate the problem as an equivalent linear programming problem. Give a physical interpretation to the objective function.

Chapter 5

Storage Systems Layout

5.1 INTRODUCTION

In Chapter 4 we addressed single-facility location problems; multiple-facility problems are addressed in Chapter 6. In this chapter we apply material from Chapter 4 in the context of designing storage systems layouts.

A number of interesting location and layout problems arise in the process of designing storage systems. Furthermore, storage systems occur in a variety of contexts, including manufacturing, warehousing, and the service sector. Initially, we assume that the material to be stored is discrete and is in the form of unit loads (e.g., pallet loads, tote boxes, cartons, cases, crates, and containers). Subsequently, we address situations in which either the material to be stored is continuous or the discrete storage problem can be considered to be continuous. Examples of the storage of continuous material include liquids in storage containers and dry, bulk materials in silos.

Among the design issues facing the storage systems designer are those related to the *size* of the storage system, the *storage methods* to be used, and the *layout* of the storage system. Among the determinations to be made are *the number of storage locations required*, *the method of storing/retrieving products*, and *the assignment of items to locations*.

Trade-offs inevitably occur between *throughput* and *storage space* in designing storage systems. The term *throughput* is used as a measure of the number of storages and retrievals performed per time period. It can be expressed directly as a rate (e.g., 320 storages per 8-hour day). Alternatively, it can be given inversely in terms of the time required to perform a storage (e.g., 1.5 minutes per storage). Space is a measure of the static nature of storage. Throughput, however, is a measure of the activity or the dynamic nature of storage; it represents the flow occurring in storage.

The *size* of the storage system depends on a number of parameters and variables. For example, the size of the storage system is influenced by storage, throughput, and cost parameters. The decision variables that influence the size of storage include the storage methods and the storage layout.

The *material characteristics* and *inventory profile* establish the storage and throughput parameters. Included in the former are the characteristics that influence the way material is stored, handled, and controlled. The material characteristics of interest include size, weight, shape, value, shelf life, stackability, toxicity, flammability, explosiveness, and environmental requirements, among others. The inventory profile includes both the amount of each product stored over time and the input/output functions that generate the activity requirements for storing and retrieving material.

The input/output functions will depend on the mission of the storage system. As an example, consider a distribution center for finished goods produced at various company-owned manufacturing plants. If a "push" system is used, the production plants "push" inventory to the distribution center and the input function will be determined by the production schedules at the plants. The output or demand function might be represented by a forecasting equation developed for the marketing department. The cumulative differences in the input and output functions will determine the storage requirements in the distribution center over time.

Among the *cost parameters* that influence the size of the storage system are the costs of providing storage versus the costs of not providing storage. The former includes the costs of providing storage internally versus leasing space or contracting with a public warehouse to provide the storage space. The costs of providing storage space include the costs associated with space, personnel, and equipment resources. The cost of not providing storage reflects the impact of a space shortage and includes the cost of lost business, goodwill costs, and the cost to the total business due to inadequate space.

The *storage method* used includes the specification of the unit load and/or container to be stored, handled, and controlled, as well as the storage/retrieval device, storage equipment, and other material handling equipment. A number of alternative methods exist for storing and retrieving material, including manually storing items on shelving, storing unit loads in pallet rack with lift trucks, storing unit loads in pallet rack with automated storage/retrieval machines, and manually storing small parts in carousel conveyors. Material can be moved to/from storage manually, or mechanically via conveyors and industrial trucks, or automatically via automatic guided vehicles and automated monorails, among others.

The storage system layout includes the height, length, and width of storage, the location of the individual items in storage, and the location and configuration of any required support functions. Both the storage capacity and the throughput capacity of the storage system will be influenced by the layout used.

In this chapter we consider the layout of the storage system. However, to accomplish our objective, it will be necessary to determine the size of the storage requirement. Storage size depends on the number of storage locations required; in turn, the number of storage locations depends on the storage location policy used.

A number of alternative storage location policies are used to determine the assignment of items to storage locations. In this chapter we consider four policies: *dedicated storage*, *randomized storage*, *class-based dedicated storage, and shared storage.* Both the requirements for storage space and the assignment of items to storage locations are included in the consideration.

To facilitate the presentation, we assume that full pallet loads of material are to be stored; also, replenishments and retrievals are assumed to occur in full-pallet-load increments. However, the principles and procedures presented are not limited to full-pallet-load applications.

Our discussion of storage systems layout concludes with a consideration of continuous layout representations of storage systems. Drawing on the discussion in Chapter 4, contour lines are used to determine the storage configuration, and a single integral expression is presented for determining the expected distance traveled.

We do not intend for the treatment of storage systems layout to be exhaustive. Rather, we simply want to demonstrate the utility of the concepts presented in other chapters in designing storage systems layouts. Also, the material presented should be of use to the storage systems designer faced with sizing and location decisions.

5.2 DEDICATED STORAGE LOCATION POLICY

Dedicated storage, also referred to as *fixed slot storage*, involves the assignment of specific storage locations or storage addresses for each product stored. Since a storage location is assigned or dedicated to a specific product, the term *dedicated storage* is used. (Reserved parking in a parking lot is an example of dedicated storage.)

Two variations of dedicated storage are commonly used. *Part number sequence storage* is often used due to its simplicity. The storage location of a product is based solely on the part number assigned to it. Low part numbers are assigned to the "best" locations in the storage region; the higher the part number, the less desirable is its location. Typically, part number assignments are made arbitrarily, without regard to activity. Hence, if a part with a very large part number experiences high demand activity, frequent trips will be made to very poor storage locations.

Throughput-based dedicated storage is an alternative to part number sequence storage. Such a storage method involves a consideration of differences in activity levels and storage requirements among products to be stored. Throughput-based dedicated storage is preferred to part number sequencing storage when there are significant differences in either the activity level or the inventory level for products being stored. Due to the increasing usage of throughput-based dedicated storage, hereafter, we refer to it simply as *dedicated storage*.

5.2.1 Space Requirements

With dedicated storage, products are assigned to specific locations. Also, one and only one product is assigned to a storage location. Hence *the number of storage locations assigned to a product must be capable of satisfying the **maximum** storage requirement for the product*. With multiproduct storage, the storage space required equals the sum of the maximum storage requirements for each of the products.

Example 5.1

Consider a simplified representation of a warehouse that stores four products: 1, 2, 3, and 4. Daily demand and replenishment quantities for the four products are given in Table 5.1. Replenishments are staggered, with product 1 being replenished on days 1, 5, 9, . . . ; product 2 being replenished on days 3, 11, 19, . . . ; product 3 being replenished on days 2, 8, 14, . . . ; and product 4 being replenished on days 6, 18, 30, ...

Due to the assumed deterministic conditions and the repeating inventory cycles for the four products, an aggregate cycle of 24 days will occur and repeat indefinitely as long as no changes occur in either the demand rates or replenishment plan. Table 5.2 depicts the end-of-period storage requirements for each of the four products. Assuming that daily demands are met prior to storing the replenishments, the storage requirements for the products are given by their replenishment quantities. Hence the storage requirement for dedicated storage equals the sum of the individual maxima, or 72 pallet loads.

5.2.2 Sizing on the Basis of Service Levels*

One approach that can be used to size storage under dedicated storage conditions is a *service-level* approach. Specifically, when demand for storage is a random variable, storage capacity can be determined on the basis of the probability of a shortage of space. With dedicated storage, Q_j storage slots are assigned to product j for $j = 1, \ldots, n$. Therefore, the probability of there being a sufficient number of storage positions for product j is simply the probability of storage demand being less than or equal to Q_j. Thus the probability is given by the cumulative distribution function $F_j(Q_j)$.

If the storage demands for the various products are statistically independent, the probability of there being one or more shortages of storage space is given by

$$\Pr(1 \text{ or more shortages}) = 1 - \Pr(\text{no shortages}) \tag{5.1}$$

* To comprehend the material presented in this section, you should have a basic understanding of probability theory, including the use of the normal and Poisson probability distributions. Also, a number of basic concepts of nonlinear optimization should be understood.

TABLE 5.1 DEMAND AND REPLENISHMENT DATA FOR EXAMPLE 5.1

Product	Daily demand (pallet loads)	Replenishment quantity (pallet loads)
1	2	8
2	5	40
3	2	12
4	1	12

where Pr(·) denotes the probability of (·) occurring. In this case the probability of no shortages is the probability of no shortage occurring for any product. Hence

$$\Pr(\text{no shortages}) = \prod_{j=1}^{n} \Pr(\text{no shortage for product } j) \tag{5.2}$$

TABLE 5.2 END-OF-PERIOD STORAGE REQUIREMENTS (STORAGE SLOTS) FOR EXAMPLE 5.1

Time period	Product 1	Product 2	Product 3	Product 4	Aggregate
1	8	10	2	5	25
2	6	5	12	4	27
3	4	40	10	3	57
4	2	35	8	2	47
5	8	30	6	1	45
6	6	25	4	12	47
7	4	20	2	11	37
8	2	15	12	10	39
9	8	10	10	9	37
10	6	5	8	8	27
11	4	40	6	7	57
12	2	35	4	6	47
13	8	30	2	5	45
14	6	25	12	4	47
15	4	20	10	3	37
16	2	15	8	2	27
17	8	10	6	1	25
18	6	5	4	12	27
19	4	40	2	11	57
20	2	35	12	10	59
21	8	30	10	9	57
22	6	25	8	8	47
23	4	20	6	7	37
24	2	15	4	6	27
Maximum	8	40	12	12	59

Since the terms on the right-hand side of equation (5.2) are the cumulative probabilities, equation (5.2) can be expressed as

$$\text{Pr(no shortages)} = \prod_{j=1}^{n} F_j(Q_j) \tag{5.3}$$

Therefore, on substituting equation (5.3) in equation (5.1), we obtain

$$\text{Pr(1 or more shortages)} = 1 - \prod_{j=1}^{n} F_j(Q_j) \tag{5.4}$$

Example 5.2

To illustrate the use of service-level considerations in sizing storage, consider a warehouse used to store five products, 1, 2, 3, 4, and 5. Suppose that the probability distribution for the number of storage slots required for each product on a given day can be approximated by a normal distribution, with the means and standard deviations given in Table 5.3. From Table 5.4, if the number of storage slots is normally distributed, the probability of the number of slots required being less than, say, 2.25 standard deviations above the mean will be equal to 0.98778. Therefore, the probability of the storage requirement being greater than the mean plus 2.25 times the standard deviation is 0.01222.

From equation (5.4), if space is assigned to each product so that each has an equal probability of a space shortage and that value is 0.01222, the overall probability of one or more shortages in the entire warehouse is equal to $1 - (0.98778)^5$ or 0.0596. Thus, having storage capacities equal to the mean plus 2.25 times the standard deviation will yield a probability of approximately 0.06 of having at least one space shortage on a given day.

The number of storage slots assigned to product j, Q_j, is determined as follows:

$$Q_j = M_j + z_j \text{SD}_j$$

where M_j = mean daily demand for storage, measured in storage slots
SD_j = standard deviation of daily demand for storage, measured in storage slots
z_j = parameter for a standard normal distribution that provides the desired probability of sufficient storage capacity (e.g., 2.25 for a probability of 0.98778)

TABLE 5.3 MEANS AND STANDARD DEVIATIONS FOR THE NUMBER OF STORAGE SLOTS REQUIRED IN EXAMPLE 5.2

Product	Mean	Standard deviation
1	30 slots	8 slots
2	40 slots	10 slots
3	50 slots	15 slots
4	50 slots	12 slots
5	40 slots	12 slots

TABLE 5.4 CUMULATIVE PROBABILITY DISTRIBUTION FOR THE STANDARD NORMAL DISTRIBUTION (MEAN OF 0.00 AND STANDARD DEVIATION OF 1.00)

z	$F(z)$	z	$F(z)$	z	$F(z)$	z	$F(z)$
0.00	0.50000	1.00	0.84135	2.00	0.97725	3.00	0.99865
0.05	0.51994	1.05	0.85314	2.05	0.97982	3.05	0.99886
0.10	0.53983	1.10	0.86433	2.10	0.98214	3.10	0.99903
0.15	0.55962	1.15	0.87493	2.15	0.98422	3.15	0.99920
0.20	0.57926	1.20	0.88493	2.20	0.98610	3.20	0.99931
0.25	0.59871	1.25	0.89435	2.25	0.98778	3.25	0.99943
0.30	0.61791	1.30	0.90320	2.30	0.98928	3.30	0.99952
0.35	0.63683	1.35	0.91149	2.35	0.99061	3.35	0.99960
0.40	0.65542	1.40	0.91924	2.40	0.99180	3.40	0.99966
0.45	0.67365	1.45	0.92647	2.45	0.99286	3.45	0.99972
0.50	0.69146	1.50	0.93319	2.50	0.99379	3.50	0.99977
0.55	0.70884	1.55	0.93943	2.55	0.99461	3.55	0.99981
0.60	0.72575	1.60	0.94520	2.60	0.99534	3.60	0.99984
0.65	0.74215	1.65	0.95053	2.65	0.99598	3.65	0.99987
0.70	0.75804	1.70	0.95543	2.70	0.99653	3.70	0.99989
0.75	0.77337	1.75	0.95994	2.75	0.99702	3.75	0.99991
0.80	0.78814	1.80	0.96407	2.80	0.99745	3.80	0.99993
0.85	0.80234	1.85	0.96784	2.85	0.99781	3.85	0.99994
0.90	0.81594	1.90	0.97128	2.90	0.99813	3.90	0.99995
0.95	0.82894	1.95	0.97441	2.95	0.99841	3.95	0.99996

Thus, for the example,

$$Q_1 = 30 + 2.25(8) = 48.00 \text{ slots}$$

$$Q_2 = 40 + 2.25(10) = 62.50 \text{ slots}$$

$$Q_3 = 50 + 2.25(15) = 83.75 \text{ slots}$$

$$Q_4 = 50 + 2.25(12) = 77.00 \text{ slots}$$

$$Q_5 = 40 + 2.25(12) = 67.00 \text{ slots}$$

Rounding off to obtain integer values yields 48 slots for the first product, 63 slots for the second product, 84 slots for the third product, 77 slots for the fourth product, and 67 slots for the fifth product—for a total storage capacity of 339 slots. There will be a 0.06 probability of at least one space shortage per day.

If the storage system is to be sized based on the probability of a space shortage occurring, two optimization problems come to mind. First, suppose that we want to be assured that the probability of a space shortage will be no greater than, say, 0.05; then we could determine the minimum number of storage slots required to provide the desired level of assurance. In general, if the minimum acceptable probability of no space shortage is **P**, we want to determine the number of storage slots to assign to each product such that we minimize the total amount of space without exceeding

P. Hence we wish to determine the values of $\{Q_1, Q_2, \ldots, Q_n\}$ such that the sum of the Q_j values is minimized. Stated mathematically, the problem reduces to

$$\text{minimize} \sum_{j=1}^{n} Q_j \tag{5.5}$$

subject to

$$\prod_{j=1}^{n} F_j(Q_j) \geq \mathbf{P} \tag{5.6}$$

$$Q_j \geq 0 \qquad \text{for all } j \tag{5.7}$$

where $F_j(Q_j)$ is the cumulative distribution function for the number of storage slots required for product j evaluated at the capacity level Q_j.

Example 5.3

A storage system is to be designed for storing three products: 1, 2, and 3. The number of storage slots required on any given day is a Poisson-distributed random variable; the storage requirements for the three products are statistically independent. The expected values of the daily space requirements are 10, 15, and 20, respectively, for the three products. (The associated cumulative probabilities are given in Table 5.5.) It is desired to minimize the total storage capacity, subject to the probability of at least one shortage occurring being no greater than 0.05.

The optimization problem can be solved using dynamic programming or, for those unfamiliar with dynamic programming, enumeration. No fewer than 70 slots will be required to assure that the probability of a space shortage will be no greater than 0.05. Although there are several feasible ways to allocate 70 slots, the allocation that yields the smallest shortage probability (0.0466) is to make the following assignment: 17 slots to product 1, 24 slots to product 2, and 29 slots to product 3.

The second optimization problem involves the allocation of a given aggregate level of storage space among multiple products in such a way that we maximize the probability of not having a space shortage. In particular, given a number of storage slots, **S**, the mathematical formulation of the resulting optimization problem is

$$\text{maximize} \prod_{j=1}^{n} F_j(Q_j) \tag{5.8}$$

subject to

$$\sum_{j=1}^{n} Q_j \leq \mathbf{S} \tag{5.9}$$

$$Q_j \geq 0 \qquad \text{for all } j \tag{5.10}$$

Example 5.4

For the situation considered in Example 5.3, suppose that an aggregate capacity of 60 is available. How should the storage space be allocated among the products if it is desired to minimize the probability of one or more shortages?

TABLE 5.5 CUMULATIVE PROBABILITY DISTRIBUTION FOR THE POISSON DISTRIBUTION

Mean	z	$F(z)$	Mean	z	$F(z)$	Mean	z	$F(z)$	Mean	z	$F(z)$
5.0	0	0.0067	10.0	5	0.0671	15.0	25	0.9938	20.0	39	1.0000
5.0	1	0.0404	10.0	6	0.1301	15.0	26	0.9967	25.0	7	0.0000
5.0	2	0.1247	10.0	7	0.2202	15.0	27	0.9983	25.0	8	0.0001
5.0	3	0.2650	10.0	8	0.3328	15.0	28	0.9991	25.0	9	0.0002
5.0	4	0.4405	10.0	9	0.4579	15.0	29	0.9996	25.0	10	0.0006
5.0	5	0.6160	10.0	10	0.5830	15.0	30	0.9998	25.0	11	0.0014
5.0	6	0.7622	10.0	11	0.6968	15.0	31	0.9999	25.0	12	0.0031
5.0	7	0.8666	10.0	12	0.7916	15.0	32	1.0000	25.0	13	0.0065
5.0	8	0.9319	10.0	13	0.8645	20.0	4	0.0000	25.0	14	0.0124
5.0	9	0.9682	10.0	14	0.9165	20.0	5	0.0001	25.0	15	0.0223
5.0	10	0.9863	10.0	15	0.9513	20.0	6	0.0003	25.0	16	0.0378
5.0	11	0.9945	10.0	16	0.9730	20.0	7	0.0008	25.0	17	0.0605
5.0	12	0.9980	10.0	17	0.9857	20.0	8	0.0021	25.0	18	0.0920
5.0	13	0.9993	10.0	18	0.9928	20.0	9	0.0050	25.0	19	0.1336
5.0	14	0.9998	10.0	19	0.9965	20.0	10	0.0108	25.0	20	0.1855
5.0	15	0.9999	10.0	20	0.9984	20.0	11	0.0214	25.0	21	0.2473
5.0	16	1.0000	10.0	21	0.9993	20.0	12	0.0390	25.0	22	0.3175
7.5	0	0.0006	10.0	22	0.9997	20.0	13	0.0661	25.0	23	0.3939
7.5	1	0.0047	10.0	23	0.9999	20.0	14	0.1049	25.0	24	0.4734
7.5	2	0.0203	10.0	24	1.0000	20.0	15	0.1565	25.0	25	0.5529
7.5	3	0.0591	15.0	2	0.0000	20.0	16	0.2211	25.0	26	0.6294
7.5	4	0.1321	15.0	3	0.0002	20.0	17	0.2970	25.0	27	0.7002
7.5	5	0.2414	15.0	4	0.0009	20.0	18	0.3814	25.0	28	0.7634
7.5	6	0.3782	15.0	5	0.0028	20.0	19	0.4703	25.0	29	0.8179
7.5	7	0.5246	15.0	6	0.0076	20.0	20	0.5591	25.0	30	0.8633
7.5	8	0.6620	15.0	7	0.0180	20.0	21	0.6437	25.0	31	0.8999
7.5	9	0.7764	15.0	8	0.0374	20.0	22	0.7206	25.0	32	0.9285
7.5	10	0.8622	15.0	9	0.0699	20.0	23	0.7875	25.0	33	0.9502
7.5	11	0.9208	15.0	10	0.1185	20.0	24	0.8432	25.0	34	0.9662
7.5	12	0.9573	15.0	11	0.1848	20.0	25	0.8878	25.0	35	0.9775
7.5	13	0.9784	15.0	12	0.2676	20.0	26	0.9221	25.0	36	0.9855
7.5	14	0.9897	15.0	13	0.3632	20.0	27	0.9475	25.0	37	0.9908
7.5	15	0.9954	15.0	14	0.4657	20.0	28	0.9657	25.0	38	0.9943
7.5	16	0.9980	15.0	15	0.5681	20.0	29	0.9782	25.0	39	0.9966
7.5	17	0.9992	15.0	16	0.6641	20.0	30	0.9865	25.0	40	0.9980
7.5	18	0.9997	15.0	17	0.7489	20.0	31	0.9919	25.0	41	0.9988
7.5	19	0.9999	15.0	18	0.8195	20.0	32	0.9953	25.0	42	0.9993
7.5	20	1.0000	15.0	19	0.8752	20.0	33	0.9973	25.0	43	0.9996
10.0	0	0.0000	15.0	20	0.9170	20.0	34	0.9985	25.0	44	0.9998
10.0	1	0.0005	15.0	21	0.9469	20.0	35	0.9992	25.0	45	0.9999
10.0	2	0.0028	15.0	22	0.9673	20.0	36	0.9996	25.0	46	1.0000
10.0	3	0.0103	15.0	23	0.9805	20.0	37	0.9998			
10.0	4	0.0293	15.0	24	0.9888	20.0	38	0.9999			

Again, using either dynamic programming or enumeration,the allocation of the 60 slots that yields the greatest probability of no space shortage will be found to be 14, 20, and 26 for products 1, 2, and 3, respectively. From the Poisson probabilities given in Table 5.6, the probability of no space shortage is (0.9165)(0.9170)(0.9221) or 0.7750.

5.2.3 Sizing on the Basis of Costs

The previous analysis of space requirements for dedicated storage was based entirely on service-level considerations. Under deterministic conditions, the size of the storage system was determined to be equal to the sum of the maximum requirements for each product. When random conditions exist, two approaches were considered: minimize the amount of space required to ensure that the probability of a shortage is no greater than a prespecified quantity and, given the storage capacity, allocate the space among the products so that the probability of no shortage is maximized.

Alternatively, storage size can be determined using cost models. Such models might reflect the costs of owning and operating space versus contracting space or incurring a space shortage. To motivate the consideration of cost models in sizing storage, consider a situation in which the cost to provide Q_j storage slots for product j is equal to the sum of a fixed cost of building Q_j slots, a variable cost of storing product j each time period, and a variable cost that occurs when the requirement for space exceeds Q_j. One formulation of such a situation, under deterministic conditions, follows:

$$\text{minimize} \quad \text{TC}(Q_1, \ldots, Q_n) =$$

$$\sum_{j=1}^{n} \left[C_0 Q_j + \sum_{t=1}^{T} \{ C_{1,t}[\min(d_{t,j}, Q_j)] + C_{2,t}[\max(d_{t,j} - Q_j, 0)] \} \right] \tag{5.11}$$

where
Q_j = "owned" storage capacity for product j
T = length of the planning horizon in time periods
$d_{t,j}$ = storage space required for product j during period t
$\text{TC}(Q_1, \ldots, Q_n)$ = total cost over the planning horizon as a function of the set of storage capacities
C_0 = discounted present worth cost per unit storage capacity owned during the planning horizon of T time periods
$C_{1,t}$ = discounted present worth cost per unit stored in owned space during time period t
$C_{2,t}$ = discounted present worth cost per unit stored in leased space or per unit of space shortage during time period t

$$\min(d_{t,j}, Q_j) = \begin{cases} d_{t,j} & \text{if } d_{t,j} < Q_j \\ Q_j & \text{if } d_{t,j} \geq Q_j \end{cases}$$

$$\max(d_{t,j} - Q_j, 0) = \begin{cases} 0 & \text{if } d_{t,j} - Q_j < 0 \\ d_{t,j} - Q_j & \text{if } d_{t,j} - Q_j \geq 0 \end{cases}$$

In Equation (5.11) the discounted present worth cost of building the space for product j is $C_0 Q_j$. The operating cost each time period is based on the amount of product j in storage, either the storage requirement ($d_{t,j}$) or the storage capacity (Q_j), whichever is the smallest. If the storage requirement is greater than the storage capacity, a space shortage occurs. Under such conditions, we assume that the excess requirement ($d_{t,j} - Q_j$) is met via leased storage at an incremental cost of $C_{2,t}$ per unit stored in leased space during period t.

Due to the separable nature of equation (5.11), the optimum storage capacity can be determined independently for each product. (For some of the homework problems given at the end of the chapter, this might not be the case. Thus, for more general formulations, search techniques and enumeration might be required to determine the optimum solution.)

The total cost function given by equation (5.11) can be shown to be piecewise linear and convex. Consequently, a simple solution procedure can be used to determine the optimum capacity. Before stating the procedure, let $C' = C_0/(C_2 - C_1)$. The optimum capacity can be obtained as follows:

1. Sequence in decreasing order the demands for space.
2. Sum the demand frequencies over the sequence.
3. When the partial sum is first equal to or greater than C', stop; the optimum capacity equals that demand level.

Example 5.5

To illustrate the use of the cost formulation, suppose that only one product is to be stored. Let $T = 10$ periods, $C_0 = \$20$ per space, $C_{1,t} = \$1$ per space period, $C_{2,t} = \$4$ per space period, and the storage requirements over the planning horizon equal 4, 6, 8, 10, 9, 8, 7, 6, 5, and 4 for time periods 1 to 10, respectively.

The ordered demands, frequencies, and partial sum are given in Table 5.6. $C' = C_0/(C_2 - C_1) = 20/(4 - 1) = 6.67$ and the partial sum exceeds 6.67 when the demand equals 6. Therefore, the optimum capacity is 6. The resulting total cost equals \$223; alternatively, providing a capacity of 5 yields a total cost of \$224 and providing a capacity of 7 yields a total cost of \$228.

Example 5.6

Suppose that $T = 50$, $C_0 = \$100$, $C_1 = \$4$, and $C_2 = \$8$, with demand for space given in Table 5.7. In this case, $C' = 100/(8 - 4) = 25$. From Table 5.8, the partial sum equals 25 when demand equals 120. Therefore, the optimum capacity is 120. (*Note*: Since the partial sum equals C', there are multiple optimum solutions, namely, $115 \le Q \le 120$.)

TABLE 5.6 DETERMINATION OF THE OPTIMUM STORAGE CAPACITY FOR EXAMPLE 5.5

Ordered demand	Frequency	Partial sum
10	1	$1 < 6.67$
9	1	$2 < 6.67$
8	2	$4 < 6.67$
7	1	$5 < 6.67$
6	2	$7 > 6.67$
5	1	$8 > 6.67$
4	2	$10 > 6.67$

TABLE 5.7 STORAGE REQUIREMENT DATA FOR EXAMPLE 5.6

Periods	Space required (pallet loads)	Periods	Space required (pallet loads)
1–5	100	26–30	120
6–10	120	31–35	115
11–15	125	36–40	110
16–20	130	41–45	105
21–25	125	46–50	100

The probabilistic counterpart to equation (5.11) is as follows:

$$\text{minimize} \quad E[\text{TC}(Q_1, \ldots, Q_n)] = \sum_{j=1}^{n} \left\{ C_0 Q_j + C_1 \sum_{t=1}^{T} \left[\sum_{d_{t,j} < Q_j} d_{t,j} p(d_{t,j}) + \sum_{d_{t,j} \geq Q_j} Q_j p(d_{t,j}) \right] + C_2 \sum_{t=1}^{T} \left[\sum_{d_{t,j} \geq Q_j} (d_{t,j} - Q_j) p(d_{t,j}) \right] \right\} \tag{5.12}$$

where $E[\text{TC}(Q_1, \ldots, Q_j)]$ = expected total cost over the planning horizon as a function of the set of storage capacities

$p(d_{t,j})$ = probability mass function for the storage requirement for product j during period t

C_1 = cost per unit stored in owned space per time period

C_2 = cost per unit stored in leased space or per unit of space shortage per time period

The first term on the right-hand side of equation (5.12) represents the cost of "building" a storage facility of size $Q_1 + \cdots + Q_n$. The second term represents the expected cost of storing material in the facility. The amount of product j stored during period t is $d_{t,j}$ when $d_{t,j} < Q_j$; otherwise, it is Q_j. The computation involves determining the expected value of the storage cost each period for each product. The third term represents the expected cost of storing the "excess" material in a leased facility. The amount of product stored in the leased facility is the difference in the

TABLE 5.8 DETERMINATION OF THE OPTIMUM STORAGE CAPACITY FOR EXAMPLE 5.6

Ordered demand	Frequency	Partial sum
130	5	$5 < 25$
125	10	$15 < 25$
120	10	$25 = 25$
115	5	$30 > 25$
110	5	$35 > 25$

storage requirement and the storage capacity for product j; the use of the leased facility occurs only when $d_{t,j} > Q_j$.

Letting $\overline{D}_{t,j}$ denote the expected value of $d_{t,j}$ and simplifying equation (5.12) gives

$$\text{minimize} \quad E[\text{TC}(Q_1, \ldots, Q_n)] = \sum_{j=1}^{n} \left\{ C_0 Q_j + \sum_{t=1}^{T} C_1 \overline{D}_{t,j} + (C_2 - C_1) \sum_{t=1}^{T} \sum_{d_{t,j} \geq Q_j} (d_{t,j} - Q_j) p(d_{t,j}) \right\} \tag{5.13}$$

Equation (5.13) is the sum of separable piecewise linear convex functions.* Hence the optimum solution can be determined for each product independently; furthermore, because of the piecewise linear nature of the cost equation, an optimum solution will equal one of the values of the random variable, storage demand. In this case, by letting δ denote the increment in demand for storage capacity and by requiring that the first forward difference of the objective function [i.e., $f(Q_j + \delta) - f(Q_j)$] be nonnegative at the optimum value of Q_j and that the first backward difference [i.e., $f(Q_j) - f(Q_j - \delta)$] be nonpositive, it can be shown that a necessary and sufficient condition for Q_j to be an optimum solution is for it to satisfy the following inequality.†

$$\sum_{t=1}^{T} \hat{F}_{tj}(Q_j) \geq C' \geq \sum_{t=1}^{T} \hat{F}_{tj}(Q_j + \delta) \tag{5.14}$$

where $\hat{F}_{tj}(x_j)$ = complementary distribution function for storage demand, evaluated at x_j

$$= \sum_{d_{t,j} \geq x_j} p(d_{t,j})$$

and $C' = C_0/(C_2 - C_1)$.

Example 5.7

To illustrate the cost modeling approach, suppose that only one product is to be stored; let $C_0 = 25$, $C_1 = 2$, and $C_2 = 6$, such that $C' = 25/(6 - 2) = 6.25$. Also, let the prob-

*To establish the convexity of equation (5.13), we show that the second difference of the cost equation with respect to Q_j is nonnegative. By showing that $f(Q_j)$ is convex, the objective function will be convex since it is the sum of convex functions. Based on incremental demands for storage capacity of δ, the second difference of $f(Q_j)$ is $f(Q_j + 2\delta) - 2f(Q_j + \delta) + f(Q_j)$; in this case, the second difference reduces to $p(d_{t,j})$ evaluated at $d_{t,j} = Q_j + \delta$; hence $f(Q_j)$ is convex since the probability is nonnegative by definition.

†Since the objective function is convex, it is sufficient to find the largest value of Q_j for which $f(Q_j) \leq f(Q_j - \delta)$; alternatively, one can find the smallest value of Q_j for which $f(Q_j) \leq f(Q_j + \delta)$. Performing both calculations yields the necessary and sufficient condition given by equation (5.14).

TABLE 5.9 PROBABILITY DISTRIBUTIONS FOR STORAGE DEMAND IN EXAMPLE 5.7

Storage demand (pallet loads)	Time periods									
	1	2	3	4	5	6	7	8	9	10
10	0.30	0.30	0.30	0.30	0.30	—	—	—	—	—
15	0.40	0.40	0.40	0.40	0.40	0.20	0.20	0.20	0.20	0.20
20	0.30	0.30	0.30	0.30	0.30	0.30	0.30	0.30	0.30	0.30
25	—	—	—	—	—	0.50	0.50	0.50	0.50	0.50

ability distribution for demand be as given in Table 5.9. The complementary cumulative distribution functions are given in Table 5.10. From Table 5.10,

$$\sum_{t=1}^{10} \hat{F}_t(15) = 8.5 > 6.25$$

$$\sum_{t=1}^{10} \hat{F}_t(20) = 5.5 < 6.25$$

Therefore, a storage capacity of size 15 should be provided. A calculation establishes that the resulting expected cost will be 900. For the example, a capacity of 10 will yield an expected cost of 945 and a capacity of 20 will yield an expected cost of 915.

In the preceding two sections we considered deterministic and probabilistic approaches to use in sizing a dedicated storage facility. Both service-level and cost-based formulations were considered. Throughout, we assumed considerable knowledge of the demand for storage space; in the case of probabilistic demand, we even assumed that we knew the underlying probability distributions for storage demand. Admittedly, it is a rare situation where such perfect knowledge of the future is available. Does this mean that the analyses are of no value if we do not have perfect knowledge concerning the future? Not at all!

Too frequently, deterministic approaches are used in the face of random conditions because of a lack of knowledge concerning the form of the underlying

TABLE 5.10 COMPLEMENTARY CUMULATIVE DISTRIBUTION FUNCTIONS FOR STORAGE DEMAND IN EXAMPLE 5.7

Storage capacity Q(pallet loads)	Probability of storage demand $\geq Q$ for time periods:										
	1	2	3	4	5	6	7	8	9	10	Sum
10	1.00	1.00	1.00	1.00	1.00	1.00	1.00	1.00	1.00	1.00	10.0
15	0.70	0.70	0.70	0.70	0.70	1.00	1.00	1.00	1.00	1.00	8.5
20	0.30	0.30	0.30	0.30	0.30	0.80	0.80	0.80	0.80	0.80	5.5
25	—	—	—	—	—	0.50	0.50	0.50	0.50	0.50	2.5

probability distributions. When this occurs, the designer either ignores the effects of randomness or attempts to compensate for random conditions by using "adjusted" deterministic estimates. In either case, zero variance is assumed. Experience indicates that the size of a storage facility is affected significantly by the variations in storage demand. For this reason, to ignore the impact of variation by taking a deterministic approach can lead to the storage facility being too small to accommodate the real requirements for storage.

One further benefit of using probabilistic models is the ability to determine the sensitivity of the storage capacity to changes in the form of the probability distribution and changes in the values of the parameters of the probability distributions. One might find, for example, that storage capacity is relatively insensitive to such changes; in such a case it is not necessary to spend considerable time and effort trying to obtain accurate representations of the probability distribution for storage demand. On the other hand, one might find that the storage capacity is quite sensitive to the form of the distribution or the values of its parameters; in this case, care should be taken to ensure that the best available estimates are used.

In the examples, we assumed that the distributions of storage requirements were known. We made the assumptions to simplify the discussion of the use of the probabilistic formulations. In actual applications, the distributions are determined by taking advantage of historical data, forecasting models, and the judgment and experience of persons in the organization who are familiar with the storage requirements. (The normal distribution is appropriate when central limit theorem conditions are met; similarly, the Poisson distribution is appropriate when an underlying Markov process is at work. For additional treatment of the warehouse sizing problem, see Refs. 65 and 66.)

5.2.4 Assigning Products to Storage/Retrieval Locations

With dedicated storage, products are assigned to storage/retrieval locations in an attempt to minimize the time required to perform the storage and retrieval operations. Of course, for dedicated storage to be feasible, you must have a sufficient number of storage slots to "dedicate" slots to products. In such a situation, the assignment problem becomes a matter of assigning products to slots according to an appropriate criterion. In our case the criterion will be to minimize some function of the distance traveled to store and retrieve the assigned products. To formulate the dedicated storage assignment problem, let

s = number of storage slots or locations
n = number of products to be stored
m = number of input/output (I/O) points
S_j = storage requirement for product j, expressed in number of storage slots
T_j = throughput requirement or activity level for product j, expressed by the number of storage/retrievals performed per unit time

$p_{i,j}$ = percent of the storage/retrieval trips for product j that are from/to input/output (I/O) point i
$t_{i,k}$ = time required to travel between I/O point i and storage/retrieval location k
$x_{j,k}$ = 1, if product j is assigned to storage/retrieval location k
= 0, otherwise
$f(x)$ = expected time required to satisfy the throughput requirement for the system

The formulation of the dedicated storage assignment problem is

$$\text{minimize} \quad f(x) = \sum_{i=1}^{m} \sum_{j=1}^{n} \sum_{k=1}^{s} \frac{T_j}{S_j} (p_{i,j} t_{i,k} x_{j,k}) \tag{5.15}$$

subject to

$$\sum_{j=1}^{n} x_{j,k} = 1, \qquad k = 1, \ldots, s \tag{5.16}$$

$$\sum_{k=1}^{s} x_{j,k} = S_j, \qquad j = 1, \ldots, n \tag{5.17}$$

$$x_{j,k} = (0, 1) \qquad \text{for all } j \text{ and } k$$

Equation (5.15) gives the expected time required to perform the required storages and retrievals during a time period. In particular, if product j is assigned to storage/retrieval location k ($x_{j,k} = 1$), then it takes $t_{i,k}$ time units to travel from input point i to storage location k and it takes $t_{i,k}$ time units to travel from retrieval location k to output point i. Since the total number of storage/retrieval locations for product j equals S_j, the probability of the storage/retrieval trip being from/to storage/retrieval location k is $1/S_j$ for those locations assigned to product j. The total number of storage/retrieval trips performed per time period for product j equals T_j; however, only $p_{i,j}$ percent of the total trips for product j are performed from/to I/O point i. Hence the expected time required to travel between storage/retrieval location k and I/O point i for product j is given by the product of T_j/S_j and $p_{i,j} t_{i,k} x_{j,k}$. Summing over all I/O points, products, and storage locations yields $f(x)$. Equation (5.16) ensures that only one product is assigned to storage/retrieval location k. Equation (5.17) ensures that the number of storage/retrieval locations assigned to product j equals S_j.

Again, our formulation of the storage/retrieval location assignment problem assumes that each of the S_j loads of product j is equally likely to be retrieved and that each of the S_j storage locations assigned to product j is equally likely to be selected for storage. If a first-in, first-out retrieval policy is used and storage is always performed at the location that has been empty the longest period of time, our assumptions will be valid.

On examining equation (5.15), notice it can be written equivalently as

$$f(x) = \sum_{j=1}^{n} \frac{T_j}{S_j} \sum_{k=1}^{s} x_{j,k} \sum_{i=1}^{m} (p_{i,j} t_{i,k}) \tag{5.18}$$

The term in parentheses represents the average amount of time required for product j to travel between storage/retrieval location k and the m I/O points. Letting

$$c_{j,k} = \sum_{i=1}^{m} p_{i,j} t_{i,k} \tag{5.19}$$

the objective function can be written as

$$f(x) = \sum_{j=1}^{n} \sum_{k=1}^{s} c_{j,k} x_{j,k} \tag{5.20}$$

where $c_{j,k} = (T_j/S_j)\check{t}_{j,k}$. Thus the dedicated storage assignment problem can be formulated as a transportation problem.

Example 5.8

To illustrate the formulation of a storage/retrieval location problem as a transportation problem, we consider a very simple situation involving six storage locations, two I/O points, and three products. Consider the storage rack face shown in Figure 5.1. A lift truck travels along the aisle, stops, and raises and lowers its forks to store and retrieve unit loads of the products at the indicated (*) locations. The times required to travel between the I/O points (*) and storage/retrieval locations (*) are as given in Table 5.11.

For the example, $m = 2$, $n = 3$, and $s = 6$. Also, we assume that $p_{i,j} = 0.5$ for all i,j combinations (i.e., the two I/O points are equally likely to be used for travel to/from the 6 storage/retrieval locations for all products). Hence $\check{t}_{j,1} = 25$ seconds, $\check{t}_{j,2} = 20$ seconds, $\check{t}_{j,3} = 15$ seconds, $\check{t}_{j,4} = 40$ seconds, $\check{t}_{j,5} = 35$ seconds, and $\check{t}_{j,6} = 30$ seconds for $j = 1$, 2, and 3.

Let the number of storages and retrievals to be performed per hour for the three products be as follows: $T_1 = 4$, $T_2 = 6$, and $T_3 = 3$. Also, let the number of storage positions required for each product be as follows: $S_1 = 2$ slots, $S_2 = 1$ slot, and $S_3 = 3$ slots.

The resulting formulation of the transportation problem will be as follows:

$$\begin{aligned} \text{minimize} \quad f(x) = {} & 50x_{1,1} + 40x_{1,2} + 30x_{1,3} + 80x_{1,4} + 70x_{1,5} + 60x_{1,6} + 150x_{2,1} \\ & + 120x_{2,2} + 90x_{2,3} + 240x_{2,4} + 210x_{2,5} + 180x_{2,6} + 25x_{3,1} \\ & + 20x_{3,2} + 15x_{3,3} + 40x_{3,4} + 35x_{3,5} + 30x_{3,6} \end{aligned}$$

subject to

$$\sum_{j=1}^{3} x_{j,k} = 1, \qquad k = 1, \ldots, 6$$

$$\sum_{k=1}^{6} x_{1,k} = 2$$

$$\sum_{k=1}^{6} x_{2,k} = 1$$

$$\sum_{k=1}^{6} x_{3,k} = 3$$

$$x_{j,k} = (0, 1) \qquad \text{for all } j \text{ and } k$$

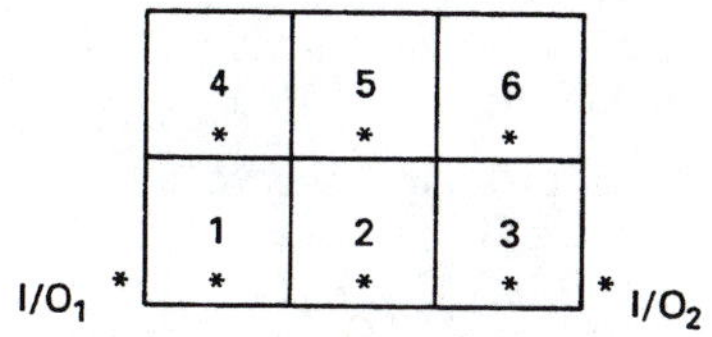

Figure 5.1 Storage rack layout for Example 5.8.

TABLE 5.11 TRAVEL TIMES FOR EXAMPLE 5.8

i	k	$t_{i,k}$	i	k	$t_{i,k}$
1	1	5 sec	2	1	45 sec
1	2	15 sec	2	2	25 sec
1	3	25 sec	2	3	5 sec
1	4	20 sec	2	4	60 sec
1	5	30 sec	2	5	40 sec
1	6	40 sec	2	6	20 sec

A tableau representation of the transportation problem is given in Figure 5.2. On solving the transportation problem, we find that product 1 should be assigned to storage locations 1 and 2, product 2 should be assigned to storage location 3, and product 3 should be assigned to storage locations 4, 5, and 6.

	1	2	3	4	5	6	Number of storage slots per product
Product 1	50 *	40 *	30	80	70	60	2
Product 2	150	120	90 *	240	210	180	1
Product 3	25	20	15	40 *	35 *	30 *	3
Maximum number of unit loads per storage slot	1	1	1	1	1	1	

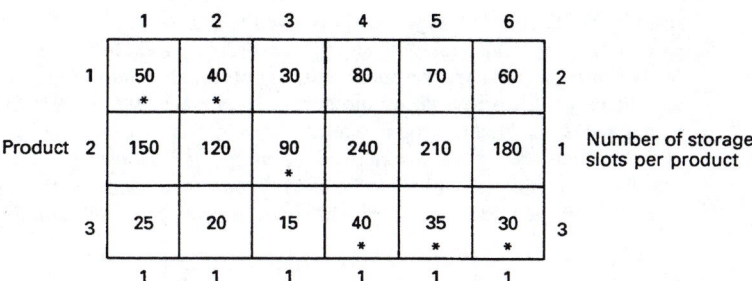

Figure 5.2 Transportation problem tableau for Example 5.8. (* denotes an optimum storage/retrieval location assignment.)

Due to a special property of the example problem, it is possible to solve the storage/retrieval location assignment problem without having to utilize one of the standard transportation problem algorithms. In particular, notice that the value of $p_{i,j}$ was the same for all values of j (i.e., $p_{i,j} = p_i$). *When the percentage of travel between I/O point i and the storage/retrieval locations is the same for all products*, the following procedure can be shown to yield an optimum solution to the dedicated storage assignment problem [26,51].

1. Number the products according to the ratio of their throughput (T_j) and storage (S_j) requirements, such that

$$\frac{T_1}{S_1} \geq \frac{T_2}{S_2} \geq \cdots \geq \frac{T_n}{S_n} \tag{5.21}$$

2. Compute the $\check{t}_k$ values for all products, where

$$\check{t}_k = \sum_{i=1}^{m} p_i t_{i,k} \tag{5.22}$$

3. Assign product 1 to the S_1 storage locations having the smallest $\check{t}_k$ values; assign product 2 to the S_2 unassigned storage locations having the next-lowest $\check{t}_k$ values; and so on.

The motivation for the ranking procedure is to put the item with the largest T_j to S_j ratio in the slots with the smallest average travel times ($\check{t}_k$ values), put the item with the next largest ratio in the slots with the next smallest travel times, and so on. As emphasized previously, the procedure is based on a critical assumption; namely, *all products being stored have the same percentage distribution of movement between the storage/retrieval locations and the I/O points*. Similarly, we assume that all storage and retrieval operations are "single command" operations (i.e., either one storage or one retrieval operation is performed per trip between storage and the I/O point).

Example 5.9

Consider the array of storage locations given in Figure 5.3. The storage slots might be openings in a storage rack with storage and retrieval via lift truck; alternatively, the storage locations might represent potential spots for stacking pallets on the floor. In this case, however, the array represents storage bays, each 20 ft × 20 ft. Rectilinear travel is used and is assumed to originate and/or terminate at the centroid of the storage bay.

Products are received from production through doors at I/O points 4 and 5; each receiving door is equally likely to be used. Products are shipped through I/O points 1, 2, and 3, with the middle door more likely to be used. Since full unit loads are assumed

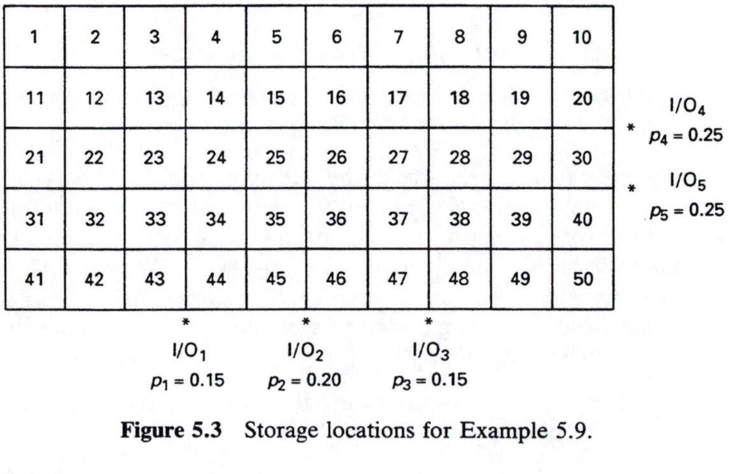

Figure 5.3 Storage locations for Example 5.9.

to be received and shipped, the number of loads received equals the number of loads shipped. Hence, 50% of the storage/retrieval activity is divided equally between I/O points 4 and 5. The remaining 50% of the movement is divided among I/O points 1, 2, and 3 as shown.

Four products are to be stored in the warehouse with only one type of product stored in a given storage bay. Products A to D have daily throughput requirements, measured in number of storages or retrievals per day, of 60, 70, 80, and 90, respectively; their storage requirements are 20, 10, 15, and 5 bays, respectively. Hence their throughput-to-storage ratios will be 60/20 = 3.0, 70/10 = 7.0, 80/15 = 5.3, and 90/5 = 18.0, respectively. Numbering the products in decreasing order of their throughput-to-storage ratios results in product D being number 1, product B being number 2, product C being number 3, and product A being number 4. (Hereafter, we refer to the products by their numbers rather than their alphabetic designations.) Product 1 requires 5 storage bays, product 2 requires 10 storage bays, product 3 requires 15 storage bays, and product 4 requires 20 storage bays.

We assume rectilinear travel at constant speeds in each dimension. Hence min-

205	185	165	148	134	124	118	115	115	115
185	165	145	128	114	104	98	95	95	95
170	150	130	113	99	89	83	80	80	80
165	145	125	108	94	84	78	75	75	75
165	145	125	108	94	84	78	75	75	75

* I/O$_4$

* I/O$_5$

* I/O$_1$ * I/O$_2$ * I/O$_3$

Figure 5.4 Expected distances traveled for Example 5.9.

4(A)	4(A)	4(A)	4(A)	4(A)	4(A)	4(A)	3(C)	3(C)	3(C)
4(A)	4(A)	4(A)	4(A)	3(C)	3(C)	3(C)	3(C)	3(C)	3(C)
4(A)	4(A)	4(A)	3(C)	3(C)	2(B)	2(B)	2(B)	2(B)	2(B)
4(A)	4(A)	4(A)	3(C)	3(C)	2(B)	2(B)	2(B)	1(D)	1(D)
4(A)	4(A)	4(A)	3(C)	3(C)	2(B)	2(B)	1(D)	1(D)	1(D)

* I/O$_4$

* I/O$_5$

* I/O$_1$ * I/O$_2$ * I/O$_3$

Figure 5.5 Dedicated storage assignments for Example 5.9.

imizing distance traveled is equivalent to minimizing travel time. The resulting $\check{t}_k$ values are given in Figure 5.4. To illustrate the computation, consider storage bay 1.

$$\check{t}_1 = (0.15)(140) + (0.20)(180) + (0.15)(220) + (0.25)(220) + (0.25)(240) = 205$$

Assigning products to storage bays in the order 1, 2, 3, and 4 yields the dedicated storage assignment given in Figure 5.5. The resulting expected distance traveled per day is obtained as follows:

$$\begin{aligned} f(x) &= (75 + 75 + 75 + 75 + 75)\frac{90}{5} \\ &\quad + (75 + 78 + 78 + 80 + 80 + 80 + 83 + 84 + 84 + 89)\frac{70}{10} \\ &\quad + (94 + 94 + 95 + 95 + 95 + 98 + 99 + 104 + 108 + 108 + 113 + 114 + 115 \\ &\quad + 115 + 115)\frac{80}{15} \\ &\quad + (118 + 124 + 125 + 125 + 128 + 130 + 134 + 145 + 145 + 145 + 148 \\ &\quad + 150 + 165 + 165 + 165 + 165 + 170 + 185 + 185 + 205)\frac{60}{20} \\ &= 29{,}823.67 \text{ ft/day} \end{aligned}$$

5.3 RANDOMIZED STORAGE LOCATION POLICY

Randomized storage, also referred to as *floating slot storage*, allows the storage location for a particular product to change or "float" over time. In practice, randomized storage is defined as follows. When a load arrives for storage it is placed in the "closest" open feasible location; retrievals occur on a first-in, first-out basis. If there is more than one input point, the storage location selected is the one "closest" to the input point through which the unit load enters the storage facility.

In modeling randomized storage it is commonly assumed that each empty storage slot is equally likely to be chosen for storage when a storage operation is performed; similarly, it is assumed that each unit of a particular product is equally likely to be retrieved when multiple storage locations exist for a product and a retrieval operation is performed. When the warehouse is relatively full, there are no significant differences in the travel distances obtained via the "equal likelihood" assumptions and those resulting from the "closest open slot" practice. However, for a "sparse warehouse" there can be significant differences in the travel distances obtained.

5.3.1 Space Requirements

With randomized storage, products can be stored in any available storage slot. Hence, the storage space requirements will equal the maximum of the aggregate storage requirements for the products.

Example 5.10

Recall Example 5.1. From Table 5.2, the maximum aggregate storage requirement is 59 storage slots in time period 20. Recall that for dedicated storage 72 storage slots were required. In this example, randomized storage reduces the storage requirements by 13 slots or 18%.

However, the randomized storage requirement depends on the aggregate inventory levels for the four products. The aggregate level can be quite sensitive to changes in the depletion and replenishment schedules of the various products. For instance, suppose that replenishments are not staggered as in Table 5.2. Instead, suppose that all four products are replenished at the same time. In this case, the randomized storage requirements would be the same as that for dedicated storage.

5.3.2 Sizing on the Basis of Service Levels and Costs

We have considered two approaches in sizing a storage facility: determining the minimum size that satisfies a service-level objective (expressed as a probability of a space shortage), and determining the size on the basis of trade-offs in the costs of providing space versus having insufficient space. In this section, both approaches are considered in sizing a storage facility for randomized storage.

Due to the dynamic conditions that typically exist in the replenishment of products, it is very difficult to predict *exactly* the storage requirements for randomized storage. For this reason, storage capacity levels are sometimes established by treating inventory levels of the products as random variables.

Example 5.11

Consider the data provided in Table 5.3 for Example 5.2. Recall that five products are to be stored. The means (and standard deviations) for the number of storage slots required were 30 (8), 40 (10), 50 (15), 50 (12), and 40 (12) for products 1 to 5, respectively. The mean aggregate number of storage slots equals the sum of the means for the five products, or 210 slots. Due to the assumed statistical independence among the products, the variance for the aggregate number of storage slots equals the sum of the variances for the five products, or 677. Hence the standard deviation for the aggregate number of storage slots equals approximately 26.02. Further, since the number of storage slots required for each product was given to be normally distributed, the aggregate number of storage slots is normally distributed.

From Table 5.4 we can determine the storage capacity having an equivalent probability of meeting the storage requirements obtained in Example 5.2 for dedicated storage. Recall from example 5.2 that the probability of a shortage of storage space was approximately 0.06. From Table 5.4, a shortage probability of 0.06 occurs at a point on the normal distribution approximately 1.56 standard deviations above the mean. Hence the storage capacity for randomized storage should be approximately 210 + 1.56(26.02), or 251 slots. Since 339 slots were required for dedicated storage, randomized storage can provide the same level of protection against a space shortage with 25.96% fewer storage slots.

Example 5.12

From Example 5.3, the storage requirements for three products were Poisson distributed with means of 10, 15, and 20. Since the sum of statistically independent Poisson distributed random variables is Poisson distributed, the aggregate storage requirement is Poisson distributed with a mean of 45.

We will use a normal distribution to approximate the Poisson distribution. Recall that the mean equals the variance for the Poisson distribution. The storage level that yields a shortage probability of 0.05, from Table 5.4, is 1.645 standard deviations above the mean. Therefore, the storage capacity should be equal to $45 + 1.645(45)^{0.5}$, or 56 storage spaces, whereas 70 spaces were required for dedicated storage.

Example 5.13

In Examples 5.11 and 5.12 we determined the storage capacity necessary to provide a given probability of a space shortage. Now we consider the randomized storage extension to example 5.4, in which 60 storage slots were available. With dedicated storage, an enumeration procedure was used to determine the allocation of space among the products that would minimize the probability of a space shortage; the resulting probability was 0.225. For randomized storage, we treat the aggregate storage requirement as a single product. Hence, to determine the shortage probability, we simply evaluate the cumulative distribution function for a Poisson-distributed random variable having a mean of 45. Using a normal approximation to the Poisson, we find that the shortage probability is 0.013, rather than 0.225 with dedicated storage.

One reason for randomized storage requiring less space than dedicated storage in the examples is the assumption of statistical independence among the products. The sum of the standard deviations of statistically independent random variables is greater than the standard deviation of the sum of statistically independent random variables. However, the assumption of statistical independence among products might not be valid. Peaks, valleys, and surges in demand for storage tend to occur together for groups of products (e.g., lawn furniture, lawn mowers, and spring wardrobe sales are highly correlated). Also, the inventory level of a product is serially correlated (i.e., correlated with itself over time).

Because of the likely presence of statistical dependence rather than statistical independence, we recommend that the results obtained using the statistical approaches be considered to be bounds or approximations on the amount of storage space required. Where possible, the distribution of the aggregate storage requirement should be developed directly rather than attempting to develop it from the distributions of the individual products.

5.3.3 Comparison with Dedicated Storage

Even though there are difficulties associated with determining exactly the storage requirements for randomized storage, it is still the case that randomized storage typically requires less space than does dedicated storage. Since this is the case, why isn't it always preferred to dedicated storage? One reason is the *necessity for having a good locator system to keep track of the location of products*; with dedicated storage you know the permanent or fixed address of the product. However, perhaps a more compelling reason relates to the travel times required to store and retrieve products under the two storage systems.

With dedicated storage, the "high-activity" products are assigned to the premium slots; "slow movers" are assigned to the least desirable slots. With randomized storage a premium slot might not be available when a high-activity product is to be stored; hence it occupies a less desirable slot. Similarly, when a slow mover arrives, if a premium slot is available, it will be taken by the slow mover. Thus it is expected that the travel times resulting from randomized storage will be greater than those resulting from dedicated storage.

In fact, Francis [24] showed that the dedicated storage layout resulting from solving the transportation problem formulated in Section 5.2.4 will yield an average distance traveled less than that which will result using randomized storage *as long as the same number of storage slots are required for both dedicated and randomized storage*. However, since randomized storage typically requires less storage space than does dedicated storage, perhaps the reduction in space will be sufficient to produce travel times less than those resulting from dedicated storage. The following example addresses the issue.

Example 5.14

Consider again the data for Example 5.9 and the average travel distances for the storage bays given in Figure 5.4. With randomized storage any product can be stored in any storage bay. Computing the average distance over the 50 storage bays yields an average travel distance of 115.4 ft. Multiplying the distance by the total number of storages and retrievals performed per day (300) gives an average travel distance of 34,620 ft/day, as opposed to 29,823.67 ft/day with dedicated storage.

Since it is anticipated that the storage requirements for the four products can be met with randomized storage by using less than 50 storage bays, it is instructive to determine the reduction in storage space required for randomized storage to yield the same daily travel distance as obtained using dedicated storage. Using enumeration, by eliminating the use of the 11 storage bays having the largest average travel distances (i.e., 1, 2, 11, 21, 31, 41, 3, 12, 22, 4, and 32) an average travel distance of 100.56 ft to/from the remaining 39 storage bays is obtained. The resulting daily travel equals 30,169 ft/day. Eliminating storage bay 42 reduces the daily travel total to 29,818 ft/day. Hence, for the example, the space savings realized from randomized storage must provide a reduction of at least 12 storage bays in order for randomized storage to yield a lower average travel distance than obtained from dedicated storage. Figure 5.6 provides the resulting storage layout for randomized storage.

As noted previously, there do not exist any simple ways to determine the storage requirements for randomized storage. A rule of thumb that is used by some firms is to multiply the average aggregate requirement by 1.70. There is no obvious justification for 1.70. From Table 5.2, the average aggregate requirement for Example 5.1 is 41; hence multiplying by 1.70 yields a level of 69.7 or 70 rather than the maximum aggregate requirement of 59. In fact, with such an approach for Example 5.1, the level established would be almost that required for dedicated storage (72).

Another rule of thumb is to set a storage level equal to 85% of that required

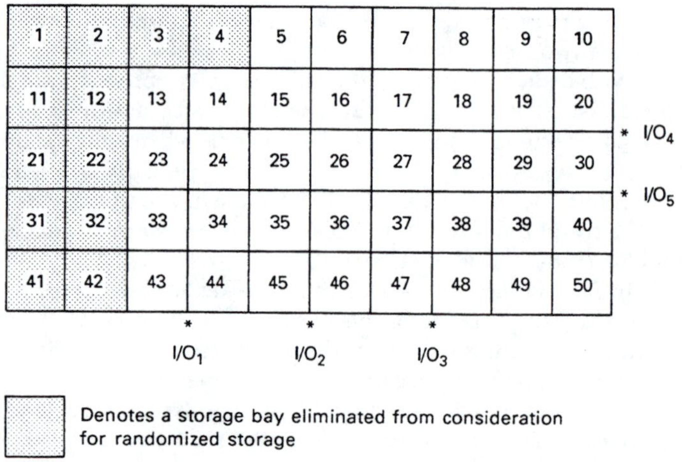

Figure 5.6 Storage space reductions required for randomized storage travel distance to be equivalent to dedicated storage distance in Example 5.14.

for dedicated storage. Again, no justification is given for using the 85% other than the mysterious claim that "it works." From Example 5.1, 85% of 72 is 61.2, which is quite close to the level of 59 established in Table 5.2. Despite the existence of rules of thumb concerning sizing for randomized storage, we strongly recommend that *the storage capacity be established on the basis of an analysis of needs* rather than by using rules of thumb.

5.4 CLASS-BASED DEDICATED STORAGE LOCATION POLICY

As a compromise between dedicated storage and randomized storage, *class-based dedicated storage* is frequently used. With class-based dedicated storage, products are divided into three, four, or five classes based on their throughput (T)-to-storage (S) ratios. The relatively few fast movers are categorized as class 1 products, next are the class 2 products, then class 3 products, and so on. Dedicated storage is used for the classes and randomized storage is used within a class.

It should be noted that the entire discussion of dedicated storage given in Section 5.2 applies to class-based dedicated storage if, instead of dealing with products, one deals with classes of products. For this reason, our treatment of class-based dedicated storage focuses on the formation of classes of products.

Example 5.15

Table 5.12 provides the throughput (T) and storage (S) requirements for 50 products to be accommodated in a storage facility, with the storage requirements representing

TABLE 5.12 THROUGHPUT AND STORAGE DATA FOR EXAMPLE 5.15

Product	T	S	T/S	Product	T	S	T/S
1	4	8	0.500	26	3	2	1.500
2	5	12	0.417	27	10	16	0.625
3	9	4	2.250	28	3	6	0.500
4	7	8	0.875	29	8	4	2.000
5	3	8	0.375	30	15	13	1.154
6	9	5	1.800	31	10	9	1.111
7	3	10	0.300	32	7	5	1.400
8	30	24	1.250	33	5	6	0.833
9	2	28	0.071	34	15	13	1.154
10	34	12	2.833	35	30	8	3.750
11	12	12	1.000	36	3	4	0.750
12	13	10	1.300	37	10	4	2.500
13	1	25	0.040	38	6	4	1.500
14	9	10	0.900	39	4	9	0.444
15	4	2	2.000	40	10	6	1.667
16	11	20	0.550	41	3	7	0.429
17	3	5	0.600	42	5	15	0.333
18	13	19	0.684	43	50	16	3.125
19	2	40	0.050	44	10	45	0.222
20	17	4	4.250	45	4	18	0.222
21	1	18	0.056	46	56	20	2.800
22	8	19	0.421	47	3	15	0.200
23	1	15	0.067	48	4	25	0.160
24	3	50	0.060	49	1	20	0.050
25	1	10	0.100	50	20	32	0.625

the number of storage slots required if dedicated storage were used. Also shown in Table 5.12 are the *T/S* ratios for the products. The *T/S* ratios range from 0.040 to 4.250.

After determining the *T/S* ratios, the products are ordered according to their *T/S* ratios, from largest to smallest. Next, the cumulative sum of the *T/S* ratios is determined (51.803). The ordinate of Figure 5.7 represents the normalized cumulative sum of *T/S* ratios; the abscissa represents the rank ordering of products. From the resulting figure it is apparent that 20% of the products (10) account for more than 50% of the cumulative throughput-to-storage ratios.

The boundary lines established for the classes are usually determined somewhat arbitrarily. As shown in Table 5.13, the top 10 products are assigned to class 1; the 15 products having the next largest *T/S* ratios are designated class 2; and the remaining 25 products are designated class 3. (Obviously, if 50 classes were established, the result would be a dedicated storage problem rather than a class-based dedicated storage problem. Similarly, if a single class is established, a randomized storage problem results.)

The throughput requirements for class 1 products total 227 and the sum of the individual storage requirements is 79. If, in fact, 85% of the dedicated storage requirement is needed for randomized storage, then 67 slots are needed for class 1. For class 2 products the throughput totals 158 and the dedicated storage requirement totals 145 slots; again, 85% of the dedicated storage total yields 123 slots for class 2. Class 3

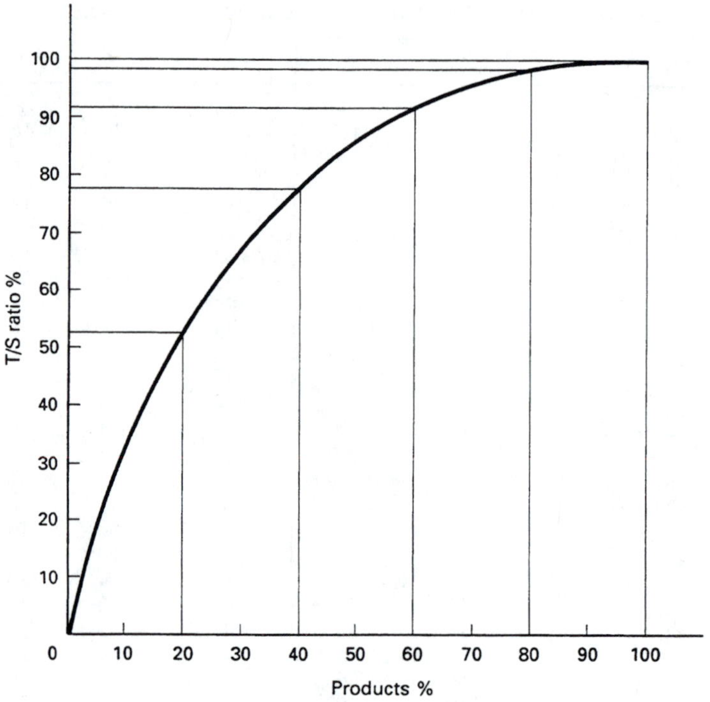

Figure 5.7 Distribution of throughput-to-storage ratio for Example 5.15.

TABLE 5.13 ASSIGNMENT OF PRODUCTS TO CLASSES IN EXAMPLE 5.15

Class	Products
1	3, 6, 10, 15, 20, 29, 35, 37, 43, 46
2	4, 8, 11, 12, 14, 18, 26, 30, 31, 32, 33, 34, 36, 38, 40
3	1, 2, 5, 7, 9, 13, 16, 17, 19, 21, 22, 23, 24, 25, 27, 28, 39, 41, 42, 44, 45, 47, 48, 49, 50

products have a throughput total of 115, with a storage total of 476 slots; a 15% reduction in space required indicates 405 slots are required for class 3. (Again, we do not recommend using the 15% reduction in dedicated storage space to determine the randomized storage requirement. We used the multiplier here for illustrative purposes.)

Given the throughput requirements for the three classes of products (227, 158, and 115) and the storage requirements for the classes (67, 123, and 405 slots), the dedicated storage assignment procedure can be used to determine the storage locations for each class. Since the sum of the storage requirements totals 595, we will use a total of 600 slots, as shown in Figure 5.8, and adjust the sizes of the three classes to be 68, 124, and 408.

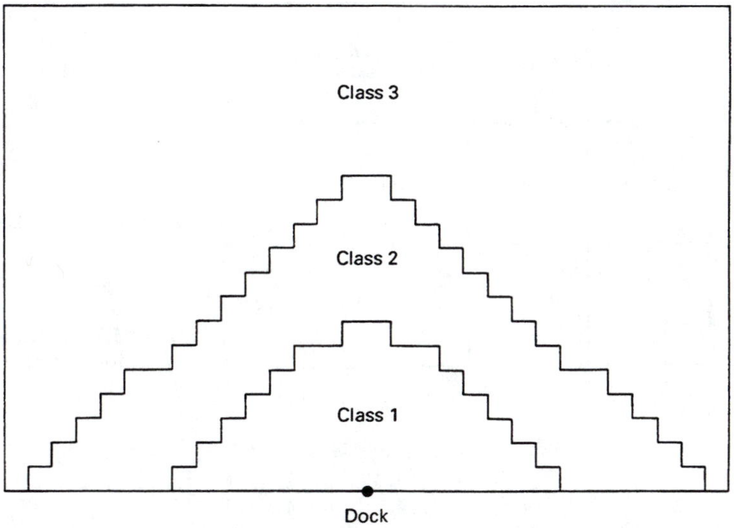

Figure 5.8 Class-based dedicated storage layout for Example 5.15.

For simplicity, we assume that rectilinear travel is performed between the single dock and the centroid of each storage slot. The warehouse is 20 slots deep and 30 slots wide. (It remains as an exercise for you to establish that the classes have been assigned to the appropriate storage slots.)

In Example 5.15 the classes were established on the basis of the throughput-to-storage ratios of the products. The boundary lines for the classes were established arbitrarily. What other possibilities exist for forming the classes? The following example addresses the issue.

Example 5.16

Consider a simplified storage problem involving the storage of pallet loads of five canned vegetables: broccoli, beans, cauliflower, corn, and peas. The number of storage bays required for each are 1, 3, 1, 2, and 3, respectively. The hourly throughput requirements are 2, 15, 3, 20, and 60, respectively. Ranking the products on the basis of throughput-to-storage ratios results in the following numerical assignments to products: peas (1), corn (2), beans (3), cauliflower (4), and broccoli (5).

The layout of the storage region is given in Figure 5.9, along with the average times required to store/retrieve loads assigned to the various storage bays. If dedicated storage is used, rather than class-based dedicated storage, one optimum assignment is to assign peas to bays 4, 5, and 6; corn to bays 9 and 10; beans to bays 2, 3, and 8; cauliflower to bay 7; and broccoli to bay 1. The resulting travel time required to perform the 100 storage/retrieval operations would be 2775 seconds.

Suppose that two classes are to be used. How should they be formed? There are 15 possible ways to form two classes without distinguishing between the classes. The

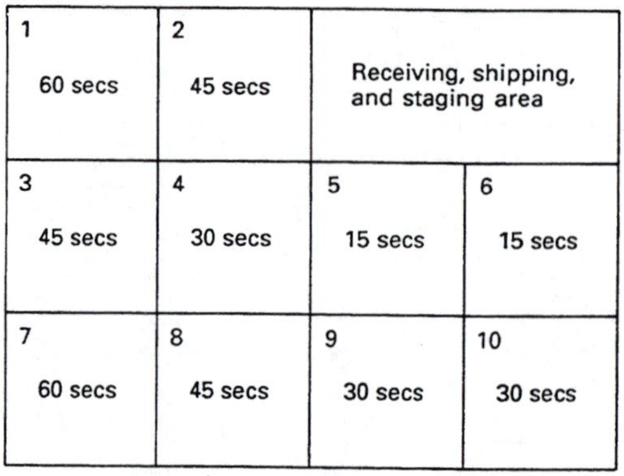

Figure 5.9 Layout of the storage region for Example 5.16.

15 combinations are listed as follows: {(1), (2, 3, 4, 5)}, {(2), (1, 3, 4, 5)}; {(3), (1, 2, 4, 5)}; {(4), (1, 2, 3, 5)}; {(5), (1, 2, 3, 4)}; {(1, 2), (3, 4, 5)}; {(1, 3), (2, 4, 5)}; {(1, 4), (2, 3, 5)}; {(1, 5), (2, 3, 4)}; {(2, 3), (1, 4, 5)}; {(2, 4), (1, 3, 5)}; {(2, 5), (1, 3, 4)}; {(3, 4), (1, 2, 5)}; {(3, 5), (1, 2, 4)}; and {(4, 5), (1, 2, 3)}. Of these, since the products were ranked $1 > 2 > 3 > 4 > 5$ on the basis of the *T/S* ratios, we will consider the following four combinations: {(1), (2, 3, 4, 5)}; {(1, 2), (3, 4, 5)}; {(1, 2, 3), (4, 5)}; and {(1, 2, 3, 4), (5)}. (You can consider the remaining possibilities if you wish.)

If product 1 is assigned to class 1 and all others are assigned to class 2, the resulting average time required to perform the 100 operations would be 3000 seconds. If products 1 and 2 are assigned to class 1, the time will be 2940 seconds. If products 1, 2, and 3 are assigned to class 1 and products 4 and 5 are assigned to class 2, the time will be 3328.125 seconds. Finally, if products 1, 2, 3, and 4 are assigned to class 1 and product 5 is assigned to class 2, the time will be 3550 seconds. Hence if two classes are used, it is best to assign products 1 and 2 to class 1 and the remaining products to class 2.

Suppose that three classes are to be used. How should they be formed? We consider the following six possibilities: {(1), (2), (3, 4, 5)}; {(1), (2, 3), (4, 5)}; {(1), (2, 3, 4), (5)}; {(1, 2), (3), (4, 5)}; {(1, 2), (3, 4), (5)}; and {(1, 2, 3), (4), (5)}. The resulting times required to perform the 100 storages/retrievals were 2820, 2865, 2935, 2895, 2917.5, and 3328.125 seconds. Thus if three classes are used, product 1 should be assigned to class 1, product 2 should be assigned to class 2, and the remaining products should be assigned to class 3.

It should be obvious that the best way to form four classes is as follows: {(1), (2), (3), (4, 5)}. The resulting time will be 2775 seconds.

Example 5.16 illustrated that the time required to perform the required storages and retrievals tends to decrease with an increase in the number of classes. However, the example was based on the storage level remaining constant. Recall that by grouping products into classes and using randomized storage within a class, reductions in space often occur.

The primary objective of the example was to illustrate that many alternatives are available concerning the formation of products into classes. In practice, the assignment of products to classes is performed somewhat arbitrarily, with consideration given to grouping products on the basis of the ranking of their *T/S* ratios.

Example 5.17

Before concluding the consideration of class-based dedicated storage, it is instructive to consider the assignment of classes to storage locations. Specifically, we consider the assignment of classes to storage locations within an automated storage and retrieval system (AS/RS) used for storing pallet loads of products. An elevation view of one side of a storage aisle is given in Figure 5.10. The storage rack has 10 storage levels vertically and 40 storage columns horizontally. The input/output (I/O) point for the system is 7.5 ft from the end of the aisle and elevated 20 ft, as shown. The storage slots are 5 ft × 5 ft in size. The storage/retrieval (S/R) machine travels horizontally at a speed of 250 ft/min; it travels vertically at a speed of 50 ft/min. Additionally, the S/R travels vertically and horizontally simultaneously; hence the time required to travel between two points is the maximum of the time to travel horizontally and the time to travel vertically. The S/R travels to the midpoint of the load support level for each opening. All travel is between a storage slot and the I/O point.

To illustrate the computation of travel time, consider the movement of the S/R from the I/O to the storage slot in the second column of storage and at the top level. The horizontal distance to be traveled to the midpoint of the support level is 15 ft; the vertical distance is 25 ft. The time required to travel horizontally is 15/250, or 0.06 minute; the time required to travel vertically is 25/50, or 0.50 minute. Hence the travel time between the I/O and the storage slot is the maximum of 0.06 and 0.50, or 0.50 minute.

The storage aisle includes 800 storage slots, since storage occurs on both sides of the aisle. Four classes of products are to be stored. One class requires 80 storage slots and has a throughput of 27 operations per hour, where an operation is defined to be either a storage or a retrieval. A second class requires 172 storage slots and has a throughput requirement of 15 operations per hour. A third class requires 248 storage slots and has a throughput of 6 operations per hour. The fourth class requires 300 slots and has a throughput of 2 operations per hour. Each operation requires 0.25 minute to either pick up or deposit (P/D) a load.

The throughput-to-storage ratios for the products indicate that the classes should be numbered in the same sequence as they were identified above. The resulting layout is given in Figure 5.11. The same layout applies on both sides of the aisle.

To determine the utilization of the S/R, note that a total of 50 operations are performed hourly and each operation requires that a load be picked up and deposited.

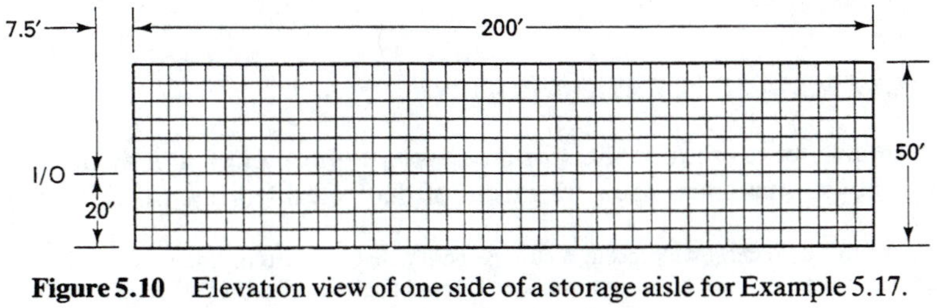

Figure 5.10 Elevation view of one side of a storage aisle for Example 5.17.

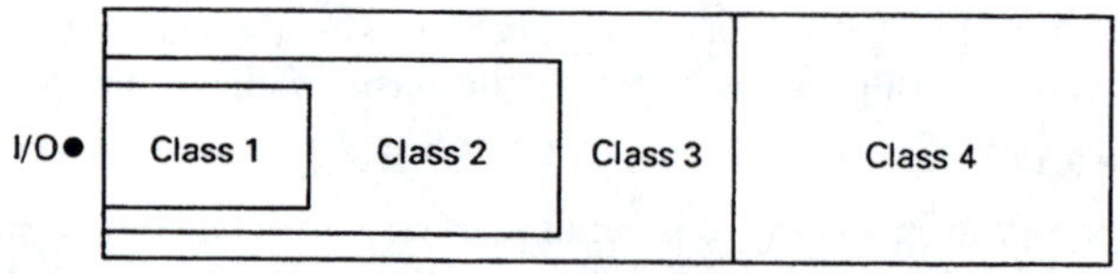

Figure 5.11 Assignment of classes to storage locations for Example 5.17.

Hence 25 minutes will be consumed each hour in performing the P/D requirements. This leaves 35 minutes each hour for traveling to and from storage locations.

We assume that each slot assigned to a class is equally likely to be selected for storage or retrieval. The round-trip travel time for class 1 products equals 0.3040 minute; for class 2 products 0.5977 minute; for class 3 products 0.9016 minute; and for class 4 products 1.36 minute. Thus the hourly travel totals 27(0.3040) + 15(0.5977) + 6(0.9016) + 2(1.360) or 25.303 minute. Thus the hourly utilization is (25 + 25.303)/60 or 83.83%.

Our consideration of class-based dedicated storage included the establishment of classes and the assignment of classes to storage locations. Before turning our attention to a fourth storage/retrieval location policy, shared storage, we should emphasize that our discussion of sizing for dedicated storage and class-based dedicated storage did not reflect fully the manner in which it is applied in practice. In particular, *it is seldom the case that the amount of space actually provided is equal to the sum of the maximum requirements for all products*. Instead, the total amount of space is reduced by some percentage, recognizing that there is the possibility that a space shortage condition might result.

One firm, for example, established five classes and five dedicated storage zones in each aisle of its automated storage and retrieval system. Each zone was sized at 90% of its maximum requirement. When a class 1 product arrived, if there was no space in zone 1, it was placed in zone 2; if sufficient space was not available in zone 2, it then tried, in consecutive order, zones 3, 4, and 5. When a class 2 product arrived, the computer searched, in consecutive order, for space in zones 2, 3, 1, 4, and 5. When a class 3 product arrived, storage space was sought, in order, in zones 3, 4, 2, 5, and 1. When a class 4 product arrived, in order, storage space was sought in zones 4, 5, 3, 2, and 1. The firm was pleased with the results; however, they did observe that it seldom occurred that space was assigned in other than the first two zone preferences.

5.5 SHARED STORAGE LOCATION POLICY*

In an attempt to reduce the storage space requirement for dedicated storage, some warehouse managers use a variation of dedicated storage in which the assignment

* In this section we present a storage policy that is based on industry practice but one that has received relatively little research attention. Goetschalckx and Ratliff [31] appear to be the first to study

of products to spaces is managed carefully. In particular, over time, different products use the same storage slot, albeit only one product occupies the slot when it is occupied. The location policy used is referred to here as *shared storage*.

To motivate our consideration of shared storage, consider the arrival of 100 pallet loads of a particular "fast mover" product to be stored in pallet rack. Pallet loads will be retrieved and shipped at a rate of 5 pallet loads per day over a 20-day period.

With randomized storage, 100 empty storage slots are selected "at random" for the product; no recognition is given to the fact that the product is a fast mover. With dedicated storage, on the other hand, at least 100 empty slots must be available among the premium locations assigned to the fast-mover product.

If randomized storage is used, each time a pallet load is removed from storage, the slot is available for use by the next product requiring storage. However, with dedicated storage, each removal of a pallet from storage creates an empty slot that will not possibly be filled until, at the earliest, the arrival of the next shipment of the same product.

Shared storage recognizes that while the product might be considered to be a fast mover, each pallet load stays in storage different lengths of time. Depending on the amount of the product in inventory at the time the shipment arrives, it is possible that five pallet loads will be in storage for only 1 day, whereas five other pallet loads within the same shipment will be in storage for 20 days. From the perspective of the storage positions in the warehouse, five pallet loads appear to be super fast movers; the remaining pallet loads are viewed as being less fast, perhaps even medium or slow movers. *Shared storage recognizes and takes advantage of the inherent differences in lengths of time that individual pallet loads remain in storage.*

5.5.1 Space Requirements

The storage space requirements for shared storage range from that required for randomized storage to that required for dedicated storage, depending on the amount of information available regarding the inventory levels over time for each product. As noted above, the distinction between shared storage and randomized storage is that the former involves total specificity regarding the storage locations for products, whereas with the latter, the locations depend solely on the emergence of empty slots within the warehouse. Shared storage and dedicated storage differ due to the distinction made by the former regarding *the time that each load of a product spends*

---

formally the shared storage policy. Relatively little experience has been gained in applying the policy to large-scale problems "optimally"; further, as will be seen, it is not immediately obvious how one might go about making optimum location decisions for large-scale applications. We present the policy because (a) multiple products "sharing" storage space is quite common in some industries, and (b) its features are sufficiently attractive to warrant our further attention. Since the material might be of less interest to those interested in "tried and true" storage policies, this section may be omitted without jeopardizing an understanding of the remaining material in the chapter.

in storage; dedicated storage assigns the total replenishment lot of a product to a number of storage positions based on *the average time spent in storage for the replenishment lot*.

A situation that naturally suggests the use of shared storage is a production line that is used to produce multiple products. Since products are produced sequentially rather than simultaneously, inventory replenishments are distributed over time. A beverage bottling line is an example of the type of situation we have in mind; other examples include production lines for paint, bleach, and industrial chemicals.

In each case cited, the same production equipment is used to produce different package sizes of the same product, as well as different products. A new setup or changeover is required between the production of different sizes or products. Hence it is not possible for the inventory levels of the various sizes and products produced on the same production line to be increasing at the same time. While one product is being produced, the inventory levels of the other products are decreasing. Not all products can be at their maximum inventory levels simultaneously. Hence the use of dedicated storage would result in some empty storage slots existing at all times.

Example 5.18

To illustrate the shared storage situation, suppose that a production line is used to produce two products, alpha and beta. The production schedule calls for alpha to be produced uniformly each of five days on odd-numbered weeks; beta is similarly produced during even-numbered weeks. Demand is constant over time and equal to one pallet load per day for each product. The daily production rate is two pallet loads, regardless of the product being produced. Demand occurs immediately following production and before storage occurs.

From Table 5.14, the storage requirement is constant over time and equal to five pallet loads. If randomized storage is used, five storage spaces are required; for dedicated storage, space must be provided for 10 loads. Shared storage will require five storage spaces; however, because of a knowledge of the time spent in storage for each

TABLE 5.14 INVENTORY DATA FOR EXAMPLE 5.18

		Beginning-of-day inventory (loads)			Production/day		End-of-day inventory inventory (loads)		
Week	Day	Alpha	Beta	Total	Alpha	Beta	Alpha	Beta	Total
Odd	1	0	5	5	2	—	1	4	5
	2	1	4	5	2	—	2	3	5
	3	2	3	5	2	—	3	2	5
	4	3	2	5	2	—	4	1	5
	5	4	1	5	2	—	5	0	5
Even	1	5	0	5	—	2	4	1	5
	2	4	1	5	—	2	3	2	5
	3	3	2	5	—	2	2	3	5
	4	2	3	5	—	2	1	4	5
	5	1	4	5	—	2	0	5	5

load, the assignment of storage locations can be performed to reduce the average travel time from that obtained using randomized storage.

Suppose that five storage positions are located along an aisle at coordinate points 1 to 5. Let the I/O point be located at the origin. A first-in, first-out inventory policy is used.

The time spent in storage by the kth load produced is shown in Table 5.15. The same distribution of times holds for both products. Recall that dedicated storage assigns the "most active" products to the "most preferred" locations. Shared storage, on the other hand, assigns the pallet load having the "shortest stay in storage" to the "most preferred" location.

Let A_i denote loads of alpha produced on day i during the 10-day cycle and let B_j denote loads of beta produced on day j during the 10-day cycle. The storage assignments for the units produced over a 4-week period are given in Table 5.16. The

TABLE 5.15 TIME SPENT IN INVENTORY FOR EACH LOAD PRODUCED FOR EXAMPLE 5.18

Unit load produced	Time spent in inventory (days)
1	0
2	1
3	1
4	2
5	2
6	3
7	3
8	4
9	4
10	5

TABLE 5.16 SHARED STORAGE ASSIGNMENTS FOR EXAMPLE 5.18

End of day:	Storage location 1	2	3	4	5	End of day:	Storage location 1	2	3	4	5
1	A1^	B4^	B4	B5	B5	11	A1^	B4^	B4	B5	B5
2	A2^	A2	B4^	B5	B5	12	A2^	A2	B4^	B5	B5
3	A3	A2^	A3	B5^	B5	13	A3	A2^	A3	B5^	B5
4	A3	A4	A3^	A4	B5^	14	A3	A4	A3^	A4	B5^
5	A3^	A4	A5	A4	A5	15	A3^	A4	A5	A4	A5
6	B1^	A4^	A5	A4	A5	16	B1^	A4^	A5	A4	A5
7	B2^	B2	A5	A4^	A5	17	B2^	B2	A5	A4^	A5
8	B3	B2^	A5^	B3	A5	18	B3	B2^	A5^	B3	A5
9	B3	B4	B4	B3^	A5^	19	B3	B4	B4	B3^	A5^
10	B3^	B4	B4	B5	B5	20	B3^	B4	B4	B5	B5

^Load to be retrieved at the beginning of the next day.

storage assignment is initialized as shown for time period 1. Since the single load of alpha (A) remains in storage for the shortest possible period of time, 1 day, we assign it to the most preferred location, coordinate 1. During period 2 we retrieve one load of alpha and beta, but we store the two loads of alpha that were just produced. We continue the process, with the results given in Table 5.16.

5.5.2 Throughput Performance

As demonstrated in the preceding section, under certain conditions, the storage space required to support shared storage can be the same as that required for randomized storage. Furthermore, as demonstrated in the following example, it is also the case that shared storage can deliver greater throughput benefits than either randomized or dedicated storage. (Here we define the throughput rate of a storage system to be the reciprocal of the average travel time required to store and retrieve material in the storage system.)

Example 5.19

The deterministic simulation of the storage system we provided in Table 5.16 illustrates that a repeating cycle exists with a duration of 10 days, beginning with day 1. During the repeating cycle, 20 loads are produced; however, only 18 loads are placed in storage. From Table 5.17, for those loads stored, the average distance traveled from the I/O point to the storage location equals 2.5 distance units. To compare the distances traveled using shared storage with those resulting from randomized storage and dedicated storage, notice that if each storage position is equally likely to be selected, the average distance traveled to perform a storage is 3.0 distance units. However, since randomized storage

TABLE 5.17 DISTANCE TRAVELED TO PERFORM RETRIEVAL OPERATIONS FOR EXAMPLE 5.19

Time period (day)	Distance traveled to perform retrieval operations	
	Using shared storage	Using randomized storage[a]
2	1 + 2 = 3	6
3	1 + 3 = 4	6
4	2 + 4 = 6	6
5	3 + 5 = 8	6
6	1	3
7	1 + 2 = 3	6
8	1 + 4 = 5	6
9	2 + 3 = 5	6
10	4 + 5 = 9	6
11	1	3
Total distance traveled	= 45	54
Number of retrievals	= 18	18
Average distance/retrieval	= 2.5	3.0

[a] Expected distance from the I/O points to a retrieval point is 3.0.

is often implemented by storing in the closest open location (COL) and retrieving on a first-in, first-out (FIFO) basis, it is instructive to consider the average travel distance using the closest open location (COL) method.

Table 5.18 depicts the COL method, beginning with all storage spaces empty and producing only alphas. A 10-day repeating cycle develops on day 5. As indicated in Table 5.19, the average distance traveled using the COL method is 2.78.

The average distance traveled to perform 18 storages using dedicated storage is 4.72. Table 5.20 depicts the 18 dedicated storages. In computing the travel distance, we gave the same benefit of COL travel to dedicated storage and assumed that storage occurred using COL storage and FIFO retrieval.

TABLE 5.18 CLOSEST OPEN LOCATION STORAGE ASSIGNMENTS FOR EXAMPLE 5.18

End	Storage location					End	Storage location				
of day:	1	2	3	4	5	of day:	1	2	3	4	5
1	A1^					11	B4^	B5	B4	A1^	B5
2	A2^	A2				12	A2^	B5	B4^	A2	B5
3	A3	A2^	A3			13	A3	B5^	A3	A2^	B5
4	A3^	A4	A3	A4		14	A3^	A4	A3	A4	B5^
5	A5	A4	A3^	A4	A5	15	A5	A4	A3^	A4	A5
6	A5	A4^	B1^	A4	A5	16	A5	A4^	B1	A4	A5
7	A5	B2^	B2	A4^	A5	17	A5	B2^	B2	A4^	A5
8	A5^	B3	B2^	B3	A5	18	A5^	B3	B2^	B3	A5
9	B4	B3^	B4	B3	A5^	19	B4	B3^	B4	B3	A5^
10	B4	B5	B4	B3^	B5	20	B4	B5	B4	B3^	B5

^Load to be retrieved at the beginning of the next day.

TABLE 5.19 DISTANCE TRAVELED TO PERFORM RETRIEVAL OPERATIONS FOR EXAMPLE 5.19, BASED ON COL STORAGE

Time period (day)	Distance traveled to perform retrieval operations using COL storage
6	3
7	2 + 3 = 5
8	2 + 4 = 6
9	1 + 3 = 4
10	2 + 5 = 7
11	4
12	1 + 4 = 5
13	1 + 3 = 4
14	2 + 4 = 6
15	1 + 5 = 6
Total distance traveled	= 50
Number of retrievals	= 18
Average distance/retrieval	= 2.78

TABLE 5.20 DISTANCE TRAVELED TO PERFORM STORAGES AND RETRIEVALS USING DEDICATED STORAGE FOR EXAMPLE 5.19

	Storage/retrieval location										Distance traveled		
Day	1	2	3	4	5	6	7	8	9	10	To perform retrievals	To perform storages	Total
1	A[a,b]					B5	B4[a]		B4	B5	8	1	9
2	A2[a,b]	A2[b]				B5			B4[a]	B5	1 + 7 = 8	1 + 2 = 3	11
3	A3[b]	A2[a]	A3[b]			B5				B5	1 + 9 = 10	1 + 3 = 4	14
4	A3[a]	A4[b]	A3	A4[b]		B5[a]					2 + 10 = 12	2 + 4 = 6	18
5	A5[b]	A4	A3[a]	A4	A5[b]						1 + 6 = 7	1 + 5 = 6	13
6	A5	A4[a]		A4	A5	B1[a,b]					3	6	9
7	A5			A4[a]	A5	B2[a,b]	B2[b]				2 + 6 = 8	6 + 7 = 13	21
8	A5				A5[a]	B3[b]	B2[a]	B3[b]			4 + 6 = 10	6 + 8 = 14	24
9	A5					B3[a]	B4[b]	B3	B4[b]		5 + 7 = 12	7 + 9 = 16	24
10						B5[b]	B4	B3[a]	B4	B5[b]	1 + 6 = 7	6 + 10 = 16	23
											85	85	170

[a] Retrieved at the beginning of the next day.
[b] Stored after retrievals are performed on this day.

Example 5.20

As a second example of shared storage, consider a storage situation involving three products: apples (A), bananas (B), and cantaloupes (C). The retrieval rate is one pallet load per day for each product. Apples are replenished every 6 days, bananas every 4 days, and cantaloupes every 3 days. Table 5.21 depicts the end-of-day inventory levels based on the replenishment schedules. The aggregate storage space requirement is eight storage slots. With dedicated storage, 10 slots are required. Table 5.18 lists the closest open location storage assignments for Example 5.18.

Table 5.22 provides the shared storage assignment for each day in the 12-day cycle. Table 5.23 depicts the dedicated storage assignment. Storage slots are numbered in increasing order of the average distance from the I/O point(s).

TABLE 5.21 END-OF-DAY INVENTORY LEVELS FOR EXAMPLE 5.20

Day	Apples	Bananas	Cantaloupes	Aggregate
1	0	3	2	5
2	5	2	1	8
3	4	1	0	5
4	3	0	2	5
5	2	3	1	6
6	1	2	0	3
7	0	1	2	3
8	5	0	1	6
9	4	3	0	7
10	3	2	2	7
11	2	1	1	4
12	1	0	0	1

TABLE 5.22 SHARED STORAGE ASSIGNMENT FOR EXAMPLE 5.20

	Storage/retrieval location							
Day	1	2	3	4	5	6	7	8
1	*B*	*C*	*B*	*C*	*B*			
2	*A*	*A*	*B*	*C*	*B*	*A*	*A*	*A*
3		*A*			*B*	*A*	A	A
4	*C*	*C*				*A*	*A*	*A*
5	*B*	*C*	*B*	*B*			*A*	*A*
6			*B*	*B*				*A*
7	*C*	*C*		*B*				
8	*A*	*C*	*A*		*A*	*A*	*A*	
9	*B*	*B*	*A*	*B*	*A*	*A*	*A*	
10	*C*	*B*	*C*	*B*	*A*	*A*	*A*	
11			*C*	*B*		*A*	*A*	
12							*A*	

TABLE 5.23 DEDICATED STORAGE ASSIGNMENT FOR EXAMPLE 5.20

	Storage/retrieval location									
Day	1	2	3	4	5	6	7	8	9	10
1	*C*	*C*	*B*	*B*	*B*					
2		*C*		*B*	*B*	*A*	*A*	*A*	*A*	*A*
3					*B*		*A*	*A*	*A*	*A*
4	*C*	*C*						*A*	*A*	*A*
5		*C*	*B*	*B*	*B*				*A*	*A*
6				*B*	*B*					*A*
7	*C*	*C*			*B*					
8		*C*				*A*	*A*	*A*	*A*	*A*
9			*B*	*B*	*B*		*A*	*A*	*A*	*A*
10	*C*	*C*		*B*	*B*			*A*	*A*	*A*
11		*C*			*B*				*A*	*A*
12										*A*

Shared storage combines the throughput-based storage assignment feature of dedicated storage with the floating slot feature of randomized storage. Savings in space are combined with reduced travel times in using the shared storage concept. Although the notion of shared storage has been used by warehouse managers for years, it has been used on an ad hoc basis; there is little evidence that it has been applied optimally.

From the foregoing it should be clear that shared storage can provide substantial benefits when precise information is available concerning the timing of storages and retrievals for individual loads of products. However, it would be incorrect to assume that the absence of such precise information negates the opportunity to benefit from using shared storage; the concept can still be applied, albeit in a modified form.

One lesson learned from our consideration of shared storage is that dedicated storage zones should be overlapped when using class-based dedicated storage. By retrieving loads first from the shared zone, space will be provided for other products. To realize the benefits from the shared zone, some firms relax the first-in, first-out retrieval rule; although they enforce FIFO for lots, they do not require FIFO within a lot.

5.6 CONTINUOUS WAREHOUSE LAYOUT

In this section we assume that the set of storage locations can be represented adequately as a "continuous" planar region, a set of positive area, rather than a set of discrete locations. The reasons for studying continuous warehouse layout are threefold. First, results from continuous formulations can provide insights concerning the underlying discrete problem. Second, many storage problems involve such a large number of storage locations that a continuous representation is quite appropriate. Third, the continuous problem may be easier to solve than the corresponding discrete problem.

5.6.1 Storage Region Configuration

Perhaps after some reflection it should be apparent that the shape of the continuous storage region for a product can be defined by a contour line. In Chapter 4 we introduced the concepts of level lines, level sets, and contour sets. In this chapter we apply these concepts in defining the boundaries of storage regions within a warehouse.

To motivate the use of contour lines, recall that in Chapter 4 we developed contour lines for the weighted distance traveled between new and existing facilities. In this chapter we let the existing facilities be the source (input) and destination (output) points for travel to/from storage; the percentage of travel between storage and the I/O points represent the "weights" between existing and new facilities.

By definition, a contour line encloses all points having an expected distance traveled less than or equal to the value of the contour line; we call the set of such points a level set or contour set. In this chapter we will be interested in constructing level sets of a given area in defining the storage regions to be used in a warehouse. A storage region defined by a level set of area A will have an expected distance less than or equal to the expected distance for any other set of equal area.

Based on the use of contour lines to define storage regions, the problem of determining storage configurations for continuous warehouse layouts reduces to a geometry problem. To illustrate the use of contour lines, we provide the following example problems.

Example 5.21

Suppose that randomized storage is to be used and a storage area of 152,000 square feet is to be provided. There is a single input/output point and it is to be located along the exterior of the storage facility. Rectilinear travel occurs between the I/O point and the centroid of a storage bay. Both discrete and continuous storage configurations are to be provided. The discrete storage situation involves the use of storage bays that are 20 ft × 20 ft. A total of 380 storage bays are required.

Applying the storage layout algorithm given in Section 5.2.4, the discrete storage layout solution to the example problem is given in Figure 5.12. The storage region is shaped like a pyramid: the base of the pyramid is defined by the exterior wall containing the I/O point and has a length equivalent to 38 storage bays; its "height" is equivalent to 19 storage bays. A computation establishes the expected distance traveled is 260 ft.

To solve the continuous version of the problem, the I/O point is considered to be the origin of an x-y coordinate system. The storage region will be required to lie in the first and fourth quadrants. From Chapter 4 the contour lines in the fourth quadrant will have slopes of $+1$ and those in the first quadrant will have slopes of -1. The resulting storage region is given in Figure 5.13. It, too, is triangularly shaped, with a base of 779.75 ft and a height of 389.87 ft; the I/O point is located centrally along the

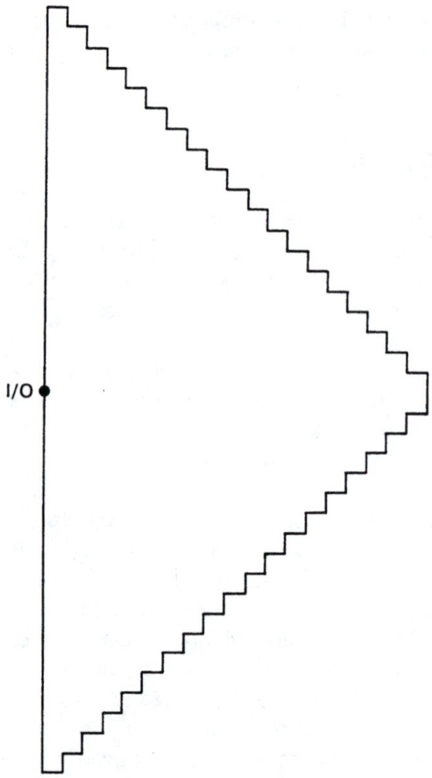

Figure 5.12 Discrete storage layout for Example 5.21.

base. A computation of the expected distance traveled yields a value of 259.9145 ft. Hence the continuous approximation is quite good.

The storage region obtained in Example 5.21 could be placed in a conventional rectangularly shaped building having dimensions at least equal to 800 ft × 400 ft. The remaining space in the facility could be used for offices, equipment storage, locker rooms, future expansion of the storage area, and the like. (Or, for that matter, it could define the walls of a triangularly shaped warehouse.) We again emphasize that the storage space configuration obtained is a benchmark solution and should be used as a design aid in obtaining the final warehouse layout. Also, the design of the storage region is based solely on the objective of minimizing the expected distance traveled between storage and the I/O point. It considers neither construction cost nor any other costs that are not proportional to expected distance traveled. Hence it might be necessary to modify the solution obtained before implementing it.

Example 5.22

In a number of practical situations a layout design is to be designed for an existing warehouse. To illustrate the approach taken, consider a warehouse of dimensions 200

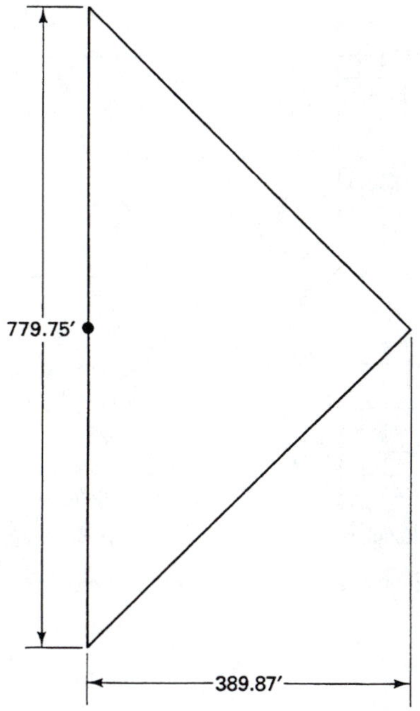

Figure 5.13 Continuous warehouse layout solution for Example 5.21.

ft × 150 ft having a single dock located as shown in Figure 5.14. Randomized storage is to be used. The storage space required will be either 18,000 or 27,000 ft^2. Assume that item movement is equally likely to occur between the dock and any point in the storage region, and assume that travel is rectilinear.

Constructing contour lines inside the existing warehouse results in three different families of geometric shapes, as shown in Figure 5.15. The smallest-valued contour line is triangularly shaped and is applicable for storage areas no greater than 10,000 ft^2; the next set of contour lines applies for storage areas between 10,000 and 20,000 ft^2; and the final set of contour lines applies for storage areas between 20,000 and 30,000 ft^2.

The storage area (A) enclosed by a contour line can be expressed as a function of the value of the contour line (k) as follows:

$$A = \begin{cases} k^2, & 0 \le k \le 100 \\ 200k - 10{,}000, & 100 \le k \le 150 \\ 30{,}000 - (250 - k)^2, & 150 \le k \le 250 \end{cases} \qquad (5.23)$$

In the first case, the contour line is triangularly shaped with a base of $2k$ and a height of k; its area is k^2 and its values vary from 0 to 100 ft as the area enclosed varies from 0 to 10,000 ft^2.

In the second case, picking the point where the contour line intersects the uppermost wall of the facility, the distance from the contour line to the I/O point is the sum of 100 ft of travel parallel to the y-axis and $(k - 100)$ ft of travel parallel to the x-axis. The contour line varies between 100 and 150 ft as the storage area ranges from 10,000 to 20,000 ft^2. The geometric shape of the contour line can be represented as the union

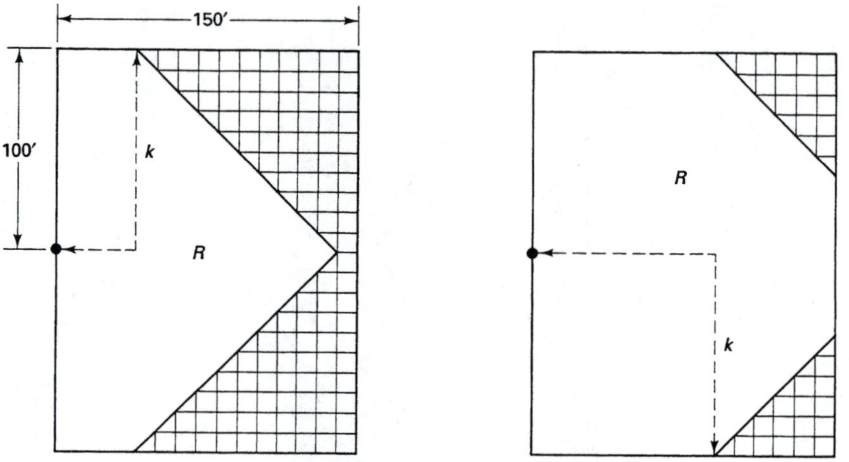

(a) Storage region of size 18,000 ft^2.

(b) Storage region of size 27,000 ft^2.

Figure 5.14 Continuous warehouse layout designs in an existing warehouse.

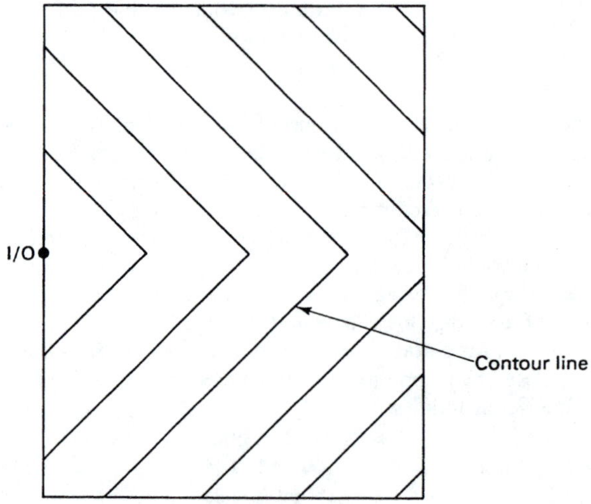

Figure 5.15 Contour lines for an existing warehouse.

of a rectangle, 200 ft × $(k - 100)$ ft in size, and a triangle 200 ft × 100 ft in size. Hence the area enclosed by a contour line is $200k - 30{,}000$.

In the third case, the area enclosed by the contour line can be obtained simply by subtracting the area not enclosed by the contour line from the total area for the building. Each corner of the building not enclosed by a contour line is triangularly shaped and of size $(250 - k) \times (250 - k)$. Hence the area enclosed by a contour line is equal to the area of the building, 30,000, less the sum of the areas of the two corners,

$(250 - k)^2$. The contour lines range in value from 150 to 250 ft as the storage area ranges from 20,000 to 30,000 ft^2.

Letting A equal 18,000 and solving for k gives a value of 140 ft and results in the design given in Figure 5.14(a). Letting A equal 27,000 and solving for k gives a value of 200 ft and results in the design given in Figure 5.14(b).

Example 5.23

Suppose that the storage region is to be in the first and fourth quadrants, two docks (P_1 and P_2) are to be located along the y-axis separated by a distance c, rectilinear travel to/from storage is equally likely to occur for each dock, and a storage area of A is to be provided. What will be the shape and dimensions of the storage region?

In Figure 5.16, r is the rectilinear distance from the intersection of the y-axis and contour line to the closest dock. The contour line encloses an area of

$$A = r(c + r) \tag{5.24}$$

Solving for r yields

$$r = 0.5[(4A + c^2)^{0.5} - c] \tag{5.25}$$

If a weight of 0.5 is associated with each dock, the relationship between r and k, the value of the contour line, is given by

$$k = 0.5r + 0.5(r + c)$$

or

$$r = k - 0.5c \tag{5.26}$$

From equations (5.24) and (5.26) we can determine the relationship between k and A. Although we do not need to know the relationship between k and A to determine the storage configuration, we will need to know the relationship subsequently when we determine the expected distance traveled.

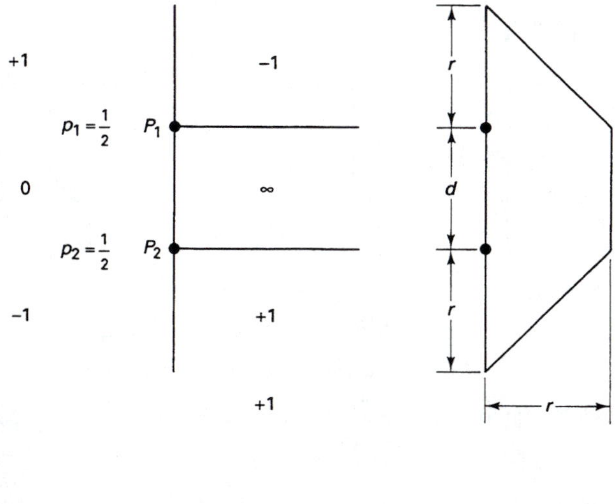

Figure 5.16 Continuous storage region for a two-dock situation.

Example 5.24

Suppose that the existing warehouse in Example 5.22 has two docks, one for truck delivery and one for rail delivery, as shown in Figure 5.17. Letting the weight for each dock be 0.5 gives the contour lines shown.

The area of the storage region enclosed by a contour line can be expressed as a function of the value of the contour line as follows:

$$A = \begin{cases} -3984.375 + 87.5k + 0.5k^2, & 87.5 \le k \le 162.5 \\ -937.5 + 150k, & 162.5 \le k \le 187.5 \\ -4453.125 + 262.5k - 0.5k^2, & 187.5 \le k \le 262.5 \end{cases} \tag{5.27}$$

for $7500 \le A \le 30{,}000$. For those contour lines for which $0 \le e \le 100$, $k = e + 87.5$. Solving for e in terms of A gives

$$e = \begin{cases} -175 + (15{,}625 + 2A)^{0.5}, & 75 \le A \le 23{,}437.50 \\ \dfrac{(A - 12{,}187.50)}{150}, & 23{,}437.50 \le A \le 27{,}187.50 \end{cases} \tag{5.28}$$

If A equals 18,000 ft^2, then e equals approximately 52.5 ft, k equals 140 ft, and the storage region is as depicted in Figure 5.17(a); if A equals 27,000 ft^2, then e equals approximately 98.75 ft, k equals 186.25 ft, and the storage region is as depicted in Figure 5.17(b).

The previous examples involved either randomized storage or a single product. As illustrated by the following three examples, the same approach can be used to determine the dedicated storage regions for multiple products and classes of products.

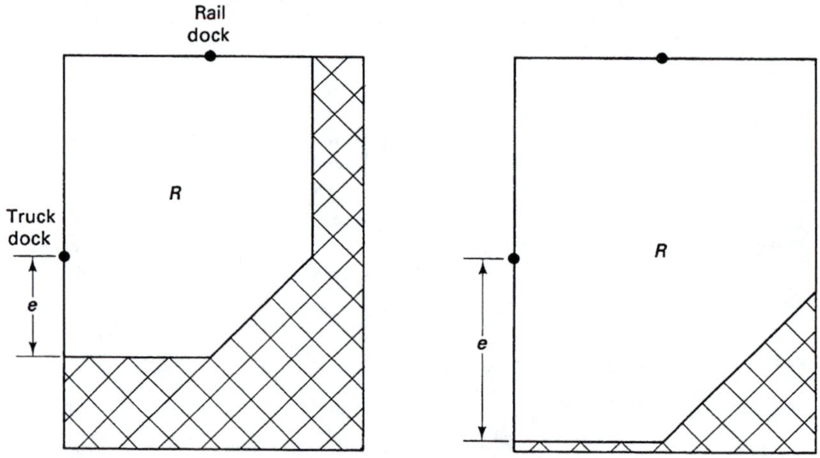

(a) Storage region of size 18,000 ft^2.

(b) Storage region of size 27,000 ft^2.

Figure 5.17 Two-dock layout design in an existing warehouse.

Example 5.25

Consider two products (1 and 2) where $S_1 = 2500\ \text{ft}^2$, $S_2 = 2400\ \text{ft}^2$, $T_1 = 100$ operations per day, and $T_2 = 50$ operations per day. Since $T_1/S_1 > T_2/S_2$, product 1 is placed in the layout first. Hence it is necessary for a contour line to be constructed enclosing an area of 2500 ft^2. A single dock is to be located along the y-axis and the storage region is to be in the first and fourth quadrants. Thus a triangular storage region results with a base of 100 ft and a depth or height of 50 ft. The union of the two storage regions will also be enclosed by a triangular contour line; since a combined area of 4900 ft^2 is to be enclosed, the base will be 140 ft and the height will be 70 ft as depicted in Figure 5.18.

Example 5.26

To illustrate a more complicated multiproduct application, consider the data from Example 5.9 involving five docks and four products. Using the contour line approach described in Chapter 4, we constructed the contour lines shown in Figure 5.19. The highest-ranking product, D, requires 2000 ft^2 for storage and is to be placed in the warehouse first. Next, product B is to be placed in the warehouse. The combined storage areas for B and D total 6000 ft^2; hence we construct a contour line that encloses 6000 ft^2. Finally, product C is to be placed in the warehouse. The combined storage areas for B, C, and D total 12,000 ft^2; we construct a contour line enclosing 12,000 ft^2. The remaining 8000 ft^2 of space is to be assigned to product A.

Example 5.27

Recall Example 5.17, involving an automated storage/retrieval system. Travel occurred simultaneously horizontally and vertically. Suppose that the I/O point is located at the origin and the storage region is to be constructed in the first quadrant. Horizontal travel occurs at a speed of 400 ft/min and vertical travel occurs at a speed of 80 ft/min. If a storage area of 18,000 ft^2 is to be provided, what should be the dimensions of the storage region to minimize the average time required to travel from the I/O point to a point in storage?

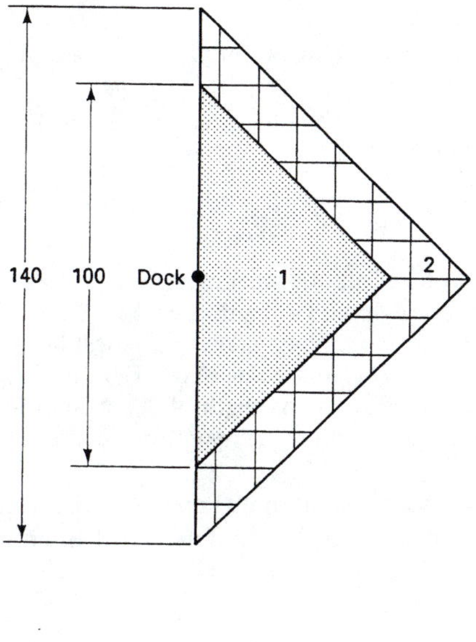

Figure 5.18 Single-dock, multiproduct storage region.

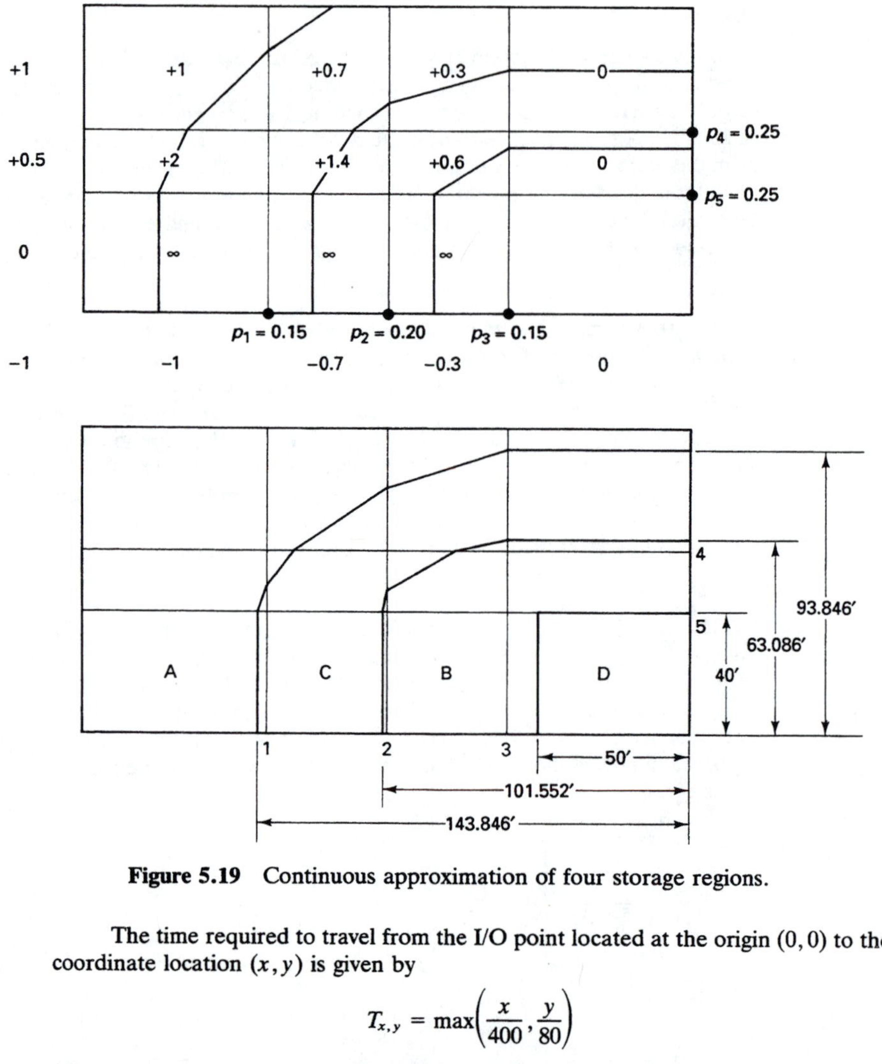

Figure 5.19 Continuous approximation of four storage regions.

The time required to travel from the I/O point located at the origin $(0, 0)$ to the coordinate location (x, y) is given by

$$T_{x,y} = \max\left(\frac{x}{400}, \frac{y}{80}\right)$$

Since the horizontal speed is five times the vertical speed, the contour line for travel from the I/O point will be a rectangle with a length equal to five times its height. Letting the height be H, the area (A) enclosed by a contour line will be $A = 5H^2$. Setting A equal to 18,000 ft^2 and solving for H gives a value of 60. Thus the storage region should be rectangular with a height of 60 ft and a length of 300 ft in order to minimize the expected travel time from the I/O to a random point in storage.

We realize that we used the popular mathematical procedure of "proof by assertion" to establish that a contour or level set of area A is the minimum-distance region. Rather than provide a formal proof of the property, we appealed to your intuition and gave a number of examples. The property has been proven formally using a special case of the Neyman–Pearson lemma from statistics [24,26].

5.6.2 Computing the Expected Distance Traveled

In the case of discrete storage locations, the expected distance traveled over the storage region can be determined by summing the expected distances for the individual products. The latter can be determined by summing the distances traveled to/from all storage locations assigned to a product, dividing the sum by the number of locations dedicated to the product, and multiplying the result by the average number of storages and retrievals performed per time period for the product. Similarly, in the case of continuous storage, the expected distance for the storage region dedicated to a product can be obtained by integrating over the storage region and multiplying the result by the throughput to space ratio for the product.

As an alternative to integrating over the spatial region, it is possible to integrate over the set of contour lines defining the region. To do so, it is necessary to develop a functional relationship between the value of a contour line and the area it encloses. To facilitate the description of the "contour line integration" approach, recall Example 5.21, involving the use of randomized storage. The storage region was served by a single dock located at the origin; the storage region was to be contained in the first and fourth quadrants; and rectilinear travel was assumed.

As depicted in Figure 5.13, the resulting contour line was a triangle. Choosing any arbitrary contour line of value k, the area it encloses (A) is equal to k^2. Therefore,

$$A = k^2 = q(k) \tag{5.29}$$

$q(k)$ is the functional relationship between A and k; specifically, it is the area of a level set of value k. Furthermore, on inverting equation (5.29),

$$k = A^{1/2} = r(A) \tag{5.30}$$

$r(A)$ is the inverse function relating k to A and is found by solving $q(k)$ for k. The inverse function $r(t)$ can be computed from $A = q(r(t))$. For example, $q(k) = k^2$ gives $A = r(A)^2$, or $r(A) = A^{1/2}$.

In general, as a contour line ranges from its minimum to its maximum value, the area it encloses changes from its minimum value to a value A. In this case the minimum value of the contour line can be obtained from equation (5.30) by letting A equal zero; its maximum value can be obtained by evaluating equation (5.30) at the storage area to be enclosed.

In Example 5.21, an area of 152,000 ft^2 is to be enclosed. Applying equation (5.30), the minimum value of the contour line will be zero and the maximum value will be 389.8718. To compute the expected distance traveled, we will use the following single integral expression:

$$E[\mathbf{R}] = \int_{\mathbf{R}} \frac{T}{A} f(\mathbf{X})\, d\mathbf{X} = \frac{T}{A} \int_{r(0)}^{r(A)} q'(k) k\, dk \tag{5.31}$$

where $E[\mathbf{R}]$ is the expected distance traveled over the storage region $\mathbf{R}$, T is the number of "trips" to or from storage performed per unit time (i.e., throughput), $f(\mathbf{X})$ is the average distance per trip to/from the point $\mathbf{X}$, and $q'(k)$ is the first derivative of $q(k)$ with respect to k.

To motivate equation (5.31), the cumulative distribution function for distance traveled is given by $q(k)/A$; hence the probability density function is given by $q'(k)/A$ for $r(0) \le k \le r(A)$. Thus the expected distance traveled is as shown. For a proof of equation (5.31), see Refs. 24 and 26.

Example 5.28

To illustrate the use of equation (5.31) in computing the expected distance traveled, recall the single product–single dock situation addressed in Example 5.21. Applying equation (5.31) yields

$$E[\mathbf{R}] = \frac{T}{A} \int_{0}^{A^{1/2}} (2k) k\, dk = \frac{2T}{3} A^{1/2} \tag{5.32}$$

Thus for a throughput of 1 per minute and an area of 152,000 ft^2, $E[\mathbf{R}] = 259.9145$ ft/min, as noted previously.

Example 5.29

Recall Example 5.22 involving two products and a single dock. Extending the results for a single product, the expected distance traveled is given by

$$E[\mathbf{R}_1, \mathbf{R}_2] = \frac{T_1}{A_1} \int_{r(0)}^{r(A_1)} (2k) k\, dk + \frac{T_2}{A_2} \int_{r(A_1)}^{r(A_1 + A_2)} (2k)\, k\, dk \tag{5.33}$$

where T_1 and T_2 are the throughput values for products 1 and 2, respectively.

The limits on the second integral result from the contour line enclosing the storage region for product 2 taking on values expressed in terms of the overall storage area enclosed. Hence for product 2, its contour lines range in value from the maximum value for product 1 to a value coinciding with the enclosure of the composite areas for the two products. Therefore,

$$E[\mathbf{R}_1, \mathbf{R}_2] = \frac{100}{2500} \int_{0}^{(2500)^{1/2}} (2k) k\, dk + \frac{50}{2400} \int_{(2500)^{1/2}}^{(4900)^{1/2}} (2k) k\, dk = 6361.11 \text{ ft}$$

Example 5.30

As an illustration of a two-dock situation, recall Example 5.23 and the associated Figure 5.16. Suppose that the storage area to be enclosed by the contour line is 10,000 ft^2, the docks are separated by a distance of 20 ft, and there are 100 storages or retrievals performed per hour. Travel is equally likely to be to/from either dock. From equation (5.26),

$$r = k - 0.5c$$

Substituting equation (5.26) in equation (5.24) gives

$$A = (k - 0.5c)(k + 0.5c)$$

or

$$A = k^2 - 0.25c^2 = q(k) \tag{5.34}$$

Furthermore, on solving for k as a function of A,

$$k = (A + 0.25c^2)^{1/2} = r(A) \tag{5.35}$$

and

$$r(0) = 0.5c$$

The expected distance traveled is given by

$$E[\mathbf{R}] = \frac{T}{A}\int_{0.5c}^{[A + 0.25c^2]^{1/2}} (2k)k\,dk = \frac{T}{12A}[(4A + c^2)^{3/2} - c^3] \tag{5.36}$$

For $c = 20$ ft, $T = 100$ per hour, and $A = 10{,}000$ ft^2, $E[\mathbf{R}] = 6760.25$ ft/hr.

Earlier we addressed the issue of randomized storage space requirements versus dedicated storage requirements. Based on the continuous space approach, it is possible to establish an upper bound on the amount of space required for randomized storage in order for it to yield an expected distance traveled equal to that for dedicated storage.

Example 5.31

To illustrate the approach, Example 5.30 is extended to include multiple product classes. Doing so yields the following results for product j, $j = 1, 2, 3$:

$$q(k_j) = k_j^2 - 0.25c^2$$

$$r(B_j) = (B_j + 0.25c^2)^{1/2}$$

where $B_j = A_1 + \cdots + A_j$. For the case of three product classes, the expected distance traveled is given by

$$E[\mathbf{R}_1, \mathbf{R}_2, \mathbf{R}_3] = \left(\frac{2}{3}\right)\left\{\frac{T_1}{A_1}[(B_1 + 0.25c^2)^{3/2} - (0.25c^2)^{3/2}] + \frac{T_2}{A_2}[(B_2 + 0.25c^2)^{3/2} - (B_1 + 0.25c^2)^{3/2}] + \frac{T_3}{A_3}[(B_3 + 0.25c^2)^{3/2} - (B_2 + 0.25c^2)^{3/2}]\right\}$$

Suppose that the total throughput is 100 operations per hour and the total storage requirement is 10,000 ft^2. Class I products represent 75% of the throughput and 15% of the storage space; class II products account for 20% of the throughput and 35% of the storage space; and class III products contribute only 5% of the throughput and represent 50% of the storage space. Letting $T_1 = 75$, $A_1 = 1500$, $T_2 = 20$, $A_2 = 3500$, $T_3 = 5$, and $A_3 = 5000$, the resulting throughput-to-space ratios will be 0.0500, 0.005714, and 0.0010 for the three product classes. Letting $c = 20$ ft, the expected distance traveled for the three classes will be 3677.49 ft/hr.

To establish an upper bound on the amount of space required for randomized storage to yield the same expected distance traveled as dedicated storage for the three product classes, set the expected distance traveled for one product class of unknown area equal to the expected distance traveled for the three product classes. Thus from equation (5.36), for $c = 20$ ft, $T = 100$ per hour, and $E[\mathbf{R}] = 3677.49$ ft/hr,

$$E[\mathbf{R}_{rs}] = \frac{100[(4A_{rs} + (20)^2)^{3/2} - (20)^3]}{12A_{rs}}$$

$$= 3677.49$$

Solving for A_{rs} by trial and error yields a value of approximately 2772 ft^2. Thus, based on the results obtained, for the two-dock case with three product classes having the given throughput-to-space ratios, in comparison with 10,000 ft^2 for dedicated storage, the space required for randomized storage would have to be no greater than 27.72% the size of the dedicated storage system to yield the same value for expected distance traveled.

Before concluding our consideration of expected distance computations, it is important to note that it is not always the case that identically shaped contour lines define the storage region of interest. In such situations it is necessary to break the integration into parts, one for each appropriate contour line. The following three examples illustrate a variety of situations involving changing shapes of contour lines.

Example 5.32

To illustrate the approach to be taken when multiple contour lines are involved, consider a storage region in the first and fourth quadrants with three I/O points located at the coordinate points $(0, c)$, $(0, 0)$, and $(0, -c)$, as depicted in Figure 5.20. Assuming that travel is equally divided among the three I/O points, the contour lines are as shown in Figure 5.20. In this case, there are two sets of contour lines; the first set is appropriate for $2c/3 < k < c$, the second set is appropriate for $k > c$.

For the first set of contour lines,

$$k = \tfrac{1}{3}(c - a) + \tfrac{1}{3}a + \tfrac{1}{3}(a + c)$$

which reduces to

$$k = \tfrac{2}{3}c + \tfrac{1}{3}a$$

Thus the parameter, a, is related to k and c as follows:

$$a = 3k - 2c$$

Also, in terms of the parameter a, the area enclosed by the first set of contour lines is

$$A = \frac{a^2}{3}$$

for $0 < A < c^2/3$, or

$$A = \frac{(3k - 2c)^2}{3} = q(k) \tag{5.37}$$

for $0 < A < c^2/3$. Thus

$$k = \frac{2c + (3A)^{1/2}}{3} = r(A) \tag{5.38}$$

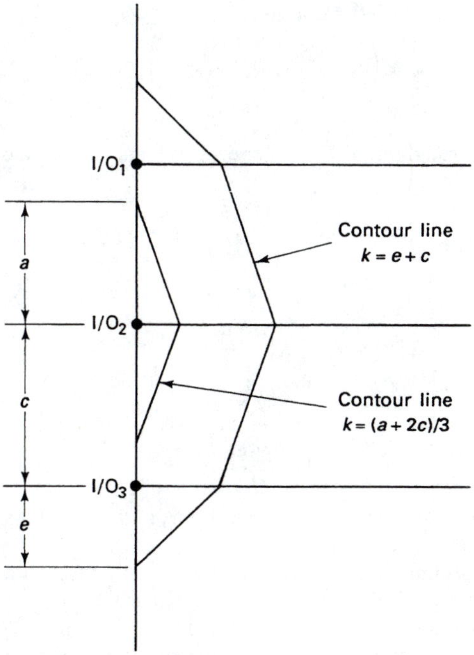

Figure 5.20 Three-dock example of contour line construction.

for $2c/3 < k < c$.

For the second set of contour lines,

$$k = \tfrac{1}{3}e + \tfrac{1}{3}(e + c) + \tfrac{1}{3}(e + 2c)$$

or

$$k = e + c$$

Thus the parameter e is related to k and c as follows:

$$e = k - c$$

The area enclosed by the second set of contour lines is given in terms of e and c as

$$A = e^2 + 2ec + \frac{c^2}{3}$$

Thus, in terms of k and c, the area is

$$A = (k - c)^2 + 2(k - c)c + \frac{c^2}{3}$$

or

$$A = k^2 - \frac{2c^2}{3} = q(k) \tag{5.39}$$

for $A > c^2/3$. Solving equation (5.39) for k gives

$$k = \left(A + \frac{2c^2}{3}\right)^{1/2} = r(A) \tag{5.40}$$

for $k > c$.

Therefore, given a throughput of T, the expected distance traveled over a storage region of area A is given by

$$E[\mathbf{R}] = \frac{T}{A}\left\{\int_{2c/3}^{c} (6k - 4c)k\,dk + \int_{c}^{[A + 2c^2/3]^{1/2}} (2k)k\,dk\right\}$$

which reduces to

$$E[\mathbf{R}] = \frac{T}{A}\left(\frac{2}{3}K_1^3 - \frac{10}{27}c^3\right) \tag{5.41}$$

where

$$K_1 = \left(A + \frac{2c^2}{3}\right)^{1/2}$$

For the case in which $c = 20$ ft, $T = 100$ per hour, and $A = 10{,}000$ ft^2, a calculation establishes that $E[\mathbf{R}] = 6905.47$ ft/hr.

It is interesting to compare the result (6905.47) with that obtained for the two-dock equivalent in Example 5.30 (6760.25). Also, compare the two-dock result with the expected distance for the one-dock equivalent, from equation (5.32), (6666.67). Although we will not give a formal proof of the property, the example demonstrates that the expected distance traveled increases with an increasing number of docks when travel to/from storage is randomly distributed over the docks. For a proof of the property, see Refs. 24 and 26.

Example 5.33

To illustrate the computation of the expected travel time for other than rectilinear distances, consider a storage aisle for an AS/RS with the I/O located at the lower left-hand corner of the storage region. Let the time required to travel horizontally from the I/O to the farthest storage point be t_h and let the time required to travel vertically from the I/O to the farthest storage point be t_v, where $t_v < t_h$.

A contour line for travel time is rectangularly shaped, as shown in Figure 5.21. Let A continue to denote the area to be enclosed by the contour line and let k denote the value of the contour line. For $k < t_v$, the contour line is square in time and of dimension $k \times k$; for $k > t_v$, the contour line is rectangular, of height t_v and length k. Therefore, the determination of the expected travel time is performed in two steps.

Consider first the square-in-time contour line. The area enclosed by the contour line is k^2. Therefore, $A = k^2 = q(k)$, $k = A^{1/2} = r(A)$, and $0 < k < t_v$. For the second set of contour lines, the value of the contour line is k; the area enclosed by the contour line is kt_v. Hence $A = kt_v = q(k)$, $k = A/t_v$, and $t_v < k < t_h$, since $A = t_v t_h$. The expected time traveled is given by

$$E[\mathbf{R}] = \frac{1}{t_v t_h}\left[\int_0^{t_v} (2k)k\,dk + \int_{t_v}^{t_h} t_v k\,dk\right] = \frac{t_v^2 + 3t_h^2}{6t_h} \tag{5.42}$$

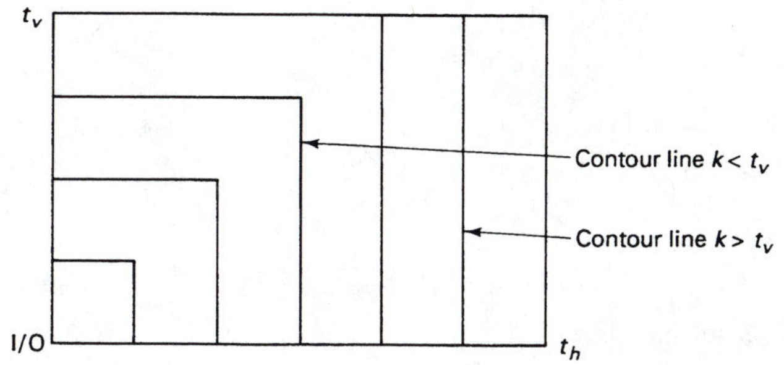

Figure 5.21 Contour lines for an automated storage/retrieval system.

Letting $t_v = bt_h$, where $0 < b < 1$ is called the shape factor, equation (5.42) reduces to

$$E[\mathbf{R}] = \frac{t_h(b^2 + 3)}{6} \tag{5.43}$$

(Recall that $E[\mathbf{R}]$ is based on one way travel. For round-trip travel, the value obtained must be doubled.)

Example 5.34

As a final example of the determination of expected distance traveled, consider Figure 5.22. A storage region is to be defined in the first quadrant; two I/O points are to be provided at the coordinates $(c, 0)$ and $(0, c)$; travel to/from storage is equally divided between the two I/O points; and travel is rectilinear. In this case there exists a minimum-distance region rather than a minimum-distance point. Hence all points in the square region shown are the same distance, c, from the I/O points. Outside the square region, the contour lines are as shown in Figure 5.22. The value of a contour line enclosing an area greater than c^2 is defined to be

$$k = 0.5e + 0.5(e + 2c)$$

or

$$k = e + c$$

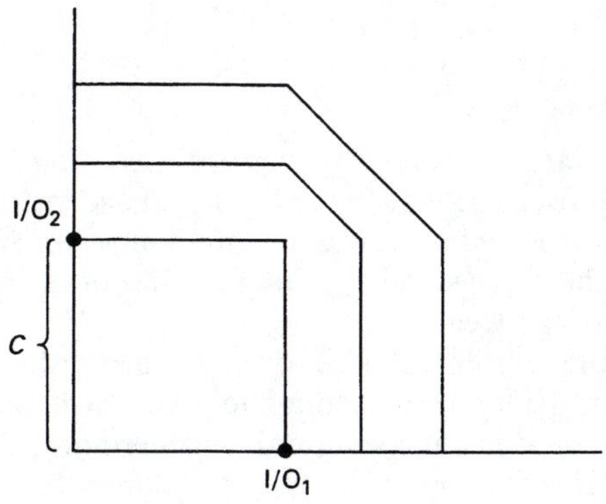

Figure 5.22 Example of a warehouse layout with a minimum storage region.

Therefore,

$$e = k - c$$

The area enclosed by the contour line is

$$A = 0.5e^2 + c^2 + 2ce$$

or

$$A = 0.5(k^2 - c^2) + kc = q(k) \tag{5.44}$$

for $A > c^2$. Therefore, on solving for k,

$$k = [2(A + c^2)]^{1/2} - c = r(A) \tag{5.45}$$

for $k > c$.

Given a throughput of T, the expected distance traveled will be

$$E[\mathbf{R}] = \frac{T}{A}\left[(c^2)c + \int_c^{[2(A + c^2)]^{1/2} - c} (k + c)k\,dk\right] \tag{5.46}$$

or

$$E[\mathbf{R}] = \frac{T}{6A}(2K_2^3 - 3K_2^2 c + 2c^3)$$

where

$$K_2 = [2(A + c^2)]^{1/2}$$

In equation (5.46) the first term reflects the value of the contour line over the square region c, applied for a region of area c^2. The second term represents the contribution made by the remaining area of $A - c^2$; the value of the contour line ranges from a value of c to a value of $[2(A + c^2)]^{1/2} - c$. Therefore, if $c = 100$ ft, $T = 100$ per hour, and $A = 25{,}000$ ft^2, then a calculation yields $E(\mathbf{R}) = 12{,}027$ ft/hr. In this case $e = 64.575$ and the maximum value of k is 164.575. The resulting storage region is almost a square region of dimension 164.575 ft $\times$ 164.575 ft, except for the removal of an area of $(0.5)(64.575)^2$ ft^2 from the upper right-hand corner of the "square."

5.7 SUMMARY

In summary, the design of material storage systems represents a fertile area for applying the results obtained from research on facility layout and location problems. Of the numerous problems that must be addressed in designing storage systems, we considered two explicitly in the chapter: determining the size of the storage system and assigning products to storage locations.

The determination of the "optimum" size of the storage system is a complex undertaking because of the difficulty of projecting the joint distribution of demand for space from the multiple products to be stored. Furthermore, the size of the storage system is strongly influenced by the storage location policy used.

We considered four storage location policies and we presented a number of mathematical models for determining the size of the storage system and assigning products to storage locations. Among the models presented were several that allowed the computation of expected distance traveled for continuous storage by integrating over the value of the contour line, rather than integrating over the spatial region.

Although the consideration of storage systems layout was abbreviated, hopefully, it served to establish the opportunity for applying a number of the location models presented in the text. With the increased emphasis on reducing (if not eliminating) the need for storage, it is essential that storage systems be designed well, taking into account space and throughput requirements and both economic and service-level considerations.

5.8 FURTHER READING

Our treatment of the subject of storage systems layout focused primarily on determining the location of stock within a warehouse. In addition to our coverage of the subject, an abundant research literature is available treating the broader subject of storage systems design. Within the overall category of storage systems design, the subject of order picking is probably one of the richest subjects from a research perspective. The References provide a sampling of the literature on order picking (e.g., Refs. 1, 4, 6, 7, 9–11, 13, 14, 19, 22, 27–30, 50, 56, and 60). Because the order picking problem is generally formulated as a version of the traveling salesman problem, most of the papers present heuristics; others present travel-time models for estimating the time required to perform a sequence of "picks" in the warehouse.

Because of the"unique" travel metrics associated with some of the storage/retrieval equipment found in a warehouse (e.g., automated storage/retrieval machines and carousel conveyors), a number of researchers have analyzed various aspects of storage/retrieval hardware performance. Among studies of the performance of storage/retrieval equipment are Refs. 8–14, 18–23, 27–29, 34, 36, 37, 40, 43, 47, 48, 50, 53, 59, 60, 63, and 64.

Another subject that has received considerable attention is that of locating items in a warehouse. The randomized, dedicated, class-based dedicated storage, and shared storage problems we treated have been studied by a number of researchers, including in Refs. 5, 17, 31, 38, 40–43, 46, 48, 52, 53, 57, 64, 67, and 68.

In addition to those who have addressed stock location issues, provided travel time models, modeled order picking problems, and analyzed automated storage/retrieval systems, several have addressed aggregate design issues, as reported in Refs. 2, 3, 15, 45, and 58.

Surprisingly, very little attention has been given to sizing warehouses. There is very little in the literature giving specific treatment to the storage capacity of a warehouse. However, there is a rich literature of inventory models, with a significant body of multiproduct literature treating various aspects of aggregate inventories.

Among the inventory models that have been developed that relate closely to the storage capacity problem are those described in Refs. 32, 33, 35, 39, 44, 49, 54, 55, 61, 62, and 69.

REFERENCES

1. Armstrong, R. D., W. D. Cook, and A. L. Saipe, "Optimal Batching in a Semiautomated Order Picking System," *Journal of the Operational Research Society,* Vol. 30, No. 8, August 1979, pp. 711–720.
2. Ashayeri, J., and L. F. Gelders, "Warehouse Design Optimization," *European Journal of Operations Research,* Vol. 21, No. 3, September 1985, pp. 285–294.
3. Azadivar, F., "Minimum-Cost Modular Design of Automated Warehousing Systems," *Material Flow,* Vol. 4, No. 3, June 1987, pp. 177–188.
4. Bachers, R., W. Dangelmaier, and H. J. Warnecke, "Selection and Use of Order Picking Strategies in a High-Bay Warehouse," *Material Flow,* Vol. 4, No. 4, March 1988, pp. 233–245.
5. Ballou, R. H., "Improving the Physical Layout of Merchandise in Warehouses," *Journal of Marketing,* Vol. 31, No. 3, July 1967, pp. 60–64.
6. Bartholdi J. J., and L. K. Platzman, "Efficient Bin Numbering Schemes for Warehouses," *Material Flow,* Vol. 4, No. 4, March 1988, pp. 247–254.
7. Bartholdi, J. J., and L. K. Platzman, "Heuristics Based on Spacefilling Curves for Combinatorial Problems in Euclidean Space," *Management Science,* Vol. 34, No. 3, March 1988, pp. 291–305.
8. Bozer, Y. A., "A Minimum Cost Design for an Automated Warehouse," master's thesis, Georgia Institute of Technology, Atlanta, GA, 1978.
9. Bozer, Y. A., "Optimizing Throughput Performance in Designing Order Picking Systems," doctoral dissertation, Georgia Institute of Technology, Atlanta, GA, 1986.
10. Bozer, Y. A., and G. P. Sharp, "An Empirical Evaluation of a General Purpose Automated Order Accumulation and Sortation System Used in Batch Picking," *Material Flow,* Vol. 2, No. 2, 1985, pp. 111–131.
11. Bozer, Y. A., M. A. Quiroz, and G. P. Sharp, "An Evaluation of Alternative Control Strategies and Design Issues for Automated Order Accumulation and Sortation Systems," *Material Flow,* Vol. 4, No. 4, March 1988, pp. 265–282.
12. Bozer, Y. A., and J. A. White, "Travel Time Models for Automated Storage/Retrieval Systems," *IIE Transactions,* Vol. 16, No. 4, December 1984, pp. 329–338.
13. Bozer, Y. A., and J. A. White, "Design and Performance Models for End-of-Aisle Order Picking Systems," *Management Science,* Vol. 36, No. 7, July 1990, pp. 852–866.
14. Bozer, Y. A., and J. A. White, "A Generalized Design and Performance Analysis Model for End-of-Aisle Order Picking Systems," *Progress in Material Handling and Logistics: Material Handling '90,* Vol. 2 (J. A. White, I. W. Pence, Jr., R. J. Graves, L. F. McGinnis, M. R. Wilhelm, and R. E. Ward, editors), Springer-Verlag, New York, pp. 205–243.
15. Dangelmaier, W. and R. Bachers, "SIMULAP: A Simulation System for Material Flow and Warehouse Design," *Material Flow,* Vol. 3, No. 4, June 1986, pp. 207–214.
16. Dearing, P. M., "On Some Minimax Location Problems Using Rectilinear Distances," doctoral dissertation, University of Florida, Gainesville, FL, 1972.

17. Dichtl, E. and W. Beeskow, "Optimal Allocation of Commodities in Storehouses by Means of Multi-dimensional Scaling," *Zeitschrift fur Operations Research,* Vol. 24, pp. B51–B64, 1980.

18. Elsayed, E. A., "Algorithms for Optimal Material Handling in Automatic Warehousing Systems," *International Journal of Production Research,* Vol. 19, No. 5, September–October 1981, pp. 525–535.

19. Elsayed, E. A. and R. G. Stern, "Computerized Algorithms for Order Picking in Automated Warehousing Systems," *International Journal of Production Research,* Vol. 21, No. 4, July–August 1983, pp. 579–586.

20. Elsayed, E. A. and O. I. Unal, "Order Batching Algorithms and Travel-Time Estimation for Automated Storage/Retrieval Systems," *International Journal of Production Research,* Vol. 27, No. 7, July 1989, pp. 1097–1114.

21. Elsayed, E. A., "Order Sequencing in Automated Storage/Retrieval Systems with Due Dates," *Progress in Material Handling and Logistics: Material Handling '90,* Vol. 2 (J. A. White, I. W. Pence, Jr., R. J. Graves, L. F. McGinnis, M. R. Wilhelm, and R. E. Ward, editors), Springer-Verlag, New York, pp. 245–267.

22. Foley, R. D., and E. H. Frazelle, "Recent Analytical Results on Miniload Performance," *Progress in Material Handling and Logistics: Material Handling '90,* Vol. 2 (J. A. White, I. W. Pence, Jr., R. J. Graves, L. F. McGinnis, M. R. Wilhelm, and R. E. Ward, editors), Springer-Verlag, New York, pp. 515–526.

23. Foley, R. D., and E. H. Frazelle, "Analytical Results for Miniload Throughput and the Distribution, of Dual Command Travel, *IIE Transactions,* to appear.

24. Francis, R. L., "On Some Optimum Facility Design Problems," doctoral dissertation, Northwestern University, Evanston, IL, 1967.

25. Francis, R. L., "On Some Problems of Rectangular Warehouse Design and Layout," *The Journal of Industrial Engineering,* Vol. 18, No. 10, 1967, pp. 595–604.

26. Francis, R. L., "Sufficient Conditions for Some Optimum-Property Facility Designs," *Operations Research,* Vol. 15, No. 3, May-June 1967, pp. 448–466.

27. Frazelle, E. H., and G. P. Sharp, "Correlated Assignment Strategy Can Improve Any Order-Picking Operation," *Industrial Engineering,* Vol. 21, No. 4, April 1989, pp. 33–37.

28. Frazelle, E. H., *Stock Location Assignment and Order Picking Productivity,* MHRC-TD-89-11, Material Handling Research Center, Georgia Institute of Technology, Atlanta, GA, June 5, 1990.

29. Goetschalckx, M., and H. D. Ratliff, "Sequencing Picking Operations in a Man-Aboard Order Picking System," *Material Flow,* Vol. 4, No. 4, March 1988, pp. 255–264.

30. Goetschalckx, M., and H. D. Ratliff, "An Efficient Algorithm to Cluster Order Picking Items in a Wide Aisle," *Engineering Costs and Production Economics,* Vol. 13, No. 4, April 1988, pp. 263–271.

31. Goetschalckx, M., and H. D. Ratliff, "Shared Storage Policies Based on the Duration of Stay" *Management Science,* Vol. 36, No. 9, September, 1990, pp. 1120–1132.

32. Goyal, S. K., "Determination of Optimum Packaging Frequency of Items Jointly Replenished," *Management Science,* Vol. 21, No. 4, December 1974, pp. 436–443.

33. Goyal, S. K., "A Note on Multi-Product Inventory Situations with One Restriction," *Journal of Operational Research Society,* Vol. 29, No. 3, March 1978, pp. 269–271.

34. Graves, S. C., W. H. Hausman, and L. B. Schwarz, "Storage Retrieval Interleaving in Automatic Warehousing Systems," *Management Science*, Vol. 23, No. 9, September 1977, pp. 935–945.

35. Hall, N. G., "A Multi-item EOQ Model with Inventory Cycle Balancing," *Naval Research Logistics Quarterly*, Vol. 35, No. 3, June 1988, pp. 319–325.

36. Han, M. H., L. F. McGinnis, J. S. Shieh, and J. A. White, "Sequencing Retrievals from Automated Storage/Retrieval Systems," *IIE Transactions*, Vol. 19, No. 1, March 1987, pp. 56–66.

37. Han, M. H., L. F. McGinnis, and J. A. White, "Analysis of Rotary Rack Operation," *Material Flow*, Vol. 4, No. 4, March 1988, pp. 283–294.

38. Harmatuck, D. J., "A Comparison for Two Approaches to Stock Location," *Logistics and Transportation Review*, Vol. 12, No. 4, 1976, pp. 282–284.

39. Hartley, R., and L. C. Thomas, "The Deterministic Two Product Inventory System with Capacity Constraint," *Journal of the Operational Research Society*, Vol. 33, No. 11, November 1982, pp. 1013–1020.

40. Hausman, W. H., L. B. Schwarz, and S. C. Graves, "Optimal Storage Assignment in Automatic Warehousing Systems," *Management Science*, Vol. 22, No. 6, February 1976, pp. 629–638.

41. Heskett, J. L., "Cube-per-Order Index: A Key to Warehouse Stock Location," *Transportation and Distribution Management*, Vol. 3, No. 4, April 1963, pp. 27–31.

42. Heskett, J. L., "Putting the Cube-per-Order Index to Work in Warehouse Layout," *Transportation and Distribution Management*, Vol. 4, No. 8, August 1964, pp. 23–30.

43. Hodgson, T. J., and T. J. Lowe, "Production Lot Sizing with Material Handling Cost Considerations," *IIE Transactions*, Vol. 14, No. 1, March 1982, pp. 44–51.

44. Homer, E. D., "Space-Limited Aggregate Inventories with Phased Deliveries," *The Journal of Industrial Engineering*, Vol. 17, No. 6, June 1966, pp. 327–332.

45. Jünemann, R., and F. Meister, "Rational Warehouse Planning," *Material Flow*, Vol. 4, No. 4, March 1988, pp. 217–224.

46. Kallina, C., and J. Lynn, "Application of the Cube-per-Order Index Rule for Stock Location in a Distribution Warehouse," *Interfaces*, Vol. 7, No. 1, November 1976, pp. 37–45.

47. Kanet, J. J., and R. G. Ramirez, "Optimal Stock Picking Decisions in Automatic Storage and Retrieval Systems," *OMEGA, The International Journal of Management Science*, Vol. 14, No. 3, May–June 1986, pp. 239–244.

48. Kim, W. B., and E. Koenigsberg, "The Efficiency of Two Groups of N Machines Served by a Single Robot," *Journal of the Operational Research Society*, Vol. 38, No. 6, June 1987, pp. 523–538.

49. Krone, L. H., "A Note on Economic Lot Sizes for Multi-Purpose Equipment," *Management Science*, Vol. 10, No. 3, April 1964, pp. 461–464.

50. Kunder, R. and T. Gudehus, "Mittlere Wegzeiten beim eindimensionalen Kommissionieren," (Average Travel Times in One-Dimensional Order Picking), *Zeitschrift fur Operations Research*, Vol. 19, No. 3, April 1975, pp. B53–B72.

51. Malette, A. J., and R. L. Francis, "A Generalized Assignment Approach to Optimal Facility Layout," *AIIE Transactions*, Vol. 4, No. 2, June 1972, pp. 144–147.

52. Malmborg, C. J., and B. Krishnakumar, "On the Optimality of the Cube per Order Index for Conventional Warehouses with Dual Command Controls," *Material Flow,* Vol. 4, No. 3, June 1987, pp. 169–176.

53. Medeiros, D. J., and B. Emamizadeh, "Optimal Container Location in Miniload AS/R Systems," *Progress in Material Handling and Logistics: Material Handling '90,* Vol. 2 (J. A. White, I. W. Pence, Jr., R. J. Graves, L. F. McGinnis, M. R. Wilhelm, and R. E. Ward, editors), Springer-Verlag, New York, pp. 303–316.

54. Page, E., and R. J. Paul, "Multiple-Product Inventory Situations with One Restriction," *Journal of the Operational Research Society,* Vol. 27, No. 4, 4th Quarter 1976, pp. 815–834.

55. Parsons, R. A., "A Note on Krone's Economic Lot Size Formula," *Management Science,* Vol. 12, No. 3, November 1965, p. 314.

56. Ratliff, H. D., and A. S. Rosenthal, "Order Picking in a Rectangular Warehouse: A Solvable Case of the Traveling Salesman Problem," *Operations Research,* Vol. 31, No. 3, May–June 1983, pp. 507–521.

57. Roll, Y., and M. J. Rosenblatt, "Shifting in Warehouses," *Material Flow,* Vol. 4, No. 3, June 1987, pp. 147–158.

58. Rosenblatt, M. J., and Y. Roll, "Warehouse Design with Storage Policy Considerations," *International Journal of Production Research,* Vol. 22, No. 5, September–October 1984, pp. 809–821.

59. Schwarz, L. B., S. C. Graves, and W. H. Hausman, "Scheduling Policies for Automatic Warehousing Systems: Simulation Results," *AIIE Transactions,* Vol. 10, No. 3, September 1978, pp. 260–270.

60. Sharp, G. P., K. I, Choe, and C. S. Yoon, "Small Parts Order Picking: Analysis Framework and Selected Results," *Progress in Material Handling and Logistics: Material Handling '90,* Vol. 2 (J. A. White, I. W. Pence, Jr., R. J. Graves, L. F. McGinnis, M. R. Wilhelm, and R. E. Ward, editors), Springer-Verlag, New York, pp. 317–341.

61. Silver, E. A., "A Simple Method of Determining Order Quantities in Joint Replenishments under Deterministic Demand," *Management Science,* Vol. 22, No. 12, August 1976, pp. 1351–1361.

62. Thomas, L. C. and R. Hartley, "An Algorithm for Limited Capacity Inventory Problems with Staggering," *Journal of the Operational Research Society,* Vol. 34, No. 1, January 1983, pp. 81–85.

63. Tompkins, J. A. and J. A. White, *Facilities Planning,* John Wiley & Sons, Inc., New York, 1984.

64. Wenzel, C. D., "Maximize Your Miniload System—Softloading: The Key to Mini AS/RS Production," *Proceedings of the 3rd International Conference on Automation in Warehousing,* November 1979, pp. 67–78.

65. White, J. A., "On Some Segregated Storage and Warehouse Sizing Problems," doctoral dissertation, The Ohio State University, Columbus, Ohio, 1970.

66. White, J. A., and R. L. Francis, "Normative Models for Some Warehouse Sizing Problems," *AIIE Transactions,* Vol. 3, No. 3, September 1971, pp. 185–193.

67. Wilson, H. G., "Order Quantity, Product Popularity and the Location of Stock in Warehouses," *AIIE Transactions,* Vol. 19, No. 3, September 1977, pp. 230–237.

68. Yang, M., *Analysis and Optimization of Class-Based Dedicated Storage Systems,* MHRC-TD-88-07, Material Handling Research Center, Georgia Institute of Technology, Atlanta, GA, November 14, 1988.

69. Zoller, K., "Deterministic Multi-Item Inventory Systems with Limited Capacity," *Management Science,* Vol. 24, No. 4, December 1977, pp. 451–455.

PROBLEMS

5.1. Consider a storage problem involving dedicated storage of four products: nuts, bolts, screws, and nails. Demands and replenishments for the products occur randomly and independently; hence the inventory levels are statistically independent random variables, with the following distributions:

Inventory level (pallet loads)	Probability			
	Nuts	Bolts	Screws	Nails
5	0.10	—	0.25	0.10
10	0.10	0.20	—	0.20
15	0.10	—	0.25	0.30
20	0.10	0.20	—	0.40
25	0.10	—	0.25	—
30	0.10	0.20	—	—
35	0.10	—	0.25	—
40	0.10	0.20	—	—
45	0.10	—	—	—
50	0.10	0.20	—	—

(a) Suppose that 100 storage positions exist. How should they be allocated to minimize the probability of a storage space shortage?

(b) What is the minimum number of storage slots necessary for the probability of a storage space shortage to be no greater than 0.80? For the solution, how many slots are assigned to each product?

5.2. Consider a storage problem involving dedicated storage of three product classes (1, 2, and 3). Suppose that the dedicated storage requirements for the product classes are statistically independent and continuously and uniformly distributed as follows.

$$f(x_1) = 0.010, \qquad 0 < x_1 < 100$$

$$f(x_2) = 0.005, \qquad 50 < x_2 < 250$$

$$f(x_3) = 0.004, \qquad 250 < x_3 < 500$$

(a) Determine the minimum amount of space required for the probability of a storage space shortage to be no greater than 0.30.

(b) Suppose that space exists for storing 500 units. Determine the assignment of slots to products that will minimize the probability of a storage space shortage.

(c) Suppose that each storage space costs \$3 per period to provide and that each space shortage costs \$5 per period. Determine the assignment of space to product classes that minimizes the expected cost per period.

5.3. Dedicated storage is to be used for three product classes (1, 2, and 3). Storage requirements for the classes are statistically independent and distributed as follows:

$$f(x_1) = 0.020, \qquad 50 < x_1 < 100$$

$$f(x_2) = 0.005, \qquad 0 < x_2 < 200$$

$$f(x_3) = 0.010, \qquad 100 < x_3 < 200$$

Determine the minimum amount of space required for the probability of a storage space shortage to be no greater than 0.25.

5.4. Dedicated storage is to be used for three product classes (1, 2, and 3). Storage demands occur randomly and independently. Inventory levels are distributed continuously and uniformly over the following ranges:

Product	Range
1	(0, 50)
2	(0, 100)
3	(50, 100)

(a) Space exists for 200 units to be stored. Assign slots to products to minimize the probability of a storage space shortage occurring.

(b) If the probability of a storage space shortage is to be no greater than 0.80, how much space should be assigned to each product to minimize the total amount of space required?

(c) Suppose that each storage space costs \$20 per day and each space shortage costs \$40. The inventory levels are daily levels; how much space should be assigned each product to minimize expected cost?

5.5. Suppose that the dedicated storage requirements for individual products are statistically independent and distributed as follows:

Storage requirement (storage positions)	Product *A*	*B*	*C*	*D*
0	0.1	0.1	0.4	—
10	—	0.1	—	—
20	0.2	0.1	0.3	0.1
30	—	0.1	—	0.3
40	0.3	0.2	0.2	0.4
50	—	0.2	—	0.2
60	0.4	0.2	0.1	—

(a) Determine the minimum number of storage positions having a probability no greater than 0.30 of a storage space shortage.

(b) Determine the allocation of 160 storage positions among the four products that will minimize the probability of a space shortage.

5.6. Consider a storage problem involving dedicated storage of five products: oats, peas, beans, barley, and corn. Demands and replenishments for the products occur randomly and independently. Hence the inventory levels are considered to be statistically independent random variables, with the following distributions:

Inventory level (thousands of bushels)	Probability				
	Oats	Peas	Beans	Barley	Corn
5	0.1	—	0.1	0.4	—
10	0.1	0.1	—	—	0.2
15	0.1	—	0.1	—	—
20	0.1	0.2	—	0.3	0.2
25	0.1	—	0.2	—	—
30	0.1	0.3	—	—	0.2
35	0.1	—	0.2	0.2	—
40	0.1	0.4	—	—	0.2
45	0.1	—	0.4	—	—
50	0.1	—	—	0.1	0.2

(a) Suppose that 200 storage slots exist, each capable of storing 1000 bushels of a product. Assign the slots to products so that the probability of a shortage of storage space is minimized.

(b) What is the minimum number of storage slots necessary for the probability of a storage space shortage to be no greater than 0.50? How many slots are assigned to each product?

5.7. In Problem 5.6, suppose that it costs $2500 per year to provide a storage position for 1000 bushels. Also, suppose that it costs $5 per day per storage position short. Determine the number of storage positions to provide to minimize expected cost. Also, indicate the optimum allocation of storage spaces among the products. Consider the inventory levels to be daily levels and consider 200 working days in a year.

5.8. Storage space is to be provided in the face of the following forecast for space. Where more than one estimate is given in a period, treat each value as equally likely to occur. For example, the figures of "55, 56, or 57" given for period 1 means that the space required is equally likely to be either 55, 56, or 57 storage positions for that particular period.

Period	Space required	Period	Space required
1	55, 56, or 57	11	34 or 38
2	48, 50, or 52	12	40, 44, or 48
3	45, 58, or 51	13	52, 55, or 58
4	40, 42, 44, or 46	14	60 or 64
5	38, 39, 40, or 41	15	62, 64, 66, or 68
6	36 or 38	16	65 or 68
7	34, 35, or 36	17	60 or 62
8	30, 32, 34, or 36	18	58, 59, 60, or 61
9	32, 34, or 36	19	56 or 58
10	33, 34, 35, or 36	20	52, 54, or 55

Suppose that it costs $30 per storage position to construct the storage space. Alternatively, space can be rented at a cost of $4 per storage position per period. Assume that constructed space will have no salvage value at the end of the planning horizon of 20 periods. How much space should be constructed to minimize total costs over the 20 periods?

5.9. Rework Problem 5.8 with the following modification. The cost required to construct storage space is given as follows:

Amount of spae built	Cost per position
[1, 35]	$35
(35, 45]	30
(45, 50]	26
(50, 55]	23
(55, 60]	21
(60, 70]	20

5.10. Storage space is to be provided to meet the forecast shown below. Space can be provided "internally" by constructing space or "externally" by leasing space. Suppose that it costs $63 per storage position to construct storage space. Alternatively, leased space can be provided at a cost of $0.30 per storage space per period. Assume that constructed space will have no salvage value at the end of 500 periods. How much space should be constructed to minimize total costs over the 500 periods? The demands are for each period over the stated interval (e.g., 2500 storage positions are required for each of the first five periods).

Period(s)	Demand	Period(s)	Demand
1– 5	2500	237–261	4000
6– 15	2750	262–302	4500
16– 25	3000	303–340	4700
26– 45	3200	341–396	4200
46– 75	4000	397–415	4000
76–100	3600	416–424	3600
101–128	3500	425–448	3500
129–147	3200	449–460	3400
148–200	3000	461–478	3000
201–236	3500	479–500	2800

5.11. Three products (hammers, drills, and wrenches) are to be stored in a warehouse. The inventory replenishment and demand patterns are as follows:

Beginning of period	Inventory level: Hammers	Drills	Wrenches
1	3	1	2
2	2	4	1
3	1	3	6
4	3	2	5
5	2	1	4
6	1	4	3
7	3	3	2
8	2	2	1
9	1	1	6
10	3	4	5
11	2	3	4
12	1	2	3

(a) Determine the storage requirements for randomized storage, dedicated storage, and shared storage.

(b) Determine the location of products along a line or storage aisle; use dedicated storage and shared storage.

5.12. Storage space is to be provided in the face of the following forecast for space.

Month	Demand	Probability	Month	Demand	Probability
Jan.	20	0.40	July	10	0.70
	25	0.40		20	0.30
	30	0.20			
Feb.	30	0.30	Aug.	20	0.60
	40	0.40		30	0.40
	50	0.30			
Mar.	60	0.20	Sept.	30	0.50
	80	0.60		40	0.50
	100	0.20			
Apr.	40	0.20	Oct.	40	0.30
	50	0.40		50	0.40
	60	0.40		60	0.30
May	30	0.50	Nov.	30	0.40
	40	0.40		40	0.50
	50	0.10		50	0.10
June	20	0.60	Dec.	20	0.70
	30	0.40		30	0.30

(a) Suppose that it costs \$23 per storage position to construct the storage space. Alternatively, space can be rented at a cost of \$4 per storage position per period.

Assume that constructed space will have no salvage value at the end of the 12 periods. How much space should be constructed to minimize total costs over the 12 periods?

(b) Considering the probability distributions for monthly demand given above and assuming that months are statistically independent, determine the probability distribution for any month in the year. Next, determine the probability of a space shortage if a storage capacity of 50 is provided.

5.13. Consider a warehouse (50 ft × 100 ft) that lies in the first quadrant. Fifty storage bays (10 ft × 10 ft) are served by three docks having coordinates $\mathbf{P}_1 = (0, 20)$, $\mathbf{P}_2 = (0, 50)$, and $\mathbf{P}_3 = (0, 80)$. $\mathbf{P}_1$ and $\mathbf{P}_2$ are used for shipping and $\mathbf{P}_3$ is used for receiving. The movement of product to–from storage is equally likely to be from receiving or to shipping. Three product classes are to be stored in the warehouse using dedicated storage. Class 1 requires 19 storage bays, class 2 requires 21 storage bays, and class 3 requires 10 storage bays. Throughput values are 55, 20, and 45 for 1, 2, and 3, respectively. Use rectilinear travel between docks and centroids of storage bays.

(a) Suppose that 80% of shipping is through $\mathbf{P}_2$ and 20% is through $\mathbf{P}_1$. Determine the layout that minimizes expected distance traveled.

(b) For (a), determine the bound on storage space reduction for randomized storage to have an expected distance traveled less than that for dedicated storage.

(c) Suppose that movement from storage to shipping is always to the closest dock, rectilinearly. Determine the layout that minimizes expected distance traveled.

(d) Suppose that one-half of shipping is to the closest dock and one-half occurs as in part (a). Determine the layout that minimizes expected distance traveled.

5.14. Dedicated storage is to be used for storing four products (A, B, C, and D) in a warehouse that consists of 30 storage bays (20 ft × 20 ft). The layout of the warehouse is shown in Figure P5.14, including the location of the receiving docks (1 and 2) and shipping docks (3 and 4). Product A requires 5 bays for storage; its daily activity averages 25 storages and 25 retrievals. Product B requires 10 bays for storage; its daily activity averages 5 storages and 5 retrievals. Product C requires 10 bays for storage; its daily activity averages 10 storages and 10 retrievals. Product D requires 5 bays for storage; its daily activity averages 10 storages and 10 retrievals. All movement within the warehouse is performed by lift trucks using single command cycles; rectilinear travel between the docks and the centroids of the storage bays can be assumed. Activity is

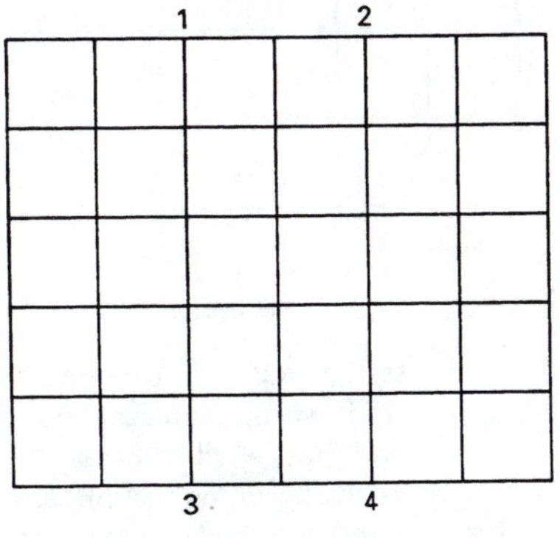

Figure P5.14

divided equally among the docks. Also, travel occurs at constant velocity throughout the warehouse.

(a) Determine the dedicated storage layout that will minimize the average distance traveled per day.

(b) Suppose that class-based dedicated storage is to be used, with only two classes (I and II). How should the two classes be formed to minimize the average distance traveled? To reduce the number of combinations, assume that class I is the highest-ranking class and will be assigned to the "best" locations. On the answer sheet specify the products to be included in class I.

5.15. Consider the layout of 32 storage bays and one I/O point shown in Figure P5.15. Suppose that four products are to be stored using dedicated storage. The storage bays are 15 ft × 15 ft. The products (oats, peas, beans, and barley) have the following storage and throughput requirements:

Product	Storage (bays)	Throughput (operations/day)
Oats	16	40
Peas	8	30
Beans	5	10
Barley	3	20

(a) Determine the dedicated storage layout for the Oats, Peas, Beans, and Barley Grow, Inc. distribution center that will minimize the distance traveled in storing/retrieving unit loads of product, based on single command cycles and rectilinear travel.

(b) Determine the bound on space required for randomized storage to yield an expected distance traveled no greater than that obtained using dedicated storage.

(c) Determine the best way to form two classes for class-based dedicated storage.

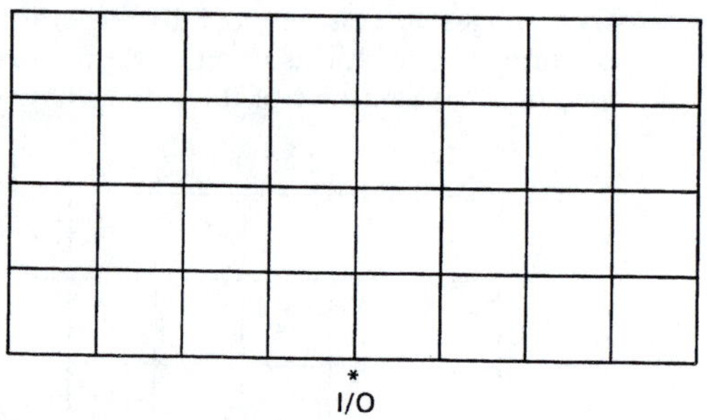

Figure P5.15

5.16. In Problem 5.15, suppose that oats and peas must be stored in bays that are 30 ft × 30 ft, rather than 15 ft × 15 ft, such that their storage space requirement is four (30 ft × 30 ft) bays for oats and two (30 ft × 30 ft) bays for peas. Solve part (a) and determine the average distance traveled, assuming rectilinear travel to the centroid of the 30 ft × 30 ft storage bay for oats and peas and to the centroid of the 20 ft × 20 ft storage bay for beans and barley.

5.17. Consider the storage locations depicted in Figure P5.17. Suppose that rectilinear travel occurs. Furthermore, let 40% of the operations be storages and the balance be retrievals. Assume that only single command cycles occur. Three products are to be stored (brand X, brand Y, and brand Z). Brand X requires 10 storage locations, brand Y requires 25 storage locations, and brand Z requires 15 storage locations. Brand X has a daily throughput requirement of 60 trips to/from storage per day; brand Y requires 10 trips to/from storage per day; and brand Z requires 30 trips to/from storage per day. Storage locations are 10 ft × 10 ft. Dedicated storage is to be used.

(a) Determine the dedicated storage layout of the products assuming that inputs to storage are always initiated from the closest input point (I_1 or I_2) and outputs from a storage location are always to the closest output point (O_1 or O_2).

(b) Determine the dedicated storage layout assuming that inputs are divided equally between I_1 and I_2 and outputs are divided equally between O_1 and O_2.

(c) Determine the upper bound on the number of storage locations for randomized storage to yield a lower average distance traveled than dedicated storage.

(d) If two classes are to be formed, how should they be formed in support of class-based dedicated storage?

(e) Solve part (b) assuming that travel occurs simultaneously in the x and y directions at constant velocities.

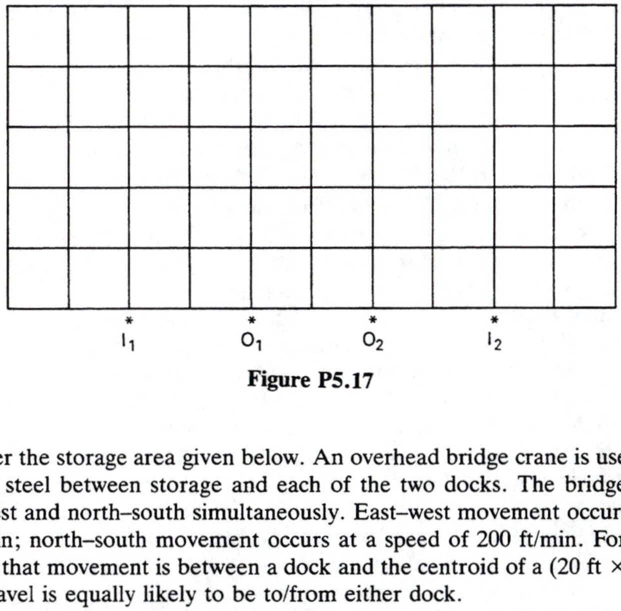

Figure P5.17

5.18. Consider the storage area given below. An overhead bridge crane is used to transport coils of steel between storage and each of the two docks. The bridge crane travels east–west and north–south simultaneously. East–west movement occurs at a speed of 80 ft/min; north–south movement occurs at a speed of 200 ft/min. For convenience, assume that movement is between a dock and the centroid of a (20 ft × 20 ft) storage bay. Travel is equally likely to be to/from either dock.

Three products (grade A, grade B, and grade C) are to be stored using dedicated storage. Grade A requires 6 storage bays, grade B requires 12 storage bays, and grade C requires 18 storage bays. Demand for the products results in there being 12 moves to/from storage per day for grade A, 12 per day for grade B, and 24 per day for grade C.

(a) Determine the expected travel times for each storage bay.

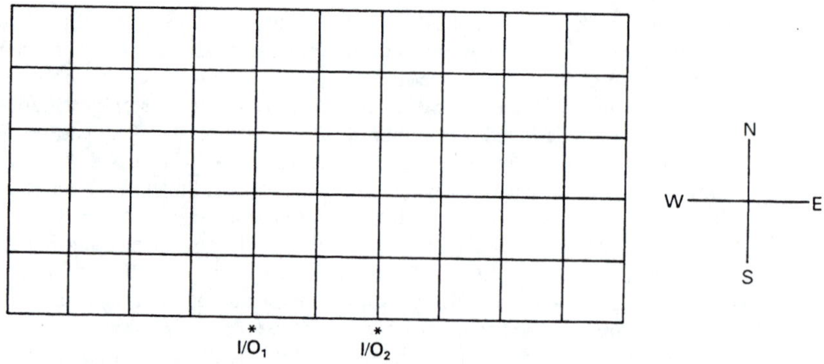

Figure P5.18

(b) Determine the dedicated storage layout that will minimize the average time to store/retrieve products/day.

(c) Based on the solution to part (b), compute the average daily workload for the bridge crane, expressed in daily travel time required.

(d) Suppose that all inputs occur at dock 1 and all outputs occur at dock 2. Also, suppose that the travel of the bridge crane is always from dock 1 to a storage location where a coil is stored, then directly to another storage location where a coil is retrieved, then to dock 2 for shipment of the coil, and then to dock 1 to pick up the next coil destined for storage. Assuming that products are selected for storage and for retrieval in proportion to their throughput values, determine the average daily workload for the bridge crane. Compare the result with that obtained in part (c).

5.19. Consider a storage region for a warehouse (50 ft × 100 ft) that lies in the first quadrant. Docks are located at the coordinate points $\mathbf{P}_1 = (0,30)$, $\mathbf{P}_2 = (0,50)$, $\mathbf{P}_3 = (0,70)$, and $\mathbf{P}_4 = (50,50)$. Fifty storage bays (10 ft × 10 ft) are to be assigned for the dedicated storage of four products: alpha, beta, gamma, and delta. Receiving occurs at $\mathbf{P}_4$ and shipping occurs at $\mathbf{P}_1$, $\mathbf{P}_2$, and $\mathbf{P}_3$. Fifty percent of the item movement is from $\mathbf{P}_4$; the remaining movement is to $\mathbf{P}_1$, $\mathbf{P}_2$, and $\mathbf{P}_3$ using the "closest dock" assignment rule. The throughput requirements for the products are measured in operations per week, where an operation is either a storage or a retrieval. The throughput values are 40, 30, 20, and 10 for alpha, beta, gamma, and delta, respectively. The storage requirements, measured in storage bays, are 5, 10, 15, and 20 for alpha, beta, gamma, and delta, respectively. Rectilinear travel occurs between the docks and the centroids of the storage bays.

(a) Determine the layout of products that will minimize expected distance traveled.

(b) Suppose that travel is twice as likely to be to $\mathbf{P}_2$ as to $\mathbf{P}_1$ and it is equally likely for travel to be to $\mathbf{P}_1$ as it is to $\mathbf{P}_3$. Solve part (a).

(c) For the conditions of part (a), what minimum reduction in space would have to occur for randomized storage to have an expected distance traveled less than that of dedicated storage?

(d) For the conditions of part (b), use contour lines to portray the continuous layout solution.

5.20. In Problem 5.19, suppose that class-based storage is to be used. For the conditions of part (a), determine the best formation of 2-classes and 3-classes among the four products. Form 2-classes and 3-classes for the conditions of part (b).

5.21. A warehouse layout is to be determined for the storage of four products, W, X, Y, and Z. The warehouse layout is to be in the first quadrant, with the x and y axes forming two of the walls of the rectangular facility. Docks are located at the coordinates $(20, 0)$ and $(0, 20)$. The warehouse consists of 24 bays, each 10 ft $\times$ 10 ft. The warehouse is located within the rectangle given by $0 < x < 60$ and $0 < y < 40$. Rectilinear travel occurs in the warehouse, with travel equally likely to be to/from either dock. Throughput and storage requirements for the four products are as follows:

Product	Throughput requirement (operations/day)	Storage requirement (storage bays)
W	15	10
X	30	5
Y	50	4
Z	5	5

(a) Determine the dedicated storage layout that will minimize the average distance traveled per day between the docks and the centroid of each storage bay.

(b) Suppose that randomized storage is to be used and the facility is unchanged from part (a). Determine the maximum number of storage bays required for storage of the aggregate product mix such that the expected distance traveled is no greater than that obtained in part (a).

(c) Suppose that class-based storage is to be used, with only 2-classes. Determine the assignment of products to classes that will minimize the expected distance traveled and compute the resulting expected distance traveled. Determine the minimum expected distance traveled for the 2-class solution.

(d) Suppose that the coordinates of the two docks are $(0, 20)$ and $(60, 20)$; let the storage requirements be 8, 8, 4, and 4 for products W, X, Y, and Z, respectively; and let travel always be to/from the closest dock. Determine the dedicated storage layout that will minimize the average distance traveled per day.

5.22. A warehouse layout is to be determined for the storage of three products, W, X, and Y. The warehouse layout is to be developed in the first quadrant, with the x and y axes forming two of the walls of the rectangular facility. Docks are located at the coordinates $(20, 0)$, $(40, 0)$, and $(0, 20)$. The warehouse consists of 24 bays, each 10 ft $\times$ 10 ft. The warehouse is located within the rectangle given by $0 < x < 60$ and $0 < y < 40$. Rectilinear travel occurs in the warehouse, with travel equally likely to be to/from either dock. The throughput and storage requirements for the three products are as follows:

Product	Throughput requirement (operations/day)	Storage requirement (storage bays)
W	10	4
X	40	8
Y	50	12

(a) Assuming that dedicated storage is to be used, determine the layout that will minimize the average distance traveled per day. Assume rectilinear travel between the docks and the centroid of each storage bay.

(b) Suppose that randomized storage is to be used and the facility is unchanged from part (a). Determine the maximum number of storage bays required for storage of the aggregate product mix such that the expected distance traveled is no greater than that obtained in part (a).

5.23. In Problem 5.22, suppose that travel is always to/from the closest dock. Determine the dedicated storage layout that will minimize the average distance traveled per day.

5.24. A warehouse layout is to be determined for the storage of three products, W, X, and Y. The warehouse layout is to be in the first quadrant, with the x and y axes forming two of the walls of the rectangular facility. Docks are located at the coordinates $(30, 0)$ and $(0, 20)$. The warehouse consists of 24 bays, each 10 ft $\times$ 10 ft. The warehouse is located within the rectangle given by $0 < x < 60$ and $0 < y < 40$. Rectilinear travel occurs in the warehouse, with travel equally likely to be to/from either dock. Throughput and storage requirements for the three products are as follows:

Product	Throughput requirement (operations/day)	Storage requirement (storage bays)
W	20	6
X	30	6
Y	10	12

(a) Determine the dedicated storage layout that will minimize the average distance traveled per day between the docks and the centroid of each storage bay.

(b) Suppose that randomized storage is to be used and the facility is unchanged from part (a). Determine the maximum number of storage bays required for storing the aggregate product mix, such that the expected distance traveled is no greater than that obtained in part (a).

(c) Suppose that travel is always to/from the closest dock. Determine the dedicated storage layout that will minimize the average distance traveled per day.

5.25. A new warehouse is being designed for storing four products. Dedicated storage is to be used. It is desired to determine the size of the warehouse as well as the assignment of products to storage locations. The number of storage locations required for individual products is independently Poisson distributed with a mean of 10 per day for product 1, 15 per day for product 2, 20 per day for product 3, and 25 per day for product 4. The fixed cost of providing space in the warehouse is \$100 per storage position per year; the daily cost for each load stored in the warehouse is \$0.50 per load; and the daily cost to store "excess" loads in a public warehouse is \$5.00 per load. There are 250 working days in a year and one load is stored in a storage position.

(a) Determine the number of storage positions to be provided for each product to ensure that the expected total annual cost is minimized. For the answers obtained, what is the probability of having to store any loads in the public warehouse?

(b) Determine the minimum number of storage positions to be provided for each product to ensure that the probability of having to store any loads in the public warehouse will be no greater than 0.10.

5.26. In Problem 5.25, suppose that the number of storage positions provided is 16, 24, 24, and 36 for products 1, 2, 3, and 4, respectively. The products are to be stored in storage bays, where a bay is defined to be a 2 × 2 cluster of four storage positions. Hence the number of storage bays required totals 4, 6, 6, and 9, respectively, for the products. The 100 storage positions are arranged in a 5 × 5 array of storage bays, as shown in Figure P5.26. Throughput requirements for the four products, measured in operations per day, are 12, 45, 7, and 36, respectively.

(a) Suppose that travel to/from storage is always from/to the closest I/O point. Travel occurs rectilinearly between the I/O points and the centroids of the 25 10 ft × 10 ft storage bays. Determine the storage assignment that will minimize the expected distance traveled.

(b) Suppose that storage and retrieval occurs using an overhead bridge crane, with simultaneous travel. Further, suppose that travel in the east–west direction is twice as fast as travel in the north–south direction. Finally, suppose that travel to/from the centroid of a 10 ft × 10 ft storage bay is equally likely to be from/to either I/O point. Determine the storage assignment that will minimize the expected distance traveled.

(c) Suppose that class-based dedicated storage is to be used. Travel occurs rectilinearly between the centroid of a storage bay and the closest I/O point. Determine the formation of two classes that will minimize the expected distance traveled and give the storage assignment for the two classes.

(d) Suppose that randomized storage is to be used and travel occurs rectilinearly between the centroid of a storage bay and the closest I/O point. Determine the maximum number of storage bays that can be required for randomized storage without the expected distance traveled being greater than that for dedicated storage. Identify the storage bays to be eliminated.

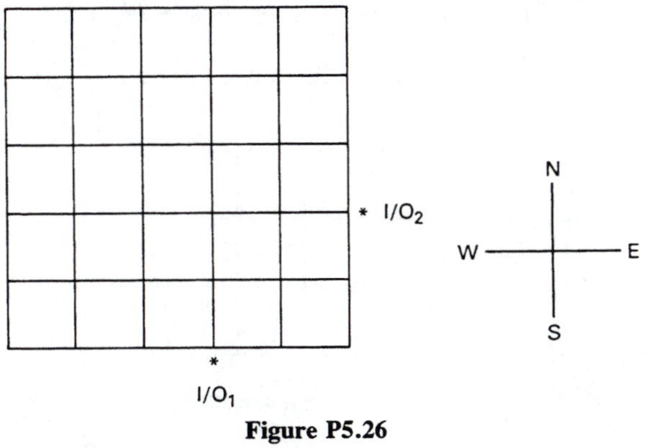

Figure P5.26

5.27. A new warehouse is being designed for storing five products. Dedicated storage is to be used. It is desired to determine the size of the warehouse as well as the assignment of products to storage locations. The number of storage locations required for individual products is independently Poisson distributed with a mean of 5 per day for product 1, 10 per day for product 2, 15 per day for product 3, 20 per day for product 4, and 25 per day for product 5. The fixed cost of providing space in the warehouse is $120 per storage position per year; the daily cost for each load stored in the warehouse is \$1.00 per load; and the daily cost to store "excess" loads in a public warehouse is \$5.00 per load. There are 300 working days in a year and one load is stored in a storage position.

(a) Determine the number of storage positions for each product to ensure that the expected total annual cost is minimized.

(b) Given the answers for part (a), suppose that the products are to be stored in storage bays, using dedicated storage. A bay is defined to be a 2×2 cluster of four storage positions. An integer number of bays must be provided for each product and no mixing of products is allowed within a storage bay. Finally, suppose that a maximum of 30 storage bays are available in the warehouse. If more than 30 bays are required, the excess must be stored in a public warehouse at a significant penalty cost based on the ratio of throughput to space required for the product. The 30 storage bays are arranged as shown in Figure P5.27. Throughput requirements for the five products, measured in operations per day, are 10, 12, 40, 7, and 31, respectively. Determine the assignment of products to storage bays that will minimize the average distance traveled daily. Assume that travel is equally likely to be to/from either dock and all travel is rectilinear between docks and storage bay centroids.

(c) Suppose that class-based dedicated storage is to be used in part (b). Determine the formation of two classes that will minimize the expected distance traveled and give the storage assignment for the two classes.

(d) Suppose that in part (b) randomized storage is to be used rather than dedicated storage. Determine the maximum number of storage bays that can be required for randomized storage without the expected distance traveled being greater than that for dedicated storage. Identify the storage bays to be eliminated.

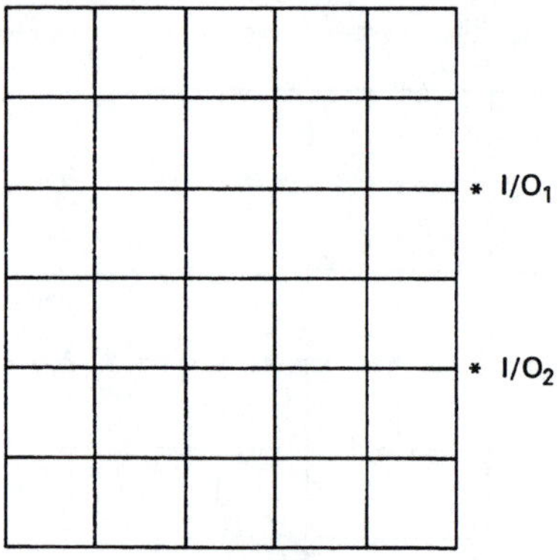

Figure P5.27

5.28. Dedicated storage is to be used for storing four products (A, B, C, and D) in a warehouse that consists of 50 storage bays (20 ft × 20 ft). The layout of the warehouse is shown in Figure P5.28, including the location of the receiving docks (1 and 2) and shipping docks (3 and 4). Product A requires 10 bays for storage; its daily activity averages 25 storages and 25 retrievals. Product B requires 8 bays for storage; its daily activity averages 10 storages and 10 retrievals. Product C requires 16 bays for storage; its daily activity averages 10 storages and 10 retrievals. Product D requires 16 bays for storage; its daily activity averages 5 storages and 5 retrievals. All movement within the

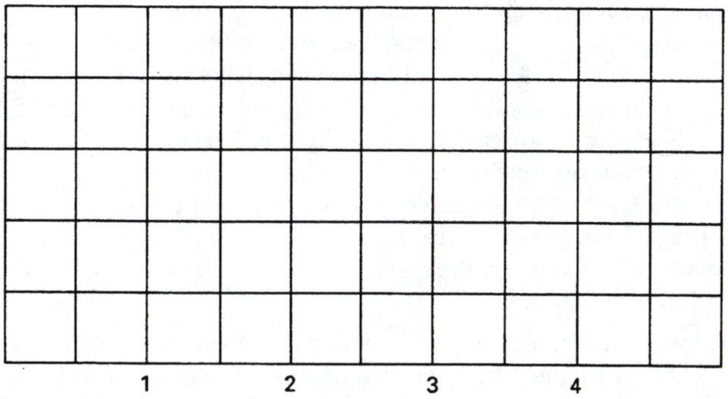

Figure P5.28

warehouse is performed by lift trucks using single command cycles; rectilinear travel between the docks and the centroids of the storage bays can be assumed. Also, travel occurs at constant velocity throughout the warehouse. Dock activity is divided equally among the docks. Determine the dedicated storage layout that will minimize the average distance traveled per day.

5.29. Suppose that shared storage is to be used in a situation involving three products: apples (A), bananas (B), and cantaloupes (C). The daily storage requirements for the three products are given below. Storage occurs along the aisle in the same manner as in the examples. Indicate the assignment of products to storage positions over a 6-day cycle.

	Number of storage positions		
End of day:	A	B	C
1	5	2	0
2	4	0	2
3	3	4	1
4	2	2	0
5	1	0	2
6	0	4	1

End of day:	1	2	3	4	5	6	7	8
1	A	A		B	B	A	A	A
2	—	—	—	—	—	—	—	—
3	—	—	—	—	—	—	—	—
4	—	—	—	—	—	—	—	—
5	—	—	—	—	—	—	—	—
6	—	—	—	—	—	—	—	—

5.30. In Example 5.18, suppose that the same production line is equipped to produce three products, alpha, beta, and gamma. Demand for gamma equals two pallet loads per day.

The production capacity is modified such that daily production is four pallet loads; four pallet loads of alpha are produced each of the first two days of the 2-week production cycle; two pallet loads of alpha and two pallet loads of beta are produced on the third day; four pallet loads of beta are produced on the fourth and fifth days; and four pallet loads of gamma are produced each of the next five days of the 2-week cycle. Production does not occur on the weekend.

Determine the shared storage assignments and the distance traveled to perform storage and retrieval operations, based on the same assumptions as used in Example 5.18. Compare the results with those that would be obtained using dedicated and randomized storage.

5.31. Consider a shared storage situation involving four products: oats, peas, beans, and barley. The daily retrieval rate is one trailer load of product. Each trailer is loaded with 10 pallet loads of oats, 15 pallet loads of peas, 5 pallet loads of beans, and 10 pallet loads of barley. Oats are replenished every 5 days; peas are replenished every 4 days; beans are replenished every 10 days; and barley is replenished every 5 days. The receipts are staggered such that the hth load of oats arrives on day $1 + 5h$; the ith load of peas arrives on day $4 + 4i$; the jth load of beans arrives on day $2 + 10j$; and the kth load of barley arrives on day $3 + 5k$.

(a) Determine the end-of-day inventory levels of each product.

(b) Based on the solution to part (a), determine the aggregate storage requirement, measured in pallet loads.

(c) Compare the dedicated storage requirement with that for randomized and shared storage.

(d) Assume that storage occurs on two sides of a single storage aisle. Also, assume that loads are stored in pallet rack having four levels of storage, with vertical travel to a level being twice the horizontal travel to a storage position. Determine the shared storage assignments and dedicated storage assignments for the products.

(e) Analyze the sensitivity of the shared storage solution to changes in the timing of receipts by considering the arrivals to occur on days $1 + 5h$, $2 + 4i$, $3 + 10j$, and $4 + 5k$.

(f) Solve part (e) for the case in which the first arrival of a replenishment occurs on day 1 for all products.

5.32. Using contour lines, determine the shape and dimensions of a continuous version of the facility that will satisfy the storage requirements given in Problem 5.22, with the docks located at the same three coordinate points. The total storage area equals 2400. The warehouse will lie in the first quadrant but is no longer constrained to be rectangularly shaped; neither is it required that $x < 60$ and $y < 40$. In addition to giving the shape and dimensions of the facility, give the shape and dimensions of the storage region for each product under dedicated storage conditions. Also determine the expected distance traveled.

5.33. Contour lines are to be used to determine the shape and dimensions of a storage region in the first quadrant enclosing an area of 9200. Two docks are used to service the storage region; they are located at the coordinate points $\mathbf{P}_1 = (100, 0)$ and $\mathbf{P}_2 = (140, 0)$. The movement to/from storage is always from/to the closest dock.

(a) Determine the shape and dimensions of the storage region that will minimize the expected rectilinear distance traveled between storage and the docks.

(b) Suppose that the distance between the two docks does not have to equal 40, but must continue to be located along the x-axis and the storage region must continue to lie in the first quadrant. Determine the spacing between the docks that will minimize the expected rectilinear distance traveled between storage and the docks and still generate a contour line that encloses an area of 9200.

(c) Using the single integral equation approach, determine the expected distance traveled for parts (a) and (b).

5.34. Contour lines are to be used to determine the shape and dimensions of a storage region in the first quadrant enclosing an area of 15,000. Two docks are used to service the storage region; they are located at the coordinate points $\mathbf{P}_1 = (50, 0)$ and $\mathbf{P}_2 = (0, 50)$.

(a) Determine the shape and dimensions of the storage region that will minimize the expected rectilinear distance traveled between storage and the docks. The movement to/from storage is equally likely to be from/to either dock.

(b) Determine the expected distance traveled for the optimum design for part (a) using the single integral expression approach described in class.

(c) Suppose that travel is always to the closest dock using rectilinear travel. Determine the shape and dimensions of the storage region that will minimize the expected rectilinear distance traveled between storage and the docks.

5.35. Contour lines are to be used to determine the configuration of a storage region in the first quadrant enclosing an area of 15,625. Three docks are used to service the storage region; they are located at the coordinate points $\mathbf{P}_1 = (50, 0)$, $\mathbf{P}_2 = (0, 50)$, and $\mathbf{P}_3 = (0, 100)$.

(a) Determine the shape and dimensions of the storage region that will minimize the expected rectilinear distance traveled between storage and the docks. Fifty percent of the movement to/from storage is from/to dock 1; the remaining travel to/from storage is equally likely to be from/to either of the remaining two docks.

(b) Determine the expected distance traveled for the optimum design for part (a) using the single integral expression approach described in class.

5.36. Using contour lines, determine the shape, area, and dimensions of the storage region located in the first quadrant that minimizes expected distance traveled using rectilinear distances. Suppose that two docks are located at the coordinate points $\mathbf{P}_1 = (30, 0)$ and $\mathbf{P}_2 = (0, 30)$. Also, suppose that travel is twice as likely to involve $\mathbf{P}_1$ as it is to involve $\mathbf{P}_2$. Finally, suppose that $\mathbf{P}_2$ is centered on the west wall or boundary of the storage region. Also, determine the expected distance traveled.

5.37. Contour lines are to be used to determine the shape and dimensions of a storage region in the first quadrant enclosing an area of 3100. Three docks are used to service the storage region; they are located at the coordinate points $\mathbf{P}_1 = (0, 20)$, $\mathbf{P}_2 = (0, 40)$, and $\mathbf{P}_3 = (30, 0)$. One-half the movement to/from storage is from/to $\mathbf{P}_3$; the remainder is equally likely to be from/to $\mathbf{P}_1$ and $\mathbf{P}_2$. Determine the expected distance traveled using the single integral approach.

5.38. Determine the shape and dimensions of the continuous regions for three product classes stored in a storage aisle served by an automatic storage and retrieval machine. The storage machine travels horizontally at a speed of 400 ft/min and simultaneously travels vertically at a speed of 80 ft/min. A storage area of 18,000 ft^2 is to be provided on each side of the storage aisle. The storage regions are to be configured so that the average time required to travel from/to the I/O point is minimized. The I/O point will be elevated 20 ft and located at the end of the storage aisle. Class-based dedicated storage is to be used.

Product class A requires 2000 ft^2 of storage, B requires 6000 ft^2 of storage, and C requires the remaining 10,000 ft^2 of storage. The throughput activity for the three classes is approximately 15 operations per hour for A, 10 per hour for B, and 5 per hour for C, where an operation is either a storage or a retrieval.

(a) For each product class, determine the configurations and dimensions of the storage regions that minimize the average time required to perform a storage or retrieval operation.

(b) Using the single integral equation approach, determine the expected time required to perform an operation.

(c) Solve parts (a) and (b) when the maximum storage height is constrained to be no greater than 50 ft.

5.39. Solve part (a) of Problem 5.38 with the following modifications. There are two input points and three output points: $\mathbf{I}_1 = (50, 20)$, $\mathbf{I}_2 = (250, 20)$, $\mathbf{O}_1 = (100, 20)$, $\mathbf{O}_2 = (150, 20)$, and $\mathbf{O}_3 = (200, 20)$. The number of storages equals the number of retrievals. Input is equally divided between the two input points; output is equally divided between the three output points.

5.40. Solve Problem 5.38 with the following modification: travel to a storage point is always from the closest input point and travel from a storage point is always to the closest output point.

5.41. A steel storage center uses a combination of overhead bridge cranes and industrial lift trucks to perform material handling. Lift truck travel is rectilinear and crane travel is Chebyshev. The number of material handling trips to/from storage is equally divided between lift trucks and cranes. The rectangular storage region lies in the first quadrant and is defined by the set of points (x, y), for which $0 < x < 400$ and $0 < y < 100$. Lift trucks travel at a speed of 400 per unit time and cranes travel at a speed of 100 per unit time; speeds are the same for travel in the x direction as in the y direction. A single dock is located at the point $\mathbf{P} = (200, 0)$.

(a) Suppose that storage bays are 20 ft × 20 ft and that the highest-ranked product requires 10 bays. Which bays should be assigned to the product?

(b) Develop the set of contour lines describing the "mixed metric" travel in the steel storage center.

5.42. Show that $x_j = P^{1/n}$ is an optimum solution for the following optimization problem:

$$\text{minimize} \quad \sum_{j=1}^{n} x_j$$

subject to

$$\prod_{j=1}^{n} x_j \geq P$$

5.43. Show that $x_j = S/n$ is an optimum solution to the following optimization problem:

$$\text{maximize} \quad \prod_{j=1}^{n} x_j$$

subject to

$$\sum_{j=1}^{n} x_j \leq S$$

5.44. Given the optimization problem of minimizing

$$\text{ETC}(Q_j) = C_0 Q_j + C_2 \int_{x > Q_j} f(x)\, dx$$

show that the optimum value of Q_j is such that

$$F_j(Q_j) = \frac{C_2 - C_1}{C_2}$$

where $F_j(Q_j)$ is the cumulative distribution function for storage demand for product j evaluated at Q_j.

5.45. Given a number $d > 0$, let $\mathbf{P}_1 = (0, 0)$ and $\mathbf{P}_2 = (d, 0)$. Let $\mathbf{L}$ consist of the "strip" in the first and fourth quadrants, parallel to the y-axis, made up of all points lying between $\mathbf{P}_1$ and $\mathbf{P}_2$; that is, $\mathbf{L} = [(x, y): 0 \le x \le d, -\infty < y < \infty]$. Denote the rectilinear distance between $\mathbf{X}$ and $\mathbf{P}_i$ by $f_i(\mathbf{X})$, and for every design R in $H(\mathbf{L}; A)$, define an average distance $E[R]$ by

$$E[R] = \sum_{i=1}^{2} \int_{R^*} \frac{1}{A} f_i(\mathbf{X})\, d\mathbf{X}$$

(a) Find and draw a design R^* in $H(\mathbf{L}: A)$ to minimize $E[R]$. [*Hint*: $q(z) = d(z - d)$.] Show that

$$E[R^*] = \frac{A + 2d^2}{2d}$$

Note that this problem may be considered as that of finding a warehouse design where all items enter at one dock and depart from the other dock.

(b) Suppose that you have the option of choosing d as well. Find the value of d to minimize $E[R^*]$). In the event that R^* turns out to be a rectangle, determine its x and y dimensions.

5.46. Given any two points $\mathbf{X}$ and $\mathbf{P}$ in the plane, denote by $g(\mathbf{X}, \mathbf{P})$ an *arbitrary* distance between $\mathbf{X}$ and $\mathbf{P}$. A famous theorem in mathematics states that $g(\mathbf{X}, \mathbf{P})$ is equal to or greater than the Euclidean distance between $\mathbf{X}$ and $\mathbf{P}$; that is, the shortest distance between two points is a straight line. Make use of this theorem to prove, for any set R of area $\mathbf{A}$, that

$$\int_{R^*} \frac{1}{A} g(\mathbf{X}, \mathbf{P})\, d\mathbf{X} \ge \frac{2}{3}\left(\frac{A}{\pi}\right)^{1/2}$$

Can you suggest any practical use of this inequality?

5.47. Let $\mathbf{L}$ be the union of the first and fourth quadrants, let b be a known positive constant, define the points $\mathbf{P}_1 = (0, 0)$, $\mathbf{P}_2 = (0, b)$, $\mathbf{P}_1 = (0, 2b)$, denote the rectilinear distance $\mathbf{X}$ and $\mathbf{P}_j$ by $d(\mathbf{X}, \mathbf{P}_j)$, and define the function $f(\mathbf{X})$ by

$$f(\mathbf{X}) = d(\mathbf{X}, \mathbf{P}_1) + 2d(\mathbf{X}, \mathbf{P}_2) + d(\mathbf{X}, \mathbf{P}_3)$$

If $q(z)$ is the area of the set $R(z)$, where $R(z) = [\mathbf{X} \in \mathbf{L}: f(\mathbf{X}) \le z]$, it is given that

$$q(z) = \frac{1}{2}\left(\frac{z}{2} - b\right)^2 \qquad \text{for } 2b \le z \le 4b$$

$$q(z) = \frac{z^2}{16} - \frac{b^2}{2} \qquad \text{for } 4b < z$$

The positive constant A is given, $H(\mathbf{L}; A)$ denotes the collection of all sets (or designs) R in $\mathbf{L}$ of area A, and for every design R in $H(\mathbf{L}; A)$ the expression $E[R]$ is defined by

$$E[R] = \int_{R^*} \frac{1}{A} f(\mathbf{X})\, d\mathbf{X}$$

(a) What physical interpretation can you give to the expression $E[R]$?
(b) Find a design R^* in $H(\mathbf{L}; A)$ that minimizes $E[R]$ when (1) $A \le b^2/2$, and (2) $A > b^2/2$. Why is it necessary to consider these two different cases?
(c) For the designs R^* found in part (b), develop single-integral expressions for $E[R^*]$, using the function $q(z)$ as given. Do not evaluate the single integrals.

5.48. Given a positive number a and any real number p, show that the value of the integral

$$\int_p^{a+p} \frac{1}{a} |x|\, dx$$

is given by $|p + (a/2)|$ when $p \ge 0$ or $p \le -a$, and by $a/4 + (1/a)[p + (a/2)]^2$ when $-a \le p \le 0$. [*Hint*: When $p \ge 0$,

$$\int_p^{a+p} \frac{1}{a} |x|\, dx = \int_p^{a+p} \frac{1}{a} x\, dx$$

When $p \le -a$, then $a + p \le 0$, so

$$\int_p^{a+p} \frac{1}{a} |x|\, dx = \int_p^{a+p} \frac{1}{a} (-x)\, dx$$

Finally, when $-a \le p \le 0$, then $a + p \ge 0$, so

$$\int_p^{a+p} \frac{1}{a} |x|\, dx = \int_p^{0} \frac{1}{a} (-x)\, dx + \int_0^{a+p} \frac{1}{a} x\, dx.]$$

5.49. Show that the procedure given for using contour lines and ranking products on the basis of the ratio of throughput to space yields a solution that minimizes the average distance traveled.

5.50. Show that use of the single integral equation is equivalent to integrating over the plane in determining the average distance traveled.

Chapter 6

Planar Multifacility Location Problems

6.1 INTRODUCTION

In this chapter we consider some multifacility location problems. As an example of one class of such problems, in the next section we present an actual application involving the location of a concrete casting facility and an assembly and storage facility in the available yard space of a construction company so as to minimize associated material handling costs to and from the two new facilities. While the problems are natural extensions of those of Chapter 4, they differ from these earlier problems in two important respects: (1) at least two new facilities are being located, and (2) costs are incurred proportional to distances between some pairs of new facilities. As a second class of multifacility location problems, in the final section we consider some location-allocation problems, where the weights as well as the new facility locations can be treated as variables, and examine an application involving locating drilling platforms for offshore oil fields and allocating wells to drilling platforms.

Figure 6.1 gives a geometric illustration of the type of problem that we consider in this chapter. Two facilities are to be located at points $\mathbf{X}_1$ and $\mathbf{X}_2$ to be determined. Four existing facilities are located at known points $\mathbf{P}_1, \ldots, \mathbf{P}_4$, respectively. Because the terms *new facility* and *existing facility* occur quite often in this chapter, we shall abbreviate them by NF and EF, respectively. In Figure 6.1 the lengths of the line

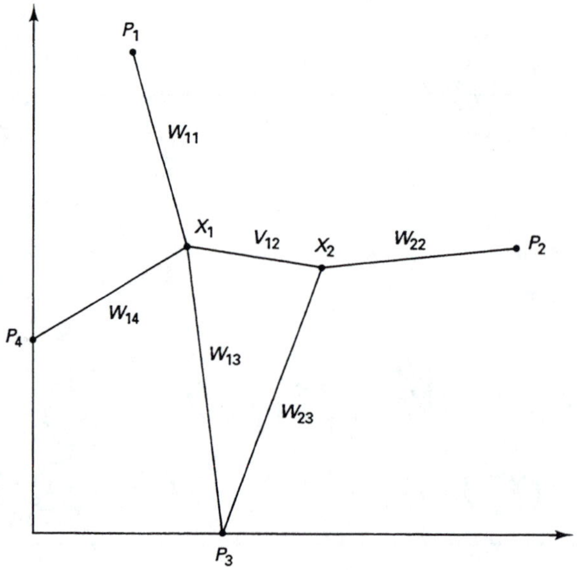

Figure 6.1 Geometric illustration of a multifacility location problem.

segments between the facilities represent the distances between the facilities; we shall denote the distance between any two points **X** and **Y** of interest by $d(\mathbf{X}, \mathbf{Y})$. Items of some sort are transported between NF 1 and EFs 1, 3, and 4; similarly, items of some sort are transported between NF 2 and EFs 2 and 3. In addition, items of some sort are transported between NFs 1 and 2. When items are transported between the facilities, costs are incurred that are proportional to the travel distances of the items. Thus the cost of transport between NFs 1 and 2 is given by

$$v_{12}\, d(\mathbf{X}_1, \mathbf{X}_2) \tag{6.1}$$

with v_{12} being the nonnegative constant of proportionality, called a *weight*. Similarly, the total transport cost involving NF 1 and EFs 1, 3, and 4 is given by the following sum:

$$w_{11}\, d(\mathbf{X}_1, \mathbf{P}_1) + w_{13}\, d(\mathbf{X}_1, \mathbf{P}_3) + w_{14}\, d(\mathbf{X}_1, \mathbf{P}_4) \tag{6.2}$$

The total transport cost involving NF 2 and EFs 2 and 3 is as follows:

$$w_{22}\, d(\mathbf{X}_2, \mathbf{P}_2) + w_{23}\, d(\mathbf{X}_2, \mathbf{P}_3) \tag{6.3}$$

In Figure 6.1 we have written the weight associated with each distance besides the line segment representing the distance. We wish to determine the NF locations, say $\mathbf{X}_1^*$ and $\mathbf{X}_2^*$, which minimize the total transport cost.

Let us denote the total transport cost, which depends on the choice of X_1 and X_2, by $f(X_1, X_2)$. We have only to add the transport costs given by (6.1), (6.2), and (6.3) to obtain the total cost of *all* transport, namely,

$$f(\mathbf{X}_1, \mathbf{X}_2) = v_{12}\, d(\mathbf{X}_1, \mathbf{X}_2) + w_{11}\, d(\mathbf{X}_1, \mathbf{P}_1) + w_{13}\, d(\mathbf{X}_1, \mathbf{P}_3)$$
$$+ w_{14}\, d(\mathbf{X}_1, \mathbf{P}_4) + w_{22}\, d(\mathbf{X}_2, \mathbf{P}_2) + w_{23}\, d(\mathbf{X}_2, \mathbf{P}_3)$$

Our problem now is to find NF locations, say $\mathbf{X}_1^*$ and $\mathbf{X}_2^*$, which minimize $f(\mathbf{X}_1, \mathbf{X}_2)$. It is the term $v_{12}\, d(\mathbf{X}_1, \mathbf{X}_2)$, involving the distance between NFs 1 and 2, which gives this problem its special character. For if we have $v_{12} = 0$, then finding an optimal choice of NF 1 would be independent of finding an optimal location of NF 2, and we could find these NF locations by solving independent single-facility location problems with costs defined by (6.2) and (6.3), respectively, using the approaches considered in Chapter 4.

If each of the given weights is 1, and distances are Euclidean, our total transport cost $f(\mathbf{X}_1, \mathbf{X}_2)$ is just the sum of all the given distances. In this case finding NF locations to minimize the total transport cost is the same as finding NF locations that minimize the total distance; in terms of Figure 6.1, we would be finding choices of $\mathbf{X}_1$ and $\mathbf{X}_2$ that minimize the total lengths of the lines drawn between facilities. Assuming each of the given weights is 1, let us choose some values for the EF locations to obtain a numerical example.

Example 6.1

Suppose that $\mathbf{P}_1 = (5, 25)$, $\mathbf{P}_2 = (25, 15)$, $\mathbf{P}_3 = (10, 0)$, $\mathbf{P}_4 = (0, 10)$. What do you think an optimal choice of $\mathbf{X}_1$ and $\mathbf{X}_2$ would be? Of course, you can make a guess at an answer (and you may wish to do so), but it is unlikely that you would guess that an approximately optimal choice of NF locations is given by

$$\mathbf{X}_1^* = (8.8388, 5.7922), \qquad \mathbf{X}_2^* = (9.1645, 5.6370)$$

with a corresponding total cost of

$$f(\mathbf{X}_1^*, \mathbf{X}_2^*) = 59.7402$$

If we had $v_{12} = 0$, you should be able to use results of Chapter 4 to see that $\mathbf{X}_2^*$ would be on the line segment joining $\mathbf{P}_2$ and $\mathbf{P}_3$, while $\mathbf{X}_1^*$ would be in the convex hull of $\mathbf{P}_1$, $\mathbf{P}_3$, and $\mathbf{P}_4$. Thus the result of having $v_{12} = 1$ is to "pull" the two NFs closer together than they would be otherwise. In fact, although this location problem is easy to understand and has a simple geometric interpretation, it is not trivial to solve. The fact that the problem is nontrivial is fortunate for us, of course, since we devote this chapter to solving such problems.

Let us consider the example again for the case when all distances are rectilinear. Before giving the solution, we give you the opportunity to try to find it yourself. In this case we find optimal solutions given by

$$\mathbf{X}_1^* = \mathbf{X}_2^* = (10, 10)$$

with a corresponding minimum total cost of

$$f(\mathbf{X}_1^*, \mathbf{X}_2^*) = 70$$

The minimum cost of movement in the x-direction is 30, while the minimum cost of movement in the y-direction is 40. Of course, we could not put both new facilities at the same location, but we could choose locations close to (10, 10) with hardly any increase in cost. For example, if we take $\mathbf{X}_1 = (8, 10)$ and $\mathbf{X}_2 = (12, 10)$, we get a total cost of $32 + 40 = 72$, which is only 2.86% above the minimum total cost.

Because we shall be interested in several types of distances, including Euclidean and rectilinear distances, it is useful to introduce a distance, called the l_p distance, which includes these distances, and others, as special cases. Given any two points $\mathbf{X} = (x_1, x_2)$ and $\mathbf{Y} = (y_1, y_2)$, the l_p distance between $\mathbf{X}$ and $\mathbf{Y}$, which we denote by $|\mathbf{X} - \mathbf{Y}|_p$, is defined as follows for $1 \le p$:

$$|\mathbf{X} - \mathbf{Y}|_p = [\,|x_1 - y_1|^p + |x_2 - y_2|^p]^{1/p}$$

We can see that when $p = 1$ the l_p distance is the rectilinear distance, while when $p = 2$ it is the usual Euclidean distance. For $1 < p < 2$, the value of the distance lies between the rectilinear distance value and the Euclidean distance value; for $2 < p$ the distance lies below the Euclidean distance value and continues to decrease as p increases, taking on the limiting value as p approaches infinity of the distance known as the Tchebychev distance:

$$|\mathbf{X} - \mathbf{Y}|_\infty = \max[\,|x_1 - y_1|, |x_2 - y_2|]$$

We saw in Chapter 5 that the Tchebychev distance is of use in modeling location problems involving the location of items moved in and out of an automated storage and retrieval system. Also, we saw the use of the Tchebychev distance in Section 4.4.

The l_p distance has the following properties for $1 \le p$ and *all* $\mathbf{X}$ and $\mathbf{Y}$:

1. Nonnegativity: $|\mathbf{X} - \mathbf{Y}|_p \ge 0$
2. Symmetry: $|\mathbf{X} - \mathbf{Y}|_p = |\mathbf{Y} - \mathbf{X}|_p$
3. Triangle inequality: $|\mathbf{X} - \mathbf{Y}|_p \le |\mathbf{X} - \mathbf{Z}|_p + |\mathbf{Z} - \mathbf{Y}|_p$ for any $\mathbf{Z}$
4. Homogeneity: $|r\mathbf{X} - r\mathbf{Y}|_p = \text{abs}(r)|\mathbf{X} - \mathbf{Y}|_p$

In property 4, r is any real number and abs(r) denotes the absolute value of r. You should be able to verify each property yourself except possibly for the triangle inequality. The origin of the term *triangle inequality* comes from the fact that if we have a right triangle with vertices $\mathbf{X}$, $\mathbf{Y}$, and $\mathbf{Z}$, with $\mathbf{Z}$ corresponding to the right angle, the length of the hypotenuse (which is just $|\mathbf{X} - \mathbf{Y}|_2$) is no greater than the sum of the lengths of the two sides (which is just $|\mathbf{X} - \mathbf{Z}|_2 + |\mathbf{Z} - \mathbf{Y}|_2$).

Let us now generalize from the example of Figure 6.1. We suppose that we have m existing facilities at known points $\mathbf{P}_1, \ldots, \mathbf{P}_m$ as in Chapter 4, and wish to determine n new facility locations $\mathbf{X}_1, \ldots, \mathbf{X}_n$. For each new facility j and existing facility i, a cost is incurred that is proportional to the l_p distance between the facilities, with the constant of proportionality being a nonnegative number, a *weight* w_{ji}. For each pair of new facilities j and k, a cost is incurred that is proportional to the l_p distance between the facilities, with the constant of proportionality being the nonnegative *weight* v_{jk}. Hence the total cost of all movement between NFs and EFs is given by

$$\sum_{j=1}^{n} \sum_{i=1}^{m} w_{ji}|\mathbf{X}_j - \mathbf{P}_i|_p \tag{6.4}$$

while the total cost of all movement between pairs of NFs is given by

$$\sum_{1 \le j < k \le n} v_{jk} |\mathbf{X}_j - \mathbf{X}_k|_p \tag{6.5}$$

Thus the total cost for new facility locations $\mathbf{X}_1, \ldots, \mathbf{X}_n$, which we denote by $f(\mathbf{X}_1, \ldots, \mathbf{X}_n)$, is simply the sum of (6.4) and (6.5). That is,

$$f(\mathbf{X}_1, \ldots, \mathbf{X}_n) = \sum_{1 \le j < k \le n} v_{jk} |\mathbf{X}_j - \mathbf{X}_k|_p + \sum_{j=1}^{n} \sum_{i=1}^{m} w_{ji} |\mathbf{X}_j - \mathbf{P}_i|_p \tag{6.6}$$

We wish to determine new facility locations that will minimize the total cost, in this case the function f. *We remark that it is the term* (6.5), *the sum of weighted distances between NFs, that gives this multifacility location problem its special character*, for without it the location problem would decompose into n single new facility problems, one for each new facility, of the type we have considered in Chapter 4.

The same seven assumptions apply to the multifacility location problem that apply to the single-facility location problem, discussed in Section 4.5. In addition, A3 is now likely to be more stringent, as is A7. The reason for A3 being more stringent is due to the fact that the multifacility model may have an optimal solution where two or more new facilities are given the same location; in such a situation we will have to modify the solution somewhat to make it usable, probably by choosing distinct locations close to those the model recommends. Further, since there are now at least two new facilities, it is more likely that some of the weights are controllable, giving rise to distribution considerations. For example, if we have a problem where each existing facility receives service from a closest new facility, the problem would not be well represented by the model (6.6), since it is based on the assumption that each existing facility receives service from a known new facility (i.e., the weights w_{ji} are known). When each existing facility receives service from a closest new facility, we obtain a location-allocation problem, which we study later in the chapter.

It will often be convenient to put the weights w_{ji} into a matrix $\mathbf{W} = (w_{ji})$ and to put the weights v_{jk} into a matrix $\mathbf{V} = (v_{jk})$, where we adopt the convention that $v_{kj} = v_{jk}$ for distinct j and k, with diagonal entries of $\mathbf{V}$ being zero. Thus we show the data for the example of Figure 6.1 in Table 6.1.

A good deal of insight into the structure of the foregoing multifacility location problem can be obtained by constructing a graph $G(\mathbf{V}, \mathbf{W})$, which we term the *weight graph*. Let the graph have nodes $E_1, \ldots, E_m$ corresponding to the existing facilities, and nodes $N_1, \ldots, N_n$ corresponding to the new facilities. For every new facility j and existing facility i for which w_{ji} is positive, construct an edge joining nodes N_j and E_i. Similarly, for every new facility j and new facility k for which v_{jk} is positive, construct an edge joining nodes N_j and N_k. Figure 6.2 illustrates the weight graph for Example 6.1.

We say two nodes of $G(\mathbf{V}, \mathbf{W})$ are *adjacent* if they are joined by an edge. For example, in Figure 6.2 nodes N_1 and E_1 are adjacent, but nodes N_1 and E_2 are not adjacent. A sequence of adjacent nodes, for example, E_1, N_1, N_2, E_3, is called a *path*. We say that the weight graph is *connected* if at least one path joins any two nodes

TABLE 6.1 EXAMPLE 6.1: DATA FOR THE EXAMPLE OF FIGURE 6.1

$$\mathbf{W} = \begin{pmatrix} 1 & 0 & 1 & 1 \\ 0 & 1 & 1 & 0 \end{pmatrix}, \qquad \mathbf{V} = \begin{pmatrix} 0 & 1 \\ 1 & 0 \end{pmatrix}$$

$$\mathbf{a} = (5 \quad 25 \quad 10 \quad 0)$$

$$\mathbf{b} = (25 \quad 15 \quad 0 \quad 10)$$

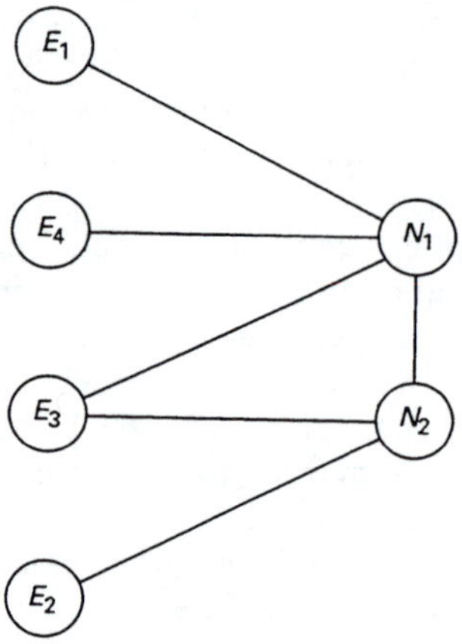

Figure 6.2 Weight graph for Example 6.1.

in the graph. Thus the graph of Figure 6.2 is connected, but it would not be connected if we remove, say, the edge joining N_1 and E_1. The graph would be disconnected if we remove only the edge joining N_2 and E_2. On the other hand, if we remove only the edge joining N_1 and E_3, the graph would still remain connected. Suppose that we remove the edge joining N_1 and N_2 and the edge joining N_1 and E_3. In this case we get two connected subgraphs that have no common edges or nodes. Such subgraphs are called components. A *component* is a connected subgraph with a maximum number of adjacent nodes and associated edges.

The structure of $G(\mathbf{V}, \mathbf{W})$ *tells us whether or not the multifacility location problem can be decomposed into smaller problems*. If the graph $G(\mathbf{V}, \mathbf{W})$ has q components, the problem decomposes into q independent problems, one for each component, with the N nodes in each component identifying the new facility locations to be determined for the problem associated with the components. For example, if we have a weight graph with at least three N nodes and nodes N_1 and N_2 are adjacent to each other but to no other N node, and are adjacent to two or more E nodes, we can find least cost locations of facilities 1 and 2 independently from those of the other new facilities. More generally, if we denote by $G(\mathbf{V})$ the graph obtained from $G(\mathbf{V}, \mathbf{W})$ by removing the E nodes, as well as the edges incident to the E and N nodes, the components of $G(\mathbf{V})$ identify the new facilities of independent location problems, one for each collection of components. The decomposed problems we obtain may likewise have structure that can be exploited; for example, if any N node is adjacent to only one other node, its optimal location is identical with that of the facility to whose node it is adjacent, and there is no point in using an algorithm to determine its location. Similarly, if some N node, say N_4, is adjacent to exactly two E nodes, say E_1 and E_2, and to no other N node, we know (by the majority theorem) that NF 4 will have an optimal location at P_1 or P_2, whichever has the larger weight with NF 4.

Thus we recommend that you do the following whenever you have a problem such as those of this chapter to solve. *Construct the weight graph* $G(\mathbf{V}, \mathbf{W})$ *and see if it decomposes into components*. If it does decompose into components, you can solve the problems associated with the components independently of one another, thus obtaining several smaller problems to solve instead of one larger problem. If $G(\mathbf{V}, \mathbf{W})$ is connected, it is still possible for $G(\mathbf{V})$ to be disconnected. Thus you should also construct $G(\mathbf{V})$ to see if it decomposes into components, and if so, solve the problems associated with the components independently of one another. For the example of Figure 6.2, if there is no arc between nodes N_1 and N_2, then $G(\mathbf{V}, \mathbf{W})$ is connected but $G(\mathbf{V})$ is not. Thus we obtain two independent location problems, one involving NF 1 with EFs 1, 3, and 4, the other involving NF 2 with EFs 2 and 3.

Some of the insights of Chapter 4 for the single-facility minisum location problem with $p = 1$ carry over to the multifacility version of the problem, as we shall see. It is again the case that there exist optimal new facility locations each of which is an intersection point. Further, each optimum location is a median location with respect to *all* the other facilities for which it has positive weights. Also, the problem decomposes into two independent location problems, one for the x locations and the other for the y locations. The two problems have identical structure, and thus any procedure to solve one problem can also be used to solve the other problem. One method of solving the x-coordinate problem is to convert it into an equivalent linear programming problem. The dual of this linear programming problem can be converted into a network transshipment problem, whose structure is quite similar to that of the weight graph, and thus can be solved by algorithms that exploit the network structure. We remark that linear constraints can be added to the linear programming formulations to represent restrictions upon locations, but in this case the problem may not decompose into two independent problems, resulting in a larger problem with less structure to exploit.

For the Euclidean distance problem ($p = 2$), provided that the graph $G(\mathbf{V}, \mathbf{W})$ is connected, it is the case that all the optimum new facility locations are in the convex hull of the existing facility locations. Further, the Weiszfeld algorithm extends in quite a direct manner to provide an algorithm commonly used for solving the problem. Similarly, the problem can be "perturbed" by placing an epsilon term under each square root to avoid the difficulties associated with division by zero. The fact that the algorithm is quite easy to program adds to its appeal. Computational experience with the algorithm appears to be generally satisfactory, although there are pathological problems for which the algorithm may not perform well in one sense or another.

The multifacility problem with squared Euclidean distances appears to be of limited interest other than to provide new facility locations to initiate the Weiszfeld algorithm for the Euclidean distance problem. The problem with squared Euclidean distances may be solved by setting partial derivatives to zero; two systems of linear equations result from setting the derivatives to zero. The two systems have the same matrix but different right-hand sides.

We now give an overview of this chapter. In Section 6.2 we consider an application of the methodology of this chapter to a real-world location problem. A multifacility model with rectilinear distances was used to solve an actual location problem for a construction company, and the solution was implemented. In Section 6.3 we consider the multifacility model with rectilinear distances. We show how to convert the model to an equivalent linear program and present a coordinate-search method for solving the model as well; next we give a way to interpret the problem as a simple network flow problem, which allows us to solve many problems by inspection and to solve large problems easily; finally, we give a motivation for the network approach, based on the duality theory of linear programming. In Section 6.4 we consider the multifacility problem with squared Euclidean distances, principally as preparation for Section 6.5, where we study the problem with Euclidean distances, state some of its properties, and develop an algorithm to solve the problem. In the final section we consider location-allocation problems, where the weights as well as the new facility locations can be treated as variables, and examine an application involving locating drilling platforms for offshore oil fields and allocating wells to the platforms.

6.2 AN APPLICATION

We consider a case study due to Love and Yerex [33]. The name and location of the company have been changed by the authors, but all other data are unchanged. In the discussion to follow we have changed the number of the figure, and deleted some of the linear program discussion, which we defer to the next section, but everything else is basically unchanged. This case study should help to convince you that there are interesting actual applications for the models of this chapter, as well as illustrate the way in which model weights are obtained.

Example 6.2

The Coastal Construction Company of Portsmouth, Virginia, operates a precast–prestressed concrete products plant. The plant was originally used to produce components for large bridges and tunnels. Coastal presently produces items such as bridge girders, pilings, and other marine construction components.

A new product, transmission poles for electrical utilities, was to be introduced and the location of two new production facilities was to be determined. Figure 6.3 shows

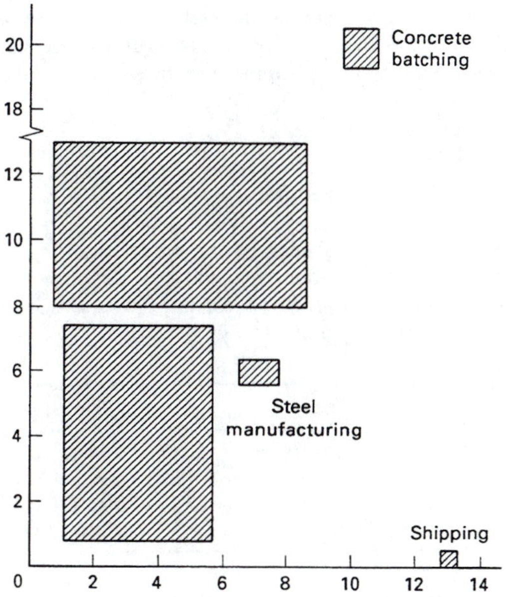

Figure 6.3 Coastal Construction Co. site.

a section of the yard area. Three existing facilities were relevant to the determination of the location of two new facilities. These facilities consisted of the following:

1. *Concrete batching plant.* The concrete batching plant supplies the premixed concrete used in pole construction. This plant also supplies the concrete for all other concrete items produced in the plant.
2. *Steel manufacturing and storage area.* A certain amount of mild steel hardware is used in pole production; about half of the steel is embedded in the concrete and forms part of the finished pole. The remainder is added in the storage area. This hardware consists of spiral wound ties, base plates, and galvanized steel support brackets.
3. *Shipping area.* All completed production moves through this area, where it is checked and weighed before leaving the plant.

Two new facilities, a concrete casting area and an assembly and storage area, were to be located in the available yard space.

1. *Concrete casting area.* In this department the poles are cast. The concrete delivered to the area is placed in holding brickets and used as required by a pouring crane, which is part of the casting area equipment. A crane also moves the cast parts into the assembly and storage area.
2. *Assembly and storage area.* Production activities in the assembly and storage area include the trimming, repair, and other finished work to be done on the poles.

In order to develop numerical data a sales estimate of 40 poles per day was used as a daily production norm. Forty poles required approximately 10 yd^3 of concrete and 8400 lb of manufactured steel. The handling equipment required was as follows:

1. *Concrete*: 3-yd^3 brickets on flatbed trucks
2. *Steel*: front-end loader in drums
3. *Poles*: Drott Travel lift crane and flatbed trailer

In order to arrive at some estimate of handling costs, the operating cost for each piece of handling equipment was obtained from the accounting department. A field accountant supplied historical costs based on studies of existing operations. The following cost–distance values were then computed:

Item	Cost	Trips required
Concrete delivery	20 cts/100 ft	4 trips req'd/40 poles
Steel delivery to casting area	8 cts/100 ft	10 trips req'd/40 poles
Crane	40 cts/100 ft	10 trips req'd/40 poles
Flatbed	20 cts/100 ft	2 trips req'd/40 poles

From the plot plan of the plant, with the origin at the lower left corner, the existing facilities are located as follows:

1. Concrete batch plant (1000, 2000)
2. Steel manufacturing area (700, 600)
3. Shipping checks (1300, 0)

Noncandidate areas are shaded in Figure 6.3; ordinates are in hundreds of feet. Based on the trip frequencies and cost estimates, the following daily costs per 100 feet between all facilities were computed.

New facilities	Existing facilities 1	2	3	New facilities	New facilities 1	2
1	0.80	0.80	0	1	0	4.0
2	0	0.80	0.40	2	0	0

Under the assumption of rectangular travel distances, a linear programming problem was formulated. The locations of the concrete casting and the assembly areas are given by (x_1, y_1) and (x_2, y_2), respectively.

For the linear program, the objective function states that the goal of the model is to minimize total daily materials handling cost between all facility pairs that are affected by the location of the new facilities. The linear programming solution procedure is thus forced to locate the new facilities in positions that minimize total materials handling cost.

The solution of the linear programming formulation, $x_1 = x_2 = 700, y_1 = y_2 = 600$, does not conflict with any existing facilities. The total daily cost is \$18.40. The solution indicates that the two new facilities are to be located together adjacent to the steel manufacturing area for maximum cost savings. The solution was implemented by the company.

It is interesting to note that management did not recognize the existence of a location problem in this form. The location study had not been specifically requested but was suggested by one of the authors, who was engaged in a more comprehensive plant study. It is not possible to state with certainty the exact savings that were achieved by using the location model since management had not reached a final decision regarding the location of the new facilities. However, actual percentage cost savings probably were considerable. For example, one possible set of sites that had been considered located the concrete casting equipment at plot plan location (10, 18) and the assembly-storage area at location (13, 10). The total daily material handling cost of these locations is \$69.60. The optimal solution represents a 73.6% cost reduction over this solution.

A final comment may be made in relation to the sales forecast of 40 poles per day. The optimal locations of the new facilities would not be different with a different sales volume, since all material flows between facilities are proportional to the flow of finished poles from the facility. For example, with a sales volume of 80 poles per day, all coefficients in the objective function of the linear model would be doubled. Since the constraints remain unchanged, the same solution would be reached.

This problem is exactly the kind of problem we are considering in this chapter. In our notation the data for it are as follows:

$$\mathbf{W} = \begin{pmatrix} 0.80 & 0.80 & 0 \\ 0 & 0.80 & 0.40 \end{pmatrix}, \qquad \mathbf{V} = \begin{pmatrix} 0 & 4.0 \\ 4.0 & 0 \end{pmatrix}$$

$$\mathbf{a} = (10 \quad 7 \quad 13)$$

$$\mathbf{b} = (20 \quad 6 \quad 0)$$

(Note that we give coordinates in hundreds of feet.) The linear program the authors refer to is identical to the one we develop in the next section. Formulating and solving this linear program gives the solution stated by the authors. We solve this problem subsequently using the network flow solution procedure.

6.3 RECTILINEAR DISTANCE MINISUM LOCATION PROBLEMS

6.3.1 Elementary Results and Solution Procedures

In this section we consider the rectilinear distance minisum location problem and present elementary results and solution procedures. In the next section we present a more advanced solution procedure, which is much more efficient.

Recall that when we have the l_p distance and take $p = 1$, we get the rectilinear distance. In particular, if NF j has location $\mathbf{X}_j = (x_j, y_j)$, and NF k has location (x_k, y_k), then

$$|\mathbf{X}_j - \mathbf{X}_k|_1 = |x_j - x_k| + |y_j - y_k| \tag{6.7}$$

Similarly, if EF i has location $\mathbf{P}_i = (a_i, b_i)$, then

$$|\mathbf{X}_j - \mathbf{P}_i|_1 = |x_j - a_i| + |y_j - b_i| \tag{6.8}$$

Thus if we substitute (6.7) and (6.8) into our total cost expression (6.6) and rearrange terms, we get

$$f(\mathbf{X}_1, \ldots, \mathbf{X}_n) = f_1(x_1, \ldots, x_n) + f_2(y_1, \ldots, y_n) \tag{6.9}$$

where

$$f_1(x_1, \ldots, x_n) = \sum_{1 \le j < k \le n} v_{jk}|x_j - x_k| + \sum_{j=1}^{n} \sum_{i=1}^{m} w_{ji}|x_j - a_i| \tag{6.10}$$

and

$$f_2(y_1, \ldots, y_n) = \sum_{1 \le j < k \le n} v_{jk}|y_j - y_k| + \sum_{j=1}^{n} \sum_{i=1}^{m} w_{ji}|y_j - b_i| \tag{6.11}$$

Here $f_1(x_1, \ldots, x_n)$ is the total cost of movement in the x-direction, while $f_2(y_1, \ldots, y_n)$ is the total cost of movement in the y-direction. Equation (6.9) says that the total cost of movement is the sum of the total cost of movement in the x-direction and the total cost of movement in the y-direction. Consequently, *we can find the minimum total transport cost by independently minimizing* $f_1(x_1, \ldots, x_n)$ *and* $f_2(y_1, \ldots, y_n)$. The choice of the x's has no effect on the choice of the y's, and vice versa. Thus we can decompose our original problem into two subproblems, each with half as many variables as the original problem. Further, if we consider equations (6.10) and (6.11) we see they *have exactly the same form*. Hence if we take equation (6.10) and replace each x by y and each a by b, we have equation (6.11). Thus any procedure we develop to minimize the total cost of movement in the x-direction can also be used to minimize the total cost of movement in the y-direction simply by replacing the x's and a's by y's and b's, respectively.

Let us now concentrate on the problem of minimizing the cost of movement in the x-direction; we shall construct an equivalent linear programming problem. The justification for the equivalence is a simple change of variables, the motivation for which is as follows. Let us interpret r_{ji} as the amount by which x_j is to the right of a_i, and s_{ji} as the amount by which x_j is to the left of a_i. In case x_j is to the right of a_i, we have

$$r_{ji} = x_j - a_i = |x_j - a_i|$$

and we *define*

$$s_{ji} = 0$$

Thus we have

$$x_j - r_{ji} + s_{ji} = a_i \qquad \text{and} \qquad r_{ji} + s_{ji} = |x_j - a_i| \tag{6.12}$$

Similarly, if x_j is to the left of a_i, we have

$$s_{ji} = a_i - x_j = |x_j - a_i|$$

and we define

$$r_{ji} = 0$$

Thus we again have (6.12). Of course, we also require that r_{ji} and s_{ji} be nonnegative. Expression (6.12) amounts to a change of variables. Similarly, we introduce nonnegative variables p_{jk} and q_{jk}, which represent the amount by which x_j is to the left or the right, respectively, of x_k.

If we now use (6.12) in our expression for $f_1(x_1, \ldots, x_n)$, the problem of minimizing $f_1(x_1, \ldots, x_n)$ becomes the following linear programming problem:

$$\text{minimize} \quad \sum_{1 \leq j < k \leq n} v_{jk}(p_{jk} + q_{jk}) + \sum_{j=1}^{n} \sum_{i=1}^{m} w_{ji}(r_{ji} + s_{ji})$$

subject to

$$\begin{aligned} x_j - x_k + p_{jk} - q_{jk} &= 0, \qquad 1 \leq j < k \leq n \\ x_j - r_{ji} + s_{ji} &= a_i, \qquad i = 1, \ldots, m \\ & \qquad\qquad j = 1, \ldots, n \end{aligned}$$

All variables but the x's are required to be nonnegative.

If we set up the original location problem so that all existing facilities are in the first quadrant, we can impose nonnegativity conditions on the x's (and y's) as well. We remark that if any weight in the objective function is zero, the corresponding constraint in the linear program can be omitted.

Example 6.3

Assume $n = 2$ new facilities and $m = 3$ existing facilities having locations of (10, 15), (20, 25), and (40, 5), respectively. All data for this example are shown in Table 6.2. If we solve the linear programming problem for the x-coordinates, we find that it is optimal to have both x-coordinates the same, any number between 10 and 20, and the minimum objective function value is 160. *To obtain the problem to solve for the y-coordinates we simply replace the x's by y's and the a's by b's.* For the example we find it is optimal to have both y-coordinates equal to 15 and get a minimum objective function value of 60. In conclusion, for this example it is optimal to give both new facilities the location $(x^*, 15)$, where $10 \leq x^* \leq 20$, and the minimum cost of movement in both x and y directions is given by $160 + 60 = 220$. Of course, it is impossible to place both new facilities at the same location, so we will have to seek locations for them close to that the linear programming solution provides. Which locations we actually choose depends on the size of the new facilities as well as the space available.

Here is the linear program to be solved to find x_1 and x_2. To check your understanding you should set up and solve the one to find y_1 and y_2:

TABLE 6.2 DATA FOR EXAMPLE 6.3

$\mathbf{W} = (w_{ji}) = \begin{pmatrix} 2 & 1 & 0 \\ 4 & 0 & 5 \end{pmatrix}$,	$\mathbf{V} = (v_{jk}) = \begin{pmatrix} 0 & 2 \\ 2 & 0 \end{pmatrix}$
$\mathbf{a} = (a_i) = (10 \quad 20 \quad 40)$	
$\mathbf{b} = (b_i) = (15 \quad 25 \quad 5)$	

$$\text{minimize} \quad 2(p_{12} + q_{12}) + 2(r_{11} + s_{11}) + 1(r_{12} + s_{12}) + 4(r_{21} + s_{21}) + 5(r_{23} + s_{23})$$

subject to:

$$\begin{aligned}
x_1 \quad & -r_{11} + s_{11} &= 10 \\
x_1 \quad & -r_{12} + s_{12} &= 20 \\
x_2 \quad & -r_{21} + s_{21} &= 10 \\
x_2 \quad & -r_{23} + s_{23} &= 40 \\
x_1 - x_2 \quad & -p_{12} + q_{12} &= 0
\end{aligned}$$

All variables but x_1 and x_2 must be nonnegative.

This example illustrates a situation that happens often with the multifacility location model with rectilinear distances. Optimal new facility locations tend to "cluster"; that is, new facility locations tend to be the same for subsets of new facilities. This situation weakens the usefulness of the model in some cases, since the locations the model provides cannot then be directly implemented. Rather, we try to find nearby locations which result in small increases in the objective function value. For example, with $\mathbf{X}_1 = (19, 15), \mathbf{X}_2 = (21, 15)$ we get $f(\mathbf{X}_1, \mathbf{X}_2) = 222$, which is only about 1% greater than the minimum cost of 220.

This example also illustrates a result that is generally true. There exists at least one optimal solution where each new facility coordinate coincides with some existing facility coordinate. Suppose that we draw horizontal and vertical lines extending to infinity through each existing facility location. Call each point that is the intersection of a horizontal and a vertical line an *intersection point*. Then the example illustrates the *intersection point property*: there is at least one optimal solution where each new facility is located at an intersection point. (This property does not always imply, however, that there is an optimal solution where each new facility is located at some existing facility, as Example 6.2 illustrates.)

Note that the linear program in Example 6.3 has 5 constraints and 12 variables. Solving such a problem by hand would be tedious, and solving any larger problem by hand would probably be unthinkable. Fortunately, this is where duality will be useful. We shall see that the dual of this example problem has so much structure that it can be solved by inspection. More generally, the dual problem will be a minimum-cost network flow problem, which can be solved very efficiently by any one of a number of special-purpose algorithms.

As an alternative to linear programming, a simple approach which sometimes finds optimal locations is known as *coordinate descent*. We shall illustrate the ideas of this approach with several examples.

Example 6.4

Consider a location problem with four existing facilities. Points $\mathbf{P}_1$ to $\mathbf{P}_4$ are given by (0, 2), (4, 0), (6, 8), and (10, 4), respectively. Table 6.3 gives all the data for the example.

We thus have

$$f_1(x_1, x_2) = 6|x_1 - x_2| + 5|x_1 - 0| + 3|x_1 - 4| + |x_2 - 4| + 8|x_2 - 6| + 4|x_2 - 10|$$

$$f_2(y_1, y_2) = 6|y_1 - y_2| + 5|y_1 - 2| + 3|y_1 - 0| + |y_2 - 0| + 8|y_2 - 8| + 4|y_2 - 4|$$

Let us first use the method to attempt to minimize $f_1(x_1, x_2)$. To get a starting point we omit the term $6|x_1 - x_2|$ in $f_1(x_1, x_2)$, so that the function is the sum of two single-facility minisum functions to which we can apply the median conditions, obtaining $(x_1, x_2) = (0, 6)$ with $f_1(0, 6) = 66$. Now we start at the point (0, 6) and try to decrease the value of f_1 by varying x_1. We minimize

$$f_1(x_1, 6) = 5|x_1 - 0| + 3|x_1 - 4| + 6|x_1 - 6| + 18$$

(You should figure out where the number 18 comes from.) *When we vary only a single variable, the function we have to minimize is just a simple one-facility minisum function, and we can use the median conditions of Chapter* 4 *to find a minimum*. We use the median conditions to minimize the latter expression, obtaining $x_1 = 4$, with $f_1(4, 6) = 50$. Next we try to decrease the value of f_1 by varying x_2. We minimize

$$f_1(4, x_2) = 6|x_2 - 4| + |x_2 - 4| + 8|x_2 - 6| + 4|x_2 - 10| + 20$$

We use the median conditions to minimize the latter expression, obtaining $x_2 = 6$, with $f_1(4, 6) = 50$. But now we have arrived at the point (4, 6) a second time using coordinate descent, so we stop. (If we did not stop, we would just be repeating the same steps over and over again.) It is known that when we stop (having found the same point twice), *if no two coordinates are the same, we have minimized* f_1. It is not essential to compute the f value at each step of coordinate descent, but doing so can provide a useful computational check, since each value should be no larger than the previous value.

Now let us apply coordinate descent to try to minimize f_2. We start at the point $(y_1, y_2) = (2, 8)$, for which $f_2(2, 8) = 66$. We first try to decrease the value of f_2 by varying

TABLE 6.3 DATA FOR EXAMPLE 6.4

$$\mathbf{W} = \begin{pmatrix} 5 & 3 & 0 & 0 \\ 0 & 1 & 8 & 4 \end{pmatrix}, \qquad \mathbf{V} = \begin{pmatrix} 0 & 6 \\ 6 & 0 \end{pmatrix}$$

$\mathbf{a} = (0 \quad 4 \quad 6 \quad 10)$

$\mathbf{b} = (2 \quad 0 \quad 8 \quad 4)$

y_1. We minimize

$$f_2(y_1, 8) = 3|y_1 - 0| + 5|y_1 - 2| + 6|y_1 - 8| + 24$$

so using the median conditions to minimize the latter expression gives $y_1 = 2$, with $f_2(2, 8) = 66$. However, we do not stop yet, since we have not tried varying y_2. We minimize

$$f_2(2, y_2) = |y_2 - 0| + 6|y_2 - 2| + 4|y_2 - 4| + 8|y_2 - 8| + 6$$

We use the median conditions and find $y_2 = 4$ with $f_2(2, 4) = 54$. Next we minimize

$$f_2(y_1, 4) = 3|y_1 - 0| + 5|y_1 - 2| + 6|y_1 - 4| + 36$$

and find $y_1 = 2$, with $f_2(2, 4) = 54$ again. Now we stop, since we have arrived at the same point again using coordinate descent. As the two y-coordinates are distinct, we can conclude that we have minimized f_2. Thus for our example we have found optimal locations $\mathbf{X}_1^* = (4, 2), \mathbf{X}_2^* = (6, 4)$, with minimum total cost $= 50 + 54 = 104$.

Our example illustrates all the ideas of coordinate descent. We pick an initial vector and choose a single coordinate to vary, using the median conditions to minimize the resulting expression. The first coordinate we choose is the first variable, the second coordinate we choose is the second variable, and so on. We continue until we obtain the same vector by coordinate descent that we have obtained previously by coordinate descent, at which point we stop. If no two entries in the vector are equal, we have found either optimal x or y coordinates, depending on whether we are minimizing f_1 or f_2, respectively. Any point can be chosen as a starting point, although we may end up doing less work with coordinate descent if our starting point is "almost" optimal.

Example 6.5

With $m = 2$, $\mathbf{P}_1 = (0, 0)$, and $\mathbf{P}_2 = (4, 4)$, we have

$$f_1(x_1, x_2) = 10|x_1 - x_2| + 8|x_1 - 0| + 2|x_1 - 4| + |x_2 - 0| + 4|x_2 - 4|$$

$$f_2(y_1, y_2) = 10|y_1 - x_2| + 8|y_1 - 0| + 2|y_1 - 4| + |y_2 - 0| + 4|y_2 - 4|$$

The data for this example are given in Table 6.4. Let us apply coordinate descent to try to find a minimum for f_1, beginning with the point $(4, 0)$, for which $f_1(4, 0) = 88$. We first try varying x_1 and thus wish to minimize

$$f_1(x_1, 0) = 18|x_1 - 0| + 2|x_1 - 4| + 16$$

TABLE 6.4 DATA FOR EXAMPLE 6.5

$$\mathbf{W} = \begin{pmatrix} 8 & 2 \\ 1 & 4 \end{pmatrix}, \qquad \mathbf{V} = \begin{pmatrix} 0 & 10 \\ 10 & 0 \end{pmatrix}$$

$\mathbf{a} = (0 \quad 4)$

$\mathbf{b} = (0 \quad 4)$

Using the median conditions gives $x_1 = 0$, and we get $(0, 0)$. Next we try varying x_2, and so minimize

$$f_1(0, x_2) = 11|x_2 - 0| + 4|x_2 - 4| + 8$$

Using the median conditions gives $x_2 = 0$, with $f_1(0, 0) = 24$.

At this point we stop coordinate descent for f_1, since we minimized $f_1(x_1, 0)$ earlier. Hence we end up with the point $(0, 0)$; *since both entries are the same, we are unable to conclude, using only coordinate descent, whether or not we have minimized* f_1. For this small example we can check our result by enumerating every intersection point for f_1 and evaluating f_1 at each point, since we know some intersection point is optimal. We find $f_1(0, 0) = 24, f_1(4, 0) = 88, f_1(0, 4) = 52, f_1(4, 4) = 36$. Hence we can conclude that $(0, 0)$ minimizes f_1.

Now consider minimizing f_2, starting at the point $(y_1, y_2) = (4, 4)$, where $f_2(4, 4) = 36$. We first apply the median conditions to minimize

$$f_2(y_1, 4) = 8|y_1 - 0| + 12|y_1 - 4| + 4$$

and obtain $y_1 = 4$ again. Next we apply the median conditions to minimize

$$f_2(4, y_2) = |y_2 - 0| + 14|y_2 - 4| + 32$$

and obtain $y_2 = 4$ again. We now stop coordinate descent. Since both coordinates of $(4, 4)$ are the same, we cannot conclude we have found a minimum of f_2. We can again check on our work by evaluating f_2 at every intersection point. We find $f_2(0, 0) = 24$, $f_2(4, 0) = 88, f_2(0, 4) = 52, f_2(4, 4) = 36$. Thus we conclude that *coordinate descent can provide a solution that is not optimal*. In fact, the solution it provides can depend on the starting point. If we had been lucky enough to start from $(y_1, y_2) = (0, 4)$, it would have provided us with the solution $(y_1, y_2) = (0, 0)$, an optimal solution.

For this second example there *is* a direction we can take from point $(4, 4)$ that will give us a minimum; namely, we can "move" along the line $y_1 = y_2$. If we require $y_1 = y_2$, then $f_2(y_1, y_2) = f_2(y_1, y_1)$, in which case we minimize

$$f_2(y_1, y_1) = 9|y_1 - 0| + 6|y_1 - 0|$$

The median conditions give $y_1 = 0$, so $y_1 = y_2$ implies that $y_2 = 0$ also, which we know is an optimal solution. To get to an optimal solution from $(4, 4)$, we had to vary both coordinates simultaneously, and coordinate descent, of course, does not do this. Checking such "diagonal" directions is impractical for large n, however, so we shall not pursue the idea beyond this example.

In conclusion, *coordinate descent is guaranteed to find a global minimum of* f_1 (of f_2) *only when no two entries in the final vector it provides are equal. If some entries in the final vector are equal, it may or it may not provide a global minimum.* The principle advantage of coordinate descent is its simplicity of use, since it is quite easy to code. It is necessary to resort to linear programming, or a related procedure as discussed in the following section, to be assured of finding global minima.

6.3.2 Minimum-Cost Network Flow Procedure

In this section we consider again the multifacility problem with rectilinear distances. Our approach is based on constructing the dual of the linear programming formulation of the preceding section. The following section provides a motivation for the approach.

Recall that the total cost of movement in the x direction is given by the following equation:

$$f_1(x_1, \ldots, x_n) = \sum_{1 \le j < k \le n} v_{jk}|x_j - x_k| + \sum_{j=1}^{n} \sum_{i=1}^{m} w_{ji}|x_j - a_i|$$

The use of the distance change of variables we developed gives the following equivalent linear program: minimize

$$\sum_{1 \le j < k \le n} v_{jk}(p_{jk} + q_{jk}) + \sum_{j=1}^{n} \sum_{i=1}^{m} w_{ji}(r_{ji} + s_{ji})$$

subject to

$$x_j - x_k + p_{jk} - q_{jk} = 0, \qquad 1 \le j < k \le n$$

$$x_j - r_{ji} + s_{ji} = a_i, \qquad i = 1, \ldots, m; \quad j = 1, \ldots, n$$

All variables but the x's must be nonnegative. If any weight in the objective function is zero, the corresponding constraint in the linear program can be omitted.

If we construct the dual of this linear programming problem for finding optimal x-coordinates and then make a change of variables so that all variables are nonnegative, we get a minimum cost network flow problem. This flow problem can be constructed using the following procedure.

Minimum-Cost Network Flow Problem Construction Procedure

1. Set up a network with a node N_j for each new facility j and a node E_i for each existing facility i.
2. Whenever $w_{ji} > 0$, construct a directed arc from node N_j to node E_i with a capacity of $2w_{ji}$ and a cost of 0. Whenever $v_{jk} > 0$ for $j < k$, construct a directed arc from node N_j to node N_k with a capacity of $2v_{jk}$ and a cost of 0.
3. Set up the matrix $\mathbf{V}^-$ with each entry v_{jk}^- defined as follows: $v_{jk}^- = v_{jk}$ for $j < k$, $v_{jk}^- = 0$ for $j = k$, $v_{jk}^- = -v_{kj}$ for $j > k$. For $t = 1, \ldots, n$, compute ω_t, defined to be the total of two sums: the sum of the entries in row t of $\mathbf{W} = (w_{ji})$, and the sum of the entries in row t of $\mathbf{V}^-$; interpret ω_t as the supply available at node N_t.
4. Compute

$$\omega_{n+1} = \sum_{j=1}^{n} \sum_{i=1}^{m} w_{ji}$$

the total of all the entries in **W**. Construct a node N_{n+1}, a sink with demand ω_{n+1}, and construct a directed arc from each node E_i to node N_{n+1} with infinite capacity and cost a_i. It is easy to verify that

$$\sum_{j=1}^{n} \omega_j = \omega_{n+1}$$

Hence the total amount available will always equal the total amount required.

Given a minimum objective function value to the network flow problem, say z^*, the minimum value of movement in the x direction, say f_1^*, can be computed as follows:

$$f_1^* = a' - z^*$$

where

$$a' = \sum_{i=1}^{m} a_i \left(\sum_{j=1}^{n} w_{ji} \right)$$

(Note that a' is the sum of the products of the entries in a and the column totals of **W**.) Any minimum cost network flow algorithm can be used to solve the problem, which provides dual variables corresponding to nodes. With π_j denoting the optimal dual variable corresponding to node N_j, $j = 1, \ldots, n + 1$, let us compute

$$x_j^* = \pi_j - \pi_{n+1}, \qquad j = 1, \ldots, n \tag{6.13}$$

It is known that x_j^* is an optimum x-coordinate for new facility j.

To get the network for finding the y-coordinates, simply replace the a's by the b's, and the x's by the y's, in the discussion above. Also replace f_1 by f_2 (and use b' instead of a').

In addition, complementary slackness is often useful. Suppose that we have an optimal solution to the network problem; let $z(N_j, N_k)$ denote an optimal flow in the arc from N_j to N_k, and $z(N_j, E_i)$ denote an optimal flow in the arc from N_j to E_i. It is known that

$$\text{if } 0 < z(N_j, N_k), \text{ then } x_k^* \le x_j^*$$
$$\text{if } z(N_j, N_k) < 2v_{jk}, \text{ then } x_j^* \le x_k^*$$

In particular,

$$\text{if } 0 < z(N_j, N_k) < 2v_{jk}, \text{ then } x_j^* = x_k^*$$

That is, if the flow from N_j to N_k is positive but less than the arc capacity, then for any optimal locations of new facilities j and k, their x-coordinates are the same. Similarly, it is known that

$$\boxed{\begin{aligned}&\text{if } 0 < z(N_j, E_i), \text{ then } a_i \le x_j^* \\ &\text{if } z(N_j, E_i) < 2w_{ji}, \text{ then } x_j^* \le a_i\end{aligned}}$$

In particular,

$$\boxed{\text{if } 0 < z(N_j, E_i) < 2w_{ji}, \text{ then } x_j^* = a_i}$$

that is, if the flow from N_j to E_i is positive but less than the arc capacity, the optimal x-coordinates of new facility j and existing facility i coincide.

We remark that it is often possible to solve small problems by inspection, simply by saturating arcs in order of increasing costs. Let us consider Example 6.2 again. For this example $a' = 10 \times 6 + 20 \times 1 + 40 \times 5 = 280$. Figure 6.4 shows the network we obtain using the foregoing construction procedure. The arc from node N_1 to N_2 has a cost of 0 and a capacity of $2v_{12} = 4$. There exist arcs from N_1 to E_1 and E_2 with costs of 0 and capacities of 4 and 2, respectively. Similarly, there exist arcs from N_2 to E_1 and E_3 with costs of 0 and capacities of 8 and 10, respectively. Node N_1 is a source with an amount available of $\omega_1 = v_{12} + w_{11} + w_{12} + w_{13} = 5$, while node N_2 is a source with an amount available of $\omega_2 = -v_{12} + w_{21} + w_{22} +$

w_{ji}				
	2	1	0	3
	4	0	5	9
	6	1	5	12

$a = (10,\ 20,\ 40)$

v_{jk}				ω_i
	0	2	2	5
	-2	0	-2	7
			0	12

$a' = 60 + 20 + 200 = 280$

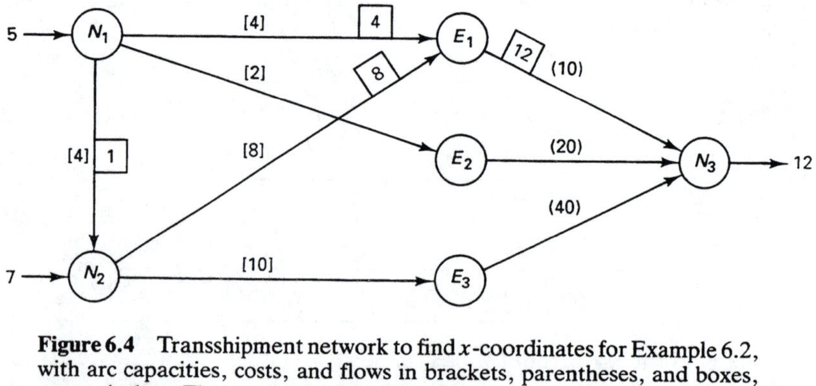

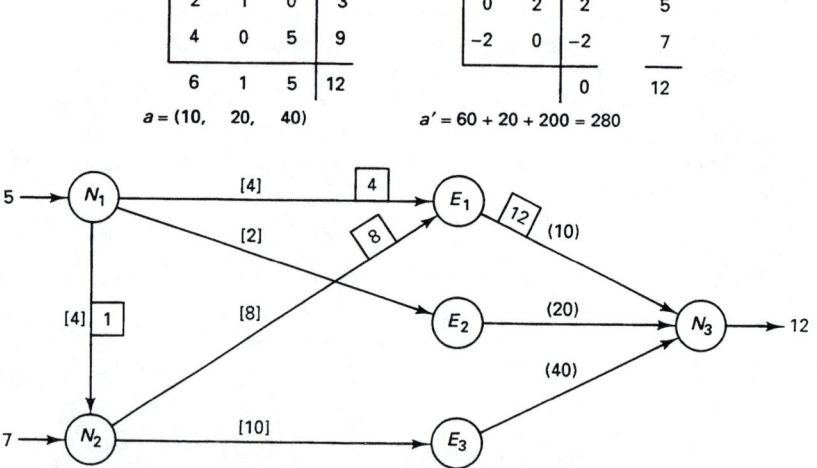

Figure 6.4 Transshipment network to find x-coordinates for Example 6.2, with arc capacities, costs, and flows in brackets, parentheses, and boxes, respectively. Flow cost $= 10 \times 12 = 120$, lower bound $= a' -$ flow cost $= 280 - 120 = 160$. Infinite capacities and zero arc costs not shown.

$w_{23} = 7$. We include a sink node N_3 with a demand of $\omega_3 = 5 + 7 = 12$. The arcs into N_3 from E_1, E_2, and E_3 have costs of 10, 20, and 40, respectively, and have capacities of infinity. We can solve this example by inspection, on noting that each unit of flow from a source to a sink must incur a cost of at least 10, the smallest of the arc costs from N to E nodes. Thus if every unit of flow incurs a cost of 10, we clearly have an optimal solution. Suppose that we send 4 units of flow from N_1 to E_1 to N_3, we send 1 unit of flow from N_1 to N_2, and we send 8 units of flow from N_2 to E_1 to N_3. Then every unit of flow incurs a cost of 10, and the minimum objective function value is given by $z^* = 10 \times 12 = 120$. We know that $a' = 280$, so $f_1^* = 280 - 120 = 160$ is the minimum cost of movement in the x-direction.

We note that many of the computations illustrated in the preceding paragraph can be organized in a tabular fashion, as follows:

w_{ji}					v_{jk}				ω_j
	2	1	0	3		0	2	2	5 (3 + 2)
	4	0	5	9		−2	0	−2	7 (9 − 2)
	6	1	5	12					12

a_i	10	20	40	$a' = 10 \times 6 + 20 \times 1 + 40 \times 5 = 280$
b_i	15	25	5	$b' = 15 \times 6 + 25 \times 1 + 5 \times 5 = 140$

To find the x-coordinates for this example, we note that complementary slackness implies that $x_1^* = x_2^* = x$, say, so we get

$$f(x) = f_1(x,x) = 6|x - 10| + 1|x - 20| + 5|x - 40|$$

Using the median conditions we conclude that $f(x)$ is minimized by any x^* between 10 and 20. Thus we take $x_1^* = x_2^*$ to be the same, any point between 10 and 20. Alternatively, complementary slackness tells us that since the flow in the arc from N_1 to E_1 is positive, that $10 \le x_1^*$, while since the arc from N_1 to E_2 is not saturated, $x_1^* \le 20$. Thus we know that $10 \le x_1^* = x_2^* \le 20$. If we compute $f_1(x_1^*, x_2^*)$ directly, we get $f_1(x_1^*, x_2^*) = 160$ for both x-coordinates identical and between 10 and 20, thus implying optimality. Alternatively (and more generally), if we solve this problem using a minimum-cost network flow algorithm, one optimal solution that we can get has optimal node labels for nodes N_1, N_2, and N_3 of 0, 0, and -10, respectively. Thus substituting into (6.13) we get $x_1^* = 0 - (-10) = 10$ and $x_2^* = 0 - (-10) = 10$.

Due to duality theory, there is a *fail-safe computational check* on this approach of solving the minimum-cost network flow problem by inspection. Given an objective function value z^* that we believe to be minimal for the flow problem, as well as locations $x_1^*, \ldots, x_n^*$, which we believe to be optimal x-coordinates, compute

$f_1(x_1^*, \ldots, x_n^*)$ directly. An *equivalent* condition for our solutions to be optimal is that

$$f_1(x_1^*, \ldots, x_n^*) = a' - z^*$$

[For our example, we have $f_1(10, 10) = 160 = 280 - 120 = a' - z^*$, so we are assured that giving new facilities 1 and 2 x-coordinates of 10 is optimal.] More generally, for *any* new facility coordinates $x_1, \ldots, x_n$ and *any* corresponding flow value z, it is true that

$$f_1(x_1, \ldots, x_n) \geq a' - z$$

An equivalent condition for optimality for both problems is that the latter inequality holds as an equality.

Returning to the example, to get the network to solve for the y-coordinates, we replace the a's by the b's in Figure 6.4, thus replacing arc costs of 10, 20, and 40 by 15, 25, and 5, respectively. We recompute a' (now called b') and find $b' = 15 \times 6 + 25 \times 1 + 5 \times 5 = 140$. Again this flow problem can be solved by inspection by saturating arcs in order of increasing costs. We send 3 units of flow from N_1 to N_2 at a cost of 0, we send 10 units of flow from N_2 to E_3 to N_3, getting a cost of $5 \times 10 = 50$, and we send 2 units of flow from N_1 to E_1 to N_3, getting a cost of $15 \times 2 = 30$, obtaining a minimum objective function value of $30 + 50 = 80$. Hence the minimum cost of movement in the y-direction is $140 - 80 = 60$. For this example we can again use complementary slackness to compute the y-coordinates. Since the flow from N_1 to E_1 is positive and less than its upper bound, we know that $y_1^* = b_1 = 15$. Since the flow from N_1 to N_2 is positive and less than its upper bound, we know that $y_1^* = y_2^*$, so we take $y_1^* = y_2^* = 15$. As a computational check, $f_2(15, 15) = 60 = 180 - 120 = b' - z^*$. Alternatively, if we solve this problem using a minimum-cost network flow algorithm, one optimal solution we can get has optimal node labels for nodes N_1, N_2, and N_3 of 0, 0, and -15, respectively. Thus substituting into (6.13) using y's instead of x's, we get $y_1^* = 0 - (-15) = 15$ and $y_2^* = 0 - (-15) = 15$.

While node N_{n+1} is always a sink, you should not conclude from the example above that every node N_j is always a source for $j = 1, \ldots, n$. In fact, in our example, if we had $v_{12} = 10$, node N_1 would still be a sink, with a supply of 13, but node N_2 would be a sink with a demand of 1, since in this case $\omega_1 = 13$, but $\omega_2 = -1$. Similarly, if we had $v_{12} = 9$, we would have $\omega_1 = 12$, but $\omega_2 = 0$, in which case N_2 would be a transshipment node. As a second example, Figure 6.5 shows the solution of the application of Section 6.2 using the network flow approach.

To summarize, we have taken a multifacility location problem with rectilinear distances, formulated it as an equivalent linear program, constructed the dual of this linear program, and then given the dual an interpretation as a minimum-cost network flow problem. We illustrated how an approach of saturating arcs in order of increasing costs often leads to an optimal solution, and also gave an approach using node labels which always gives an optimal solution. There are several other computational approaches to solving the problem, due to Picard and Ratliff [43] and to Kolen [25] which, if a computer is employed, can have computational advantages,

w_{ji}				
	0.8	0.8	0.0	1.6
	0.0	0.8	0.4	1.2
	0.8	1.6	0.4	2.8

v_{jk}				w_i
	0.0	4.0	4.0	5.6
	−4.0	0.0	−4.0	−2.8
			0.0	2.8

$a = (10,\ 7,\ 13)$ $\qquad a' = 0.8 \times 10 + 1.6 \times 7 + 0.4 \times 13 = 24.4$

$b = (20,\ 6,\ 0)$ $\qquad b' = 0.8 \times 20 + 1.6 \times 6 + 0.4 \times 0 = 25.6$

(a) Data and initial computations

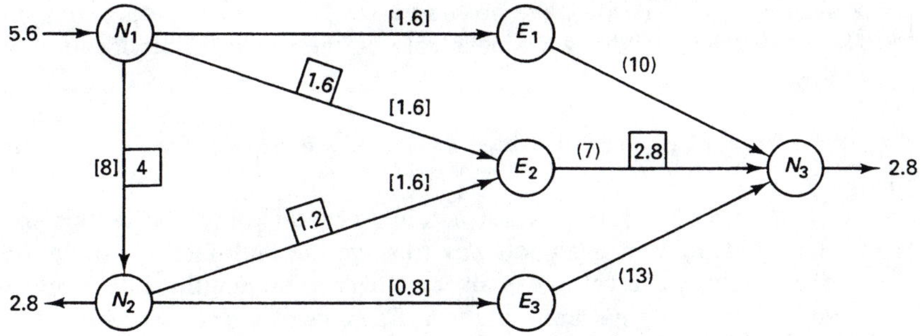

(b) Transhipment network to find X-coordinates.
Flow cost: $2.8 \times 7 = 19.6$, lower bound $= 24.4 - 19.6 = 4.8$
$O < z(N_1, N_2) < 8 \Rightarrow X_1 = X_2$. $O < z(N_1, E_2) < 1.6 \Rightarrow X_1 = 7$.
$f_1(7, 7) = 4.8 =$ lower bound.

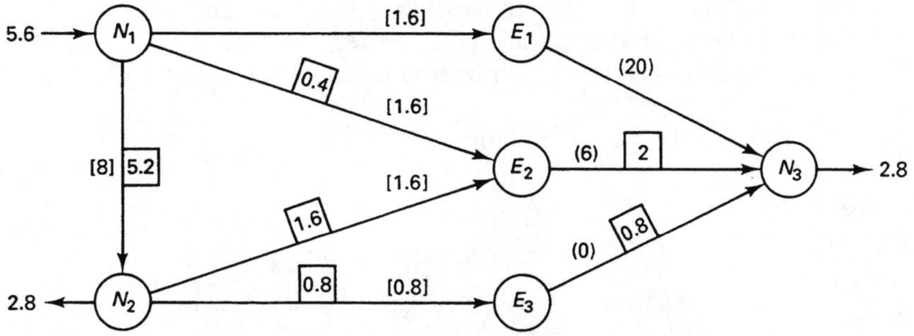

(c) Transhipment network to find Y-coordinates.
Flow cost: $6 \times 2 = 12$, lower bound $= 25.6 - 12 = 13.6$
$O < z(N_1, N_2) < 5.2 \Rightarrow Y_1 = Y_2$. $O < z(N_1, E_2) < 1.6 \Rightarrow Y_1 = 6$.
$f_2(6, 6) = 13.6 =$ lower bound.

Figure 6.5 Transshipment network solution to Coastal Construction problem: $X_1 = X_2 = (7,6)$, with arc capacities, costs, and flows shown in brackets, parentheses, and boxes, respectively. Infinite capacities and zero arc costs not shown.

in terms of a lower order of computational effort, to the approach we presented. Our approach works well, however, for obtaining insight into the problem; further, if a computer code of a transshipment algorithm is used, it can easily solve problems with hundreds of new facilities and thousands of existing facilities.

6.3.3 Motivation for the Network Solution Procedure

In this section we develop a motivation for the network solution procedure of the preceding section, utilizing the duality theory of linear programming. Accordingly, this section requires some understanding of the theory of linear programming and should be skipped if you have not had a course in linear programming.

To develop insight, we concentrate on the single-facility location problem on the line:

$$\text{minimize} \quad f(x) = w_1|x - a_1| + \cdots + w_m|x - a_m| = \sum_{i=1}^{m} w_i|x - a_i|$$

Linear programming is not actually a good way to solve this *single*-facility problem. However, it *is* a good way to solve the multifacility problem of interest, and the insights we develop will all carry over to the multifacility problem. The key to our approach is what amounts to a *Distance change of variables*. If

$$x - a = r - s, \qquad r \geq 0, \quad s \geq 0, \quad rs = 0 \tag{6.14}$$

then

$$|x - a| = r + s$$

We interpret $r(s)$ as the amount by which x is to the right (left) of a. Since x cannot be both to the right and the left of a, we require that $rs = 0$.

Hence an equivalent optimization problem we have

$$\text{minimize} \quad \sum_{i=1}^{m} w_i(r_i + s_i)$$

subject to

$$x - r_i + s_i = a_i, \qquad i = 1, \ldots, m$$

$$r_i, s_i \geq 0, \qquad i = 1, \ldots, m$$

and $r_i s_i = 0$ for $i = 1, \ldots, m$. Note that if we disregard the last set of conditions, which we call the *product conditions*, we have a linear programming problem. Fortunately, if we solve the linear program, any optimal solution will satisfy the product conditions, so we need not worry about them. Let us denote this linear program by P.

In the next three paragraphs we establish the correctness of this linear programming formulation. In effect, we introduce variables r and s in order to make a change of variables, replacing the distance $|x - a|$ by $r + s$, given that a, x, r, and s satisfy the conditions (6.14). To justify the change of variables, we observe that r and s nonnegative with a product of zero guarantees that at least one of r and s is zero. In case $s = 0$, we then have

$$s = 0 \leq r = x - a$$

Since $x - a \geq 0$, we have

$$x - a = |x - a|$$

so

$$r = |x - a| = r + 0 = r + s$$

Similarly, if $r = 0$, we conclude that

$$r = 0 \leq s = a - x = |x - a| = 0 + s = r + s$$

Hence the change of variables is valid.

Now consider problem P. Given any feasible solution, if for some i we have $r_i s_i > 0$, with, say, $r_i \geq s_i$, and we define $r_i' = r_i - s_i$, $s_i' = s_i - s_i = 0$, we have $x - r_i' + s_i' = a_i$ and we thus have a new feasible solution with a reduction of the objective function value by $2w_i s_i > 0$. Hence if we have an *optimal* solution, each product term must be zero (so that the product conditions are satisfied), as otherwise we can construct a new feasible solution with an objective function value less than the minimum objective function value—an impossibility.

We remark that an alternative way to show that the product conditions are satisfied is to observe that the variables r_i and s_i are linearly dependent. Thus at most one will be positive in any basic feasible solution, which means that their product will always be zero (nonbasic variables are zero). Since some basic feasible solution is an optimal solution, if we solve the problem by the simplex algorithm, we are guaranteed that the optimal solution so obtained will satisfy the product conditions.

Let us now construct the dual, which we denote by D, of the linear program P. We assume a familiarity with elementary linear programming duality theory. For convenience, we write out P and D for $m = 3$. If you understand this special case, you should be able to obtain the dual for any value of m. Linear program P is

$$\text{minimize} \quad w_1(r_1 + s_1) + w_2(r_2 + s_2) + w_3(r_3 + s_3)$$

subject to

				dv's
$x - r_1 + s_1$			$= a_1$	u_1
x	$- r_2 + s_2$		$= a_2$	u_2
x		$- r_3 + s_3$	$= a_3$	u_3

$$r_i, s_i \geq 0, \qquad i = 1, \ldots, 3$$

Note that we do not restrict x to be nonnegative. In the right margin we have written the dual variables corresponding to each of the three equality constraints. Dual problem D is as follows:

$$\text{maximize} \quad a_1 u_1 + a_2 u_2 + a_3 u_3$$

subject to

$$\begin{array}{rrrll} & & & & \underline{pv\text{'s}} \\ u_1 & + u_2 & + u_3 & = 0 & x \\ -u_1 & & & \le w_1 & r_1 \\ u_1 & & & \le w_1 & s_1 \\ & - u_2 & & \le w_2 & r_2 \\ & u_2 & & \le w_2 & s_2 \\ & & - u_3 & \le w_3 & r_3 \\ & & u_3 & \le w_3 & s_3 \end{array}$$

The first constraint is an equality constraint because x is unrestricted in sign, and 0 is on the right-hand side because x does not appear in the objective function of original linear program P.

To make problem D look more conventional, we make a change of variables, $z_i = -u_i + w_i$, giving $u_i = -z_i + w_i$. The inequalities $-w_i \le u_i \le w_i$ now become $0 \le z_i \le 2w_i$. The constraint $u_1 + u_2 + u_3 = 0$ becomes $z_1 + z_2 + z_3 = \omega$, where $\omega = w_1 + w_2 + w_3$. The objective function becomes

$$-a_1 z_1 - a_2 z_2 - a_3 z_3 + a'$$

where

$$a' = a_1 w_1 + a_2 w_2 + a_3 w_3$$

Hence we have the equivalent problem, denoted by D′, as follows:

$$\text{minimize} \quad a_1 z_1 + a_2 z_2 + a_3 z_3$$

subject to

$$z_1 + z_2 + z_3 = \omega$$

$$0 \le z_1 \le 2w_i, \qquad i = 1, 2, 3$$

Once we solve problem D′, subtracting its minimum objective function value from a' gives the minimum objective function value for original linear program P.

Part of the complementary slackness conditions involving any optimal solutions to the two problems P and D are as follows:

$$r_i(w_i + u_i) = 0, \qquad s_i(w_i - u_i) = 0$$

Also, for any feasible solution to P we know that $x - r_i + s_i = a_i$. Recall that we know that $-w_i \le u_i \le w_i$. Thus if $-w_i < u_i$, then $0 < w_i + u_i$, so by the complementary slackness conditions $r_i = 0$, and thus $x = a_i - s_i \le a_i$. Similarly, if $u_i < w_i$, then by the complementary slackness conditions $s_i = 0$, so $x = a_i + r_i \ge a_i$.

In conclusion, given any optimal solutions to P and D,

$$\text{if } -w_i < u_i, \text{ then } x \le a_i$$

$$\text{if } u_i < w_i, \text{ then } a_i \le x$$

From our change of variables, $z_i = -u_i + w_i$, we now conclude from complementary slackness, given any optimal solutions to P and D′, that

$$\boxed{\begin{array}{c}\text{if } z_i < 2w_i, \text{ then } x \le a_i \\ \text{if } 0 < z_i, \text{ then } a_i \le x\end{array}}$$

In particular,

$$\boxed{\text{if } 0 < z_i < 2w_i, \text{ then } x = a_i} \tag{6.15}$$

The importance of (6.15) is that it often enables us to identify an optimal solution to P given we know an optimal solution to D′, as the following example will illustrate.

Consider an example, with $a_1 = 10$, $a_2 = 20$, $a_3 = 40$, $w_1 = 5$, $w_2 = 6$, $w_3 = 4$. Let $\omega = w_1 + w_2 + w_3$. Since $\omega = 15$, $\omega/2 = 7.5$, and the use of the median conditions implies that $x^* = a_2 = 20$ is a best location, with $f(x^*) = 130$. Hence conclusions we draw from the problems (D′) and (P) should be consistent with use of the median conditions. Figure 6.6 shows a network interpretation of problem D′. We

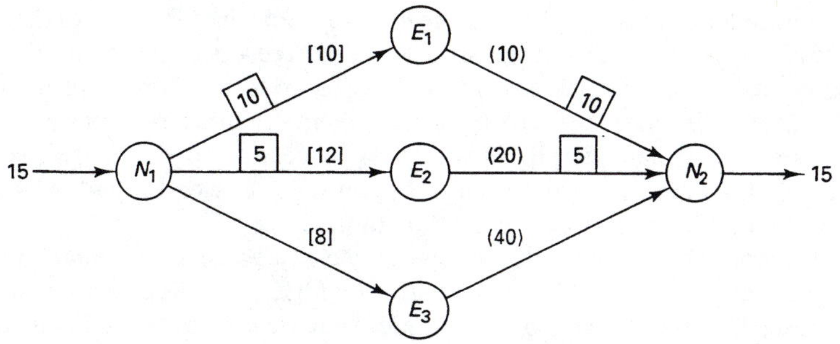

Figure 6.6 Example of a single-source transshipment network to find a single x-coordinate to minimize $f_1(x) = 5\ |x - 10| + 6\ |x - 20| + 4\ |x - 40|$. Flow cost $= 10 \times 10 + 5 \times 20 = 200$, $a' = 5 \times 10 + 6 \times 20 + 4 \times 40 = 330$, lower bound $= 330 - 200 = 130$. $0 < z(N_1, E_2) < 12 \Rightarrow x_1 = 20$. $f_1(20) = 130 =$ lower bound.

have a single source with amount available of $\omega = 15$ and a single sink with a requirement of $\omega = 15$. We have three arcs, with flows of z_1, z_2, and z_3, with capacities of 10, 12, and 8, and with costs of 10, 20, and 40, respectively; the arcs go from the source to the sink. The problem is now one of determining the arc flows so as to send the amount available at the source to the sink at a minimum cost while satisfying the arc capacity conditions. Note the condition that the amount sent from the source is equal to the sum of the flows is just $\omega = z_1 + z_2 + z_3$. Similarly, for the sink we get the equivalent condition, $-z_1 - z_2 - z_3 = -\omega$. For each arc i, $0 \leq z_i \leq 2\,w_i$ becomes the conditions of the flow being nonnegative and no more than the arc capacity. It should be intuitively clear how to get an optimal solution to this flow problem. First we use as much of the cheapest arc flow as we need; then, if necessary, we use as much of the next-cheapest arc flow as we need, and so on, until all of the amount available, ω, is used up. Hence we get $z_1^* = 10$, $z_2^* = 5$, $z_3^* = 0$, with a corresponding objective function value of 200. A computation gives $a' = 330$, and thus $330 - 200 = 130$ is the minimum objective function value of P. If you recall the median conditions, you can see that we are, in effect, using them to solve this example of problem D, since the cheapest arc has cost $a_1 = 10$, the next cheapest arc has cost $a_2 = 20$, and the third cheapest arc has cost $a_3 = 40$. Considering these arcs in order is essentially the same as considering the points a_1, a_2, and a_3 from left to right in order; comparing $z_1^* = 2w_1$ to ω is equivalent to comparing w_1 to $\omega/2$; and so on. Further, we can use the complementary slackness conditions to conclude that for any optimal arc flow z_i^* that is positive and less than its capacity, the cost of the arc identifies an optimal location; in the example, $0 < z_2^* < 12$, so $x^* = a_2 = 20$ is an optimal location, due to (6.15).

6.4 SQUARED EUCLIDEAN DISTANCE MINISUM LOCATION PROBLEMS

In this section we consider the minisum problem when, instead of distances, we have squared Euclidean distances. Our principal reason for studying the problem with squared Euclidean distances is to help develop insight for the problem with regular Euclidean distances; we do not know of too many situations where there are physical reasons for using squared Euclidean distances. Accordingly, this section is rather short. However, we recommend that you read it prior to reading the next section, which deals with the case of Euclidean distances.

Denote the function obtained when we replace each l_p distance in (6.6) by the associated squared Euclidean distance by $f^2(\mathbf{X}_1, \ldots, \mathbf{X}_n)$. Our use of the number 2 as a superscript is simply as a reminder that we are squaring distances; the number is *not* an exponent. Thus $f^2(\mathbf{X}_1, \ldots, \mathbf{X}_n)$ is the sum of the weighted squared Euclidean distances between all pairs of new facilities plus the sum of the weighted squared Euclidean distances between all pairs of new and existing facilities. We wish to find new facility locations that minimize $f^2(\mathbf{X}_1, \ldots, \mathbf{X}_n)$.

For $j = 1, \ldots, n$ it will be convenient to say that new facility j *has an exchange* with existing facility i if w_{ji} is positive. Consequently, recalling that all weights are nonnegative, new facility j has an exchange with some existing facility if the sum of the entries in row j of the matrix $\mathbf{W}$ is positive.

Recall from Chapter 4 that each term

$$w_{ji}[(x_j - a_i)^2 + (y_j - b_i)^2]$$

for which $w_{ji} > 0$, is a parabaloid, and thus a strictly convex function. Similarly, each term

$$v_{jk}[(x_j - x_k)^2 + (y_j - y_k)^2]$$

for which $v_{jk} > 0$ is a strictly convex function; a graph of the term $(x_j - x_k)^2$ [as well as the completely analogous term $(y_j - y_k)^2$] resembles a "parabolic trough" and has the property that the line segment joining any two points on the surface of the trough lies above the surface except at the endpoints of the line segment. It follows that $f^2(\mathbf{X}_1, \ldots, \mathbf{X}_n)$ is a convex function. Convexity guarantees that any point at which all partial derivative of $f^2(\mathbf{X}_1, \ldots, \mathbf{X}_n)$ is zero is a global minimum. In case each NF has an exchange with at least one EF it is known that $f^2(\mathbf{X}_1, \ldots, \mathbf{X}_n)$ is a strictly convex function. As we shall see, all partial derivatives of $f^2(\mathbf{X}_1, \ldots, \mathbf{X}_n)$ exist everywhere, so when the function is strictly convex it has a unique global minimum at the point for which the partial derivatives are all zero.

We conclude that we can find the global minima of $f^2(\mathbf{X}_1, \ldots, \mathbf{X}_n)$ by solving the equations resulting from computing the partial derivative of $f^2(\mathbf{X}_1, \ldots, \mathbf{X}_n)$ and setting them to zero. To compute the partial derivative of $f^2(\mathbf{X}_1, \ldots, \mathbf{X}_n)$ with respect to x_t, we first note that the sum of all terms in $f^2(\mathbf{X}_1, \ldots, \mathbf{X}_n)$ involving NF t is as follows:

$$\sum_{j=1}^{t-1} v_{jt}[(x_j - x_t)^2 + (y_j - y_t)^2 + v_{tt}[(x_t - x_t)^2 + (y_t - y_t)]^2$$
$$+ \sum_{j=t+1}^{n} v_{tj}[(x_t - x_j)^2 + (y_t - y_j)^2] + \sum_{i=1}^{m} w_{ti}[(x_t - a_i)^2 + (y_t - b_i)^2]$$

Recall that $v_{jt} = v_{tj}$ for all j and t. Thus we can represent the first three sums in the foregoing expression by a single sum over j from 1 to n, and obtain

$$\sum_{j=1}^{n} v_{tj}[(x_t - x_j)^2 + (y_t - y_j)^2] + \sum_{i=1}^{m} w_{ti}[(x_t - a_i)^2 + (y_t - b_i)^2]$$

Because the latter term includes all the terms involving x_t, it follows that the partial derivative of f^2 with respect to x_t is the same as the partial derivative of the latter term with respect to x_t, namely,

$$\sum_{j=1}^{n} 2v_{tj}(x_t - x_j) + \sum_{i=1}^{m} 2w_{ti}(x_t - a_i)$$

If we set the partial derivative of f with respect to x_t to zero, we set the latter expression to zero, and we can rewrite the result as

$$\sum_{j=1}^{t-1}(-v_{tj})x_j + \sum_{j=t+1}^{n}(-v_{tj})x_j + \sum_{j=1}^{n} v_{tj}x_t + \sum_{i=1}^{m} w_{ti}x_t = \sum_{i=1}^{m} w_{ti}a_i$$

or, alternatively, as

$$\sum_{j=1}^{t-1}(-v_{tj})x_j + \left(\sum_{j=1}^{n} v_{tj} + \sum_{i=1}^{m} w_{ti}\right)x_t + \sum_{j=t+1}^{n}(-v_{tj})x_j = \sum_{i=1}^{m} w_{ti}a_i$$

If you solve the latter equation for x_t, you will see that it expresses x_t as a weighted average of the x-coordinates of *all* other facility x-coordinates; we get a completely analogous conclusion for y_t. Accordingly, *each optimal new facility location will be in the convex hull of the existing facility locations for a well-formulated problem.*

Considering the latter equation again, note that the coefficient of each x_j for j not equal to t is $-v_{tj}$, while the coefficient of x_t is the sum of row t of the **V** matrix plus the sum of row t of the **W** matrix. Further, the right-hand side of the latter equation is the product of row t of the **W** matrix with the a vector. Thus let us define the matrix **A** with n rows and n columns as follows. *Each row t of* **A** *is minus one times row t of* **V** *except for the entry in column t of row t of* **A**, *which is the sum of the entries of row t of* **V** *plus the sum of the entries of row t of* **W**. Then we can write the latter equation for all t as

$$(\text{row } t \text{ of } \mathbf{A})\mathbf{X} = (\text{row } t \text{ of W})\mathbf{a}, \qquad t = 1, \ldots, n$$

That is, $\mathbf{AX} = \mathbf{Wa}$.

In conclusion, *setting all the partial derivatives of f to zero t is equivalent to the following two linear systems of equations*:

$$\mathbf{AX} = \mathbf{Wa}, \qquad \mathbf{AY} = \mathbf{Wb} \tag{6.16}$$

When at least one term in each row of **W** is positive, each linear system will have a unique solution, for in this case $f^2(\mathbf{X}_1, \ldots, \mathbf{X}_n)$ will be a strictly convex function and thus have a unique minimum.

Consider an example. We shall use the data of Example 6.1. The sum of the entries in row 1 of **W** is 3, and the sum of the entries in row 1 of **V** is 1, so the first diagonal entry in **A** is $3 + 1 = 4$. Similarly, the second diagonal entry in **A** is $2 + 1 = 3$. The nondiagonal entries of **A** are the negatives of the corresponding entries in **V**. Thus we have

$$\mathbf{A} = \begin{pmatrix} 4 & -1 \\ -1 & 3 \end{pmatrix}$$

To compute **Wa** and **Wb**, we consider the vectors **a** and **b** to be column vectors, and obtain

$$\mathbf{Wa} = \begin{pmatrix} 15 \\ 35 \end{pmatrix}, \qquad \mathbf{Wb} = \begin{pmatrix} 35 \\ 15 \end{pmatrix}$$

If we put **A**, **Wa**, and **Wb** into the partitioned matrix $(\mathbf{A}|\mathbf{Wa}|\mathbf{Wb})$ and perform elementary row operations, the partitioned matrix transforms to the following matrix:

$$\begin{pmatrix} 1 & 0 & \frac{80}{11} & \frac{120}{11} \\ 0 & 1 & \frac{155}{11} & \frac{95}{11} \end{pmatrix}$$

We read the optimal x values from the next-to-last column, and the optimal y values from the last column. Hence we have

$$\mathbf{X}_1^* = (7\tfrac{3}{11}, 10\tfrac{10}{11}), \qquad \mathbf{X}_2^* = (14\tfrac{1}{11}, 8\tfrac{7}{11})$$

Let us use the data of Example 6.3, assuming squared Euclidean distances, to solve a second problem. Computations establish that

$$\mathbf{A} = \begin{pmatrix} 5 & -2 \\ -2 & 11 \end{pmatrix}, \qquad \mathbf{Wa} = \begin{pmatrix} 40 \\ 240 \end{pmatrix}, \qquad \mathbf{Wb} = \begin{pmatrix} 55 \\ 85 \end{pmatrix}$$

The partitioned matrix $(\mathbf{A}|\mathbf{Wa}|\mathbf{Wb})$ transforms to the following matrix:

$$\begin{pmatrix} 1 & 0 & 18\frac{2}{51} & 15\frac{10}{51} \\ 0 & 1 & 25\frac{5}{51} & 10\frac{25}{51} \end{pmatrix}$$

Thus we have

$$\mathbf{X}_1^* = (18\tfrac{2}{51}, 15\tfrac{10}{51}), \qquad \mathbf{X}_2^* = (25\tfrac{5}{51}, 10\tfrac{25}{51})$$

6.5 EUCLIDEAN DISTANCE MINISUM LOCATION PROBLEMS

6.5.1 Euclidean Distance Minisum Location Problems: Introduction

In this section we consider the minisum location problem with Euclidean distances, obtained by taking $p = 2$ in (6.6). We shall develop the multifacility version of Weiszfeld's algorithm, referred to as HAP, for reasons to be discussed. HAP comes from computing partial derivatives of the function f, setting them to zero, and solving for new facility locations. We refer you to Chapter 4 for a discussion of situations where the use of Euclidean distances may be appropriate, as well as for a discussion of applications. In the following section we develop some theory to justify our approach in this section.

We wish to develop an algorithm to minimize our total cost function, $f(\mathbf{X}_1, \ldots, \mathbf{X}_n)$. To help develop a computationally stable algorithm, we shall "perturb" the problem by including a very small positive number, say ϵ, underneath each square-root term used in computing Euclidean distances. The result of this inclusion will be to avoid attempts to divide by zero and to make the perturbed cost function differentiable everywhere. Of course, we are now minimizing a different function, but if we choose ϵ sufficiently close to zero, there is no practical difference between the two functions. A graph of the function

$$[(x_t - a_i)^2 + (y_t - b_i)^2]^{1/2}$$

is a cone with its tip at the point (a_i, b_i), whereas a graph of the function

$$[(x_t - a_i)^2 + (y_t - b_i)^2 + \epsilon]^{1/2}$$

is a hyperboloid, so we are approximating a cone by a hyperboloid. For this reason the approach we develop will be referred to as HAP, an abbreviation for *hyperboloid approximation procedure*.

To simplify the notation we define terms $\alpha_t(\mathbf{X}_1, \ldots, \mathbf{X}_n)$, $\beta_t(\mathbf{X}_1, \ldots, \mathbf{X}_n)$, for $t = 1, \ldots, n$, as follows:

$$\alpha_t(\mathbf{X}_1, \ldots, \mathbf{X}_n) = \sum_{j=1}^{n} v_{tj} \frac{x_j}{[(x_t - x_j)^2 + (y_t - y_j)^2 + \epsilon]^{1/2}} + \sum_{i=1}^{m} w_{ti} \frac{a_i}{[(x_t - a_i)^2 + (y_t - b_i)^2 + \epsilon]^{1/2}}$$

$$\beta_t(\mathbf{X}_1, \ldots, \mathbf{X}_n) = \sum_{j=1}^{n} v_{tj} \frac{y_j}{[(x_t - x_j)^2 + (y_t - y_j)^2 + \epsilon]^{1/2}} + \sum_{i=1}^{m} w_{ti} \frac{b_i}{[(x_t - a_i)^2 + (y_t - b_i)^2 + \epsilon]^{1/2}}$$

Note it is only necessary to replace the x's by y's and the a's by b's in the numerators of the terms in α_t to get β_t. Also, define $\Gamma_t(\mathbf{X}_1, \ldots, \mathbf{X}_n)$ by

$$\Gamma_t(\mathbf{X}_1, \ldots, \mathbf{X}_n) = \sum_{j=1}^{n} \frac{v_{tj}}{[(x_t - x_j)^2 + (y_t - y_j)^2 + \epsilon]^{1/2}} + \sum_{i=1}^{m} \frac{w_{ti}}{[(x_t - a_i)^2 + (y_t - b_i)^2 + \epsilon]^{1/2}}$$

Note that the term $\Gamma_t(\mathbf{X}_1, \ldots, \mathbf{X}_n)$ is always positive.

For $t = 1, \ldots, n$, recall that $\mathbf{X}_t = (x_t, y_t)$. Computations establish that the partial derivative of f with respect to x_t is as follows:

$$\frac{\partial f}{\partial x_t} = \Gamma_t(\mathbf{X}_1, \ldots, \mathbf{X}_n)x_t - \alpha_t(\mathbf{X}_1, \ldots, \mathbf{X}_n)$$

Similarly, the partial derivative of f with respect to y_t is given by

$$\frac{\partial f}{\partial y_t} = \Gamma_t(\mathbf{X}_1, \ldots, \mathbf{X}_n)y_t - \beta_t(\mathbf{X}_1, \ldots, \mathbf{X}_n)$$

Thus if we set the partial derivatives with respect to x_t and y_t to zero and solve for x_t and y_t, we obtain

$$x_t = \frac{1}{\Gamma_t(\mathbf{X}_1, \ldots, \mathbf{X}_n)} \alpha_t(\mathbf{X}_1, \ldots, \mathbf{X}_n), \qquad t = 1, \ldots, n$$

$$y_t = \frac{1}{\Gamma_t(\mathbf{X}_1, \ldots, \mathbf{X}_n)} \beta_t(\mathbf{X}_1, \ldots, \mathbf{X}_n), \qquad t = 1, \ldots, n$$

The latter two equations are the basis for the iterative equations that specify the algorithm HAP, as follows:

$$x_t^{(q+1)} = \frac{1}{\Gamma_t(\mathbf{X}_1^{(q)}, \ldots, \mathbf{X}_n^{(q)})} \alpha_t(\mathbf{X}_1^{(q)}, \ldots, \mathbf{X}_n^{(q)}), \qquad t = 1, \ldots, n$$

$$y_t^{(q+1)} = \frac{1}{\Gamma_t(\mathbf{X}_1^{(q)}, \ldots, \mathbf{X}_n^{(q)})} \beta_t(\mathbf{X}_1^{(q)}, \ldots, \mathbf{X}_n^{(q)}), \qquad t = 1, \ldots, n$$

Thus we make an initial choice of NF locations, say $(\mathbf{X}_1^{(0)}, \ldots, \mathbf{X}_n^{(0)})$, use the algorithm to compute new NF locations, $(\mathbf{X}_1^{(1)}, \ldots, \mathbf{X}_n^{(1)})$, insert these into the right-hand side to compute $(\mathbf{X}_1^{(2)}, \ldots, \mathbf{X}_n^{(2)})$, and so on.

Table 6.5 illustrates the use of HAP to solve Example 6.3 with different starting points and choices of delta. Runs terminated when the rectilinear distance between two successive sets of trial locations did not exceed delta. All but the third run gave locations for both NFs very close to the point (10, 15). The third run gave both NFs at the point (11.20, 14.81). We observe, except for run 3, that all solutions had much the same cost, and making delta smaller results in a more accurate solution at the expense of a higher number of iterations. The starting point for runs 3 and 6 was the solution to the analogous squared distance problem, computed in Section 6.3.

In all these runs, the solution HAP provides is only approximately optimal (although the approximation is usually quite good) due to the fact that it does not converge to an exactly optimal solution in a finite number of iterations. Such a situation is often the case when solving nonlinear optimization problems.

TABLE 6.5 RESULTS FROM USING HAP TO SOLVE EXAMPLE 6.3 WITH DIFFERENT STARTING POINTS AND CHOICES OF DELTA, THE STOPPING CRITERION

Starting points	Delta	Number of Iterations	f value
(10, 15), (10, 15)	10^{-3}	3	172.256757
(15, 15), (15, 15)	10^{-3}	680	172.741964
(18.04, 15.20), (25.1, 10.49)	10^{-3}	708	172.880802
(10, 15), (10, 15)	10^{-6}	23	172.257049
(15, 15), (15, 15)	10^{-6}	5969	172.257078
(18.04, 15.20) (25.1, 10.49)	10^{-6}	7294	172.257079

6.5.2 Euclidean Distance Minisum Location Problems: Theory

In this section we develop some theory for the problems of interest, which also gives some justification for HAP. If you are willing to accept HAP, it is unnecessary to read this section.

It is interesting to note that the Euclidean distance model essentially includes the rectilinear distance model as a special case when all the existing facility locations are collinear, that is, all lie on a single line. For the collinear case we can choose our axes, rotated if necessary, so that all existing facility locations lie on the x-axis. Thus the existing facility locations may be represented as some points $(a_1, 0), \ldots, (a_m, 0)$ for some values of a_1 to a_m, with $a_1 < \cdots < a_m$. Recall from Chapter 4 that CH denotes the convex hull of the existing facility locations; in this case CH is just the line segment joining $(a_1, 0)$ to $(a_m, 0)$. In seeking optimal locations of the new facilities, it suffices to consider locations in CH. Further, we shall see if we restrict our new facilities to be in CH that our objective function becomes identical to the objective function $f_1(x_1, \ldots, x_n)$ in equation (6.10) encountered in solving the rectilinear distance problem. Suppose that we have any locations $\mathbf{X}_1, \ldots, \mathbf{X}_n$ of the new facilities, and we move each new facility parallel to the y-axis toward the x-axis until it lies on the x-axis. Then each new facility at its new location is at least as close to every other facility as before the moves occurred, so our total transport cost decreases. Next, identify those new facilities located on the x-axis which are not in CH, and move them to the closest point in CH. Again it is the case that each new facility moved is at least as close to every other facility as before being moved for the second time, so our total transport cost decreases again. But now every new facility is located in CH, so we only need to consider locations in CH.

Now with every NF in CH, so that its y-coordinate is zero, consider the Euclidean distance between any NF j and EF i, namely, $[(x_j - a_i)^2 + (y_j - 0)^2]^{1/2}$. Since $(y_j - 0)^2 = 0$, we have

$$[(x_j - a_i)^2 + (y_j - 0)^2]^{1/2} = [(x_j - a_i)^2 + 0]^{1/2} = |x_j - a_i|$$

Similarly, for any NF j and NF k we have

$$[(x_j - x_k)^2 + (y_j - y_k)^2]^{1/2} = [(x_j - x_k)^2 + 0]^{1/2} = |x_j - x_k|$$

Thus we can replace the Euclidean distances by absolute value distances, so we have

$$f(\mathbf{X}_1, \ldots, \mathbf{X}_n) = f_1(x_1, \ldots, x_n)$$

That is, *when every NF is restricted by being located in CH, a line segment, our total cost equation is simply the equation for the total cost of movement in the x-direction*, equation (6.10).

Example 6.6

All existing facilities are located on the x-axis. CH is just the line segment joining (10, 0) and (40, 0). The data are listed in Table 6.6. All we need to do to find optimal locations

TABLE 6.6 DATA FOR EXAMPLE 6.6

$\mathbf{w} = (w_{ji}) = \begin{pmatrix} 2 & 1 & 0 \\ 4 & 0 & 5 \end{pmatrix}$,	$\mathbf{V} = (v_{jk}) = \begin{pmatrix} 0 & 2 \\ 2 & 0 \end{pmatrix}$
$\mathbf{a} = (a_i) = (10 \quad 20 \quad 40)$	
$\mathbf{b} = (b_i) = (0 \quad 0 \quad 0)$	

is set up the function $f_1(x_1, x_2)$, find optimal x-coordinates that minimize it, and then put these coordinates together with y-coordinates of zero to get optimal locations. If you construct $f_1(x_1, x_2)$, you will find that it is simply the cost of movement in the x-direction we encountered in Section 6.2 for Example 6.3, where we found that it was optimal to have x_1 and x_2 the same, any number between 10 and 20. Hence if we take $\mathbf{X}_1^* = (x, 0) = \mathbf{X}_2^*$, with x any number between 10 and 20, we have optimal locations.

To summarize, *the cost model with Euclidean distances includes the cost model with rectilinear distances as a special case when all* EF *locations are collinear*. When each EF is on the x-axis, we can solve our Euclidean distance problem simply by solving the associated problem with absolute value distances and then attaching a y-coordinate of zero to each x-coordinate solution we obtain. You should also be able to see that any algorithm for solving the Euclidean distance problem can also be used to solve the absolute value problem simply by setting up an associated Euclidean distance problem with every existing facility on either the x-axis or the y-axis. [For example, if we wanted to minimize $f_1(x_1, x_2)$ for Example 6.6, we could construct $f(\mathbf{X}_1, \mathbf{X}_2)$ as given by Example 6.6, minimize $f(\mathbf{X}_1, \mathbf{X}_2)$ to get $(\mathbf{X}_1^*, \mathbf{X}_2^*)$, and then pick out optimal x-coordinates from $(\mathbf{X}_1^*, \mathbf{X}_2^*)$ to get an optimal solution (x_1^*, x_2^*) to the absolute value distance problem.]

There is another way in which the multifacility problems with Euclidean and rectilinear distances are related. Given a Euclidean distance problem, suppose that we construct the associated rectilinear distance problem and solve it, with f_1^* and f_2^* representing the minimum total costs of movement in the x-direction and the y-direction, respectively. Let f^* represent the minimum total cost of movement for the Euclidean distance problem, an unknown number. It is easy to compute lower and upper bounds on f^*; it is known that

$$[(f_1^*)^2 + (f_2^*)^2]^{1/2} \leq f^* \leq f(\mathbf{X}_1, \ldots, \mathbf{X}_n) \qquad (6.17)$$

where $(\mathbf{X}_1, \ldots, \mathbf{X}_n)$ is *any* choice of NF locations. The upper bound on f^* is trivial, since the cost of any choice of locations is at least as large as the minimum cost. The lower bound is not so obvious but is not difficult to derive using the triangle inequality, a result we sketch in a homework problem for Chapter 4. The consequence of (6.17) is that we can sometimes get acceptable solutions to the Euclidean distance problem without ever solving it, provided that we have an optimal solution to the associated rectilinear distance problem.

As an example of (6.17), suppose that we use the data of Example 6.3, which we have already solved for the rectilinear distance case in Section 6.2, where we found $f_1^* = 160, f_2^* = 60$, and optimal locations of (10, 15), (10, 15). We get a lower bound of

$$[(f_1^*)^2 + (f_2^*)^2]^{1/2} = [(160)^2 + (60)^2]^{1/2} = 170.88$$

To get our upper bound we compute $f((10, 15), (10, 15))$ using Euclidean distances and obtain a value of 172.28. We chose ((10, 15), (10, 15)) simply because it was convenient; it was available after solving the rectilinear distance problem. You can see that there is a difference of only 1.4 between the upper and lower bounds, so very likely ((10, 15), (10, 15)) is an acceptable solution for this example. As a second example of (6.17), we use the data of Example 6.1, where we found that $f_1^* = 30$, $f_2^* = 40$, with optimal locations of (10, 10) (10, 10) for the rectilinear distance problem. In this case the lower bound value is 50. To get an upper bound we compute $f((10, 10), (10, 10)) = 61.63$. In fact, we have already given the solution for the Euclidean distance problem in Section 6.1, where we found that $f^* = 59.71$. Thus you can see for this second example that the upper bound is better than the lower bound, and the lower bound is not too satisfactory. If we knew *only* the lower bound and the upper bound, we would probably be reluctant to use (10, 10), (10, 10) as a solution, although in fact it may well be acceptable. The second example illustrates a phenomenon often found with the Euclidean distance problem: The surface of the total cost is relatively "flat" in the neighborhood of the optimal solution. This phenomenon is based only on computational experience, but is nevertheless useful, since if we cannot use the optimal solution it indicates we can use nearby locations with only a small increase in the cost.

Let us now consider the general Euclidean distance cost model. In Chapter 4 we saw that each function of the type $|\mathbf{X}_j - \mathbf{P}_i|_2$ is a convex function; that is, the distance between any NF j and EF i is a convex function in the location of the NF. We shall now establish that any distance function $g(\mathbf{X}_j, \mathbf{X}_k) = |\mathbf{X}_j - \mathbf{X}_k|_2$ is convex. (You can skip to the next paragraph if you are willing to accept this convexity result.) For with λ any number between 0 and 1, and given any $(\mathbf{Y}_j, \mathbf{Y}_k), (\mathbf{Z}_j, \mathbf{Z}_k)$ with

$$(\mathbf{X}_j, \mathbf{X}_k) = \lambda(\mathbf{Y}_j, \mathbf{Y}_k) + (1 - \lambda)(\mathbf{Z}_j, \mathbf{Z}_k) = (\lambda\mathbf{Y}_j + (1 - \lambda)\mathbf{Z}_j, \lambda\mathbf{Y}_k + (1 - \lambda)\mathbf{Z}_k)$$

we have

$$g(\mathbf{X}_j, \mathbf{X}_k) =$$
$$|(\lambda\mathbf{Y}_j + (1 - \lambda)\mathbf{Z}_j) - (\lambda\mathbf{Y}_k + (1 - \lambda)\mathbf{Z}_k)|_2 = |\lambda(\mathbf{Y}_j - \mathbf{Y}_k) + (1 - \lambda)(\mathbf{Z}_j - \mathbf{Z}_k)|_2$$

If we use the triangle inequality we have

$$|\lambda(\mathbf{Y}_j - \mathbf{Y}_k) + (1 - \lambda)(\mathbf{Z}_j - \mathbf{Z}_k)|_2 \le |\lambda(\mathbf{Y}_j - \mathbf{Y}_k)|_2 + |(1 - \lambda)(\mathbf{Z}_j - \mathbf{Z}_k)|_2$$

Using the homogeneity property now gives

$$|\lambda(\mathbf{Y}_j - \mathbf{Y}_k)|_2 + |(1 - \lambda)(\mathbf{Z}_j - \mathbf{Z}_k)|_2 = \lambda|\mathbf{Y}_j - \mathbf{Y}_k|_2 + (1 - \lambda)|\mathbf{Z}_j - \mathbf{Z}_k|_2$$
$$= \lambda g(\mathbf{Y}_j, \mathbf{Y}_k) + (1 - \lambda)g(\mathbf{Z}_j, \mathbf{Z}_k)$$

Thus we conclude that

$$g(\mathbf{X}_j, \mathbf{X}_k) \leq \lambda g(\mathbf{Y}_j, \mathbf{Y}_k) + (1 - \lambda) g(\mathbf{Z}_j, \mathbf{Z}_k)$$

which means that the function g is convex. We note this result actually goes through for the l_p distance for any value of p at least 1, since we never used the fact that p was 2 in our derivation. A very similar approach establishes the convexity of the function in $\mathbf{X}$ defined by $|\mathbf{X} - \mathbf{P}_i|_p$, the l_p distance between $\mathbf{X}$ and $\mathbf{P}_i$, for any value of p at least 1.

The consequence of our analysis is the conclusion that $f(\mathbf{X}_1, \ldots, \mathbf{X}_n)$ is a sum of convex functions and thus is a convex function. For $p = 2$, an equivalent condition for the function to be *strictly* convex is also known: For *every* NF j, the set of locations of EFs with which NF j has an exchange (a positive weight) cannot be collinear. Equivalently, every NF j must have an exchange with at least three EFs that do not lie on the same line. This condition for strict convexity is independent of the terms involving distances between pairs of NFs, even though it is these terms that give the multifacility minisum location model its special character.

Let us now revisit HAP. In the preceding section we saw that HAP was defined by the following iterative equations:

$$x_t = \frac{1}{\Gamma_t(\mathbf{X}_1, \ldots, \mathbf{X}_n)} \alpha_t(\mathbf{X}_1, \ldots, \mathbf{X}_n), \qquad t = 1, \ldots, n$$

$$y_t = \frac{1}{\Gamma_t(\mathbf{X}_1, \ldots, \mathbf{X}_n)} \beta_t(\mathbf{X}_1, \ldots, \mathbf{X}_n), \qquad t = 1, \ldots, n$$

Equivalently, since $\mathbf{X}_t = (x_t, y_t)$ we have

$$\mathbf{X}_t = \frac{1}{\Gamma_t(\mathbf{X}_1, \ldots, \mathbf{X}_n)} (\alpha_t(\mathbf{X}_1, \ldots, \mathbf{X}_n), \beta_t(\mathbf{X}_1, \ldots, \mathbf{X}_n)), \qquad t = 1, \ldots, n$$

With

$$\Omega_t(\mathbf{X}_1, \ldots, \mathbf{X}_n) = (\alpha_t(\mathbf{X}_1, \ldots, \mathbf{X}_n), \beta_t(\mathbf{X}_1, \ldots, \mathbf{X}_n)), \qquad t = 1, \ldots, n$$

we thus have

$$\mathbf{X}_t = \frac{1}{\Gamma_t(\mathbf{X}_1, \ldots, \mathbf{X}_n)} \Omega_t(\mathbf{X}_1, \ldots, \mathbf{X}_n), \qquad t = 1, \ldots, n \qquad (6.18)$$

Equation (6.18) leads to the following iterative procedure:

$$\mathbf{X}_t^{(q+1)} = \frac{1}{\Gamma_t(\mathbf{X}_1^{(q)}, \ldots, \mathbf{X}_n^{(q)})} \Omega_t(\mathbf{X}_1^{(q)}, \ldots, \mathbf{X}_n^{(q)}), \qquad t = 1, \ldots, n \qquad (6.19)$$

To use (6.19) we choose any initial NF locations, $(\mathbf{X}_1^{(0)}, \ldots, \mathbf{X}_n^{(0)})$, substitute into equation (6.19) to compute a new set of locations, $(\mathbf{X}_1^{(1)}, \ldots, \mathbf{X}_n^{(1)})$, substitute again into equation (6.19) to compute another new set of locations, $(\mathbf{X}_1^{(2)}, \ldots, \mathbf{X}_n^{(2)})$, and so on. Thus we obtain an algorithm for finding optimal NF locations. For convenience, we refer to this algorithm as "HAP."

We remark that (6.18) expresses each NF location as a weighted average of all other facility locations. When the weight graph $G(\mathbf{V}, \mathbf{W})$ is connected, this expression is known to imply that each optimal NF location is in CH, the convex hull of the existing facility locations.

We also used HAP to solve Example 6.6, which has collinear EF locations. HAP provided solutions $\mathbf{X}_1^* = (12.5669, 0.0005)$, $\mathbf{X}_2^* = (12.5666, 0.0005)$ with a total cost of 160.000645. We know that an exact solution for this problem is given $\mathbf{X}_1^* = \mathbf{X}_2^* = (x, 0)$ with $10 \leq x \leq 20$, and a cost of 60.00, so HAP comes very close to the exact solution.

The HAP algorithm defined by (6.19) is very similar to the Weiszfeld algorithm; we can think of it as a generalization of the Weiszfeld algorithm applied to the multifacility minisum location function, $f(\mathbf{X}_1, \ldots, \mathbf{X}_n)$. When HAP computes two successive vectors of locations to be almost identical, it stops. We would like to know the justification for HAP stopping. Because the analysis is essentially the same, and the algebra is not as messy, we shall instead examine the justification for the Weiszfeld algorithm stopping. All the insight we obtain into the Weiszfeld algorithm will also apply to HAP.

Our purpose now is to establish that if the Weiszfeld algorithm computes the same point for two successive iterations, it has found a global minimum of the function f.

Recall that from Chapter 4 we defined

$$\gamma_i(\mathbf{X}) = \frac{w_i}{[(x - a_i)^2 + (y - b_i)^2 + \epsilon]^{1/2}}, \qquad i = 1, \ldots, m$$

(We include ϵ, a very small positive number, underneath the square root to avoid division by zero.) On letting $\nabla f(\mathbf{X})$ denote the gradient of f evaluated at X, we derived the following formula for the gradient in Chapter 4:

$$\nabla(\mathbf{X}) = \sum_{i=1}^{m} \gamma_i(\mathbf{X})(\mathbf{X} - \mathbf{P}_i)$$

For convenience, we define $\Gamma(\mathbf{X})$ and $\Omega(\mathbf{X})$ by

$$\Gamma(\mathbf{X}) = \sum_{i=1}^{m} \gamma_i(\mathbf{X})$$

$$\Omega(\mathbf{X}) = \sum_{i=1}^{m} \gamma_i(\mathbf{X})\mathbf{P}_i$$

[Note that $\Gamma(\mathbf{X})$ is a positive real number but $\Omega(\mathbf{X})$ is a vector.] We can now rewrite our formula for the gradient as follows:

$$\nabla f(\mathbf{X}) = \Gamma(\mathbf{X})\mathbf{X} - \Omega(\mathbf{X}) \tag{6.20}$$

Thus the equation $\nabla f(\mathbf{X}) = 0$ is equivalent to

$$\mathbf{X} = \frac{1}{\Gamma(\mathbf{X})}\Omega(\mathbf{X})$$

The latter equation leads to the Weiszfeld algorithm, where we compute a sequence of points $\{\mathbf{X}^{(q)}\}$ using the equation

$$\mathbf{X}^{(q+1)} = \frac{1}{\Gamma(\mathbf{X}^{(q)})}\Omega(\mathbf{X}^{(q)}) \tag{6.21}$$

On letting $\mathbf{X} = \mathbf{X}^{(q)}$ in (6.20), solving for $\Omega(\mathbf{X}^{(q)})$, and substituting into (6.21), we get

$$\mathbf{X}^{(q+1)} = \frac{1}{\Gamma(\mathbf{X}^{(q)})}[\Gamma(\mathbf{X}^{(q)})\mathbf{X}^{(q)} - \nabla f(\mathbf{X}^{(q)})]$$

We conclude (6.21) is equivalent to

$$\mathbf{X}^{(q+1)} = \mathbf{X}^{(q)} + \frac{1}{\Gamma(\mathbf{X}^{(q)})}[-\nabla f(\mathbf{X}^{(q)})] \tag{6.22}$$

If the algorithm computes two successive points to be the same, this means that for some q we have $\mathbf{X}^{(q+1)} = \mathbf{X}^{(q)}$, so that (6.22) now becomes

$$0 = \mathbf{X}^{(q+1)} - \mathbf{X}^{(q)} = \left[\frac{1}{\Gamma(\mathbf{X}^{(q)})}\right]\left[-\nabla f(\mathbf{X}^{(q)})\right]$$

But the term $1/\Gamma(\mathbf{X}^{(q)})$ is always positive, so the latter equation is equivalent to $\nabla f(\mathbf{X}^{(q)}) = 0$; that is, the gradient of f evaluated at $\mathbf{X}^{(q)}$ is the zero vector. Because the function f is a convex function, any point for which the gradient of f is zero is a global minimum. We can thus conclude the following: *An equivalent condition for some point $\mathbf{X}^{(q)}$ computed by the algorithm to be a global minimum is that $\mathbf{X}^{(q+1)} = \mathbf{X}^{(q)}$.* We thus have a justification for stopping the algorithm whenever it computes two successive identical points. In computational practice, we consider two successive points to be identical when the distance between the points is no greater than some positive term, say delta, which is sufficiently close to zero (e.g., delta = 0.001).

If you are familiar with steepest descent algorithms you should recognize (6.22) as being a case where we compute $\mathbf{X}^{(q+1)}$ from $\mathbf{X}^{(q)}$ using the steepest-descent direction $-\nabla f(\mathbf{X}^{(q)})$, with the term $1/\Gamma(\mathbf{X}^{(q)})$ being directly proportional to the step size, and computed automatically by the algorithm. We can view the Weiszfeld algorithm, and HAP, as being steepest descent algorithms where the step size is determined automatically.

We remark that there is a duality for the problem of this section which is quite similar to that for the rectilinear distance problem of Section 6.2. The duality leads to some interesting theoretical results, but unfortunately does not seem of much computational value, particularly as compared to the duality of Section 6.2.

6.6 LOCATION-ALLOCATION PROBLEMS

We assumed for the multifacility location problems we treated earlier in this chapter that all the weights were *known*. In fact, such an assumption can be a limitation, since sometimes the weights are adjustable within some limits, particularly if all the NFs are of the same type, and all of the EFs are of the same type, and it does not matter which NF serves which EF. Hence in this section we shall treat a class of problems, called location-allocation problems, in which the weights are decision variables, as well as the locations of the NFs. Our presentation concentrates on continuous problems with an infinite number of possible locations, and either rectilinear or Euclidean distances. Analogous discrete problems are discussed in Chapter 8.

A common example of a location-allocation problem involves the location of distribution centers, or warehouses, which receive products from production facilities and distribute products to customers such as retail or wholesale outlets; we also want to determine an allocation of customers to warehouses. Another example of a location-allocation problem occurs when a number of branch banks are to be located in a metropolitan area. Instead of branch banks or warehouses, the new facilities could easily be, say, post offices or grocery stores. In the case of branch banks, grocery stores, or post offices, the existing facilities would include the residences of the consumers (the clients), and an allocation of the existing facilities to the new facilities would be of interest.

Example 6.7

We consider a *very* small example to illustrate some basic ideas occurring in location-allocation problems. The *only* purpose of this example is to illustrate the ideas. Real-world problems are much larger, as we shall see, and are tractable only when a computer is employed to do the computations. Suppose that we have EFs 1 to 7 at locations $(0,5)$, $(8,5)$, $(5,4)$, $(2,2)$, $(3,2)$, $(0,0)$, and $(7,0)$, respectively. We wish to locate two NFs so that each EF is served by a nearest NF, and the grand total of the sum of rectilinear distances between the NFs and EFs they serve is minimal. Thus there are two aspects to this example: (1) determining the NF locations, and (2) allocating EFs to NFs. A commonly employed means of solving such a problem is the following heuristic approach: (a) With the NFs at *given* locations, allocate each EF to a closest NF; (b) *given* the allocation of EFs to NFs, minimize the total distance between each NF and its allocated EFs. Repeat (a) and (b) sequentially until no further reduction in the total distance is possible. Note that if we are working with rectilinear distances, we can use the median conditions of Chapter 4 to solve each single NF location problem; *alternatively, if we were working with Euclidean distances, we could use the Weiszfeld algorithm*.

Figure 6.7(a) shows an initial location of NFs at $\mathbf{X}_1 = (0,5)$ and $\mathbf{X}_2 = (2,2)$, respectively (we start with bad initial locations intentionally, for reasons seen later). The EFs allocated to NF_1 are given by $A_1 = \{\mathbf{P}_1, \mathbf{P}_2\}$, and to NF_2 are given by $A_2 = \{\mathbf{P}_3, \ldots, \mathbf{P}_7\}$ (points constituting allocations have lines drawn around them in the future). The total distance for NF_1 is given by $T_1 = 8$, and for NF_2 by $T_2 = 17$, giving a grand total of $T = 25$. If we now minimize the sum of distances between each NF and its allocated EFs, we find that $\mathbf{X}_1 = (0,5)$ again, while $\mathbf{X}_2 = (3,2)$; again we get $T_1 = 8$, while $T_2 = 16$, giving $T = 24$, as shown in Figure 6.7(b). Let us move NF_1 to $\mathbf{X}_1 = (3,5)$, as doing so will not cause an increase in the grand total and give NF_1 a more "favorable" location. Allocating each EF to a closest NF, we now get $A_1 = \{\mathbf{P}_1, \mathbf{P}_2, \mathbf{P}_3\}$, $A_2 = \{\mathbf{P}_4, \ldots, \mathbf{P}_7\}$; for this new allocation we get $T_1 = 11$ and $T_2 = 12$, with $T = 23$; see Figure 6.7(c). We now find that $\mathbf{X}_1 = (5,5)$ and $\mathbf{X}_2 = (3,2)$ minimize the sum of distances between the NFs and allocated EFs, with $T_1 = 9$, $T_2 = 12$, and $T = 21$ [Fig-

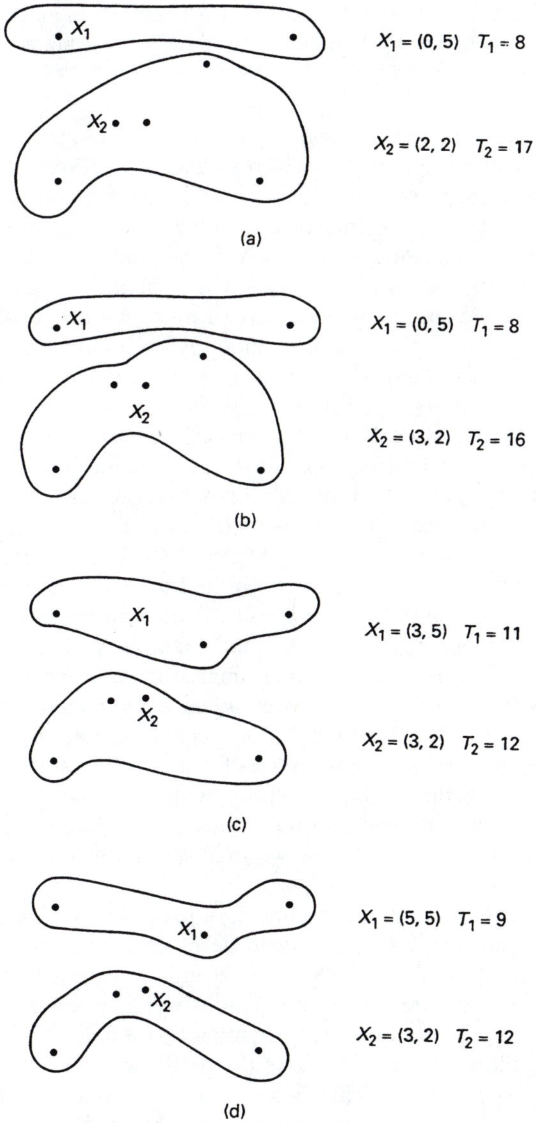

Figure 6.7 Location-allocation example with rectilinear distances.

ure 6.7(d)]. If we now reallocate, we get the same NF allocations as before, and the same NF locations as before, and so we stop. Do we have a best solution at this point? Try and answer the question yourself before reading further.

In fact, this example illustrates that *the* NF *locations we start from affect the final solution we obtain*, for there is a better solution to the problem than the one we stopped with. If we take $\mathbf{X}_1 = (2,2)$ and $\mathbf{X}_2 = (7,4)$, we get allocations of $A_1 = \{\mathbf{P}_1, \mathbf{P}_4, \mathbf{P}_5, \mathbf{P}_6\}$

and $A_2 = \{\mathbf{P}_2, \mathbf{P}_3, \mathbf{P}_7\}$; we get $T_1 = 9$, $T_2 = 8$, and $T = 17$. Thus it is essential, when solving location-allocation problems, to try a number of different starting points and run the algorithm repeatedly. We then use the best of all the solutions we find.

Let us consider some heuristic approaches to generate starting points. Let R denote a smallest rectangle, with each side parallel to an axis, that contains all the EF locations. One way to generate starting points is to locate them randomly in R. Alternatively, we could space them uniformly throughout R, or space them uniformly around the perimeter of R, locating the first one on the perimeter randomly. A related approach would be to work with the convex hull of the EF locations, denoted by CH, instead of R. For example, we could space starting points uniformly around the perimeter of CH, locating the first one randomly; we could space them equally around the perimeter of CH, and try keeping the same relative orientation while "rotating" the starting points (e.g., for $n = 3$ NFs, we could try starting locations at 12, 4, and 8 o'clock, 2, 6 and 10 o'clock, etc.). The number of possible ways of choosing starting points is limited only by one's imagination. What is important, in any case, is to try many different starting-point solutions.

Location-allocation problems are in fact rather difficult to find optimal solutions to, since they can have local minima that are not global minima. We might be able to avoid some local minima if we could treat both the location and allocation aspects simultaneously, but doing so makes the analysis and optimization much more difficult—in fact, beyond the scope of this book. Our example with rectilinear distances was simplified by the fact that there is an optimal solution where each NF is at an intersection point (recall Chapter 4) but if we had used Euclidean distances instead of rectilinear distances this intersection point property would not have been true, complicating the problem further. Thus we shall have to be satisfied with heuristic approaches. Such approaches, if used carefully, with a knowledge of their limitations and an effort made to minimize the limitations by obtaining a number of different solutions, can still be quite useful, as we shall see in our discussion to follow.

Let us now consider a real-world location-allocation problem. We shall consider some work done by Devine and Lesso [14], who developed some models for the minimum-cost development of offshore oil fields. They utilize approaches developed earlier by Cooper [7–9] for solving conventional industrial location-allocation problems, to obtain insight into the development of an offshore oil field. As we shall see, in their work the new facilities to be located are the platforms from which the drilling for oil will be done, while the existing facilities are the wells. Instead of locating warehouses and allocating customers to warehouses, they consider locating oil platforms and allocating oil wells to platforms. A little background about offshore oil fields is useful first.

After an oil field has been discovered by an exploratory well, step-out wells are then drilled to define the size and other characteristics of the field. Based on this initial information, locations are decided upon for the production wells. These well locations are specified by two map coordinates and a depth coordinate. With present technology, most fields are developed by drilling directionally from fixed platforms. The drilling and completion costs for each production well will depend upon the length and angle of the hole drilled from the platform to the target. The platform costs represent tremendous investments, usually from one-fourth to ten million dollars. For a given geographical region, the platform costs depend critically on the water depth and on the number of wells to be drilled from the platform. Thus, for a large number of production wells (25 to 300), an optimization problem that arises is to find the number, size, and location of drilling platforms and the allocation of wells to platforms, so as to minimize the sum of platform and drilling costs. This problem will be referred to as the "platform location problem."

There are many important factors which can affect the development cost for a given field. In order to better understand the assumptions inherent in the platform location model, these factors are classified into three categories: (1) decision variables, (2) prespecified variables (assumed fixed by the operators) and (3) uncontrollable variables. The decision variables are (a) the number of platforms, (b) the size of each platform, (c) the location of each platform, and (d) the assignment of wells to platforms. The variables that are assumed to be prespecified include such things as well location, type of platform, type of rig, single or dual completions, etc. The uncontrollable variables include depth of targets, water depth, geological characteristics, wind and wave forces, bottom conditions, and remoteness of location.

Another important cost incurred in developing offshore fields is the pipeline cost. While these costs are large enough to be important, they should not generally affect the platform location model because they would not vary significantly with small variations in platform location. . . . However, if the pipeline costs were deemed to be significant in a particular problem, then they could be expressed as part of the platform cost which is a function of location.

We can now formulate our problem analytically. We let m denote the number of wells, with i the well index; n denotes the number of platforms, with j the platform index. We define a zero–one variable z_{ij} as follows:

$$z_{ij} = \begin{cases} 1 & \text{if well } i \text{ is drilled from platform } j \\ 0 & \text{otherwise} \end{cases}$$

We let S_j denote the number of wells drilled from platform j (we can think of S_j as the size of platform j, since the size of the platform is directly proportional to S_j). For $i = 1, \ldots, m$, (a_i, b_i, c_i) denotes the location coordinates of well i. Similarly, for $j = 1, \ldots, n$, (x_j, y_j) denotes the location of platform j. For all values of i and j, we let

$$d_{ij} = [(x_j - a_i)^2 + (y_j - b_i)^2]^{1/2}$$

the horizontal Euclidean distance between well i and platform j. It is important to recognize that d_{ij} depends on the location of platform j, which must be determined. For all i and j, $g(d_{ij})$ denotes the drilling cost function, a function of the horizontal (Euclidean) distance between well i and platform j, and depends on (x_j, y_j). For each platform j, $P_j(S_j, x_j, y_j)$ denotes the cost of platform j as a function of the "size" of the platform size S_j, and its location (x_j, y_j), minus any salvage cost.

It is important to observe that a platform not need be directly over a well in order to drill it; in fact, wells can be drilled "at an angle," thus allowing several wells (depending on the "size" of the platform) in different locations to be drilled from the same platform. This fact is exploited when the cost to drill well i from platform j is expressed as a function of the *horizontal* distance d_{ij} between well i and platform j; if the platform had to be directly above the well, the horizontal distance between the well and the platform would be zero, and only one well could be drilled from a platform.

We can now state the problem of interest as follows:

$$\text{minimize} \quad \sum_{i=1}^{m} \sum_{j=1}^{n} z_{ij} g(d_{ij}) + \sum_{j=1}^{n} P_j(S_j, x_j, y_j)$$

subject to

$$\sum_{j=1}^{n} z_{ij} = 1, \qquad i = 1, \ldots, m$$

$$\sum_{i=1}^{m} z_{ij} = S_j, \qquad j = 1, \ldots, n$$

$$z_{ij} = 0 \text{ or } 1, \qquad i = 1, \ldots, m \qquad j = 1, \ldots, n$$

The objective function is the sum of the drilling costs and the platform costs. The first set of constraints guarantees that each well i is assigned to exactly one platform, while the second set of constraints guarantees that exactly S_j wells are assigned to platform j. The zero–one constraints state that either a well is assigned to a platform or it is not. Thus the problem is to find an allocation of wells to platforms so that each well is allocated to exactly one platform, each platform j has exactly S_j wells allocated to it, and the total of the drilling costs and the platform cost is minimized.

We remark that n is fixed in the model, and is typically small, in the range 3 to 5. Because n is small it is practical when doing sensitivity analysis to treat it as a parameter of the model.

Devine and Lesso [14] make some interesting comments about a model of such a large-scale problem:

> It is appropriate . . . to comment briefly on the scope and use of a model such as described above. Because of the real world complexity, an exact analog of the oil field development problem is not possible. Many factors add to this such as the stochastic nature of the drilling costs, uncertainty about the exact nature of the field, current economic conditions, pollution standards, etc. Thus, an "optimal" solution to the model does not imply that the analyst has an "optimal" solution ready to be implemented for the real world problem. Rather, the results from the model should be thought of as an aid to, but not as a replacement for, the analyst's intuition. The greatest attribute of the model should be to provide the decision makers with a better understanding of how the total development cost depends upon the various parameters. Of

course, these considerations should be weighed when developing techniques for the model.

The approach taken to solving the model is known as the *alternate location-allocation* (ALA) *method*. The idea of the approach is as follows: (a) *given* fixed platform locations, find a minimum cost allocation of wells to platforms; (b) *given* fixed allocations of wells to platforms, find a minimum total cost location for each platform. We alternate between steps (a) and (b) until we get convergence. It is important to recognize that ALA is a *heuristic* procedure, with no guarantee of optimality. The solution obtained is dependent on the initial starting platform locations, and it is a good idea to try solving the problem several times using different sets of initial starting locations. The solution obtained satisfies the following *necessary* conditions for optimality: (a) for the *given* assignment of targets to platform locations, the solution cannot be improved by changing locations; (b) for the *given* locations, the solution cannot be improved by altering the assignment of targets to platforms.

We now consider

I. Location subproblem for platform *j*. Let A_j denote the set of indices of the wells assigned to platform j; then—since we know the allocation—we have $z_{ij} = 1$ for $i \in A_j$, $z_{ij} = 0$ otherwise. The part of the objective function for platform j then becomes

$$\sum_{i=1}^{m} z_{ij} g(d_{ij}) + P_j(S_j, x_j, y_j) = \sum_{i \in A_j} g(d_{ij}) + P_j(x_j, y_j)$$

since the platform cost no longer depends on the platform size. Thus the problem for platform j is as follows:

$$\text{minimize} \quad \sum_{i \in A_j} g([(x_j - a_i)^2 + (y_j - b_i)^2]^{1/2}) + P_j(x_j, y_j)$$

Note that the platform cost now depends only on (x_j, y_j), since S_j is assumed to be known.

You should recognize this subproblem as being related to the minisum single-facility location problems we studied in Chapter 4; it is a generalization of these problems.

> Examination of drilling cost data and study of the drilling procedures indicate that drilling costs are a convex function of horizontal distance. In this case, since the sum of convex functions is convex, that portion of the objective function attributable to the drilling will be convex.
>
> If the platform cost function due to position is negligible because the water depth and bottom conditions do not vary significantly over the field (which appears to be the usual case) or is convex, then the total cost function to be minimized is convex. Thus one is assured that any local minimum is also the global minimum.
>
> A simple gradient search method, utilizing "golden-section search" for the one-

dimensional optimization, was used to solve the location subproblem. A more sophisticated unconstrained minimization technique could be used, but the simplicity of the problem did not seem to warrant it. In most test problems, the solution converged in two or three iterations. However, in a few observations the procedure did "zig-zag" requiring eight or nine iterations before convergence.

Of course, if the platform cost function due to position ($P_j(x_j, y_j)$) were nonconvex such that the total cost function is nonconvex, then the problem is more difficult. A local minimum can be found, but global optimality cannot be guaranteed. The best approach in this situation is to try several starting points and pick the solution (if more than one is found) which gives the lowest value of the objective function.

Next we consider the allocation subproblem.

II. Allocation subproblem. Because the platform locations are now given, with $c_{ij} \equiv g(d_{ij})$, c_{ij} is a constant, so we have the following problem:

$$\text{minimize} \quad \sum_{i=1}^{m}\sum_{j=1}^{n} c_{ij} z_{ij} + \sum_{j=1}^{n} P_j(S_j)$$

subject to

$$\sum_{j=1}^{n} z_{ij} = 1, \qquad i = 1, \ldots, m$$

$$\sum_{i=1}^{m} z_{ij} = S_j, \qquad j = 1, \ldots, n$$

$$z_{ij} = 0 \text{ or } 1, \qquad i = 1, \ldots, m \qquad j = 1, \ldots, n$$

Note the platform cost now depends only on S_j for platform j, since the platform locations are assumed to be known. The solution procedure for the allocation subproblem depends on the structure of each platform cost $P_j(S_j)$. The authors examined five different structures, obtaining Problems 1 to 5, respectively, depending on the assumptions. We shall consider the first three problems; the last two are beyond the scope of our discussion.

Problem 1: Single Fixed Cost, with No Capacity Constraints: $P_j(S_j) = a_j$. The problem can now be written as follows:

$$\text{minimize} \quad \sum_{i=1}^{m}\sum_{j=1}^{n} c_{ij} z_{ij} + \sum_{j=1}^{n} a_j$$

subject to

$$\sum_{j=1}^{n} z_{ij} = 1, \qquad i = 1, \ldots, m$$

$$z_{ij} = 0 \text{ or } 1, \qquad i = 1, \ldots, m \qquad j = 1, \ldots, n$$

Because the platform cost is constant, each feasible solution to this problem represents choosing for each well i a single platform, say $j(i)$, to which to allocate the well. Obviously, the platform to choose for each well i is the one with the smallest entry in row i of the matrix $C = (c_{ij})$. That is, find the smallest entry in each row i of the matrix C, say $c_{ij(i)}$. Set $z_{ij(i)} = 1$, and set $z_{ij} = 0$ for j distinct from $j(i)$. We allocate each well i to a "cheapest" platform $j(i)$. (You should recognize that we used the same approach in our initial rectilinear distance example when we found a closest new facility to each existing facility.)

Problem 2: Single Fixed Cost, with Capacity Constraints. Writing the capacity constraint as an inequality constraint, the problem is now as follows:

$$\text{minimize} \quad \sum_{i=1}^{m}\sum_{j=1}^{n} c_{ij}z_{ij} + \sum_{j=1}^{n} a_j$$

subject to

$$\sum_{j=1}^{n} z_{ij} = 1, \qquad i = 1, \ldots, m$$

$$\sum_{i=1}^{m} z_{ij} \le S_j, \qquad j = 1, \ldots, n$$

$$z_{ij} = 0 \text{ or } 1, \qquad i = 1, \ldots, m \qquad j = 1, \ldots, n$$

We now view S_j as a *maximum* capacity for platform j. With this interpretation Problem 2 is a transportation problem and can be solved by any algorithm for solving transportation problems. Note that for each platform it is necessary to include a slack variable, a "dummy well," to take up any excess capacity at zero cost, in order to convert the problem to the standard form of the transportation problem with all equality constraints.

Problem 3: Linear Platform Cost: $P_j(S_j) = a_j + b_jS_j$. In this case we use the fact that

$$\sum_{i=1}^{m} z_{ij} = S_j, \qquad j = 1, \ldots, n$$

It is then direct to verify that

$$\begin{aligned}\sum_{i=1}^{m}\sum_{j=1}^{n} c_{ij}z_{ij} + \sum_{j=1}^{n}(a_j + b_jS_j) &= \sum_{i=1}^{m}\sum_{j=1}^{n}(c_{ij} + b_j)z_{ij} + \sum_{j=1}^{n} a_j \\ &= \sum_{i=1}^{m}\sum_{j=1}^{n} c'_{ij}z_{ij} + \sum_{j=1}^{n} a_j\end{aligned}$$

where we define c'_{ij} by $c'_{ij} = c_{ij} + b_j$. The problem now becomes:

$$\text{minimize} \quad \sum_{i=1}^{m}\sum_{j=1}^{n} c'_{ij}z_{ij} + \sum_{j=1}^{n} a_j$$

subject to

$$\sum_{j=1}^{n} z_{ij} = 1, \qquad i = 1, \ldots, m$$

$$z_{ij} = 0 \text{ or } 1, \qquad i = 1, \ldots, m \qquad j = 1, \ldots, n$$

Note that this problem now has the same form as Problem 1 and can be solved in the same way. For each well i, examine row i of the matrix (c'_{ij}) to find a smallest entry, say $c'_{ij(i)}$, set $z_{ij(i)} = 1$, and set $z_{ij} = 0$ for j distinct from $j(i)$. Again, we allocate each well i to a "cheapest" platform $j(i)$.

The authors considered two other problems. In Problem 4, each $P_j(S_j)$ is a piecewise linear and convex function of S_j, in which case they transform the problem to a transshipment problem. In Problem 5, each $P_j(S_j)$ is a step function of S_j, in which case they transform the allocation subproblem to a linear integer programming problem and develop a heuristic to solve it approximately. The last two problems are beyond the scope of our discussion.

The first four problems are solvable by the ALA heuristic, and the authors wrote programs to solve Problems 1, 2, and 5, stating that "programs for Problems 3 and 4 can be easily obtained by making slight modifications to the programs for Problems 1 and 2, respectively." It should be emphasized that these problems are tractable only when a computer is used. Problems 1 and 3 would be easy to code, and Problem 2 would not be difficult to code if a transportation problem subroutine is available.

The authors applied their codes to two test problems. Before considering the test problems we first consider the drilling cost function they used in the first test problem. Let C_i denote the depth of target well i in thousands of feet, and recall that d_{ij} denotes the horizontal distance from platform j to target well i, again in thousands of feet. The drilling cost function was as follows.

$$g(d_{ij}) = 122.6 - 21.43C_i + 2.39C_i^2 + 12.24d_{ij} + 5.0\left(\frac{d_{ij}}{C_i - 1.5}\right)^{10}$$

the units of the function are in thousands of dollars. Note this function is strictly increasing and convex in d_{ij} and exhibits substantial diseconomies of scale.

Test Problem 1 had 60 wells randomly distributed in a 3-mile by 2.5-mile area, with depths between 4500 and 6000 ft. There were seven different size platforms available, ranging in size from a 6-well platform to a 25-well platform.

Test Problem 2 was "based on the data from an actual offshore field that has been developed. There were 102 wells distributed over an area of about 3.25 miles by 2 miles, with depths between 7900 and 9000 ft. The drilling cost function had the same general form as that of Test Problem 1, with the coefficients in each term calculated so that the cost function would simulate the actual drilling costs that were incurred when developing the field. The platform cost function had five platform sizes, ranging from a 12-well platform to a 36-well platform."

Table 6.7 shows a summary of the results for the Problem 5 algorithm. All runs

TABLE 6.7 RESULTS OF ALGORITHM FOR PROBLEM 5

	Test problem 1: Number of platforms		Test problem 2: Number of platforms		
	4	5	3	4	5
Number of runs (starting points)	6	6	5	5	5
Best solution found (millions of dollars)	12.658	12.679	32.004	32.226	32.356
Worst solution found (millions of dollars)	12.794	13.025	32.004	32.327	32.598
Average computation time per run (seconds)	9	12	13	19	20

were made on a CDC 6600 computer. Note that n was treated as a parameter, and we can pick a best value of n (e.g., 3 for Test Problem 2). The authors state that "using our estimated cost functions of Test Problem 2, the total cost of the policy actually used for developing the field was calculated to be \$33.346 million. Thus the computer solution for this test problem indicated a savings of \$1.342 million."

6.7 FURTHER READING

Miehle [37] appears to be the earliest author to have written on the topic of minisum multifacility problems; his work includes some interesting models related to the string model of Chapter 4. Since then, as our list of chapter references shows, there have been numerous papers on the topic, which can be categorized roughly as dealing with (1) Euclidean distance, (2) rectilinear distance, (3) l_p distance, and (4) duality. In addition, there are similar problems with (5) minimax objectives. All these papers can be easily identified by looking for the appropriate terms in the titles. As for (6) location-allocation problems, heuristic algorithms to solve them appear to have been introduced by Cooper [7]; his work has had a substantial impact in geography and in regional science; see the book edited by Ghosh and Rushton [21] for an extensive discussion of the location-allocation literature, as well as the paper by Rushton [47] for a discussion of applications of location-allocation models. Our brief location-allocation discussion is based on Cooper's work and that by Devine and Lesso [14].

The network flow approach we discuss for the rectilinear distance problem is from Cabot et al. [2], while the direct search method is from Pritsker and Ghare [44] and Rao [46]. For other network approaches see Picard and Ratliff [43] and Kolen [25]. The squared distance material is from White [56] and Eyster and White [17]. The algorithm HAP is based on work by Eyster et al. [18]. Some of the results in Section 5.2 are from Francis and Cabot [20]. The minisum multifacility problem continues to receive attention, as the papers by Michelot [36] and by Rado [45] indicate.

The text by Love et al. [32] has a chapter devoted to minisum multifacility problems, while the paper by Brandeau and Chiu [1] has an overview discussion of a number of location problems, including ones we have studied.

REFERENCES

1. Brandeau, M. L., and S. S. Chiu, "An Overview of Representative Problems in Location Research," *Management Science,* Vol. 35, 1989, pp. 645–673.
2. Cabot, A. V., R. L. Francis, and M. A. Stary, "A Network Flow Solution to a Rectilinear Distance Facility Location Problem," *AIIE Transactions,* Vol. 2, No. 2, 1970, pp. 132–141.
3. Calamai, P. H., and A. R. Conn, "A Projected Newton Method for l_p Norm Location Problems," *Mathematical Programming,* Vol. 38, 1987, pp. 75–109.
4. Cavalier, T. M., and H. D. Sherali, "Euclidean Distance Location-Allocation Problems with Uniform Demands over Convex Polygons," *Transportation Science,* Vol. 20, 1986, pp. 107–116.
5. Charalambous, C., "An Iterative Algorithm for the Multifacility Location Problem," *Naval Research Logistics Quarterly,* Vol. 32, 1981, pp. 373–389.
6. Chatelon, J. A., D. W. Hearn, and T. J. Lowe, "A Subgradient Algorithm for Certain Minimax and Minisum Problems," *Mathematical Programming,* Vol. 15, 1978, pp. 130–145.
7. Cooper, L., "Location-Allocation Problems," *Operations Research,* Vol. 11, No. 3, 1963, pp. 331–344.
8. Cooper, L., "Heuristic Methods for Location-Allocation Problems," *SIAM Review,* Vol. 6, No. 1, 1964, pp. 37–52.
9. Cooper, L., "Solutions of Generalized Locational Equilibrium Models," *Journal of Regional Science,* Vol. 7, 1964, pp. 1–18.
10. Cooper, L., "An Extension of the Generalized Weber Problem," *Journal of Regional Science,* Vol. 8, 1969, pp. 181–197.
11. Cooper, L., "The Transportation-Location Problem," *Operations Research,* Vol. 20, No. 1, 1972, pp. 94–108.
12. Dax, A., "A Note on Optimality Conditions for the Euclidean Multifacility Location Problem," *Mathematical Programming,* Vol. 36, 1976, pp. 628–642.
13. Dearing, P. M., and R. L. Francis, "A Network Flow Solution to a Multifacility Minimax Location Problem Involving Rectilinear Distances," *Transportation Science,* Vol. 8, 1974, pp. 126–141.
14. Devine, M. D., and W. G. Lesso, "Models for the Minimum Cost Development of Offshore Oil Fields," *Management Science,* Vol. 18, 1972, pp. B378–B387.
15. Drezner, Z., "A New Method for the Multifacility Minimax Location Problem," *Journal of the Operational Research Society,* Vol. 29, 1978, pp. 1095–1101.
16. Elzinga, J., D. W. Hearn, and W. D. Randolph, "Minimax Multifacility Location with Euclidean Distances," *Transportation Science,* Vol. 10, 1976, pp. 321–336.
17. Eyster, J. W., and J. A. White, "Some Properties of the Squared Euclidean Distance Location Problem," *AIIE Transactions,* Vol. 5, No. 3, 1973, pp. 275–280.

18. Eyster, J. W., J. A. White, and W. W. Wierwille, "On Solving Multifacility Location Problems Using a Hyperboloid Approximation Procedure," *AIIE Transactions,* Vol. 5, No. 1, 1973, pp. 1–6.
19. Francis, R. L., "On the Location of Multiple New Facilities with Respect to Existing Facilities," *Journal of Industrial Engineering,* Vol. 15, No. 2, 1964, pp. 106–107.
20. Francis, R. L., and A. V. Cabot, "Properties of a Multifacility Location Problem Involving Euclidean Distances," *Naval Research Logistics Quarterly,* Vol. 19, No. 2, 1972, pp. 335–353.
21. Ghosh, A., and G. Rusthon, eds., *Spatial Analysis and Location-Allocation Models,* Van Nostrand Reinhold Company, New York, 1987.
22. Juel, H., and R. F. Love, "An Efficient Computational Procedure for Solving the Multi-facility Rectilinear Facilities Location Problem," *Operational Research Quarterly,* Vol. 27, No. 3, 1976, pp. 697–703.
23. Juel, H., "Sufficient Conditions for Optimal Facility Locations to Coincide," *Transportation Science,* Vol. 14, 1980, pp. 125–129.
24. Juel, H., "On the Dual of the Linearly Constrained Multi-facility Location Problem with Arbitrary Norms," *Transportation Science,* Vol. 15, 1981, pp. 329–337.
25. Kolen, A., "Equivalence between the Direct Search Approach and the Cut Approach to the Rectilinear Distance Location Problem," *Operations Research,* Vol. 29, No. 3, 1981, pp. 616–620.
26. Love, R. F., "Locating Facilities in 3-Dimensional Space by Convex Programming," *Naval Research Logistics Quarterly,* Vol. 16, No. 4, 1969, pp. 503–516.
27. Love, R. F., "The Dual of a Hyperbolic Approximation to the Generalized Constrained Multi-facility Location Problem with l_p Distances," *Management Science,* Vol. 21, 1974, pp. 22–33.
28. Love, R. F., and H. Juel, "Properties and Solution Methods for Large Location-Allocation Problems," *Journal of the Operational Research Society,* Vol. 33, 1982, pp. 443–452.
29. Love, R. F., and S. Kraemer, "A Dual Decomposition Method for Minimizing Transportation Costs in Multi-facility Location Problems," *Transportation Science,* Vol. 7, 1975, pp. 297–316.
30. Love, R. F., and J. G. Morris, "Solving Constrained Multi-facility Location Problems Involving l_p Distances Using Convex Programming," *Operations Research,* Vol. 23, 1975, pp. 581–587.
31. Love, R. F., and J. G. Morris, "A Computational Procedure for the Exact Solution of Location-Allocation Problems with Rectangular Distances," *Naval Research Logistics Quarterly,* Vol. 22, 1975, pp. 441–453.
32. Love, R. F., J. G. Morris, and G. O. Wesolowsky, *Facilities Location: Models and Methods,* North-Holland Publishing Company, New York, 1988.
33. Love, R. F., and Yerex, L., "An Application of a Facilities Location Model in the Prestressed Concrete Industry," *Interfaces,* Vol. 6, No. 4, 1976, pp. 45–49.
34. Love, R. F., G. O. Wesolowsky, and S. A. Kraemer, "A Multi-facility Minimax Location Method for Euclidean Distances," *International Journal of Production Research,* Vol. 11, 1973, pp. 37–45.
35. Marucheck, A. S., and A. A. Aly, "An Efficient Algorithm for the Location-Allocation

Problem with Rectangular Regions," *Naval Research Logistics Quarterly,* Vol. 28, 1981, pp. 309–323.

36. Michelot, C., "Localization in Multifacility Location Theory," *European Journal of Operational Research,* Vol. 36, 1987, pp. 485–492.
37. Miehle, W., "Link-Length Minimization in Networks," *Operations Research,* Vol. 6, 1958, pp. 232–243.
38. Morris, J. G., "A Linear Programming Approach to the Solution of Constrained Multifacility Minimax Location Problems Where Distances Are Rectangular," *Operational Research Quarterly,* Vol. 24, 1973, pp. 419–435.
39. Morris, J. G., "Convergence of the Weiszfeld Algorithm for Weber Problems Using a Generalized 'Distance' Function," *Operations Research,* Vol. 29, 1981, pp. 37–48.
40. Morris, J. G., and W. A. Verdini, "A Single Iterative Scheme for Solving Minisum Facility Location Problems Involving l_p Distances," *Operations Research,* Vol. 27, No. 1, 1979, pp. 1180–1188.
41. Ostresh, L. M., Jr., "An Efficient Algorithm for Solving the Two Center Location-Allocation Problem," *Journal of Regional Science,* Vol. 15, 1975, pp. 209–216.
42. Ostresh, L. M., Jr., "The Multifacility Location Problem: Applications and Descent Theorems," *Journal of Regional Science,* Vol. 17, 1977, pp. 409–419.
43. Picard, J., and H. D. Ratliff, "A Cut Approach to the Rectilinear Distance Facility Location Problem," *Operations Research,* Vol. 26, No. 3, 1978, pp. 422–433.
44. Pritsker, A. A. B., and P. M. Ghare, "Locating New Facilities with Respect to Existing Facilities," *AIIE Transactions,* Vol. 2, No. 4, 1970, pp. 290–298.
45. Rado, F., "The Euclidean Multifacility Location Problem," *Operations Research,* Vol. 36, 1988, pp. 485–492.
46. Rao, M. R., "On the Direct Search Approach to the Rectilinear Facilities Problem," *AIIE Transactions,* Vol. 5, 1973, pp. 256–264.
47. Rushton, G., "Applications of Location Models," *Annals of Operations Research,* Vol. 18, 1989, pp. 25–42.
48. Sherali, H. D., and C. M. Shetty, "Rectilinear Distance Location-Allocation Problem: A Simplex Based Algorithm," *Lecture Notes in Economics and Mathematical Systems* (Extremal Methods and Systems Analysis), Vol. 174, 1977, pp. 442–464.
49. Sherali, H. D., and C. M. Shetty, "The Rectilinear Distance Location-Allocation Problem," *AIIE Transactions,* Vol. 9, No. 2, 1977, pp. 136–143.
50. Sherali, H. D., and W. P. Adams, "A Decomposition Algorithm for a Discrete Location-Allocation Problem," *Operations Research,* Vol. 32, 1984, pp. 878–900.
51. Vergin, R. C., and J. D. Rogers, "An Algorithm and Computational Procedure for Locating Economic Facilities," *Management Science,* Vol. 13, No. 6, 1967, pp. 3240–3254.
52. Wendell, R. E., and A. P. Hurter, Jr., "Minimization of a Non-separable Objective Function Subject to Disjoint Constraints," *Operations Research,* Vol. 24, 1976, pp. 643–657.
53. Wesolowsky, G. O., and R. F. Love, "The Optimal Location of New Facilities Using Rectangular Distances," *Operations Research,* Vol. 19, No. 1, 1971, pp. 124–130.
54. Wesolowsky, G. O., and R. F. Love, "A Nonlinear Approximation Method for Solving a Generalized Rectangular Distance Weber Problem," *Management Science,* Vol. 18, No. 11, 1972, pp. 656–663.

55. Wesolowsky, G. O., and W. G. Truscott, "The Multiperiod Location-Allocation Problem with Relocation of Facilities," *Management Science,* Vol. 22, 1975, pp. 57–65.
56. White, J. A., "A Note on the Quadratic Facility Location Problem," *AIIE Transactions,* Vol. 3, No. 2, 1971, pp. 156–157.

PROBLEMS

6.1. Another way to specify a weight graph is to list the nodes and the arcs, with the arcs listed as pairs of nodes. For example, the weight graph of Figure 6.2 is as follows.

Nodes: E_1, E_2, E_3, E_4, N_1, N_2, N_3
Arcs: (E_1, N_1), (E_2, N_2), (E_3, N_2), (E_4, N_1), (E_4, N_2), (N_1, N_2)

Draw the weight graph for the following nodes and arcs, and identify any special structure.

Nodes: E_1, E_2, E_3, N_1, N_2, N_3
Arcs: (E_1, N_1), (E_2, N_2), (E_3, N_3), (N_1, N_2)

6.2. With reference to Problem 6.1, draw the weight graph for the following nodes and arcs, and identify any special structure.

Nodes: E_1, E_2, E_3, N_1, N_2, N_3
Arcs: (E_1, N_1), (E_2, N_1), (E_3, N_1), (N_1, N_2), (N_2, N_3)

6.3. With reference to Problem 6.1, draw the weight graph for the following nodes and arcs, and identify any special structure.

Nodes: E_1, E_2, E_3, N_1, N_2, N_3
Arcs: (E_1, N_1), (E_2, N_1), (E_3, N_1), (N_2, N_3)

6.4. A small machine shop has five existing machines located at coordinate locations $\mathbf{P}_1 = (8, 20)$, $\mathbf{P}_2 = (10, 10)$, $\mathbf{P}_3 = (16, 30)$, $\mathbf{P}_4 = (30, 10)$, and $\mathbf{P}_5 = (40, 20)$. Two new machines are to be located in the shop. Item movement is rectilinear. It is anticipated that there will be four trips per day between the new machines. The number of trips per day between each new machine and each existing machine is

$$\mathbf{W} = \begin{pmatrix} 8 & 6 & 5 & 4 & 3 \\ 2 & 3 & 4 & 6 & 7 \end{pmatrix}$$

Determine the optimum locations for the new machines.

6.5. In Problem 6.4, determine the optimum locations when cost is proportional to the square of the Euclidean distance between new machines and between new and existing machines.

6.6. Three existing facilities are located at $\mathbf{P}_1 = (3, 4)$, $\mathbf{P}_2 = (8, 7)$, and $\mathbf{P}_3 = (15, 2)$. Two new facilities are to be located with respect to the existing machines. The cost data for the location problem are

$$v_{12} = 3$$

$$\mathbf{W} = \begin{pmatrix} 2 & 6 & 0 \\ 4 & 5 & 1 \end{pmatrix}$$

Determine the optimum locations for the new facilities, assuming cost is proportional to (a) rectilinear distance, (b) squared Euclidean distance, and (c) Euclidean distance.

6.7. Two new facilities are to be located with respect to three existing facilities located at $\mathbf{P}_1 = (8, 15)$, $\mathbf{P}_2 = (10, 20)$, and $\mathbf{P}_3 = (30, 10)$. The cost data for the location problem are

$$v_{12} = 8$$

$$\mathbf{W} = \begin{pmatrix} 6 & 3 & 5 \\ 0 & 7 & 2 \end{pmatrix}$$

Determine the optimum locations for the new facilities, assuming cost is proportional to (a) rectilinear distance, (b) squared Euclidean distance, and (c) Euclidean distance.

6.8. The Columbia Broadcasting Company (CBC) has decided to transmit its radio and television programs from its central studios by laser beams. The beams are unidirectional, and transmission power varies as the square of the distance transmitted. CBC has four central studios located at Los Angeles, Chicago, Houston, and New York. However, owing to the curvature of the earth (a laser beam travels only in straight lines), it is not possible to construct broadcasting towers high enough for Los Angeles to be "visible" from New York. Thus CBC has decided to install two transmitting stations. The first will be in a position to communicate with Los Angeles, Chicago, Houston, and the second station. The second station communicates with all others except Los Angeles. The transmission loads from each station are unequal due to their location characteristics. All space programs originate at Houston; financial and cultural news comes from Los Angeles and New York. Crime news originates from Chicago. The coordinate locations of the existing studios are $\mathbf{P}_1 = (1, 2)$, $\mathbf{P}_2 = (2, 0)$, $\mathbf{P}_3 = (3, 3)$, and $\mathbf{P}_4 = (4, 2)$. Other data are

$$v_{12} = 8$$

$$\mathbf{W} = \begin{pmatrix} 4 & 2 & 2 & 0 \\ 0 & 2 & 2 & 4 \end{pmatrix}$$

Determine the optimum location for the transmitting stations.

6.9. For Example 6.2, compute optimal x-coordinates that give an alternative to $x_1^* = x_2^* = 20$.

6.10. For Example 6.2, compute optimal y-coordinates.

6.11. Construct and solve an example multifacility rectilinear distance problem using linear programming directly.

6.12. Construct an example multifacility rectilinear distance location problem having additional linear constraints on the x's and y's which you believe is a realistic one.

6.13. Given numbers a, b, p, and q such that

$$a - b = p - q \tag{i}$$

$$p \geq 0, \qquad q \geq 0 \tag{ii}$$

$$pq = 0 \tag{iii}$$

show that $|a - b| = p + q$.

[*Hint*: Consider three cases; $a - b > 0$, $a - b = 0$, and $a - b < 0$. As a start, here is the proof for the first case. From (ii), $p \geq 0$ and $q \geq 0$; either $p > 0$ or $q > 0$, for otherwise (i) implies that $a - b = 0$, which cannot be. If $q > 0$, (iii) implies that $p = 0$, so (i) implies that $a - b = -q < 0$, which cannot be; thus $q = 0$. Since $a - b > 0$, $|a - b| = a - b$, and $a - b = p$ by (i); since $q = 0$, $|a - b| = a - b = p + q$.]

6.14. As well as attaching additional linear constraints to the x's and y's of the linear programming problem equivalent to the multifacility rectilinear distance problem, suppose that linear constraints involving other variables are attached as well. In addition to not making any physical sense in most cases, doing so will usually destroy the equivalence of the resultant linear program to the actual location problem you wish to solve. Explain why.

6.15. Consider the problem of plotting contour lines of the function $f_1(x_1, x_2) = v_{12}|x_1 - x_2| + \sum_{j=1}^{p} C_j|x_1 - c_j| + \sum_{i=1}^{q} D_i|x_2 - d_i|$, where x_1 and x_2 are identified with the x and y axes, respectively; $c_1 < \cdots < c_p$, $d_1 < \cdots < d_q$, and $v_{12} > 0$. This problem is identical with that of plotting contour lines of the one-facility rectilinear distance problem with the exception of the inclusion of the term $v_{12}|x_1 - x_2|$. Show that for any contour line passing through that part of a region (s, t) lying on or above the line $x_2 = x_1$ the slope is given by $-(M_t - v_{12})/(N_s + v_{12})$, and that for any contour line passing through that part of region (s, t) lying on or below the line $x_2 = x_1$ the slope is given by $-(M_t + v_{12})/(N_s - v_{12})$.

6.16. Use the procedure of Problem 6.15 to construct contour lines for the functions f_1 and f_2 defined by (a) Example 6.2; (b) the Coastal Construction Co. problem.

6.17. Actions are to be taken at points $x_1, \ldots, x_n$ in time, which must be determined. These actions will be affected by events occurring at known points in time $a_1, \ldots, a_m$, and by whether or not one action precedes another. More exactly, a cost per time unit w''_{ji} is incurred if $x_j \leq a_i$, a cost per time unit w'_{ji} is incurred if $x_j > a_i$, a cost per time unit v''_{jk} is incurred if $x_j \leq x_k$, and a cost per time unit v'_{jk} is incurred if $x_j > x_k$. Construct a linear programming problem that, if solved, would determine the points in time at which the n actions should be taken so as to minimize the total cost incurred.

6.18. Let $x_1, \ldots, x_n$ be the x coordinates of the new facilities. Prove that $x_1, \ldots, x_n$ minimizes f_1 if and only if there exists a feasible solution to D such that the complementary slackness conditions hold.

6.19. Three new facilities are to be located relative to six existing facilities. The cost of item movement is proportional to the square of the straight-line distance between facilities. The weighting factors are

$v_{12} = 6$,	$v_{13} = 1$,	$v_{23} = 4$,	$w_{11} = 4$,	$w_{12} = 0$,	$w_{13} = 2$
$w_{14} = 0$,	$w_{15} = 6$,	$w_{16} = 5$,	$w_{21} = 3$,	$w_{22} = 6$,	$w_{23} = 0$
$w_{24} = 0$,	$w_{25} = 8$,	$w_{26} = 1$,	$w_{31} = 0$,	$w_{32} = 0$,	$w_{33} = 8$
$w_{34} = 10$,	$w_{35} = 2$,	$w_{36} = 6$,			

The coordinate locations of the existing facilities are $\mathbf{P}_1 = (0, 12)$, $\mathbf{P}_2 = (2, 1)$, $\mathbf{P}_3 = (10, 2)$, $\mathbf{P}_4 = (6, 12)$, $\mathbf{P}_5 = (20, 10)$, and $\mathbf{P}_6 = (5, 20)$. Determine the optimum locations for the new facilities.

6.20. The following data are given for a multifacility location problem with rectilinear distances.

$$\mathbf{W} = \begin{pmatrix} 3 & 4 & 2 \\ 1 & 2 & 3 \\ 2 & 1 & 1 \end{pmatrix}, \qquad \mathbf{V} = \begin{pmatrix} 0 & 8 & 5 \\ 8 & 0 & 1 \\ 5 & 1 & 0 \end{pmatrix}$$

$$\mathbf{a} = (2 \quad 4 \quad 6)$$

$$\mathbf{b} = (0 \quad 0 \quad 0)$$

(a) State optimal values of the y-coordinates by inspection.
(b) Find optimal values of the x-coordinates using the network approach.

6.21. For Example 6.3 write out the linear programming problem for finding the x-coordinates, and then use linear programming duality theory with a change of variables to derive the network dual problem.

6.22. The following data are given for a multifacility rectilinear distance location problem:

$$v_{12} = 4$$

$$\mathbf{W} = \begin{pmatrix} 4 & 0 & 5 \\ 2 & 1 & 0 \end{pmatrix} = (w_{ji})$$

$$\mathbf{P}_1 = (10, 15), \qquad \mathbf{P}_2 = (20, 25), \qquad \mathbf{P}_3 = (40, 5)$$

(a) Find optimum x and y coordinates of new facilities 1 and 2 by using linear programming directly.
(b) Find optimum x and y coordinates of new facilities 1 and 2 by plotting contour lines of f_1 and f_2.

6.23. The following data are given for a multifacility rectilinear distance location problem: $m = 3$, $n = 5$, $\mathbf{P}_1 = (1, 0)$, $\mathbf{P}_2 = (2, 0)$, $\mathbf{P}_3 = (3, 0)$, $v_{12} = 2$, $v_{13} = 2$, $v_{14} = 2$, $v_{15} = 2$, $v_{23} = 20$, $v_{24} = 1$, $v_{25} = 0$, $v_{34} = 0$, $v_{35} = 0$, $v_{45} = 40$, and

$$\mathbf{W}' = (w_{ij}) = \begin{pmatrix} 10 & 4 & 4 & 4 & 4 \\ 0 & 1 & 1 & 5 & 5 \\ 0 & 4 & 4 & 4 & 4 \end{pmatrix}$$

(a) Show that $(x_1, y_1) = (1, 0)$ and $(x_j, y_j) = (2, 0)$ for $j = 2, \ldots, 5$ is a median location.
(b) Show that the solution in part (a) is not an optimum solution to the rectilinear problem.

6.24. For the data in Problem 6.23, assume Euclidean distance and determine the optimum locations for the five new facilities.

6.25. Given the data of Problem 6.22, construct and solve the analogous squared distance problem.

6.26. A minimax version of the multifacility Euclidean distance location problem may be formulated as follows. Locations of existing facilities at points $(a_1, b_1), \ldots, (a_m, b_m)$ are known; locations of new facilities at points $(x_1, y_1), \ldots, (x_n, y_n)$ are to be determined. Nonnegative constants v_{jk} for $1 \le j < k \le n$, and w_{ji}, for $i = 1, \ldots, m$ and $j = 1, \ldots, n$, are given. The function $g((x_1, y_1), \ldots, (x_n, y_n))$ is defined by

$$g(x_1, y_1), \ldots, (x_n, y_n)) = \max_{1 \le j < k \le n} v_{jk}[(x_j - x_k)^2 + (y_j - y_k)^2]^{1/2}$$

The function $h((x_1,y_1),\ldots,(x_n,y_n))$ is defined by

$$h((x_1,y_1),\ldots,(x_n,y_n)) = \max_{\substack{1\le i\le m\\ 1\le j\le n}} w_{ji}[(x_j-a_i)^2+(y_j-b_i)^2]^{1/2}$$

and the function $f((x_1,y_1),\ldots,(x_n,y_n))$ is defined by

$$f((x_1,y_1),\ldots,(x_n,y_n)) = \max[g((x_1,y_1),\ldots,(x_n,y_n)), h((x_1,y_1),\ldots,(x_n,y_n))]$$

The minimax location problem is to find locations of the new facilities that minimize the function $f(x_1,y_1),\ldots,(x_n,y_n))$. Identify some possible applications for which you believe the minimax location model would be more appropriate than the multifacility location models considered in Chapter 6.

6.27. The general HAP formulation can be given as

$$\hat{f}(\mathbf{X}_1,\ldots,\mathbf{X}_n) = \sum_{1\le j<k\le n} v_{jk}\hat{d}(\mathbf{X}_j,\mathbf{X}_k) + \sum_{j=1}^{n}\sum_{i=1}^{m} w_{ji}\hat{d}(\mathbf{X}_j,\mathbf{P}_i)$$

where

$$\hat{d}(\mathbf{X}_j,\mathbf{X}_k) = \begin{cases}[(x_j-x_k)^2+(y_j-y_k)^2+\epsilon]^{1/2}, & \text{if Euclidean}\\ [(x_j-x_k)^2+\epsilon]^{1/2}+[(y_j-y_k)^2+\epsilon]^{1/2}, & \text{if rectilinear}\end{cases}$$

$$\hat{d}(\mathbf{X}_j,\mathbf{P}_i) = \begin{cases}[(x_j-a_i)^2+(y_j-b_i)^2+\epsilon]^{1/2}, & \text{if Euclidean}\\ [(x_j-a_i)^2+\epsilon]^{1/2}+[(y_j-b_i)^2+\epsilon]^{1/2}, & \text{if rectilinear}\end{cases}$$

Obtain the iterative expressions that would hold regardless of the distance measure used.

6.28. Consider a multifacility rectilinear location problem for which $n = 2$ and $m = 4$, with

$$v_{12} = 6$$

$$\mathbf{W} = \begin{pmatrix}5 & 3 & 0 & 0\\ 0 & 1 & 8 & 4\end{pmatrix}$$

$$\mathbf{P}_1 = (0,2),\quad \mathbf{P}_2 = (4,0),\quad \mathbf{P}_3 = (6,8),\quad \mathbf{P}_4 = (10,4)$$

Is the optimum solution to this problem given by $(x_1,y_1) = (4,2)$ and $(x_2,y_2) = (6,4)$? Justify your answer.

6.29. Three existing facilities are located at $\mathbf{P}_1 = (3,4)$, $\mathbf{P}_2 = (8,7)$, and $\mathbf{P}_3 = (15,2)$. Two new facilities are to be located with respect to the existing machines. The cost data for the location problem are

$$v_{12} = 3$$

$$\mathbf{W} = \begin{pmatrix}2 & 6 & 0\\ 4 & 5 & 1\end{pmatrix}$$

Determine the optimum locations for the new facilities, assuming cost is proportional to (a) rectilinear distance, and (b) squared Euclidean distance.

6.30. Three new facilities are to be located relative to five existing facilities. The data for the problem are

$$v_{12} = 0, \qquad v_{13} = 2, \qquad v_{23} = 1$$

$$\mathbf{W} = \begin{pmatrix} 6 & 1 & 2 & 0 & 0 \\ 0 & 0 & 1 & 3 & 4 \\ 0 & 5 & 2 & 0 & 2 \end{pmatrix}$$

$$\mathbf{P}_1 = (0,0), \quad \mathbf{P}_2 = (2,8), \quad \mathbf{P}_3 = (5,4), \quad \mathbf{P}_4 = (7,6), \quad \mathbf{P}_5 = (8,2)$$

(a) Determine the optimum locations for the new facilities, assuming (1) rectilinear travel, and (2) straight-line travel.

(b) Determine the optimum locations for the new facilities, assuming cost is proportional to the square of the straight-line distance.

(c) Using (6.17), compute the upper and lower bounds on the minimum-cost solution to the Euclidean location problem.

6.31. Three existing facilities are located at the points $\mathbf{P}_1 = (0,0)$, $\mathbf{P}_2 = (20,15)$, and $\mathbf{P}_3 = (40,15)$. Each new facility costs \$1200 per month. The cost of item movement is \$5 per unit of distance. The number of trips per month between some new facility and each existing facility equals 10. Determine the optimum number, location, and allocation of new facilities, assuming at least one new facility is required and item movement occurs between an existing facility and only one new facility. Base your calculations on (a) rectilinear distance and (b) Euclidean distance.

6.32. Four plants are located at the points (0, 0), (10, 10), (0, 50), and (20, 40). The shipment of goods from the plants to a distribution center per month, measured in truckloads, is 100, 60, 120, and 50 loads, respectively. The cost of transporting a truckload of product from the plant to a distribution center is \$20 per unit of distance. The monthly cost of owning and operating a distribution center is \$40,000. Determine the optimum number, location, and allocation of distribution centers, assuming at least one distribution center is required and a plant ships to only one distribution center (use a Euclidean distance approximation).

6.33. Five special-purpose machines are located in a plant at the points (0, 0), (0, 10), (20, 0), (40, 10), and (10, 10). The machines require maintenance at expected frequencies of 10, 6, 12, 10, and 5 times per month, respectively. Due to the nature of the maintenance, all maintenance must be performed at the maintenance center. The cost of transporting the machines to the maintenance center is \$5 per unit of distance. The monthly cost of owning and operating a maintenance center is \$5,000. (a) Determine the optimum number, location, and allocation of maintenance centers. Assume a machine is serviced by only one maintenance center and travel is rectilinear. (b) What values of travel cost result in one maintenance center being the optimum number?

6.34. Solve the location-allocation problem of Example 6.7 when n is changed from 2 to (a) 3; (b) 4; (c) 5; (d) 6.

6.35. A location-allocation problem with rectilinear distances like Example 6.7 is given with existing facilities having coordinates of (0, 0), (10, 0), (10, 10), and (0, 10). Solve the problem for values of n of (a) 2; (b) 3; (c) 4; (d) 5.

6.36. A location-allocation problem with rectilinear distances like Example 6.7 is given with existing facilities having coordinates of (0, 3), (1, 0), (3, 6), (6, 2), (9, −1), (9, 8), and (11, 5). Solve the problem for values of n of (a) 2; (b) 3; (c) 4.

6.37. A location-allocation problem with rectilinear distances like Example 6.7 is given with existing facilities having coordinates of (0, 0), (3, 0), (5, 5), (5, −5), (7, 0), and (10, 0). Solve the problem for values of n of (a) 2; (b) 3; (c) 4.

6.38. Using the approach illustrated in Example 6.7, write a computer code to solve location-allocation problems with rectilinear distances. Use the code to solve Problems 6.34 to 6.37. Report the best solutions you find; try several starting points in each case.

6.39. Using the approach illustrated in Example 6.7, write a computer code to solve location-allocation problems with Euclidean distances. Use the code to solve Problems 6.34 to 6.37 but assuming Euclidean distances. Report the best solutions you find; try several starting points in each case.

Chapter 7

Network Location Problems

7.1 INTRODUCTION

Most of the location problems we have considered in previous chapters involve locating one or more new facilities so as to minimize some function of transport distance, time, or cost. The transport of interest has almost always occurred on transport *networks*, such as aisle networks or road networks. The distances we have been computing often are approximations to network distances, with one example being the use of rectilinear distances when working with rectilinear aisle or road networks. Further, it is usually realistic to require the facilities to be located on or adjacent to the network of interest. Hence it seems reasonable at this point to consider the following question: Instead of somehow approximating the transport network of interest, *why not work directly with the network itself*? To give a positive response to this question, we shall consider a number of network location problems of interest, many of which are directly analogous to planar location problems we have studied earlier, where the appropriate "network distances" replace the planar distances we have used previously. In addition to aisle and road networks yielding location problems, other networks resulting in location problems include river networks, air transport networks made up of flight corridors, and ocean transport networks made up of shipping lanes.

While network location models can be more realistic than analogous planar models, the price we pay for this additional realism is the effort needed to construct the network. It is only necessary to imagine the effort needed to construct, say, a network model of the interstate network of the United States, to realize that constructing such a model can be quite time consuming. Thus it is a good idea to balance the benefits of having a network model with the cost needed to construct it, prior to making a firm decision to use a network model.

The network location problems we study in this chapter are principally those called n-median and n-center problems. The n-median problem is to locate n new facilities, called *medians*, on a network with respect to m existing facilities with vertex locations, so as to minimize a sum of weighted distances between each existing facility and a *closest* median. An example of an n-median problem would be to locate mailboxes with respect to neighborhoods so as to minimize the total travel distance between mailboxes and neighborhoods when neighborhood residents use a closest mailbox. The 1-median problem is, by the way, entirely analogous to the single-facility minisum location problems we considered in Chapter 4 and, indeed, includes the location of a new facility on the line (for which we used the **median** conditions), as a special case.

The n-center problem is to locate n new facilities, called *centers*, on a network with respect to m existing facilities with vertex locations, so as to minimize a maximum of weighted distances between each existing facility and a closest center. An example of an n-center problem would be to locate n fire stations with respect to m neighborhoods, when each neighborhood is served by a closest fire station, so as to minimize the *maximum* time to respond to a fire. The one-center problem is a direct network analog of the single-facility minimax problems we studied in Chapter 4.

Initially in this chapter we concentrate on a special kind of network called a *tree*. (Figure 7.1 shows an example of a tree.) Such a network has no cycles and has a unique shortest path between any two points in the network. Location problems on tree networks are much more tractable than are location problems on general networks, and as we shall see, 1-median and 1-center problems on such networks can be easily solved with pencil and paper and the occasional use of a small pocket calculator. In addition to median and center problems, we consider a location problem called a *covering* problem, which involves locating the minimum number of new facilities so that the distance between each of m existing facilities and a closest new facility does not exceed a specified upper limit. Covering problems may occur when some corporation or government agency must locate a number of essentially identical new facilities and wishes to minimize the number of new facilities it locates, while meeting "coverage constraints" that require at least one new facility to be within a specified travel time or distance of each existing facility to receive service. The covering problem is also of interest because we can, and will, solve the n-center problem by solving a sequence of covering problems.

We now give an overview of the chapter. In Section 7.2 we introduce tree networks, giving necessary definitions and some basic properties that we use in following sections. In Section 7.3 we study the 1-median problem for a tree network, giving very efficient algorithms to solve the problem. Next, in Section 7.4, we consider the 1-center problem for a tree network and give a "dual algorithm" for solving the problem. In Section 7.5 we develop an algorithm for solving covering problems when the network is a tree. Such problems are of interest in themselves, and also help to solve the n-center problem, which we consider in Section 7.6. In Section 7.7 we develop basic ideas we need to solve center and median problems when the network is not a tree but has "cycles" and is referred to as a cyclic network. In Section 7.8 we employ the ideas developed in Section 7.7 in studying median and center problems for cyclic networks. We shall give algorithms for solving the 1-median and 1-center problems, and establish, for the n-median and n-center problems, that only a finite number of possible locations need be considered in order to find optimal locations: the finiteness results will allow us later, in Chapter 8, to construct integer programming models for solving the n-center and n-median problems on general networks.

7.2 TREE NETWORKS

In this section we introduce tree networks, together with a few definitions we need in order to develop the notion of distances on a tree network. Figure 7.1 illustrates a tree network T. You can see from the figure that the network has the property that it has no "cycles," a distinguishing property of a tree network. Similarly, there is a unique shortest "path" between any two points in T. As you might guess, the name "tree" is used because a tree network can be drawn to resemble a real tree. If you have ever seen a "family tree' chart, you have seen an example of a tree network, with the vertices representing the family members and the arcs indicating the parent–child relationship. Family trees obviously have no cycles, for a cycle would mean that each person represented by a vertex in the cycle is both his or her own ancestor and own descendant as well!

There are a number of reasons for considering location problems on tree networks. Large portions of very expensive road networks, such as interstate highways, are often trees. If you consider the interstate highway system of a specific state in the United States you will find that in about half of the states the system is a tree. Usually, a state where the system is not a tree is one of the more populous states, such as California or New York. At the other extreme, road networks in sparsely settled areas may well be trees, since tree road networks are usually the cheapest to construct. Some tree networks, which resemble the spokes of a wheel, occur naturally in distribution problems, with the source of the distribution typically being located at the "hub" of the wheel: a classic example of such a network is the railroad network of Hungary, which has Budapest as its hub. Another quite similar example is the interstate system of the state of Georgia, which has Atlanta as the hub. Two large American airlines, with headquarters in Atlanta and Chicago, have route systems often referred to as "spoke and hub" systems. Other networks, such as river networks, are quite naturally trees. Some new industrial plants resemble a tree, in

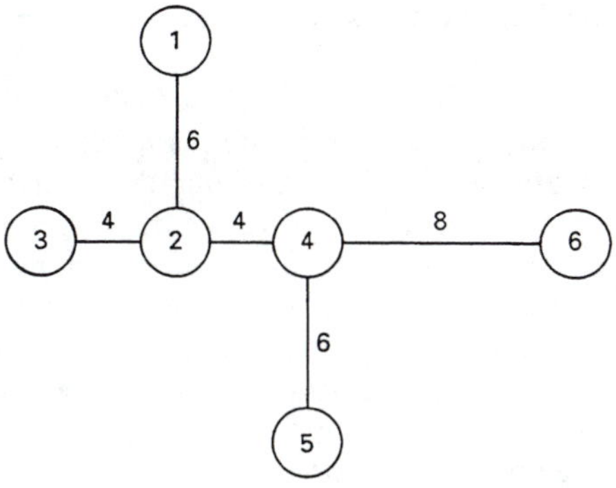

Figure 7.1 Example of a tree network.

the sense that there is a main "trunk" together with "branches"; such a structure is quite flexible, as branches can easily be added if the plant must be expanded in size. Many airport terminals also have tree structures.

Even if a network is not a tree, we can consider *approximating* it by a tree. One such approximation is called a *spanning tree*, which is a tree subnetwork that includes every "vertex" of the original network, but omits some of the "arcs." Such approximations will allow us to use algorithms for solving various location problems on tree networks to get "quick and dirty" solutions to the problems on more general networks.

Because it is easier to get insight into tree network problems, it is logical to consider tree networks prior to considering more general networks with cycles. Also, one particularly simple type of tree is a *path*, which for our purposes is essentially equivalent to a line segment, so that location problems on the line, such as those considered in Chapter 4, may be considered special cases of location problems on a tree. In fact, one nice feature of trees is that most of the insight we can gain from studying location problems on the line carry over to location problems on a tree. Finally, we remark that some problems we shall consider later for cyclic networks are in fact known to be provably difficult, in the sense that if they could always be well solved, so could an enormous class of other problems which has proven, to date, to be computationally intractable. Some day we may hope to find that cyclic network location problems can always be well solved, but that day has not yet arrived. This is not to say that *specific* cyclic network location problems may not be well solved. How well solved the problem is often depends on its data and its structure. Some problems are in fact easily solvable "most" of the time; it is just the case that certain types of such problems, which may well not occur in reality, are not well solved.

With the above by way of background, at this point we will develop some definitions we need in order to be more precise. Having such definitions will also allow us to utilize a number of results in network theory and graph theory. One word of warning: *network definitions are not standardized*! Although we shall be consistent in our use of terminology, you cannot expect authors of other books and papers to use the same terminology we do. This lack of standardization is an unfortunate fact of life.

Figure 7.1 illustrates a network called a tree, with six vertices and five arcs. For our purposes each vertex will usually be a point in the plane, and each arc will be a line segment. We let $[v_i, v_j]$ denote the arc connecting v_i and v_j, with its length being that of the line segment; the length of each arc is always a positive number because the vertices are all distinct. An arc length usually has units of distance, (e.g., miles or kilometers) but can just as well have units of travel time, energy expenditure, travel cost, or whatever is appropriate. (All arcs must have the same units, however, or else we shall end up trying to "add apples and oranges.") When we say that a point is *in* or *on* a tree T, we shall mean that it lies on one of the arcs; the point may or may not be a vertex. When a tree network represents a road network, the vertices usually represent road intersections, and the arcs represent road segments between adjacent intersections.

We can use Figure 7.1 to introduce the notion of a path. A *path* is an ordered sequence of arcs, with adjacent arcs in the sequence sharing the same ending and beginning vertices; thus $P = \{[v_1, v_2], [v_2, v_4], [v_4, v_5]\}$ is a path joining v_1 and v_5, and$P' = \{[v_3, v_2], [v_2, v_4], [v_4, v_2], [v_2, v_3]\}$ is a path joining v_3 and v_3. The path P is an example of a *simple path* because no vertex in the path repeats. The path P' is a nonsimple path and is an example of a *cycle*, a path for which the initial and terminal vertices are the same. As long as we deal with trees, we shall only be interested in simple paths, so any path we discuss subsequently will be a simple path unless we specify otherwise. We note that a distinguishing feature of any tree is that there is a unique (simple) path between any two vertices. The *length* of a path is the sum of the lengths of the arcs in the path; thus the length of the path P is $6 + 4 + 6 = 16$.

With the foregoing basic ideas, we can define a *tree* $T = (V, A)$ to be a collection of m vertices $V = \{v_1, \ldots, v_m\}$ and a collection A of exactly $m - 1$ arcs, with the property that there are no cycles. The *distance* between any two vertices v_i and v_j is denoted by $d(v_i, v_j) = d(v_j, v_i)$ and is the length of the path joining the two vertices; for example, $d(v_1, v_5) = 16$ for the tree of Figure 7.1. We also need the notion of the distance $d(x, y)$ between any two points x and y in a tree, when x and y may or may not be vertices. The latter distance is defined in exactly the same way as the distance between two vertices, simply by *imagining* x and y to be vertices, which partition the arcs on which they lie into "subarcs," each of which is a line segment with its length defined to be the length of the line segment. (For example, if x and y are midpoints of the arcs $[v_4, v_5]$ and $[v_4, v_6]$, respectively, for the tree of Figure 7.1, then $d(x, y) = 3 + 4 = 7$.) Naturally, in the case when $x = y$, we have $d(x, y) = 0$.

We now have distances adequately defined. For any x and y in a tree, we note that the following important distance properties are true.

1. *Symmetry*: $d(x, y) = d(y, x)$
2. *Positivity*: $d(x, y) \geq 0$, with $d(x, y) = 0$ meaning $x = y$
3. *Triangle inequality*: $d(x, y) \leq d(x, z) + d(z, y)$ for any points x, y, and z.

A principal reason that tree network location problems are tractable is because of convexity. If v is any vertex, it is a fact that the function in x defined by $d(x, v)$ is *convex* on the entire network, in the sense that if we plot the function as it varies along any path, and then join any two points on the graph of the function by a straight line, the straight line always lies on or above the graph of the function. Similarly, when the network is a tree, the function in x *and* y defined by $d(x, y)$ is convex. In fact, we can state the following

Convexity Property

(a) The function in x defined by $d(x, v)$ is convex on the network for every choice of a vertex v if and only if the network is a tree.

(b) The function in x and y defined by $d(x, y)$ is convex on the entire network if and only if the network is a tree.

To gain some insight into the convexity property, consider Figure 7.1. If we graph $d(x, v_4)$ along the path $P(v_3, v_6)$, we obtain a V-shaped graph that decreases linearly from 8 to 0 as x varies from v_3 to v_4, and then increases linearly from 0 to 8 as x varies from v_4 to v_6. Similarly, if we graph $d(x, v_4)$ along the path $P(v_3, v_1)$, we obtain a V-shaped graph again, with the graph decreasing linearly from 8 to 4 as x varies from v_3 to v_2, and then increasing linearly from 4 to 10 as x varies from v_2 to v_1. We shall see later that if a network has cycles, $d(x, v)$ can actually be *concave*, increasing linearly and then decreasing linearly along a path; for an example of such nonconvex behavior, refer to the graph of $d(x, v_5)$ in Figure 7.8 (page 428). Hence we conclude that if we are to have convex distances when working with a network, the network must be a tree.

Why is convexity useful? In the following sections we shall consider the problem of finding an *absolute 1-median*, a point x^* that minimizes

$$f(x) = w_1 d(x, v_1) + \cdots + w_m d(x, v_m)$$

and of finding an *absolute 1-center*, a point y^* that minimizes

$$g(y) = \max\{w_1 d(y, v_1), \ldots, w_m d(y, v_m)\}$$

These two problems are entirely analogous to the single-facility minisum and minimax location problems we discussed in Chapter 4, where, as usual, the w_i are nonnegative weights. It is known that a sum of convex functions is a convex function, and a maximum of convex functions is a convex function. Hence, when the network is a tree, the 1-median function $f(x)$ is a convex function, as is the 1-center function $g(y)$.

A property of convex functions of fundamental value in solving minimization problems is that any relative minimum of a convex function is a global minimum. Hence if we are solving a convex location minimization problem and can find a location, say x^*, which is the minimum value of the function being minimized in a small neighborhood of the point, it will be a minimizing point over the entire tree.

As we shall see subsequently, this will allow us to conclude that a minimum of the 1-median function over some single arc incident to a "tip" of some subtree of the original tree is a global minimum of the function.

Just as we have convex functions of distances for trees, we also have convex subsets of trees. A subset S of a tree is said to be *convex* if for any two points x and y in S every point on the unique shortest path joining x and y, denoted by $P(x,y)$, is in S; that is, $P(x,y)$ is contained in S. Thus a tree is itself a convex set, as is the subtree obtained by deleting any tip vertex v_t of a tree together with the interior points of the arc connecting the tip vertex to the tree. Another example of a convex set is any *neighborhood with center u and radius r*, denoted by $N(u,r)$; the collection of all points in a tree whose distance from u is at most r.

7.3 1-MEDIAN PROBLEM

The 1-median problem is the problem of locating a new facility at some point x^* in a network so as to minimize a sum of weighted distances between the new facility and a subset of the vertices: such a point x^* is called an *absolute median*. We may imagine the new facility at x (e.g., a mailbox) provides services to existing facilities placed at the vertices (so defined as to include homes and businesses), and the new facility is to be placed so as to minimize the total travel distance (for the time period of interest) between the new and existing facilities. Thus if w_i denotes the nonnegative number of round trips per time period between x and vertex v_i, then $w_i d(x, v_i)$ is the total travel distance per time period between x and v_i. Hence, letting $f(x)$ denote the total travel distance between x and all m vertices, we have

$$f(x) = w_1 d(x, v_1) + \cdots + w_m d(x, v_m)$$

There are other interpretations of the weights, of course, such as total transport cost per unit distance, or total travel time per unit distance. The weights can be whatever nonnegative numbers are appropriate; we require only that all weights have the same units.

When the tree is a path, we note that the 1-median problem is equivalent to the location problem on the line of Chapter 4, which is solved by using the median conditions: hence the choice of the term *median*.

We shall examine two related algorithms for solving the 1-median problem. As we shall see, each algorithm is quite simple and also quite efficient, being of linear order in m, the number of vertices; that is, the effort to use the algorithm is linearly proportional to m. Each algorithm considers only vertices as locations: we shall see that at least one vertex is an absolute 1-median due to a vertex optimality property. Hence subsequently we shall not distinguish between absolute 1-medians and 1-medians, and shall use only the term 1-*median*.

We call the first algorithm we consider the *Chinese algorithm*, since it was published in a Chinese journal, its authors being identified as "Hua Lo-Keng and others"; the names of the others are not given. Apparently, the problem that motivated the development of the Chinese algorithm was the location of a threshing floor, used to separate the wheat from the chaff after a wheat harvest. In this case the tree network represents a road network, with the vertices being the locations of wheat fields. The weight for each vertex represents the amount of wheat to be transported to and from the field and the threshing floor. Locating the threshing floor at the 1-median causes the total cost of transporting the wheat to and from the threshing floor to be minimized. The authors state their algorithm in the form of a poem! Prior to giving their poem, we must tell you that the authors use the word "loop" instead of "cycle" and use the word "end" instead of "tip." Their poem is as follows.

When the network has no loops,
Take all the ends into consideration:
The smallest advances one station.

We shall illustrate the algorithm using the network of Figure 7.2, with the weights for vertices 1 through 6 being 1, 4, 2, 2, 2, and 1, respectively. A tip with a smallest weight is v_1, so we add its weight to the weight of v_2, making 5 the new

$$\mathbf{B} = (b_{ij}) = \begin{pmatrix} 0 & 4\frac{4}{5} & 6\frac{2}{3} & 6\frac{2}{3} & 10\frac{2}{3} & 9 \\ 4\frac{4}{5} & 0 & 5\frac{1}{3} & 5\frac{1}{3} & 13\frac{1}{3} & 9\frac{3}{5} \\ 6\frac{2}{3} & 5\frac{1}{3} & 0 & 8 & 14 & 10\frac{2}{3} \\ 6\frac{2}{3} & 5\frac{1}{3} & 8 & 0 & 6 & 5\frac{1}{3} \\ 10\frac{2}{3} & 13\frac{1}{3} & 14 & 6 & 0 & 9\frac{1}{3} \\ 9 & 9\frac{3}{5} & 10\frac{2}{3} & 5\frac{1}{3} & 9\frac{1}{3} & 0 \end{pmatrix}$$

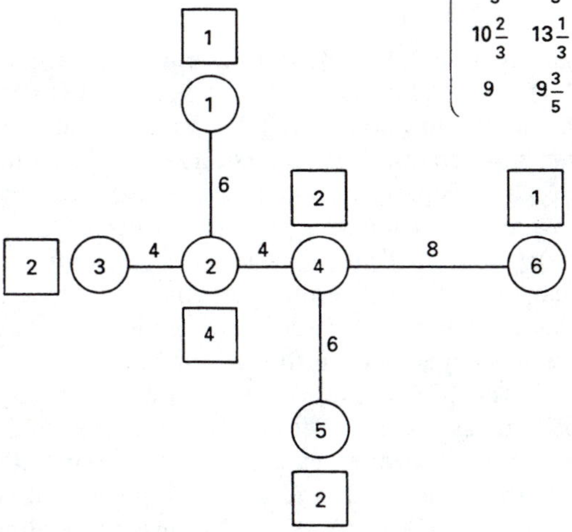

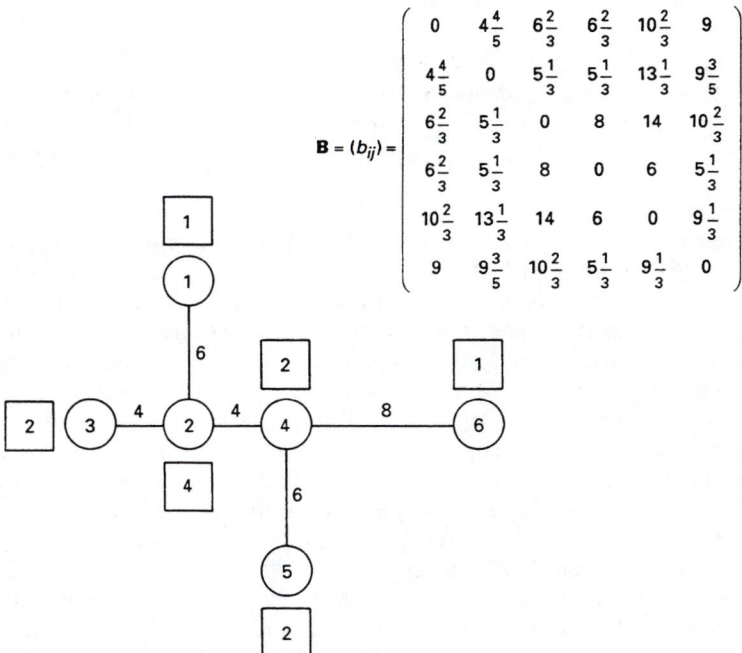

Figure 7.2 Example of a tree network with vertex weights inside squares, and bound matrix $B = (b_{ij})$.

weight for v_2; also, we now delete v_1 from the tree, as well as the arc joining it to v_2. At this point v_1 has "advanced one station." For the revised tree v_6 is the tip with the smallest weight, so the new weight for v_4 becomes $3 = 2 + 1$, and we remove v_6 and its incident arc from the tree. The new tree has two tip vertices of the same weight, v_3 and v_5, so we choose v_3 to advance one station, deleting it and its incident arc from the network and changing the new weight of v_2 to $7 = 5 + 2$. Next v_5 advances one station and we remove it and its incident arc from the tree, making $5 = 3 + 2$ the new weight for v_4. At this point the tree consists of the single arc joining vertices v_2 and v_4, which have weights of 7 and 5, respectively. Vertex 4 now advances one station, the new weight of v_2 is 12, and the remaining tree consists of v_2 alone. The conclusion at this point is that v_2 is a 1-median.

The example illustrates another way of thinking about the algorithm. We can imagine that the tree network is represented by a corresponding string network, with the vertices corresponding to knots of the string network, and the weights of the vertices corresponding to the sizes of the knots. With the algorithm, what we do is "roll up" the string network one (smallest) knot and string at a time, adjusting the size of the knot to which the "rolled-up" knot was adjacent. When we roll up the string network into a single knot, we stop with a 1-median.

We remark the algorithm can be easily augmented to compute the minimum value of $f(x)$, say $f(x^*)$, as follows:

0. Initialize a sum, say S, to zero.

1. Each time a tip is chosen to advance, add the product of the tip weight and the length of the arc the tip is on to S.

3. The final value of S is $f(x^*)$, and each intermediate value of S is a lower bound on $f(x^*)$.

Thus for the example of Figure 7.2, the successive values of S are $0, 0 + 1(6) = 6$, $6 + 1(8) = 14$, $14 + 2(4) = 22$, $22 + 2(6) = 34$, and $34 + 5(4) = 54$. Therefore, $f(x^*) = f(v_2) = 54$ is the minimum value of $f(x)$. We suggest that you compute $f(v_2)$ directly to verify its value is 54. To test your understanding of the Chinese algorithm, you should ask yourself why this modification we have given to compute the minimum value of $f(x)$ is correct. The Chinese algorithm illustrates an important point about the 1-median: *the choice of a 1-median does not depend on the lengths of the arcs*! The arc lengths can be *any* positive numbers and the 1-median will remain unchanged, provided that the tree structure does not change. Thus we can conclude that the 1-median, say x^*, is dependent on the weights and the tree structure, but *not* on the lengths of the arcs. [Of course, $f(x^*)$ depends on the arc lengths.]

The Chinese algorithm has one minor disadvantage: it can require unnecessary work. Referring to the example above, we can see when the remaining tree consists of vertices 2, 4, and 5, with weights 7, 3, and 2, respectively, that vertex 2 will be the 1-median, a fact the algorithm does not recognize, just proceeding until the tree consists of a single vertex. The next algorithm we consider, which we call the *majority algorithm*, is "smarter" and will terminate in the case that a tip vertex, vertex 2 in the example above, has a weight (7 in the example) that is at least half—*a majority*—of the sum of the weights ($6 = \frac{12}{2}$ in the example).

In order to state the majority algorithm, it is convenient to let W denote the sum of all the weights.

Majority Algorithm

1. Choose any tip v_t with weight w_t. If w_t is at least as large as $W/2$, take the tip to be a 1-median and stop.
2. Denote by v_a the vertex adjacent to v_t; add w_t to w_a to obtain the new weight for v_a. Delete all the points in the arc $[v_t, v_a]$ but v_a from the tree, and return to step 1.

If we apply the majority algorithm to the example we solved using the Chinese algorithm, choosing the tips in the same order, the two algorithms proceed in exactly the same manner until they reach the point where v_2 is a tip with a weight of 7. Since 7 is greater than $W/2 = 6$, the majority algorithm, since it can choose any tip, can choose v_2 as a tip and terminate at this point. The name we have chosen for the algorithm derives from the fact that the algorithm compares weights at tips to $W/2$, stopping when a weight is at least as large as $W/2$, that is, at least as large as a *majority* of the weights. Note that unlike the Chinese algorithm, with the majority algorithm, tips can be chosen in *any* order.

We can utilize the majority algorithm to develop a useful property which we call the

Majority Subtree Property. Given a tree T and an arc $[v_p, v_q]$ in the tree, suppose that we delete the points in the interior of the arc from T to obtain two subtrees corresponding to the vertices of the arc, say T_p and T_q. If the sum of the weights of the vertices in the subtree T_p is at least $W/2$, the subtree T_p will contain a 1-median.

We can illustrate the property above using the tree of Figure 7.2. If we remove the arc connecting v_2 and v_4, the subtree T_2 containing v_1, v_2, and v_3 has a total weight of 7, which is greater than $W/2 = 6$. The other subtree, T_4, containing v_4, v_5, and v_6, has a total weight of 5. We can use the algorithm to illustrate the property is true for this example. Suppose that we apply the algorithm to the original tree of Figure 7.2, applying it to v_5, v_6, and v_4 in succession. Because the sum of the weights in the subtree with these vertices is less than $W/2$, the algorithm will not terminate until after v_4, v_5, and v_6 are eliminated, implying then that a 1-median must be in the subtree T_2 with vertices 1, 2, and 3, which we know to be the case for this example. *The majority subtree property gives a useful qualitative insight: the subtree containing the majority of the weights should contain a 1-median.* Thus if the vertices were towns and the weights the populations of the towns, the property would state that a subtree containing the majority of the population should contain a 1-median.

We now give and motivate two properties that provide the basic justification for the majority algorithm, and for the Chinese algorithm as well.

Local Minimum Property

(a) If v_t is a tip of the tree of interest and we have $W - 2w_t \leq 0$, then v_t is a local minimum. (When equality holds, every point on the arc with v_t as its tip is a local minimum.)

(b) If $W - 2w_t > 0$, we can exclude from further consideration all points in the arc $[v_t, v_a]$ except the vertex v_a.

We illustrate this property using Figure 7.2, supposing that $a^* = [v_t, v_a] = [v_6, v_4]$, and letting x be in a^* with $s = d(v_4, x)$. For vertices v_i, $i = 1, \ldots, 5$ we note that $d(x, v_i) = s + d(v_4, v_i)$, while $d(x, v_6) = 8 - s = L - s$ (where $L = 8$). Thus we have

$$f(x) = w_1[s + d(v_4, v_1)] + \cdots + w_5[s + d(v_5, v_4)] + w_6(L - s)$$
$$= (w_1 + \cdots + w_5 - w_6)s + K$$
$$= (W - 2w_6)s + K = (W - 2w_t)s + K$$

where K consists of constant terms. Hence $W - 2w_t$ is *the slope of f as s varies between 0 and L*. Thus part (a) of the local minimum conclusion represents the case where the slope is either negative (implying that v_t is the unique relative minimizing point of f in the arc a^*) or zero (implying every point in a^* is a relative minimum). Similarly, part (b) of the conclusion represents the case where the slope is positive as s varies between 0 and L, in which case v_a is the unique relative minimum of f on the arc, so that no other point in a^* can be a relative minimum. For the specific example of Figure 7.2, $W - 2w_6 = 12 - 2(1) = 10$, so we can exclude from further consideration all points in the arc a^* but v_4.

Recall that we mentioned earlier that the 1-median function is a convex function, with the property that any relative minimum is a global minimum. Hence in the local minimum conclusion above we can replace "relative minimum" by "global minimum" and the conclusion is still true. Therefore, the conclusion above justifies step 1 of the majority algorithm.

We still need to motivate step 2 of the majority algorithm. As an illustration, consider the arc $a^* = [v_1, v_2]$ of Figure 7.2. If x is *not* in a^*, then $d(x, v_1) = d(x, v_2) + d(v_2, v_1)$, in which case, with $w_2' = w_1 + w_2$, we have

$$f(x) = f'(x) + w_1 d(v_2, v_1)$$

where

$$f'(x) = w_2' d(x, v_2) + w_3 d(x, v_3) + \cdots + w_6 d(x, v_6)$$

If we let T' denote the tree obtained from T by removing all points in a^* but v_2, we note that $f(x)$ and $f'(x)$ *differ only by a constant on T'*, so that *minimizing f on T*

is equivalent to minimizing f' on T'. Note we have obtained f' by adding the weight of v_1 to that of v_2, no longer considering the arc a^* in minimizing f' on T'. Thus at this point we have illustrated the

Equivalent 1-Median Problem Property. In step 2 of the majority algorithm, by adding the weight for vertex t to that for vertex a, and by deleting all points on the arc $[v_t, v_a]$ but v_a from the tree, we construct an equivalent 1-median problem to solve on the remaining tree.

We have now provided the basic insights needed to justify the majority algorithm. You can test your understanding of our discussion by developing on your own the (closely related) insights needed to justify the Chinese algorithm.

7.4 1-CENTER PROBLEM

The 1-center problem is the problem of locating a center (a new facility) at an *absolute center*, a point y^* in the tree network that minimizes a maximum of weighted distances between the center and a subset of the vertices. So as to identify the subset of the vertices, we now assume that there are *m or more* vertices in the tree, and number the vertices so that the first m vertices have positive weights, while any other vertices have zero weights. Letting $g(y)$ be the maximum of the weighted distances between y and the first m vertices gives

$$g(y) = \max\{w_1 d(y, v_1), \ldots, w_m d(y, v_m)\}$$

An absolute center y is a point in the tree minimizing $g(y)$.

As with the 1-median problem, the weights have a variety of interpretations, including time per unit distance, cost per unit distance, and loss per unit distance. If the new facility provides service to the vertices, then seeking an absolute center implies that we wish to minimize a maximum time, cost, or loss: we are concerned about the *worst case* and want to make the worst case as "good" as possible. Another way to view the 1-center problem is that of providing "quick service." For example, the fire insurance rate of a commercial firm will probably depend on the response time of the nearest fire station. If, say, we want to know if a station can be located so that its time to respond to a fire alarm at vertex v_i, say $w_i d(y, v_i)$, does not exceed 7 minutes, then we want y to satisfy $w_i d(y, v_i) \le 7$ for $i = 1, \ldots, m$; that is, we want y to satisfy $g(y) \le 7$. The existence of a y satisfying $g(y) \le 7$ is equivalent to an absolute center y^* satisfying $g(y^*) \le 7$. Similar quick service concerns may be relevant in locating police stations or ambulance rescue services.

We now begin to analyze the 1-center problem. We first observe that an equivalent way to write the 1-center problem is to

$$\text{minimize} \quad z$$

subject to

$$w_i d(y, v_i) \leq z, \qquad i = 1, \ldots, m$$

The equivalence follows from the fact that another way to write the constraints of the problem is $g(y) \leq z$, so since we minimize z we minimize $g(y)$. Because another way to express the constraints is $d(y, v_i) \leq z/w_i$ for $i = 1, \ldots, m$; yet another way to write the constraints is to require the collection of neighborhoods $N(v_1, z/w_1), \ldots, N(v_m, z/w_m)$ to have a nonempty intersection. These neighborhoods are convex subsets of a tree.

An important property of convex subsets of a tree we shall use subsequently when studying the 1-center problem is the following:

Pairwise Intersection Property. An equivalent condition for any finite collection of convex (and closed) subsets of a tree to have a nonempty intersection is that every pair of the sets in the collection has a nonempty intersection.

As an illustration of this property, consider a collection of neighborhoods $N(u_1, r_1), \ldots, N(u_m, r_m)$ which are subsets of a tree. We first make an initial observation that any two neighborhoods $N(u_i, r_i)$ and $N(u_j, r_j)$ have a nonempty intersection if and only if $d(u_i, u_j) \leq r_i + r_j$; that is, the distance between the centers of the two neighborhoods does not exceed the sum of their radii. Hence if we apply the pairwise intersection property, we obtain the

Neighborhood Intersection Property. Any collection of m neighborhoods of a tree has a nonempty intersection if and only if for every pair of such neighborhoods, the distance between their centers does not exceed the sum of their radii.

You can get some insight into this property by drawing a collection of line segments, which are neighborhoods of their center points with radii half their lengths. An equivalent condition for the collection of line segments to have a nonempty intersection is for each pair of the line segments to have a nonempty intersection. Roughly speaking, all the line segments overlap if and only if every pair of the line segments overlap.

But now we can use the neighborhood intersection property to conclude that an equivalent condition for the foregoing constraints to hold is that the following inequalities hold for all i and j satisfying $1 \leq i < j \leq m$:

$$d(v_i, v_j) \leq \frac{z}{w_i} + \frac{z}{w_j}$$

$$w_i w_j d(v_i, v_j) \leq z w_j + z w_i$$

$$b_{ij} \equiv \frac{w_i w_j d(v_i, v_j)}{w_i + w_j} \leq z$$

Thus an equivalent condition for the inequalities above to hold for all values of i and

j is that

$$b_{st} \equiv \max\{b_{ij}: 1 \le i < j \le m\} \le z$$

Hence it follows that b_{st} *is the minimum value of* z, which in turn is the minimum value of the 1-center function.

As the terms b_{ij} defined above occur often in analyzing the 1-center problem, it is convenient to give them a name; we shall call them *bounds*. What we have concluded above is that the largest bound, namely b_{st}, is the minimum value of $g(y)$. To find the absolute 1-median y^* for which $g(y^*) = b_{st}$, it is convenient to demonstrate in another manner that $b_{st} \le g(y)$ for every y in T. Given any center location y, we have

$$\begin{aligned} b_{st} &= \frac{(w_s w_t)d(v_s, v_t)}{w_s + w_t} \\ &\le \frac{w_s w_t[d(v_s, y) + d(y, v_t)]}{w_s + w_t} \\ &= \frac{w_t}{w_s + w_t}[w_s d(y, v_s)] + \frac{w_s}{w_s + w_t}[w_t d(y, v_t)] \\ &\le \frac{w_t}{w_s + w_t} g(y) + \frac{w_s}{w_s + w_t} g(y) = g(y) \end{aligned}$$

Note we use the triangle inequality and the distance symmetry property to obtain the conclusion that $b_{st} \le g(y)$. Further, by examining the inequalities in the string of terms above beginning with b_{st} and ending with $g(y)$, we note we can also conclude that an equivalent condition for $b_{st} = g(y)$ is to have

$$d(v_s, v_t) = d(v_s, y) + d(y, v_t) \tag{7.1}$$

and

$$w_s d(y, v_s) = g(y) = w_t d(y, v_t) \tag{7.2}$$

Now let us find the absolute 1-center y^* corresponding to the value b_{st}. From (7.1) and (7.2) above we conclude that $b_{st} = g(y^*)$ if and only if

$$d(v_s, v_t) = d(v_s, y^*) + d(y^*, v_t)$$

and

$$w_s d(y^*, v_s) = g(y^*) = w_t d(y^*, v_t)$$

As we concluded above that $b_{st} = g(y^*)$, we can replace the two equations just above by

$$w_s d(y^*, v_s) = b_{st} = w_t d(y^*, v_t)$$

giving

$$d(y^*, v_s) = \frac{w_t d(v_s, v_t)}{w_s + w_t} \tag{7.3}$$

and

$$d(y^*, v_t) = \frac{w_s d(v_s, v_t)}{w_s + w_t} \tag{7.4}$$

We can now summarize our procedure to solve the absolute 1-center problem as follows. First compute b_{st}. *Then take as y^* the (unique) point in the path joining vertices s and t that satisfies (7.3) and (7.4).*

After having used the procedure above to find the absolute 1-center y^*, to solve the *vertex* 1-center problem (if y^* is not a vertex) it is only necessary to evaluate g at the vertices of the arc containing y^* and choose as the vertex 1-center one of the two vertices with a smaller g value.

Let us now consider an example of finding an absolute 1-center for the tree of Figure 7.3, with all vertex weights being 1; we call such a 1-center problem an *unweighted* 1-center problem. For convenience, we display the lower bounds, the b_{ij}, in a matrix shown in the figure; the example illustrates a general fact that such a matrix will always be symmetric. We shall refer to the matrix $\mathbf{B} = (b_{ij})$ subsequently as the *bound matrix*. The largest entry in the bound matrix is 5, so the minimum value of $g(y)$ is also 5. Since each b_{ij} is half the distance between the vertices i and j, $5 = b_{13}$ is half the distance between vertices 1 and 3. To construct the 1-center y^*, we use (7.3) and (7.4) and conclude that y^* is the midpoint of the path joining vertices 1 and 3.

Another point of view for the 1-center problem is illustrated in Figure 7.3(c), which shows each graph $w_i d(y, v_i)$ along the edge $[v_2, v_3]$, where y is expressed in terms of its distance from v_2. Note that $g(y)$ is just the pointwise maximum of the four graphs and that its minimum value is 5, which occurs at the point y^* where $w_1 d(y^*, v_1) = w_3 d(y^*, v_3)$, so that y^* is on the arc $[v_2, v_3]$ at a distance of 2 from v_2. Note further that where the graphs intersect the vertical axis values of 3, 4, and 5 corresponds to b_{23}, b_{43}, and b_{13}, respectively.

This example illustrates a result that is true in general: *for any unweighted 1-center problem the unique* 1-*center is at the midpoint of any longest path in the tree*. This result can be exploited to give a particularly simple and elegant algorithm for solving the unweighted 1-center problem. The algorithm is as follows:

1. Choose any vertex, say v (e.g., v_4 for Figure 7.3) of the tree.
2. Find a tip that is farthest from v, say v' (e.g., v_3).
3. Find a tip that is farthest from v', say v'' (e.g., v_1)). Take the midpoint of the path joining v' and v'' to be the unique absolute 1-center.

Next we consider an absolute 1-center problem defined by Figure 7.2, where the weights are not all 1. Again, the bound matrix is shown in the figure, where we see that the largest bound is $b_{35} = 14$. Thus $g(y^*) = 14$, where y^* is the unique 1-center. To find y^*, we use (7.3) and (7.4), and since $w_3 = 2 = w_5$, we conclude that y^* is the midpoint of the path joining vertices 3 and 5. Since y^* is on the arc joining

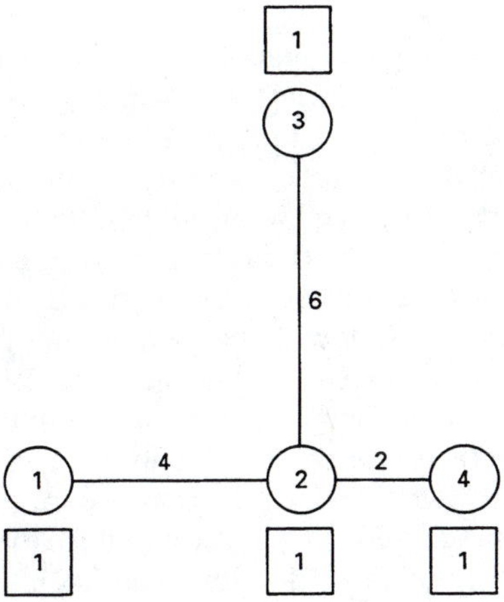

(a)

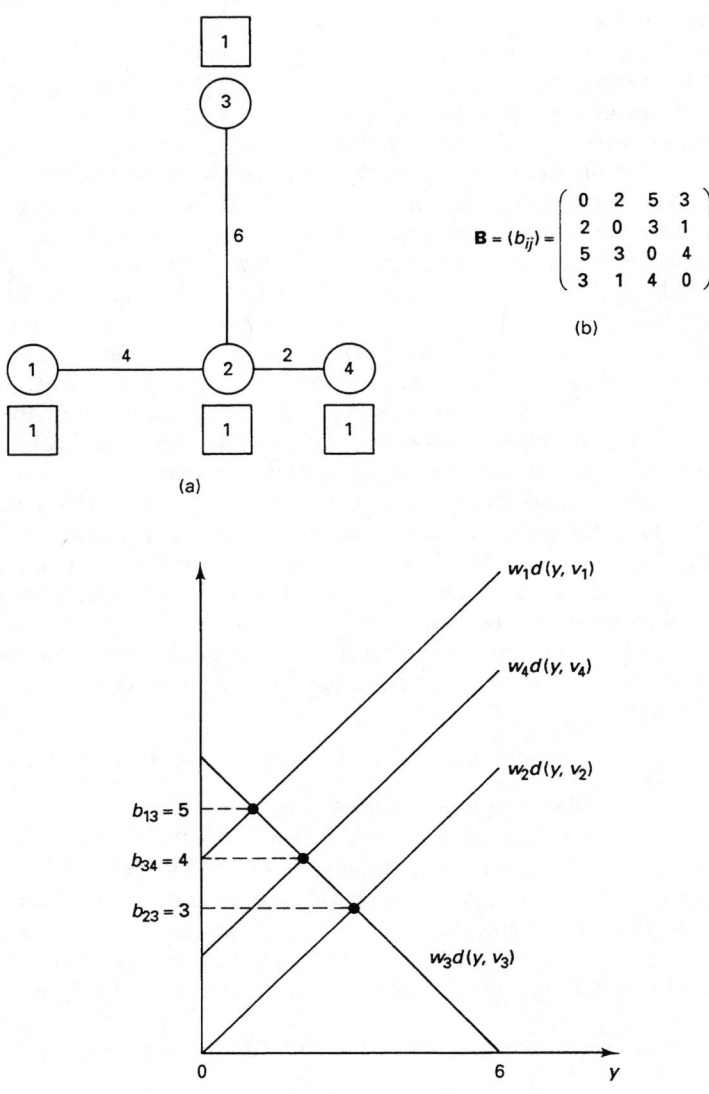

$$\mathbf{B} = (b_{ij}) = \begin{pmatrix} 0 & 2 & 5 & 3 \\ 2 & 0 & 3 & 1 \\ 5 & 3 & 0 & 4 \\ 3 & 1 & 4 & 0 \end{pmatrix}$$

(b)

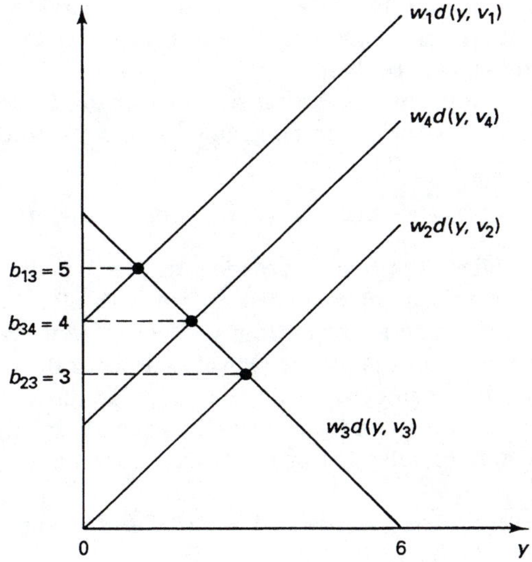

(c)

Figure 7.3 (a) Example tree, with (b) bound matrix B, and (c) graph of weighted distances on arc $[v_2, v_3]$.

vertices 2 and 4, we can also conclude that either vertex 2 or vertex 4 will be a *vertex* 1-center. Calculations establish that $g(v_2) = 20$ and that $g(v_4) = 16$, so vertex 4 is a vertex 1-center.

Even for the relatively small example of Figure 7.2, we can see that computing all the entries in the bound matrix **B** is a nontrivial task and is of order m^2, the number of entries in the matrix [actually, because of the symmetry of **B** we only need to compute $m(m-1)/2$ entries, but this number of entries is still quadratic in m]. Hence for problems of any substantial size, we would probably want to use a calculator or computer to solve the problem. It has been shown, however, that it is not always necessary to compute all the entries in the matrix **B**. A procedure that will always find a largest entry in **B** is as follows. Choose any row of **B**, say $r(1)$, and find a largest entry in row $r(1)$, say in column $c(1)$. Now find a largest entry in column $c(1)$, which occurs, say, in row $r(2)$. Next find a largest entry in row $r(2)$, and so on, continuing until for the first time the same entry is found twice in succession; this number will be a largest entry in **B**. To illustrate this procedure using Figure 7.2, suppose that we start in row 6 and compute a largest number, $10\frac{2}{3}$, which we find in column 3. A largest number in column 3 is 14, which occurs in row 5. A largest number in row 5 is 14, so as we have found the same number twice in succession, we can stop and conclude that 14 is the largest entry in **B**. (Note that when we use this approach it makes sense to compute the entries in **B** as we proceed rather than compute them all before starting.) You should be able to see that this procedure for finding a largest entry in **B** is closely related to the algorithm for solving the unweighted 1-center problem.

As a last topic in this section we consider briefly solving the 1-center problem with "addends," that is, finding an absolute 1-center y^* that minimizes $g(y)$ defined by

$$g(y) = \max\{w_1 d(y, v_1) + h_1, \ldots, w_m d(y, v_m) + h_m\}$$

For this problem, the h_i are the addends, so that the 1-center problem we have considered to this point is a special case with all zero addends. If we interpret $w_i d(y, v_i)$ as the time to travel from y to vertex i, we can interpret h_i as the time to prepare to travel plus the time to provide service at the vertex, plus the time to travel from vertex i to some other known point in the network, such as a hospital. (Naturally, depending on the circumstances, we may take some of these times to be zero.) One way to solve the problem with addends is as follows:

1. For every i and j satisfying $1 \le i < j \le m$, compute b_{ij}, now defined to be

$$b_{ij} = \frac{w_i w_j d(v_i, v_j) + w_j h_i + w_i h_j}{w_i + w_j}$$

   and then compute b_{st}, defined to be the maximum of the b_{ij}.
2. Compute h_p, defined to be the maximum of the h_i.
3. Compute b, defined to be the maximum of b_{st} and h_p.

4. If $b = h_p$, take $y^* = v_p$ as the unique 1-center, and stop.
5. Else, when $b = b_{st}$, take y^* to be the unique point on the path joining vertices s and t that satisfies

$$d(y^*, v_s) = \frac{w_t d(v_s, v_t) + h_t - h_s}{(w_s + w_t)} \tag{7.5}$$

$$d(y^*, v_t) = \frac{w_s d(v_s, v_t) + h_s - h_t}{(w_s + w_t)} \tag{7.6}$$

The derivation of the approach above is much like the derivation of the approach we used to solve the 1-center problem without addends, and makes use of the neighborhood intersection property. You might like to try deriving the approach as an exercise. Note the principal difference in solving the problem with addends is that if any addend is sufficiently large (i.e., $b = h_p$), the optimum location is at the vertex v_p, so that no travel time at all will be incurred to respond to an emergency at v_p. Otherwise, the approach to solving the problem is much the same as before.

7.5 COVERING PROBLEM

In this section we consider a location problem called a *covering problem*. There are many kinds of covering problems, and the covering problem we consider is one of the simplest. It is also one of the more useful covering problems for obtaining insight, and as we shall see, can be solved by a rather simple algorithm. Later, in Chapter 8, we consider more complicated covering problems, as well as procedures for solving them when the network has cycles.

So as to provide a context in which to discuss the problem, we shall imagine that a single service organization is to locate a number of new facilities, which we shall call *centers*, to provide service to a subset of the vertices, numbered 1 through m. Because of its criticality in this section and subsequently, we state the following assumption formally.

Closest Center Assumption. For $i = 1, \ldots, m$, each vertex v_i receives service from a closest center to v_i: if X is any collection of centers, $D(X, v_i)$ denotes the distance between v_i and a closest center in X.

Each vertex i to be served has a *coverage constraint*, stating that the distance between vertex i and a *closest* center to vertex i shall not exceed a known positive constant, say s_i. The centers are all essentially the same, so it is reasonable for each vertex i to receive service from a closest center. The service organization would like to locate the centers so that every coverage constraint is satisfied, while *minimizing* the number of centers it must locate. A typical example of such a coverage problem might be the location of a number of branch post offices with, say, a coverage condition imposed that each neighborhood in a town must have at least one branch post office within 8 kilometers of the neighborhood. Other application contexts include the location of gasoline stations, garbage dumpsters, fast-food outlets, and branch banks.

We now state the covering problem analytically. Given a collection of centers, X, we let $|X|$ denote the number of centers in the collection, called the *cardinality* of X. The covering problem of interest, denoted by C, is as follows:

(C) $$\text{minimize} \quad |X|$$

subject to

$$D(X, v_i) \le s_i \qquad \text{for } i = 1, \ldots, m$$

That is, we want to minimize the number of centers we locate while constraining the distance between each vertex i and its closest center to be at most s_i.

Figure 7.4 shows an example of a covering problem. We shall suppose that the wavy lines emanating from vertices 1 through 5 are strings whose lengths are the upper bounds in the coverage constraints: thus $s_1 = 10, s_2 = 5$, and so on. Any center x whose location in the network satisfies $d(x, v_i) \le s_i$ is said to *cover* vertex i. At least one center must be within a distance of 10 of vertex 1, within 5 of vertex 2, and so on; that is, at least one center must cover each vertex.

An overview of the covering algorithm is as follows. We choose a tip of the tree and check to see if the strings at the tip can reach the vertex adjacent to the tip (if there are no strings at the tip, the tip and the arc having the tip may be removed from the tree). If a shortest string at the tip cannot reach the vertex adjacent to the tip, a center is located at the end of the string and all strings that can "reach" the center are removed from the model. On the other hand, if all strings at the tip can reach the adjacent vertex, the arc with the tip is removed from the model and the remaining lengths of strings now emanate from the vertex adjacent to the tip. The algorithm continues in this way until all strings are removed from the model or until the tree degenerates to a single vertex, in which case one final center is located at the vertex if any string remains at the vertex. Also, the algorithm keeps track of vertices, called *distinguished vertices*, whose strings cause centers to be located.

To illustrate the algorithm, suppose that we choose the string at v_4 and pull it toward v_6. The string reaches v_6, so we remove the arc with v_4 as a tip from the model and are left with a string of remaining length $14 - 10 = 4$ emanating from v_6. (To represent the arc being removed from the model, we say we "color it brown," imagining the entire original tree to be green.) Next, suppose that we choose v_5 and pull the string of length 15 toward v_6. As the string reaches v_6 we color brown the arc with v_5 as a tip, and are left with a string of length $15 - 12 = 3$ emanating from v_6. We now have two strings emanating from v_6, so we delete the longest from the model. Next suppose that we choose as a tip of the remaining green tree v_6, which has a string of length 3. As this string does not reach v_2, we locate a center at a point y_1 on the arc joining vertices 2 and 6, which is a distance of 3 from v_6, and color the entire arc brown. Since the string originating at v_5 caused center 1 to be located when

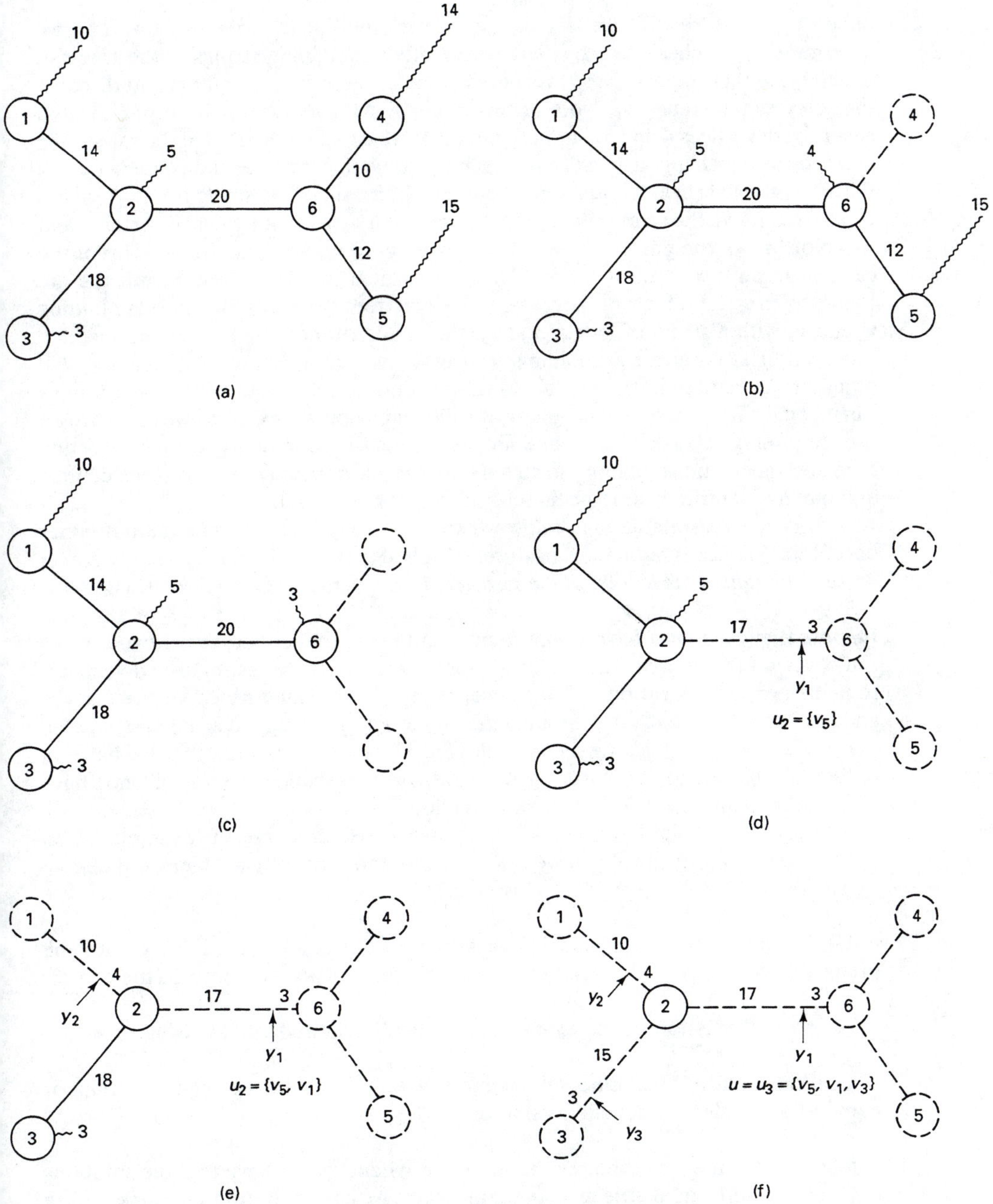

Figure 7.4 Example of COVER, a covering algorithm.

pulled tight, we record the fact that v_5 is a *distinguished vertex*. At the time we determine y_1 we check to see if any of the other remaining strings in the tree can reach y_1 (so that the associated vertices can be covered by the center) and discover that they cannot reach y_1. Note at this point that the center we have just located covers both vertices 4 and 5. Suppose next that we choose v_1 as a tip of the remaining green tree: the string at v_1 will not reach v_2, so we locate a second center at y_2 on the arc joining vertices 1 and 2 at a distance of 10 from v_1. The string from v_2 reaches y_2, so we remove the string. We record the fact that v_1 is a distinguished vertex. Next we color brown the part of the arc $[v_1, v_2]$ from v_1 to y_2, and y_2 becomes a tip of the current green tree with no string. Thus at the next iteration we can omit the arc joining y_2 to v_2. At this point the remaining green tree consists of the single arc joining v_2 and v_3 with a string of length 3 at v_3: this string cannot reach v_2, so we locate a third center at y_3 on the arc joining vertices 2 and 3 at a distance of 3 from v_3. No other strings remain in the model, so we need not check to see if any other strings can reach y_3. We record the fact that v_3 is a distinguished vertex. Next we color brown the arc joining vertices 2 and 3 and are left with a degenerate green tree consisting of v_2 and having no remaining strings. Hence we stop, having located three centers and having identified three distinguished vertices, v_1, v_3, and v_5.

It is now reasonable to ask how we know we have located a minimum number of centers. We observe that *the distance between any two of the distinguished vertices exceeds the sum of the lengths of the associated string lengths*, that is, $32 = d(v_1, v_3) > s_1 + s_3 = 13$, $46 = d(v_1, v_5) > s_1 + s_5 = 25$, and $50 = d(v_3, v_5) > s_3 + s_5 = 18$. Now *suppose* that we do not have an optimum solution, so that either one or two centers can cover all the vertices. In this case, since there are three distinguished vertices, some center, with location x, must cover at least two distinguished vertices, say v_i and v_j, so that $d(x, v_i) \leq s_i$ and $d(x, v_j) \leq s_j$, implying $d(v_i, v_j) \leq d(v_i, x) + d(x, v_j) \leq s_i + s_j$, and giving a contradiction. Hence we conclude that we have an optimum solution to the covering problem, because supposing we did not have an optimum solution led us to a contradiction.

The distinguished vertices we have just considered are an example of an optimum solution to a problem we call the *dual* of the covering problem and denote by DC:

(DC) Find a subset of the first m vertices of maximum cardinality with the property that $d(v_i, v_j) > s_i + s_j$ for *any* two distinct vertices v_i and v_j in the subset.

In the discussion of the example above we illustrated what is called the

Weak Duality Theorem. Any objective function value of DC is less than or equal to any objective function value of C.

It follows from the weak duality theorem that whenever we have feasible solutions to C and DC with equal objective function values, we have optimum solutions to each problem; the example above illustrates this fact.

Not only is DC useful for verifying optimality, as we saw in the example, but it is also useful for doing a sensitivity analysis. Any vertex v_i in an optimum solution to DC identifies a "tight constraint" in the primal problem, that is, a constraint for which the inequality holds as an equality. *Knowing which constraints are tight is useful in seeking alternative optimum solutions*. For example, with reference to the optimum solution shown in Figure 7.4, we see that any location of center 3 within a distance of 3 units of v_3 will be optimal, as will any location of center 2 within a distance of 9 to 10 of v_1. Any location of center 1 within a distance of 15 of v_5 and a distance of 14 of v_4 will be optimum. It is often the case that there are alternative optimum solutions to covering problems, and the distinguished vertices help to identify alternative optimum solutions.

Further, it is often true that *lengthening strings sufficiently which originate from distinguished vertices will allow the minimum objective function value to be reduced*. However, lengthening strings originating from "undistinguished" vertices will have no effect whatever on the minimum objective function value. For example, with reference to Figure 7.4, lengthening the string originating at v_3 from 3 to 22 will allow the minimum objective function value to be reduced from 3 to 2, while lengthening the strings originating from vertices 2 and 4 will have no effect on the minimum objective function value.

One thing to notice about problem DC is that it makes no sense if the minimum objective function value of problem C is 1; this is no reason for concern, however, for if the covering problem objective function value is 1, it is clearly the minimum value (there must be at least one center), so we do not need problem DC to verify optimality. Whenever problem DC is defined, however, we have the following relationship between the two problems:

Strong Duality Theorem. Whenever problem DC is defined, the optimal objective function values of problems C and DC are equal.

The example of Figure 7.4 illustrates the strong duality theorem, with each optimum objective function value being 3.

Now that we have provided insight into the covering problem, its solution algorithm, and the dual of the covering problem, we shall give a formal statement of the algorithm, which we call COVER. To state COVER a few definitions are convenient. We imagine that the tree is represented by inscribing straight-line segments on a planar surface so that each segment represents an arc and all the segments together represent the tree of interest. We *fasten* strings of length s_i to each vertex v_i of the inscribed tree (by convention, we allow strings of zero length): we say that the string s_i *originates* from vertex v_i for $i = 1, \ldots, m$. Every fastened string has one end permanently affixed to the planar surface. In addition, during the use of the algorithm we *engage* previously fastened strings at various points on the tree: when a string is engaged, some point of the string is permanently affixed to the tree such that there is no slack in the portion of the string so far engaged. We say that any string engaged or fastened at a vertex *emanates* from the vertex. When we *remove* strings, we imagine that they are physically deleted from the model.

During each iteration of the algorithm, we partition the original tree into two subsets: one green, the other brown. The green subset is always a tree, denoted by GT (for treen tree), while the brown subset consists of a collection of subtrees of the original tree. By convention, the point joining each brown subtree to the current green tree is in both the green and the brown subtrees, and thus is both green and brown.

COVER

0. Initially, take GT to be T and let $k = 0$. We suppose that each tip vertex of T is both brown and green. For $i = 1, \ldots, m$ we fasten a string of length s_i to v_i. We let U_0 be the empty set to begin with.
1. Choose a tip t of GT. If GT consists of a single tip and has no strings emanating from it, go to step 7, while if it has strings emanating from it, go to step 6. Else (when GT is not a single tip) find $a(t)$, the vertex in GT adjacent to t, and, for convenience, let a^* denote the arc $[t, a(t)]$.
2. If no strings emanate from t, color a^* brown and go to step 1. Else continue.
3. Pull all strings at t tight toward $a(t)$. If all strings reach $a(t)$, remove all but a shortest such string, engage this string at $a(t)$, color a^* brown, and go to step 1. Else continue.
4. Add 1 to k. Delete all but a single shortest string, say $s_{(k)}$, emanating from t. Find the unique vertex, say $v_{(k)}$, from which $s_{(k)}$originates. Adjoin $v_{(k)}$ to U_{k-1} to obtain U_k. Locate center k at the farthest point on a^* that $s_{(k)}$ reaches, and remove $s_{(k)}$. Color a^* brown. Go to step 5.
5. Remove all strings from GT that reach center k (center k can cover the vertices from which the strings originate). If no strings remain, let $U = U_k$ and go to step 7. Else return to step 1.
6. Add 1 to k. Locate center k at t. Choose any one of the strings emanating from center k and find the vertex $v_{(k)}$ from which the chosen string originates. Adjoin $v_{(k)}$ to U_{k-1} to get U. Go to step 7.
7. With q denoting the total number of centers that COVER places, X, consisting of the locations of centers 1 through q, is optimal to problem C. If $q = 1$, no feasible solution to problem DC exists. If $q \geq 2$, U is optimal to problem DC. STOP.

We now illustrate COVER using Figure 7.4. For convenience, we shall refer to the various parts of the figure simply as (a), (b), and so on. Part (a) illustrates step 0, (b) illustrates the first execution of steps 1 to 3, (c) illustrates the second execution of steps 1 to 3, and (d) illustrates the first execution of steps 1 to 5. Part (e) illustrates the second execution of steps 1 to 5, and (f) illustrates the third execution of steps 1 to 5, together with step 7. Step 6, which represents the situation where the green tree degenerates to a single tip with strings emanating from it, never occurs. Note that in step 7, COVER identifies an optimum solution to problem DC, as well as an optimum solution to problem C.

As a final topic in this section, we consider briefly a *vertex-restricted version of the covering problem*, denoted by CV, for which every center must be at a vertex. Such a situation might be of interest when the network represents an air transport network, in which case vertices typically represent airports. Alternatively, for a street network, vertices often represent intersections of streets, which are often preferable locations for such things as mailboxes or gasoline stations, due to their additional accessibility to customers.

We can use COVER with a very minor modification to solve the vertex-restricted covering problem. The modification is as follows: in step 4, we replace the sentence "Locate center k . . . and remove $s_{(k)}$" by the following sentence: "Locate center k at t and remove $s_{(k)}$." Thus we only need to change a single sentence in COVER to solve problem CV. When a shortest string emanating from t does not reach the adjacent vertex $a(t)$, we simply locate a center at the vertex t rather than in the interior of the arc a^*.

To illustrate the use of the modified algorithm, suppose that we solve a vertex-restricted version of the covering problem of Figure 7.4(a), choosing tips in the same order. After coloring brown the arc with v_4 as a tip, and then the arc with v_5 as a tip, we locate center 1 at v_6 and record the fact that v_5 is a distinguished vertex. Then we color brown the arc with v_6 as a tip. Next, when we choose v_1 as a tip, we locate center 2 at v_1, record the fact that v_1 is a distinguished vertex, and color the associated arc brown. Note that we do *not* remove the string emanating from v_2, as the string cannot reach center 2.

At this point the remaining green tree consists of the arc of length 18 joining vertices 2 and 3, with strings of lengths 5 and 3, respectively. If we choose v_3 as a tip, since the string at the vertex cannot reach v_2 we locate a center at v_3 and color the arc brown, recording the fact that v_3 is a distinguished vertex. The remaining green tree now consists solely of v_2 with a string of length 5. Hence we execute step 6 of the algorithm, placing center 4 at v_2 and recording v_2 as a *fourth* distinguished vertex. Next we execute step 7 of COVER, concluding that an optimum solution to problem CV is given by $X = \{v_6, v_1, v_3, v_2\}$ and obtaining a set of distinguished vertices given by $U = \{v_5, v_1, v_3, v_2\}$. We shall see that U solves the dual of the problem CV once we define the dual problem.

We denote the dual of problem CV by DCV. The latter problem is as follows:

(DCV) Find a subset of the first m vertices, say U, of maximum cardinality, with the property that for *any* two vertices v_i and v_j in U, there is no *vertex* v for which $d(v, v_i) \leq s_i$ and $d(v, v_j) \leq s_j$.

We suggest that you verify that the set of four vertices U in the example above has the property indicated; roughly speaking, no two strings originating from vertices in U can "reach" any vertex v in the graph. Note that for the example we can use the set U to verify that the given X is optimal. *If* there could be 3 or less vertices in an optimum solution to the example problem CV, some center in X at a vertex v would have to cover at least two vertices in U (as U has four members), and we know that this is impossible.

We remark that the same duality theorems apply to problems CV and DCV that apply to problems C and DC, with no changes in wording being required, and that the example we have gone over illustrates both theorems. Also, in addition to being useful for verifying optimality, it is again the case that the distinguished vertices can help in carrying our a sensitivity analysis. To reduce the minimum number of centers in an optimum solution to problem CV we must increase the lengths of strings originating from distinguished vertices. For example, making s_2 at least 14 would permit an optimum solution requiring only three centers.

While this completes our discussion of covering problems for tree networks as such, in the next section we shall see how we can solve a problem called the n-center problem by solving a sequence of covering problems. Thus you should be sure you understand the discussion of this section well before going on to the next section.

7.6 *n*-CENTER PROBLEM

In this section we formulate and develop a method for solving the n-center problem when the network is a tree. As we mentioned at the end of Section 7.5, we shall solve the n-center problem by solving a sequence of covering problems, so you also need to understand the discussion on covering problems.

We are given a collection $Y = \{y_1, \ldots, y_n\}$ of center locations, with each center in some arc. As with the covering problem, for each vertex location v_i that is an existing facility location, $i = 1, \ldots, m$, we make a *closest center assumption* as follows: for any n-center Y,

$$D(Y, v_i) = \min\{d(y_1, v_i), \ldots, d(y_n, v_i)\}$$

That is, each existing facility receives service from a closest center.

As with the 1-center problem, we have positive weights for the existing facilities, which together with the closest distances $D(Y, v_i)$, allow us to define our objective function of interest as follows:

$$G(Y) = \max\{w_1 D(Y, v_1), \ldots, w_m D(Y, v_m)\}$$

Our problem is a natural extension of the 1-center problem. The application contexts of the two problems are identical, and the weights typically have the same interpretations for the two problems. We wish to find an *absolute n-center*, an n-center, say Y^*, which minimizes $G(Y)$. We shall assume that $n < m$, for if $n \geq m$, the problem has a trivial solution (put at least one center at each vertex).

To find an absolute n-center we shall use some of our results about the 1-center problem. In particular, we shall use the fact that there is *some* 1-center problem whose minimal objective function value is equal to the minimal value of $G(Y)$. Thus we can limit our search for the minimal objective function value to the positive entries in the bound matrix $\mathbf{B} = (b_{ij})$. Since this fact is rather important, we shall state it formally as follows.

Minimal Objective Function Value Property. The minimal objective function value of the n-center problem, denoted by z_n, is some entry in the bound matrix $\mathbf{B} = (b_{ij})$.

Let us motivate the property. If you are willing to accept the fact that the property is true, you can skip ahead to the next paragraph. Suppose that we have some 2-center problem with $m = 6$ existing facilities, and with $Y = \{y_1, y_2\}$ an absolute 2-center. Each center location will typically be closest to some existing facilities. In particular, we shall suppose that y_1 is closest to existing facility locations 1, 2, and 3, while y_2 is closest to existing facility locations 4, 5, and 6. Let $g_1(y_1)$ denote the 1-center function for vertices 1, 2, and 3, and let $g_2(y_2)$ denote the 1-center function for vertices 4, 5, and 6. It now follows that $G(Y) = \max\{g_1(y_1), g_2(y_2)\}$. If we minimize the two given 1-center problems defined by the functions g_1 and g_2, we find locations, say y_1^* and y_2^*, for which $g_1(y_1^*) \le g_1(y_1)$ and $g_2(y_2^*) \le g_2(y_2)$, so that

$$\max\{g_1(y_1^*), g_2(y_2^*)\} \le \max\{g_1(y_1), g_2(y_2)\} = G(Y)$$

Defining the 2-center $Y^* = \{y_1^*, y_2^*\}$, the closest center assumption implies that

$$G(Y^*) \le \max\{g_1(y_1^*), g_2(y_2^*)\}$$

Hence we conclude that $G(Y^*) \le G(Y)$. But since Y^* is a 2-center and Y is an absolute 2-center, we also have $G(Y) \le G(Y^*)$. It now follows that

$$G(Y) = G(Y^*) = \max\{g_1(y_1^*), g_2(y_2^*)\}$$

From our analysis of the 1-center problem we know that

$$g_1(y_1^*) = \max\{b_{12}, b_{13}, b_{23}\}$$

and

$$g_2(y_2^*) = \max\{b_{45}, b_{46}, b_{56}\}$$

Thus we have

$$z_2 \equiv G(Y) = \max\{b_{12}, b_{13}, b_{23}, b_{45}, b_{46}, b_{56}\}$$

so z_2 is one of the entries in the bound matrix $B = (b_{ij})$.

To gain some insight into the solution procedure we shall develop, let us consider the 2-center problem defined by Figure 7.3, with four vertices and all unit weights. Even though all the weights are 1, we shall write out the weights explicitly, as the insights we obtain from this example will be true in the more general situation where the weights may differ from 1.

One way to write the 2-center problem is as follows:

$$\text{minimize} \quad G(Y)$$

subject to

$$G(Y) = \max\{w_1 D(Y, v_1), \ldots, w_4 D(Y, v_4)\}$$
$$|Y| = 2$$

Equivalently, we can minimize z subject to $G(Y) \le z$ and $|Y| = 2$. Thus, again equivalently, we can

$$\text{minimize} \quad z$$

subject to

$$D(Y, v_i) \le \frac{z}{w_i}, \qquad i = 1, \ldots, 4 \tag{7.7}$$

$$|Y| = 2 \tag{7.8}$$

When we consider the formulation (7.7) and (7.8), recalling that z_2 denotes the minimal objective function value, we get *insight 1*: z_2 is the smallest value of z such that there exists a 2-center Y for which Y and z satisfy (7.7) and (7.8).

With reference again to Figure 7.3, let R consist of the set of entries in the **B** matrix, excluding the smallest and the largest entry, so that $R = \{1, 2, 3, 4, 5\}$. Due to the minimal objective function value property above, we have the following conclusion, called *insight 2*: *z_2 will be one of the numbers in the set R*. Thus z_2 will be either 1, 2, 3, 4, or 5. (We cannot have $z_2 = 0$ obviously, for then we would need four centers.)

Combining insights 1 and 2 above we can now obtain a basic insight, called *insight 3*: *z_2 is the smallest number z in R for which there exists a Y for which Y and z satisfy* (7.7) *and* (7.8). Hence what we need at this point is a way to check whether or not some Y satisfies conditions (7.7) and (7.8) when z is any one of the numbers in R.

We now make use of our results about covering problems. Noticing that (7.8) resembles the constraints of a covering problem, we let C(z) denote the following covering problem:

C(z) $\qquad$ minimize $|Y|$

subject to

$$D(Y, v_i) \le \frac{z}{w_i}, \qquad i = 1, \ldots, m$$

(In this example $m = 4$.) We can think of z as a parameter of the covering problem: each time we change z we get a different covering problem. Let $q(z)$ denote the minimal objective function value of C(z).

We shall now explain why z_2 is the *minimal* element z in R for which $q(z) \le 2$. If you are willing to accept this fact, you can skip to the next paragraph. We note the following: if Y and z satisfy conditions (7.7) and (7.8), they satisfy the constraints of the covering problem (with $m = 4$); since $|Y| = 2$ the minimal objective function of the covering problem, $q(z)$, is at most 2. Conversely, suppose that $q(z)$ is at most 2, and let Y^* be an optimal solution to C(z). If $|Y^*| = 2$, we observe from C(z) that Y^* and z satisfy conditions (7.7) and (7.8). If $|Y^*| = 1$ and we construct a new 2-center, say Y', by appending *any* center to Y, we note that we have $D(Y', v_i) \leq D(Y^*, v_i)$, since the closest center in Y' to v_i is at least as close as the closest center in Y^* to v_i. Hence we conclude that Y' and z satisfy (7.7) and (7.8). To summarize, we can conclude that an equivalent condition for there to exist a Y such that Y and z satisfy conditions (7.7) and (7.8) is that $q(z)$, the minimal objective function value of C(z), not exceed 2. Thus z_2 is the minimal z in R for which $q(z) \leq 2$.

We now have a way to solve the example. We choose a value of z from R, and use the value to solve problem C(z), obtaining the minimal objective function value $q(z)$: if $q(z) \leq 2$, we conclude that z is an upper bound on z_2; otherwise, z is a lower bound on z_2. Since there are five possible values of z (1 to 5), there are five corresponding covering problems. One of these five problems we need not solve explicitly; since the largest entry in R is the maximum entry in the bound matrix **B** is the minimal objective function value of the 1-center problem, say z_1, it is always true that the associated covering problem C(z_1) has a minimal objective function value of 1, and the optimal location for the problem is the location of the absolute 1-center, which we have already found in Section 7.3, and know also that $q(z_1) = q(5) = 1$. We show solutions to C(z) for $z = 1, 2, 3, 4$ in Figure 7.5. We can see from the figure that $q(4) = 2$, $q(3) = 2$, $q(2) = 3$, and $q(1) = 3$. Hence we conclude that $z_2 = 3$; further, we can see that the solution shown in Figure 7.5(b) is an optimal solution to the 2-center problem. Note we can also use the results illustrated in Figure 7.5 to solve the 3-center problem defined by Figure 7.3. Because z_3 is the minimal z in R for which $q(z) \leq 3$, with reference to Figure 7.5 we conclude that $z_3 = 1$, and that the solution shown in Figure 7.5(d) is an optimal 3-center.

Let us now generalize from insights the example provides to draw generally valid conclusions about finding an n-center in a tree network to minimize

$$G(Y) = \max\{w_1 D(Y, v_1), \ldots, w_m D(Y, v_m)\}$$

We first note that we can assume n is between 1 and $m - 1$, for the n-center problem can be solved trivially by placing a center at each of the first m vertices if n exceeds $m - 1$. Let the set R consist of the distinct positive entries in the bound matrix $\mathbf{B} = (b_{ij})$, where for $1 \leq i < j \leq m$ we have

$$b_{ij} = \frac{w_i w_j d(v_i, v_j)}{w_i + w_j} \tag{7.9}$$

Further, letting z_n denote the minimal objective function value of the n-center problem, and letting $q(z)$ denote the minimal objective function value of the covering problems C(z), we conclude that *z_n is the minimal element r in R for which $q(r) \leq n$*. Hence we can solve the n-center problem by solving a *sequence* of covering problems.

Because R can have as many as $m(m - 1)/2$ elements, we need a good procedure for searching through the set R to find z_n. One such procedure, which is most

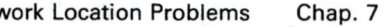

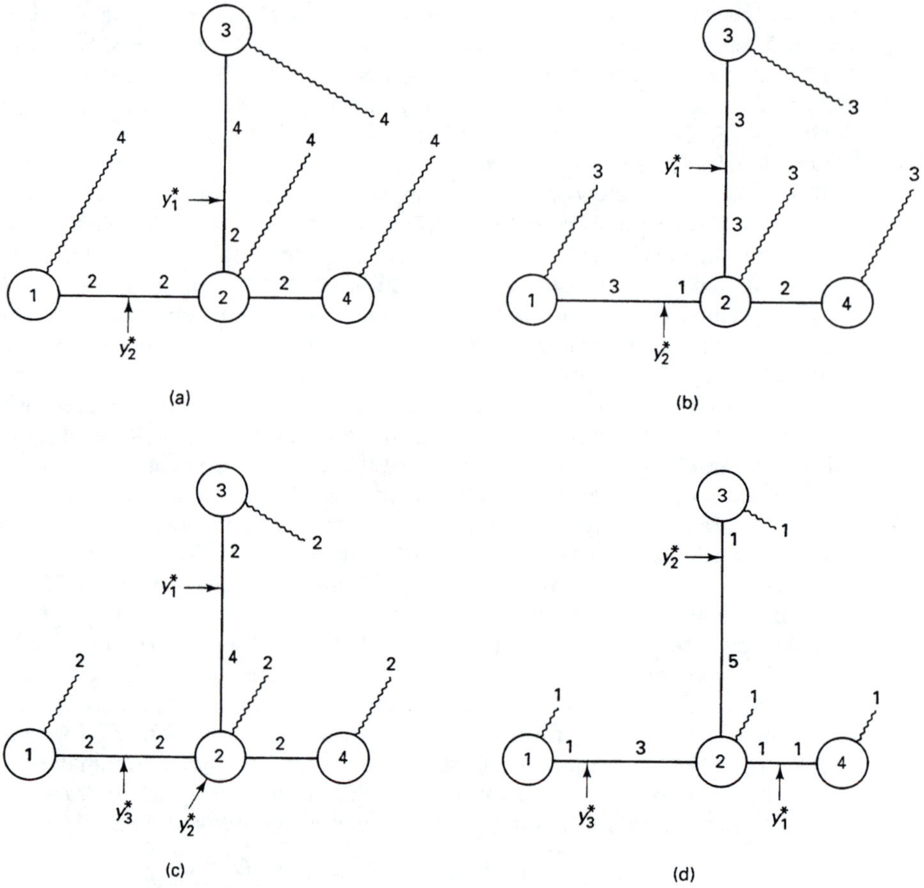

Figure 7.5 Solutions to a sequence of covering problems used to solve 2-center and 3-center problems defined from Figure 7.3.

often used, is bisection search. The exact way in which we shall employ bisection search is specified by the following algorithm, which we call NCENT, to solve the n-center problem.

NCENT

0. Construct the set R consisting of all the distinct positive b_{ij} defined by (7.9).
1. Repeat step 2 below until R consists of a single element.
2. Find a median entry in R, denoted by r. Solve the problem C(r) and compute its minimal objective function value $q(r)$. If $q(r) \le n$, delete from R all ele-

ments strictly greater than r: else delete from R all elements no greater than r (including r).

3. Let z_n be the single element in R. Solve $C(z_n)$ (if it is not already solved) and denote its optimal solution by Y^*. If $|Y^*| = n$, take Y^* as an optimal n-center and stop. In case $|Y^*| = p < n$, choose any convenient $n - p$ centers to append to Y^* to obtain an optimal n-center, say Y', and stop.

To illustrate NCENT, with reference to Figure 7.4, let us use it to solve the example 2--center problem. Initially, we have $R = \{1, 2, 3, 4, 5\}$. If we choose $r = 3$ as a median of R, we solve C(3), find $2 = q(3) = q(r) = n = 2$, so we delete 4 and 5 from R and have $R = \{1, 2, 3\}$. If next we choose $r = 2$ as a median of R, we solve C(2), find $3 = q(2) = q(r) > n = 2$, so we delete 1 and 2 from R and have $R = \{3\}$. We thus conclude that $z_2 = 3$ and that the 2-center shown in Figure 7.4(b) is optimal.

We can now see that bisection search reduces the number of elements in R by (roughly) half at each iteration (each performance of step 2), allowing R to be searched quite efficiently. For example, if initially R consists of 1000 elements, you should be able to verify that NCENT would have to solve at most 12 covering problems. Even if R initially consists of 1,000,000 elements, NCENT would have to solve at most 21 covering problems! Hopefully, we have now convinced you that using bisection search is a good idea.

The example 2-center problem we solved above may have led you to conclude that we can always exclude the largest entry in the bound matrix **B** from the set R of positive entries in **B** when searching for z_n. Unfortunately, if you came to this conclusion, you made a mistake, as the somewhat pathological example of Figure 7.6 shows, where vertices 1 to 4 are the existing facility locations, and every weight is 1. If we compute the entries in **B**, we find that every positive entry is 1, giving $R = \{1\}$. If we solve the associated covering problem C(1), we obtain a unique single center, located at vertex 5, so that $q(1) = 1$. (Note that another way to obtain v_5 is to find the absolute 1-center.) Thus with reference to step 3 of NCENT, to solve the 2-center problem we take v_5 and any other location as an absolute 2-center. Similarly, to solve the 3-center problem we take v_5 and any other two distinct locations as an absolute 3-center. Hence this example illustrates that the last part of step 3 of NCENT is in fact necessary.

As a last topic in this section we consider briefly the vertex-restricted n-center problem, for which each center must be a vertex. Reasons for considering such a problem are entirely analogous to those for considering the vertex-restricted covering problem, as discussed in the preceding section. We can use the algorithm NCENT with only minor changes to solve the vertex-restricted n-center problem. As previously, we shall assume that n does not exceed $m - 1$. Letting h denote the total number of vertices in the tree, since each of the centers must be some vertex we note that z_n must be one of the positive entries in the matrix $\mathbf{B}' = (w_j d(v_i, v_j))$ with h rows and m columns, and thus take R to consist of the set of all distinct positive entries in $\mathbf{B}'$. We let CV(r) denote the vertex-restricted version of problem C(r), and let $qv(r)$ denote the minimal objective function value of the problem CV(r). We then

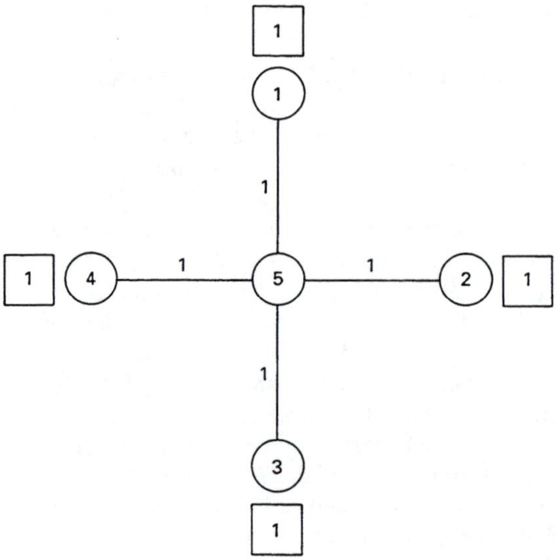

$$\mathbf{B} = (b_{ij}) = \begin{pmatrix} 0 & 1 & 1 & 1 \\ 1 & 0 & 1 & 1 \\ 1 & 1 & 0 & 1 \\ 1 & 1 & 1 & 0 \end{pmatrix}$$

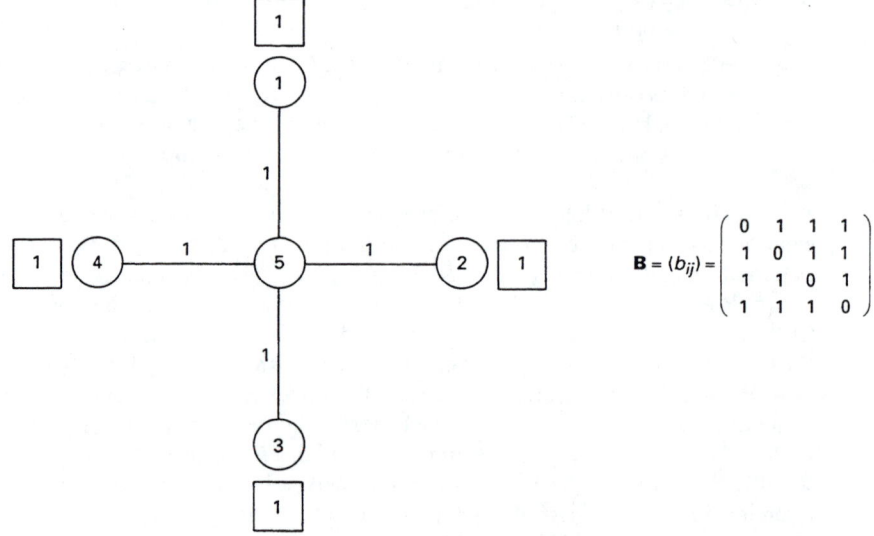

Figure 7.6 Example of a tree with $R = \{1\}$.

use the vertex-restricted version of the algorithm COVER to solve $CV(r)$ and compute $qv(r)$. With the foregoing changes we then use NCENT as before, except that in step 3, in case $|Y^*| = p < n$, the $n - p$ centers we choose to append to Y^* must be any convenient *vertices*.

Consider now examples where we solve the vertex-restricted n-center problems defined by Figure 7.3 for $n = 1, 2$, and 3. When we construct the set R we obtain $R = \{2, 4, 6, 8, 10\}$, leading to corresponding covering problems $CV(2), \ldots, CV(10)$. When we use the vertex-restricted version of NCENT we find that $\{v_2\}$ obtained by solving $CV(6)$ is an optimal 1-center, that $\{v_2, v_3\}$ obtained by solving $CV(4)$ is an optimal 2-center, and that $\{v_1, v_2, v_3\}$ and $\{v_1, v_3, v_4\}$ obtained by solving $CV(2)$ are optimal 3-centers. We recommend that you solve these example problems to be certain that you understand how to solve the vertex-restricted n-center problem.

7.7 CYCLIC NETWORKS

In the remainder of this chapter we consider network location problems for general planar networks, called *cyclic* networks for the obvious reason that they contain cycles. Such networks lead to more realistic location models than those for tree networks but are also more difficult to work with, as we shall see. As is very often the case with models, there is a trade-off between realism and tractability: the more realistic a model is, the less tractable it is.

Figure 7.7 illustrates a typical cyclic network. We can imagine that the figure illustrates a road network or any other type of network, such as the ones we discussed when introducing tree networks. We shall use Figure 7.7 to illustrate a number of definitions we need. Thus any time you come across "(e.g., . . .)," you should refer to Figure 7.7 for an example.

A *cyclic network* is a network $N = (V, A)$ which has at least one cycle. $V = \{v_1, \ldots, v_m\}$ is a vertex set of distinct points in the plane (e.g., 1, 2, 3, 4, and 5 in Figure 7.7) and A is an arc set corresponding to two-tuples (i, j) with $i < j$ in the set I of pairs of indices of vertices joined by arcs [e.g., $I = \{(1,2), (1,5), (2,3), (2,5), (3,4), (3,5), (4,5)\}$]. Thus $[v_i, v_j]$ is the arc corresponding to the two-tuple (i, j) and is represented either by a line segment or by a continuous planar curve such as an arc of a circle. Each arc has a given positive length. At most one arc joins any two vertices. Recall that we defined a *path* as a sequence of adjacent arcs, with each pair of adjacent arcs sharing a common vertex (e.g., $\{[v_1, v_2], [v_2, v_5], [v_5, v_4]\}$ is a path joining v_1 and v_4 of length $10 + 12 + 20 = 42$). If the initial and final vertices in the path are identical, the path is called a *cycle* (e.g., $\{[v_3, v_4], [v_4, v_5], [v_5, v_3]\}$). A path not containing a cycle is called a *simple path*, and a cycle in which no arc repeats is called a *simple cycle*. Note that it is the fact that a cyclic network contains one or more simple cycles that distinguishes it from a tree, which contains no simple cycles. Whenever we speak of paths and cycles subsequently, we shall mean simple paths and simple cycles. As with a tree, the *length of a path* is the sum of the lengths of the arcs in the path, and the *distance* between any two vertices v_i and v_j, denoted by $d(v_i, v_j) = d(v_j, v_i)$, is the length of a *shortest* path joining v_i and v_j [e.g., $d(v_1, v_4) = 10 + 14 + 10 = 34$].

If you are unfamiliar with algorithms for computing shortest paths, we refer you to any good book on operations research, or to the appendix of this chapter, where we illustrate Floyd's algorithm for finding a shortest path between every pair of vertices. It is true, of course, that for small networks, such as the one shown in Figure 7.7, we can identify shortest paths by inspection. However, for large networks

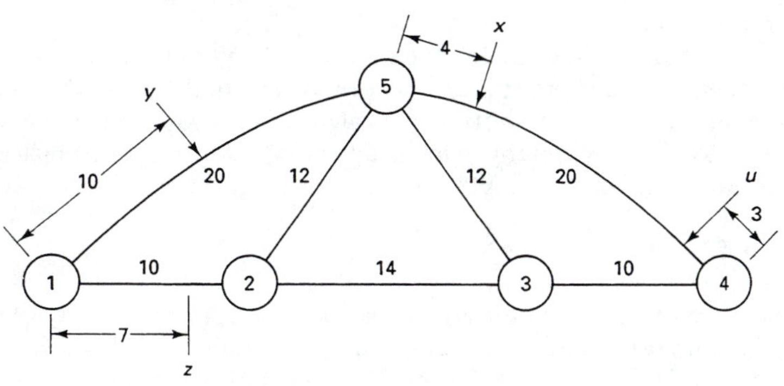

Figure 7.7 Example of a cyclic network.

it is convenient, and sometimes essential, to have an algorithm for finding the paths that can be programmed and run on a computer. (If you doubt this, try finding by inspection a shortest path between every pair of U.S. cities with populations of at least 20,000.) We remark that a network is called *connected* if at least one path joins any two nodes, and is *disconnected* otherwise. All the networks we consider will be connected, unless we indicate otherwise.

So far we have defined distances between vertices. Given any two points x and y in the network that may not be vertices, we simply *imagine* them to be vertices in order to define $d(x, y)$, the distance between x and y, to be the length of a shortest path between x and y. The definition requires, of course, that the lengths of the "subarcs" defined by x and y be well defined: with reference to Figure 7.7, as examples of the definition we have $d(x, y) = 14, d(y, z) = 17, d(x, z) = 19, d(x, u) = 13$. Our resulting distance, as defined, has the symmetry, positivity, and triangle inequality properties we stated in Section 7.2 when we considered tree networks.

We now have all the definitions we need and shall use the definitions in the remainder of the chapter. The principal thing for you to gather from this section, in conclusion, is that *in the network models to follow we shall always assume that travel distances are lengths of shortest paths*. Thus applications of the results to follow should be to problems where the shortest-path-length assumption is valid.

7.8 MEDIAN AND CENTER PROBLEMS ON CYCLIC NETWORKS

In this section we shall consider the 1-median and the 1-center problems on cyclic networks, as well as extensions of the problems involving the location of n new facilities, referred to as the *n-median* and *n-center* problems, respectively. We shall give procedures to solve the 1-median and 1-center problems. For the n-median and n-center problems we shall identify a finite collection of points in the network, say P, which must contain optimal new facility locations. This identification of the set P allows us to formulate the n-median and the n-center problems as discrete optimization problems in Chapter 8, where we shall give procedures for solving the two problems.

Throughout this section the distance $d(x, y)$ between any two points x and y is the length of a shortest path between x and y in the network. Such distances can be computed using any shortest-path algorithm. If you are unfamiliar with shortest-path algorithms, we refer you to the appendix of the chapter, which presents Floyd's algorithm, one of the simpler shortest-path algorithms.

7.8.1 Median Problems

The 1-median problem for a cyclic network is identical to that for a tree network, except that the distances are defined more generally so as to allow the consideration of cycles. With the 1-median problem we wish to minimize a sum of weighted distances between the new facility and a subset of the vertices: such a point is called an *absolute median*, and minimizes the function $f(x)$ defined by

$$f(x) = \sum_{i=1}^{m} w_i d(x, v_i)$$

Consider as an example the location of a distribution center (DC) for a parcel delivery service for a town. Trucks drive from the DC loaded with parcels to their delivery areas. They then drive about in their delivery areas and eventually return empty to the DC. The location of the DC affects the total driving distance of the trucks. If we are free to choose the location of the DC, we have what is essentially a 1-median problem. Each v_i can be the centroid of a delivery area, while x is the location of the DC, and the simplest choice of the w_i is to take each w_i to be 2. In this case a 1-median is the location that minimizes total driving distance to and from the delivery areas and the DC. By having a list of locations of interest ranked in increasing order of their $f(x)$ values, we are in a position to do a trade-off analysis between total driving distance and real estate cost.

In seeking an absolute median, a point x^* to minimize $f(x)$, we shall see, just as was the case for a tree network, that some vertex location is optimal, a result called the *vertex-optimality property*. The property is useful because it reduces our search for x^* to a finite number of points rather than an infinite number of points.

To obtain insight into the vertex-optimality property, consider Figure 7.8, which shows graphs of $d(x, v_i)$, $i = 1, \ldots, 5$, as x varies along the points in the arc $[v_2, v_3]$ of the network of Figure 7.7. As x moves from v_2 to v_3, we can see that $d(x, v_3)$ decreases linearly from 14 to 0, $d(x, v_2)$ increases linearly from 0 to 14, $d(x, v_4)$ decreases linearly from 24 to 10, and $d(x, v_1)$ increases linearly from 10 to 24. The graph of particular interest is $d(x, v_5)$, which increases linearly from 12 to 19 as x varies from v_2 to the midpoint of $[v_2, v_3]$, and then *decreases* linearly as x varies from the midpoint of $[v_2, v_3]$ to v_3. The reason for this behavior of $d(x, v_5)$ is that the shortest path between x and v_5 is either $\{[v_5, v_2], [v_2, x]\}$ or else $\{[v_5, v_3], [v_3, x]\}$, depending on whether x is closer to v_2 or to v_3, respectively. We can generalize from the insight we have obtained from this example as follows.

Piecewise Linearity and Concavity Property. For any given network $N = (V, A)$, let $[v_p, v_q]$ be any arc in A, and let v_i be any vertex in N. The function in x defined by $d(x, v_i)$:

1. Is continuous on $[v_p, v_q]$
2. As x varies from v_p to v_q in $[v_p, v_q]$, either
   (a) increases linearly, or
   (b) decreases linearly, or
   (c) first increases linearly and then decreases linearly
3. Is *concave*, in the sense that a line segment joining any two points on the graph of the function lies on or below the graph of the function.

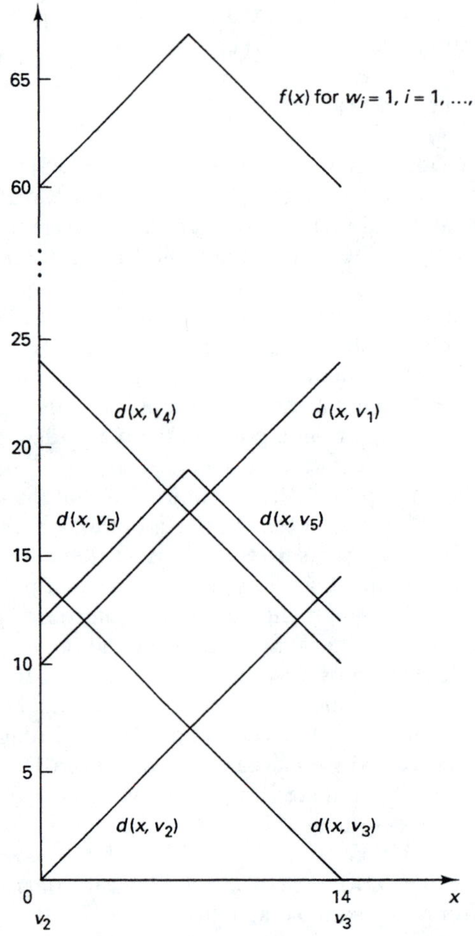

Figure 7.8 Graph of $d(x, v_i)$, $i = 1, \ldots, 5$, x in $[v_2, v_3]$, for the network of Figure 7.7.

We note that if w_i is any positive number, the property above also applies to the function in x defined by $w_i d(x, v_i)$. In particular, $w_i d(x, v_i)$ is concave on $[v_p, v_q]$. Because a sum of concave functions is a concave function, it follows that the 1-median function, $f(x)$, is concave on the arc $[v_p, v_q]$. The function $f(x)$ is illustrated in Figure 7.8 for the case where $w_i = 1$ for $i = 1, \ldots, 5$, and is clearly concave. Further, because of the concavity of x, we can see that the minimal value of $f(x)$ on the arc occurs at some vertex of the arc.

We can generalize from the example above as follows.

Vertex Optimality Property. Because the 1-median function is concave on each arc, it has at least one minimizing point in the arc occurring at a vertex of the arc. Thus we need only consider vertex locations in order to find an absolute 1-median.

We now have a simple algorithm to find an absolute 1-median: evaluate $f(x)$ at each vertex and choose a vertex with the smallest value of $f(x)$. An implementation of the algorithm is s follows. Let $\mathbf{D} = (d_{ij})$ be the m by m matrix whose entry in row i and column j is $d(v_i, v_j)$. Let $\mathbf{W} = (w_i)$ be the m by 1 vector whose entry in row i is the weight w_i. Compute **DW**. The entry in row i of **DW** is $f(v_i)$, so a smallest entry in **DW**, say $f(v_s)$ in row s, identifies an absolute median, say v_s, and its corresponding minimal cost, $f(v_s)$.

We illustrate the 1-median algorithm using the network of Figure 7.7, taking every w_i to be 1. We use an algorithm such as Floyd's algorithm, given in the appendix of the chapter, to compute the matrix **D**. We find that

$$\mathbf{DW} = \begin{pmatrix} 0 & 10 & 24 & 34 & 20 \\ 10 & 0 & 14 & 24 & 12 \\ 24 & 14 & 0 & 10 & 12 \\ 34 & 24 & 10 & 0 & 20 \\ 20 & 12 & 12 & 20 & 0 \end{pmatrix} \begin{pmatrix} 1 \\ 1 \\ 1 \\ 1 \\ 1 \end{pmatrix} = \begin{pmatrix} 88 \\ 60 \\ 60 \\ 88 \\ 64 \end{pmatrix}$$

Hence both vertices v_2 and v_3 are absolute medians, and the minimal value of $f(x)$ is 60. As a second example, when the weights for the vertices v_1 through v_5 are 2, 10, 3, 5, and 1, respectively, if we again compute **DW** we find its entries in rows 1 to 5 are 362, 194, 250, 358, and 296, respectively, giving v_2 as an absolute median. You should solve this second example yourself to be sure you understand the algorithm.

We next consider the n-median problem, with $n \geq 2$. We have a collection of *medians* $X = \{x_1, \ldots, x_n\}$, with each median in some arc of the network. As with the covering and n-center problems, we make a *closest median assumption*; we have

$$D(X, v_i) = \min\{d(x_1, v_i), \ldots, d(x_n, v_i)\} \qquad \text{for } i = 1, \ldots, m$$

We should keep in mind the closest median assumption in studying the n-median problem, for if the assumption is invalid, the results for the problem may be of little use to us. For example, if we consider using the model to locate mailboxes in a city, we would need to justify the assumption that each customer uses a closest mailbox. If, in fact, customers combine a trip to the mailbox with other stops as well, such as at a grocery store and drugstore, it may be inconvenient for them to stop at a closest mailbox to their home, and the n-median model may not be applicable.

As with the 1-median problem, we are given nonnegative weights $w_1, \ldots, w_m$ for vertices 1 to m, respectively, and compute the total cost for any n-median X as follows:

$$F(X) = \sum_{i=1}^{m} w_i D(X, v_i)$$

Our interest is now to find an n-median that minimizes $F(X)$, that is, to find an *absolute n-median*.

We shall now motivate the

Vertex Optimality Property. There exists at least one absolute n-median for which each median is at a vertex. Hence it suffices to consider only vertex locations in order to find an absolute n-median.

If X is an absolute n-median, and we can find *another* n-median X^* with each median at a vertex for which $F(X^*) \leq F(X)$, we will have illustrated the property, for X being an absolute median implies that $F(X) \leq F(X^*)$, so we conclude that $F(X^*) = F(X)$ and thus X^* is an absolute n-median with each median at a vertex. To be more concrete, let us suppose that $n = 2$, $X = \{x_1, x_2\}$ is an absolute 2-center, that there are $m = 5$ vertices, and that

$$d(x_1, v_i) = D(X, v_i) \qquad \text{for } i = 1, 2, 3$$

$$d(x_2, v_i) = D(X, v_i) \qquad \text{for } i = 4, 5$$

That is, x_1 is a closest median to vertices 1, 2, and 3, while x_2 is a closest median to vertices 4 and 5 (ties for closeness can be broken arbitrarily). Defining the 1-median functions $f_1(x_1)$ and $f_2(x_2)$, respectively, by

$$f_1(x_1) = \sum_{i=1}^{3} w_i d(x_1, v_i)$$

$$f_2(x_2) = \sum_{i=4}^{5} w_i d(x_2, v_i)$$

we then have

$$f_1(x_1) + f_2(x_2) = \sum_{i=1}^{5} w_i \mathbf{D}(X, v_i) = F(X)$$

But now by the absolute 1-median property there are absolute 1-medians x_1^* and x_2^* at vertex locations for which

$$f_1(x_1^*) \leq f_1(x_1) \qquad \text{and} \qquad f_2(x_2^*) \leq f_2(x_2)$$

Letting $X^* = \{x_1^*, x_2^*\}$ be a 2-median, the closest median assumption implies that

$$D(X^*, v_i) \leq d(x_1^*, v_i) \qquad \text{for } i = 1, 2, 3$$

$$D(X^*, v_i) \leq d(x_2^*, v_i) \qquad \text{for } i = 4, 5$$

Thus

$$\begin{aligned} F(X^*) &= \sum_{i=1}^{5} w_i D(X^*, v_i) \\ &\leq \sum_{i=1}^{3} w_i d(x_1^*, v_i) + \sum_{i=4}^{5} w_i d(x_2^*, v_i) \end{aligned}$$

$$= f_1(x_1^*) + f_2(x_2^*) \le f_1(x_1) + f_2(x_2) = F(X)$$

so we conclude that $F(X^*) \le F(X)$. Since each median in X^* is at a vertex, we have now illustrated the property.

We remark that for the n-median problem to be nontrivial, we need to have $n < m$, for otherwise we can construct an n-median X with at least one median at each of the m vertices, and have $F(X) = 0$. When $n < m$, there are m combinations taken n at a time, that is, m-choose-n, possible assignments of medians to vertices, so that the number of possible assignments can be very large; for example, 40-choose-20 is approximately 137,846,528,800. Hence although the vertex optimality property reduces the n-median problem to a finite problem, it usually does not facilitate solution by total enumeration. In Chapter 8 we shall see how the vertex optimality property can be exploited by an algorithm to solve the n-median problem.

7.8.2 Center Problems

Recall the 1-center problem is the problem of finding an absolute center, a point y^* in the given network that minimizes a maximum of weighted distances between the center and a subset of the vertices. So as to identify the subset of the vertices, we assume that there are m or more vertices in the network, and number the vertices so that only the first m vertices have positive weights. Letting $g(y)$ denote the maximum of the weighted distances involving y and the first m vertices, we then have

$$g(y) = \max\{w_1 d(y, v_1), \ldots, w_m d(y, v_m)\}$$

An absolute center is then a point in the network that minimizes $g(y)$.

Let us now consider an example problem of finding an absolute center. We shall use for the example the network of Figure 7.7 with $w_i = 1$ for $i = 1, \ldots, 5$. Throughout the example we shall write out the w_i explicitly rather than use their values of 1, because the insights of the example apply for any positive values of the weights. In Figure 7.8, each of the weighted distances $w_i d(y, y_i)$, is displayed for y in $[v_2, v_3]$. From the figure we can readily determine $g(y)$ to be as follows:

$$g(y) = \begin{cases} w_4 d(y, v_4), & 0 \le y \le 6 \\ w_5 d(y, v_5), & 6 \le y \le 8 \\ w_1 d(y, v_1), & 8 \le y \le 14 \end{cases}$$

where y is expressed as a distance from v_2. There are two local minima of $g(y)$, one defined by

$$w_4 d(y, v_4) = w_5 d(y, v_5)$$

and one defined by

$$w_5 d(y, v_5) = w_1 d(y, v_1)$$

The locally minimum values are $g(6) = 18$ and $g(8) = 18$, respectively.

For this particular case, each local minimum of $g(y)$ occurs at an *intersection point*, a point that is defined by exactly two distinct weighted distance functions being equal, with the additional property that one of the two functions increases locally as y moves away from the point in one direction, while the other increases locally as y moves away from the point in the other direction. For example, $y = 6$ is an intersection point, since $w_4 d(6, v_4) = w_5 d(6, v_5)$, while $w_4 d(y, v_4)$ increases as y decreases locally from 6, and $w_5 d(y, v_5)$ increases as y increases locally from 6.

It is also possible for $g(y)$ to have local minima at the endpoints of the arc. For example, suppose that we modify the vertex weights, so that $w_3 = w_4 = 0$. Now $g(0) = 12$ is a local minimum, not defined by an intersection point.

It should be clear now that to minimize $g(y)$ on an arc $[v_p, v_q]$, we need to identify the relevant intersection points. Furthermore, to do this, we must construct, explicitly or implicitly, certain weighted distance functions. In the following, we give a simple procedure for computing all the values required to construct the distance functions, and then describe a rudimentary algorithm for enumerating the intersection points.

Refer to Figure 7.9. With L the length of arc $[v_p, v_q]$ and s and $L - s$ the lengths of subarcs $[v_p, y]$ and $[y, v_p]$, respectively, for y in $[v_p, v_q]$, since $d(y, v_i)$ is the length of a shortest path between y and v_i, we have

$$\begin{aligned} d(y, v_i) &= \min\{d(v_i, v_p) + s, d(v_i, v_q) + L - s\} \\ &= \min\{P_i + s, Q_i + L - s\} \end{aligned}$$

where we let

$$P_i = d(v_i, v_p), \qquad Q_i = d(v_i, v_q)$$

With reference to Figure 7.10, if t is the point satisfying $P_i + t = Q_i + L - t$, then $t = (Q_i + L - P_i)/2$ and $P_i + t = Q_i + L - t = u \equiv (Q_i + L + P_i)/2$. We can

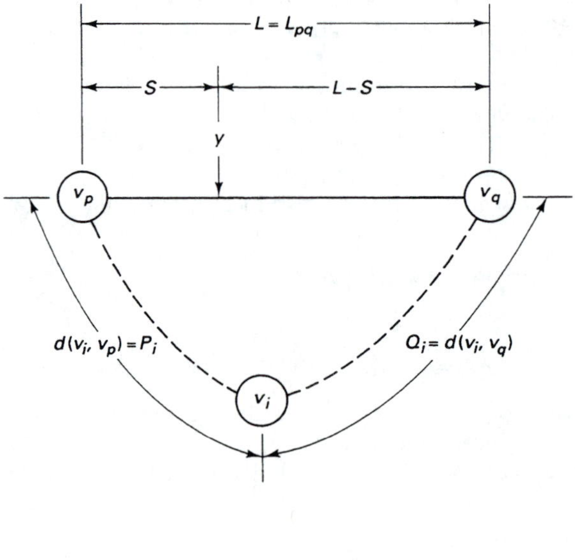

Figure 7.9 Illustration of $d(y,v_i) = \min\{P_i + s, Q_i + L - s\}$ for y in $[v_p, v_q]$.

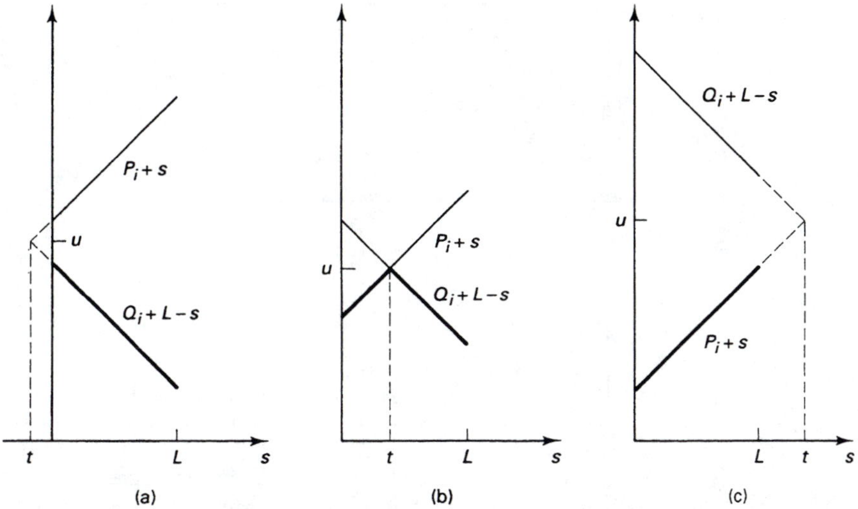

Figure 7.10 Illustration of the three cases for $d(x, v_i)$ based on Figure 7.9.

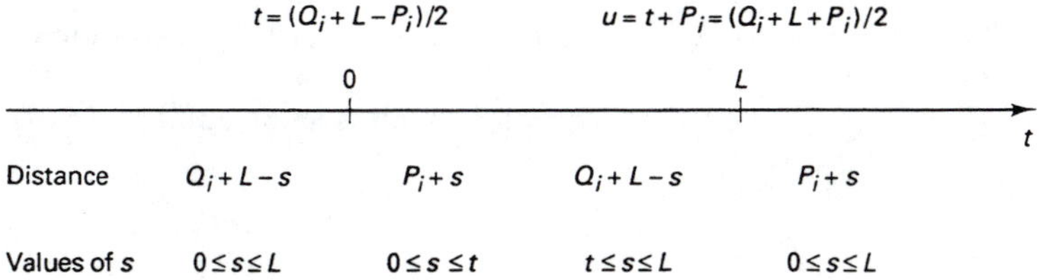

Figure 7.11 Summary of Figure 7.10.

see, as summarized in Figure 7.11, that cases (a), (b), and (c) of Figure 7.10 are as follows:

(a) If $t \le 0$, then $d(y, v_i) = Q_i + L - s$ for $0 \le s \le L$.

(b) If $0 \le t \le L$, then

$$d(y, v_i) = \begin{cases} P_i + s & \text{for } 0 \le s \le t \\ Q_i + L - s & \text{for } t \le s \le L \end{cases}$$

(c) If $L \le t$, then $d(y, v_i) = P_i + s$ for $0 \le s \le L$.

Note that since $t = (Q_i + L - P_i)/2$, $t \le 0$ means that $Q_i - P_i \le -L$, while $L \le t$ means that $L \le Q_i - P_i$. Thus if we define t_i by

$$t_i = \begin{cases} 0 & \text{if } Q_i - P_i < -L \\ \dfrac{L + Q_i - P_i}{2} & \text{if } -L \le Q_i - P_i \le L \\ L & \text{if } L < Q_i - P_i \end{cases}$$

it follows that

$$d(y, v_i) = \begin{cases} P_i + s & \text{for } 0 \le s \le t_i \\ Q_i + L - s & \text{for } t_i \le s \le L \end{cases}$$

Since $d(y, v_i)$ increases as s increases from 0 to t_i and then decreases as s increases from t_i to L, *t_i defines the farthest point in $[v_p, v_q]$ from v_i*. The expression we have for $d(y, v_i)$ is a simple two-piece (or one-piece) linear function in s, and allows us to work with s, the length of arc $[v_p, y]$, instead of y. Further, it gives us a way of constructing the graph of $d(y, v_i)$ for all y in $[v_p, v_q]$.

Table 7.1 illustrates the computation of t_i for arcs $[v_1, v_5]$, $[v_2, v_3]$, and $[v_3, v_4]$ of the example network. See Figure 7.12 for a further illustration of computations for arc $[v_1, v_5]$.

Let us now consider searching for an intersection point for i and j on arc $[v_p, v_q]$. We know that

$$w_i d(y, v_i) = \begin{cases} w_i(P_i + s) & \text{for } 0 \le s \le t_i \\ w_i(Q_i + L - s) & \text{for } t_i \le s \le L \end{cases}$$

$$w_j d(y, v_j) = \begin{cases} w_j(P_j + s) & \text{for } 0 \le s \le t_j \\ w_j(Q_j + L - s) & \text{for } t_j \le s \le L \end{cases}$$

Any intersection point y' for i and j, represented by a distance s' on $[v_p, v_q]$ from v_p, must be an interior point in $[v_p, v_q]$ where one of the two weighted distance functions decreases locally and the other increases locally as s increases in the neighborhood of s'; thus s' must satisfy

$$t_i \ne s' \ne t_j$$

$$w_i(Q_i + L - s') = w_j(P_j + s'), \qquad t_i \le s' \le L, \quad 0 \le s' \le t_j$$

or

$$w_i(P_i + s') = w_j(Q_j + L - s'), \qquad 0 \le s' \le t_i, \quad t_j \le s' \le L$$

These last three conditions imply that s' must satisfy

$$w_i(Q_i + L - s') = w_j(P_j + s'), \qquad t_i < s' < t_j$$

TABLE 7.1 EXAMPLE OF COMPUTING DISTANCE FUNCTION PARAMETERS

		$p = 1,\ q = 5,\ L = 20$			
i	w_i	P_i	Q_i	$Q_i - P_i$	t_i
1	1	0	20	20	20
2	1	10	12	2	11
3	1	24	12	−12	4
4	1	34	20	−14	3
5	1	20	0	−20	0
		$p = 2,\ q = 3,\ L = 14$			
i	w_i	P_i	Q_i	$Q_i - P_i$	t_i
1	1	10	24	14	14
2	1	0	14	14	14
3	1	14	0	−14	0
4	1	24	10	−14	0
5	1	12	12	0	7
		$p = 3,\ q = 4,\ L = 10$			
i	w_i	P_i	Q_i	$Q_i - P_i$	t_i
1	1	24	34	10	10
2	1	14	24	10	10
3	1	0	10	10	10
4	1	10	0	−10	0
5	1	12	20	8	9

or

$$w_i(P_i + s') = w_j(Q_j + L - s'), \qquad t_j < s' < t_i$$

But these two conditions imply s' must satisfy

$$s' = \frac{w_i(Q_i + L) - w_j P_j}{w_i + w_j}, \qquad t_i < s' < t_j$$

or

$$s' = \frac{w_j(Q_j + L) - w_i P_i}{w_i + w_j}, \qquad t_j < s' < t_i$$

We can now state a procedure that either finds an intersection point for i and j on arc $[v_p, v_q]$ or concludes no intersection point exists. The inputs are i, j, p, q, L, the length of arc $[v_p, v_q]$, and the data that depend on these indices. The output is an intersection point distance s' (when an intersection point exists) along with a Boolean variable, IPExists, which is either true or false.

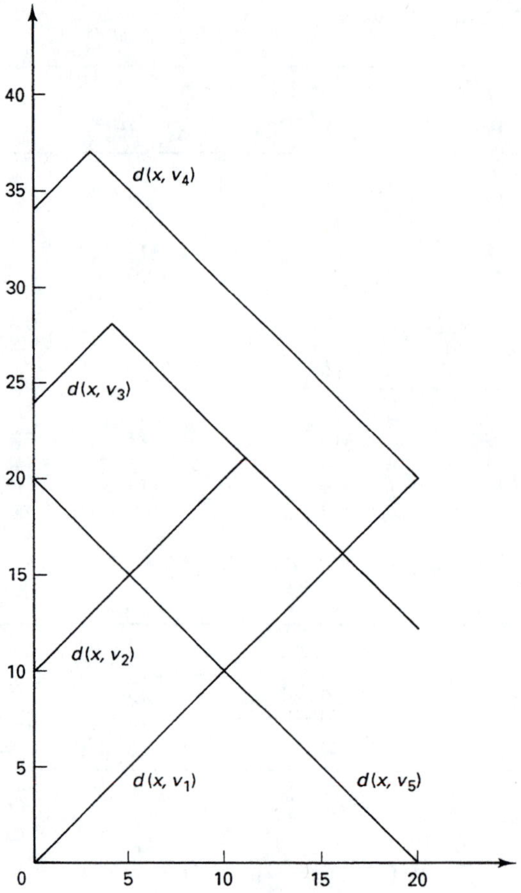

Figure 7.12 Graphs of distances on the arc $[v_1, v_5]$.

Procedure Check_IP_For(i,j,p,q,s',IPExists)

$s1 := [w_i(Q_i + L) - w_j P_j]/(w_i + w_j)$;
$s2 := [w_j(Q_j + L) - w_i P_i]/(w_i + w_j)$;
case1 := FALSE; case2 := FALSE;
IF $t_i < s1 < t_j$, THEN
    case1 := TRUE; $s' := s1$;
IF $t_j < s2 < t_i$, THEN
    case2 := TRUE; $s' := s2$;
IPExists := case1 OR case2;

Note that when $t_i = t_j$ the procedure will always conclude that no intersection point exists, so it is unnecessary to call the procedure in this case. Note also that to compute $d(y', v_i)$ for y' in $[v_p, v_q]$ we use the formula $d(y', v_i) = \min\{d(v_i, v_p) + s', d(v_i, v_q) + L - s'\}$, with s' provided by the procedure when IPExists is true.

We shall now present a pseudocode to find a best intersection point. After executing this code we would still need to compute $\min\{g(v_i): i = 1, \ldots, m\}$ to find a best vertex solution, and then compare this best vertex solution to the best intersection point to get an absolute 1-center.

Procedure FindBestIP

```
MinIPValue := big;
FOR each arc [v_p, v_q] DO
   ArcMinIPValue := big;
   FOR i := 1 TO m - 1 DO
      FOR j := i + 1 TO m DO
         IF t_i ≠ t_j, THEN
            Check_IP_For(i,j,p,q,s',IPExists);
            IF IPExists, THEN
               IPValue := g(y');
            IF IPValue < ArcMinIPValue, THEN
               ArcMinIPValue := IPValue;
               ArcBestIPSoFar := s';
   IF ArcMinIPValue < MinIPValue, THEN
      MinIPValue := ArcMinIPValue;
      BestArcSoFar := [v_p, v_q];
      BestIPSoFar := ArcBestIPSoFar;
```

Hopefully the variable names are mostly self-explanatory; the variable "big" is a very big number used for initialization purposes. What the procedure does is to find a best intersection point on each arc (when one exists), and then choose the best of these. At the end it gives the minimum g value found for an intersection point, the arc that contains this intersection point, and the distance along the arc of this best intersection point.

Table 7.2 illustrates the computation of intersection points for every i and j on arc $[v_p, v_q] = [v_2, v_3]$. In the table, the intersection point distance from v_2 is denoted by s'_{ij} (this is the s' provided by the Check_IP procedure); for each i and j the table shows t_i, t_j, case 1 or case 2 (when an intersection point exists), and the value of IPExists.

For our example problem, after enumerating all the intersection points on every arc and choosing a best one, and after evaluating $g(y)$ at every vertex v_i and choosing a best one, we find that the absolute center is on arc $[v_2, v_3]$ with value 18. Note that there are two optimal solutions.

The solution to the absolute center problem can be characterized by the following:

TABLE 7.2 EXAMPLE OF CALCULATION OF INTERSECTION POINTS

			$p = 2, q = 3$			
i	j	t_i	t_j	Case	IPExists	s'_{ij}
1	2	14	14	—	False	—
	3	14	0	2	True	2
	4	14	0	2	True	7
	5	14	7	2	True	8
2	3	14	0	2	True	7
	4	14	0	2	True	12
	5	14	7	2	True	13
3	4	0	0	—	False	—
	5	0	7	1	True	1
4	5	0	7	1	True	6

Vertex and Intersection Point (VIP) Property. There exists an optimal absolute center that is a vertex or an intersection point.

This property may seem obvious, but it will prove quite useful when we consider the n-center problem.

Let us examine very briefly a technique that can improve our enumeration approach for finding an absolute center, by allowing us to avoid enumerating intersection points on some arcs. Suppose, in the course of the enumeration, that we have a current best "trial" solution, $y^\wedge$, with function value $g(y^\wedge)$. We now consider arc $[v_p, v_q]$; if we know a number, say β_{pq}, so that $g(y) \geq \beta_{pq}$ for all y in $[v_p, v_q]$, and also $\beta_{pq} > g(y^\wedge)$, *no* location on $[v_p, v_q]$ can be an absolute center. We may be able to use the following property, which gives one possible choice of β_{pq}, to avoid enumerating all the intersection points:

Arc Exclusion-Bounding Property. Define

$$\beta_{ipq} = \min\{w_i d(v_p, v_i), w_i d(v_q, v_i)\}$$

$$\beta_{pq} = \max\{\beta_{ipq}: i = 1, \ldots, m\}$$

If $\beta_{pq} > g(y^\wedge)$, no point in arc $[v_p, v_q]$ is an absolute 1-center.

For our example we have $\beta_{15} = 20$, $\beta_{23} = 12$, and $\beta_{34} = 24$. Thus, in our enumeration, *if* we considered arc $[v_2, v_3]$ before arc $[v_3, v_4]$ we would have found a $y^\wedge$ with $g(y^\wedge) = 18$, and we could use the arc exclusion-bounding property to avoid enumerating the intersection points in $[v_3, v_4]$.

The arc-exclusion-bounding property is a result of the concavity of distance on any arc: $\beta_{ipq} \leq w_i d(y, v_i)$ for all y in $[v_p, v_q]$, so that $\beta_{pq} \leq g(y)$ for all y in $[v_p, v_q]$.

Finally, recall from our discussion of the 1-center problem on tree networks that we defined a bound matrix $\mathbf{B} = (b_{ij})$. We can calculate bounds in this fashion for cyclic networks, but the largest entry in this bound matrix may be strictly less than

$g(y^*)$, the optimum value. If we consider as an example the 1-center problem with unit weights obtained from the network of Figure 7.7, we have $b_{ij} = d(v_i, v_j)/2$, and find that $b_{st} = b_{14} = d(v_1, v_4)/2 = 17$, half the distance between the two most distant vertices, v_1 and v_4. As we know the minimal value of $g(y)$ to be 18, we can see that b_{14} is one less than the minimal value. Another way to verify that the lower bound b_{14} is not attained is to verify that the conditions (7.1) and (7.2) for the bound to be attained are violated: the conditions become $d(v_1, v_4) = d(v_1, y) + d(y, v_4)$ and $d(y, v_1) = g(y) = d(y, v_4)$, implying that y must be the midpoint, say y', of the shortest path joining v_1 and v_4. With reference to Figure 7.8, $g(y') = 19$, while $d(y', v_1) = 17 = d(y', v_4) < g(y') = 19$, so the lower bound is not attained. In general, the utility of the bound is dependent on how close it is to the minimal value of $g(y)$. If we have a value of $g(y)$ very close to the lower bound, we may decide we are "close enough" and not bother to find the minimal value of $g(y)$ if we still have many arcs remaining to search.

Before turning to the n-center problem, we consider the vertex-center problem, choosing from among all vertex locations one which minimizes $g(y)$. The vertex-center problem is much easier to solve than the absolute-center problem and in some cases is more realistic as well. If, for example, we only have a finite number of locations on the given network to choose from, we can define each to be a vertex, and then choose from among them a best location. Alternatively, if the network represents an air transport network, vertices typically represent airports, while the arcs represent air corridors, so that it is realistic to consider only vertices for locations.

To find a vertex center, we proceed as follows. Compute the matrix $\mathbf{D} = (d_{ij})$ whose entry in row i and column j is $d(v_i, v_j)$. (One way to compute the entries in $\mathbf{D}$ is to use Floyd's algorithm.) For each column j of $\mathbf{D}$, multiply every entry in the column by w_j, the weight for v_j, to obtain the matrix denoted by $\mathbf{D}''$. For each row i of $\mathbf{D}''$, compute the largest row i entry to obtain $g(v_i)$, and place this entry in the right margin. Any row s with a smallest right margin entry identifies a vertex center, say v_s.

Consider an example. We use the matrix $\mathbf{D}$ of the appendix based on the network of Figure 7.7, with weights of 1, 2, 1, 3, and 1 for vertices 1 to 5, respectively. The matrix $\mathbf{D}''$, the weights, and the $g(v_i)$ values are shown in Table 7.3. We can see

TABLE 7.3 EXAMPLE OF A VERTEX-CENTER PROBLEM

	Weight					
$\mathbf{D}''$	1	2	1	3	1	g Value
	0	20	24	102	20	102
	10	0	14	72	12	72
	24	28	0	30	12	30
	34	48	10	0	20	48
	20	24	12	60	0	60

from the table that v_3 is a vertex center, with $g(v_3) = 30$. As an exercise, we suggest that you solve the vertex-center problem for the same network when all the weights are 1; you should find that v_5 is a vertex center.

As a final topic in this section we consider again the n-center problem, for which $Y = \{y_1, \ldots, y_n\}$ is a collection of centers, with each center on some arc, and $n < m$. As is customary, for each vertex v_i we let $\mathbf{D}(Y, v_i)$ denote the distance between v_i and a closest center in Y, so that we are again assuming that travel is between each vertex and a closest center. As with the 1-center problem, we let the first m vertices have positive weights, and we wish to find an absolute n-center Y^* that minimizes

$$G(Y) = \max\{w_1 \mathbf{D}(Y, v_1), \ldots, w_m \mathbf{D}(Y, v_m)\}$$

We remind you that the n-center is a natural extension of the 1-center problem, and that the application contexts of the two problems are identical, with weights typically having similar interpretations for the two problems.

Our principal conclusion about the n-center problem in this section is the following:

Vertex and Intersection Point (VIP) Property. There exists at least one absolute n-center for which each center is at a vertex or an intersection point. Thus it suffices to consider only vertex and intersection point locations in order to find an absolute n-center.

As a motivation for the property, consider a situation with $m = 5$ vertices, and $Y = \{y_1, y_2\}$ an absolute 2-center. Suppose that y_1 is closest to vertices 1, 2, and 3, while y_2 is closest to vertices 4 and 5. If $g_1(y_1)$ denotes the 1-center function for vertices 1, 2, and 3, while $g_2(y_2)$ denotes the 1-center function for vertices 4 and 5, it now follows that $G(Y) = \max\{g_1(y_1), g_2(y_2)\}$. If we solve the 1-center problems of minimizing g_1 and minimizing g_2, there are absolute center locations y_1^* and y_2^*, each a vertex or an intersection point, for which $g_1(y_1^*) \leq g_1(y_1)$ and $g_2(y_2^*) \leq g_2(y_2)$, implying that

$$\max\{g_1(y_1^*), g_2(y_2^*)\} \leq \max\{g_1(y_1), g_2(y_2)\} = G(Y)$$

Now if we define the 2-center $Y^* = \{y_1^*, y_2^*\}$, the closest center assumption implies that $G(Y^*) \leq \max\{g_1(y_1^*), g_2(y_2^*)\}$. Hence we conclude that $G(Y^*) \leq G(Y)$. But since Y is an absolute 2-center and Y^* is a 2-center, we also know that $G(Y) \leq G(Y^*)$. Thus we conclude $G(Y^*) = G(Y)$, so that Y^* is an absolute 2-center with *each center being a vertex or an intersection point*.

The principal value of the VIP property is to reduce the absolute n-center problem from a problem with an infinite number of points to consider to a problem with a finite number of points to consider. In Chapter 8 we shall see how to exploit this finiteness result to solve the n-center problem by solving a sequence of covering problems formulated as integer programs.

7.9 FURTHER READING

The convexity discussion is from Dearing et al. [34]. The Chinese algorithm is from Hua Lo Keng et al. [85] and the median algorithm is due to Goldman [53]. The lower bound approach to the 1-center problem is from Dearing and Francis [33] and the algorithm for the unweighted problem is from Handler [71], with the analogous approach for the weighted problem being from Hakimi et al. [62]. The covering and n-center discussion is from Tansel et al. [144].

For the cyclic 1-median and 1-center discussion the sources are two papers of fundamental importance by Hakimi [59,60], which were largely instrumental in stimulating modern-day interest in network location problems. The amount of research on network location problems since the publication of these two papers really seems quite remarkable, as a glance at our list of references, which is by no means exhaustive, indicates.

The references mentioned above are all journal papers, which require a basic knowledge of network theory; they are typically not self-contained discussions. Thus for a good self-contained introduction to network location theory, see the book by Handler and Mirchandani [76], which is currently the only book devoted entirely to the topic. For books with some material on the topic, see Christofides [28] and Minieka [109]. For a book of edited readings devoted entirely to the subject of this and the next two chapters, see Mirchandani and Francis [118].

There are a number of review papers that cover network locations problems to some extent. In chronological order, the ones we are aware of are ReVelle [125], Halpern and Maimon [69], Tansel et al. [144], Francis et al. [45], Dearing [35], Moon and Chaudry [120], Hansen et al. [79], and Brandeau and Chiu [15].

For solving cyclic network location problems of any size, it is essential to have computerized shortest-path algorithms. The book by Syslo et al. [140] has codes of such algorithms, as well as of some related algorithms, such as ones for solving covering problems. It also has a discussion of computer representations of networks.

Space limitations and priorities have unfortunately prevented us from discussing several important areas of network location work, including location problems with queueing, obnoxious facility location problems, public-sector location problems, and competitive location problems; there are too many references in these areas to itemize here, but most of them can be identified easily in our list of references. Often a new facility to be located provides some service that results in a queue of servers, so that queuing aspects can be quite important. Sometimes a facility to be located, such as a garbage dump or a power generating plant, is "obnoxious" and one wants it to be far away instead of close; such obnoxious location problems have begun to receive attention. Locating facilities to provide public services can involve different objectives than ones used for private-sector problems, and the book edited by Thisse and Zoller [147] is notable in this regard. For private-sector location problems involving, say, retail stores, where the competition locates their stores can be an important consideration in choosing locations, so that competitive location problems are now receiving attention.

REFERENCES

1. Andreatta, G., and F. Mason, "k-Eccentricity and Absolute k-Centrum for a Probabilistic Tree," *European Journal of Operational Research,* Vol. 19, 1985, pp. 114–117.
2. Andreatta, G., and F. Mason, "Properties of the k-Centra in a Tree Network," *Networks,* Vol. 15, 1985, pp. 21–29.
3. Barthe'lemy, J. P., "Médians dans les graphes et localisation," *Cahiers du Centre d'Etudes de Recherche Operationnelle,* Vol. 25, 1983, pp. 163–182.
4. Barthe'lemy, J. P., "Characterisationmédiane de arbres," *Annals of Discrete Mathematics,* Vol. 17, 1983, pp. 39–46.
5. Batta, R., "Single-Server Queueing Location Models with Rejection," *Transportation Science,* Vol. 22, 1988, pp. 209–216.
6. Batta, R., R. C. Larson, and A. R. Odoni, "A Single-Server Priority Queueing-Location Model," *Networks,* Vol. 8, 1988, pp. 87–103.
7. Berman, O., "Locating a Facility on a Congested Network with Random Lengths," *Networks,* Vol. 15, 1985, pp. 275–294.
9. Berman, O., and R. C. Larson, "Optimal 2-Facility Network Districting in the Presence of Queueing," *Transportation Science,* Vol. 19, 1985, pp. 261–277.
10. Berman, O., R. C. Larson, and S. Chiu, "Optimal Server Location on a Network Operating as an M/G/1 Queue," *Operations Research,* Vol. 33, 1985, pp. 746–771.
11. Berman, O., and A. R. Odoni, "Locating Mobile Servers on a Network with Markovian Properties," *Networks,* Vol. 12, 1982, pp.73–86.
12. Berman, O., and D. Simchi-Levi, "Minisum Location of a Travelling Salesman," *Networks,* Vol. 16, 1986, pp. 239–254.
13. Brandeau, M. L., and S. S. Chiu, "Parametric Facility Location on a Tree Network with an l_p Norm Cost Function," *Transportation Science,* Vol. 22, 1988, pp. 59–69.
14. Brandeau, M. L., and R. Batta, "Finding the Two-Median of a Tree Network with Continuous Link Demands," *Annals of Operations Research,* Vol. 6, 1986, pp. 223–253.
15. Brandeau, M. L., and S. S. Chiu, "An Overview of Representative Problems in Location Research," *Management Science,* Vol. 35, 1989, pp. 645–673.
16. Chan, A., and R. L. Francis, "A Round-Trip Location Problem on a Tree Graph," *Transportation Science,* Vol. 10, 1976, pp. 35–51.
17. Chandrasekaran, R., and A. Daughety, "Location on Tree Networks: p-Center and N-Dispersion Problems," *Mathematics of Operations Research,* Vol. 6, 1981, pp. 50–57.
18. Chandrasekaran, R., and A. Tamir, "An $O(n \log p)^2$ for the Continuous p-Center Problem on a Tree," *SIAM Journal on Algebraic and Discrete Methods,* Vol. 1, 1980, pp. 370–375.
19. Chandrasekaran, R., and A. Tamir, "Polynomially Bounded Algorithms for Locating p-Centers on a Tree," *Mathematical Programming,* Vol. 22, 1982, pp. 304–315.
20. Chandrasekaran, R., and P. Viola, "The Optimum Location of Multi-centers on a Graph," *Operational Research Quarterly,* Vol. 22, 1971, pp. 145–154.
21. Chen, C. E., and R. S. Garfinkel, "The Generalized Diameter of a Graph," *Networks,* Vol. 12, 1982, pp. 335–340.

22. Chen, M.-L., R. L. Francis, J. F. Lawrence, T. J. Lowe, and S. Tufekci, "Block-Vertex Duality and the One-Median Problem," *Networks,* Vol. 15, 1985, pp. 395–412.
23. Chen, M.-L., R. L. Francis, and T. J. Lowe, "The 1-Center Problem: Exploiting Block Structure," *Transportation Science,* Vol. 22, 1988, pp. 259–269.
24. Chiu, S. S., "Optimal M/G/1 Server Location on a Tree Network with Continuous Link Demands," *Computers and Operations Research,* Vol. 13, 1986, pp. 653–669.
25. Chiu, S. S., "The Minisum Location Problem on an Undirected Network with Continuous Link Demands," *Computers and Operations Research,* Vol. 14, 1987, pp. 369–383.
26. Chiu, S. S., O. Berman, and R. C. Larson, "Locating a Mobile Server Queueing Facility on a Tree Network," *Management Science,* Vol. 31, 1985, pp. 767–772.
27. Chiu, S. S., and R. C. Larson, "Locating an *N*-Server Facility in a Stochastic Environment," *Computers and Operations Research,* Vol. 12, 1985, pp. 509–516.
28. Christofides, N., *Graph Theory: An Algorithmic Approach,* Academic Press, New York, 1975.
29. Church, R. L., and R. S. Garfinkel, "Locating an Obnoxious Facility on a Network," *Transportation Science,* Vol. 12, 1978, pp. 107–118.
30. Church, R. L., and M. E. Meadows, "Location Modelling Utilizing Maximum Service Distance Criteria," *Geographical Analysis,* Vol. 11, 1979, pp. 358–373.
31. Cunningham-Green, R. A., "The Absolute Center of a Graph," *Discrete Applied Mathematics,* Vol. 7, 1984, pp. 275–283.
32. Dearing, P. M., "Minimax Location Problems with Nonlinear Costs," *Journal of Research of the National Bureau of Standards,* Vol. 82, 1977, pp. 65–72.
33. Dearing, P. M., and R. L. Francis, "A Minimax Location Problem on a Network," *Transportation Science,* Vol. 8, 1974, pp. 333–343.
34. Dearing, P. M., R. L. Francis, and T. J. Lowe, "Convex Location Problems on Tree Networks," *Operations Research,* Vol. 24, 1976, pp. 628–642.
35. Dearing, P. M., "Location Problems," *Operations Research Letters,* Vol. 4, 1985, pp. 95–98.
36. Dobson, G., and U. S. Karmarkar, "Competitive Location on a Network," *Operations C,* Vol. 35, 1987, pp. 565–574.
37. Domschke, W., and A. Drexl, *Location and Layout Planning: An International Bibliography,* Springer-Verlag, Berlin, 1985.
38. Domschke, W., and A. Drexl, *Logistik: Standorte* (in German), R. Oldenbourg Verlag GmbH, Munich, 1985.
39. Eiselt, H. A., and G. Pederzoli, "A Location Problem in Graphs," *New Zealand Journal of Operational Research,* Vol. 12, 1984, pp. 49–53.
40. Erkut, E., R. L. Francis, and T. J. Lowe, "A Multimedian Problem with Interdistance Constraints," *Environment and Plannng B: Planning and Design,* Vol. 15, 1988.
41. Erkut, E., R. L. Francis, and T. J. Lowe, "Locating Two Facilities on a Tree Subject to Distance Constraints," *Transportation Science,* Vol. 22, 1988, pp. 199–208.
42. Erkut, E., R. L. Francis, T. J. Lowe, and A. Tamir, "Mathematical Programming Formulations of Monotonic Tree Network Location Problems," *Operations Research,* Vol. 37, 1989, pp. 447–461.

43. Farley, A. M., "Vertex Centers of Trees," *Transportation Science,* Vol. 16, 1982, pp. 265–280.

44. Francis, R. L., T. J. Lowe, and H. D. Ratliff, "Distance Constraints for Tree Network Multifacility Location Problems," *Operations Research,* Vol. 26, 1978, pp. 570–596.

45. Francis, R. L., L. F. McGinnis, and J. A. White, "Locational Analysis," *European Journal of Operational Research,* Vol. 12, 1983, pp. 220–252.

46. Frank, H., "Optimum Locations on a Graph with Probabilistic Demands," *Operations Research,* Vol. 14, 1966, pp. 409–421.

47. Frank, H., "Optimum Locations on a Graph with Correlated Normal Demands," *Operations Research,* Vol. 15, 1967, pp. 552–556.

48. Frank, H., "A Note on a Graph Theoretic Game of Hakimi's," *Operations Research,* Vol. 15, 1967, pp. 567–570.

49. Garfinkel, R. S., A. W. Neebe, and M. R. Rao, "An Algorithm for the m-Median Plant Location Problem," *Transportation Science,* Vol. 8, 1974, pp. 217–236.

50. Garfinkel, R., A. Neebe, and M. Rao, "The m-Center Problem: Minimax Facility Location," *Management Science,* Vol. 23, 1977, pp. 1133–1142.

51. Gavril, R., "Algorithms for Minimum Coloring, Maximum Clique, Minimum Covering by Cliques, and Maximum Independent Set of a Chordal Graph," *SIAM Journal of Computing,* Vol. 1, 1972, pp. 180–187.

52. Goldman, A. J., "Optimum Locations for Centers in a Network," *Transportation Science,* Vol. 3, 1969, pp. 352–360.

53. Goldman, A. J., "Optimal Center Location in Simple Networks," *Transportation Science,* Vol. 5, 1971, pp. 212–221.

54. Goldman, A. J., "Approximate Localization Theorems for Optimal Facility Placement," *Transportation Science,* Vol. 6, 1972, pp. 195–201.

55. Goldman, A. J., "Minimax Location of a Facility in a Network," *Transportation Science,* Vol. 6, 1972, pp. 407–418.

56. Goldman, A. J., and C. G. Witzgall, "A Localization Theorem for Optimal Facility Placement," *Transportation Science,* Vol. 4, 1970, pp. 406–409.

57. Guelicher, H., "Einige Eigenschaften optimaler Standorte in Verkehrsnetzen," *Schriften des Vereins fuer Sozialpolitik,* Neue Folge 42, 1965, pp. 111–137.

58. Gurevich, Y., L. Stockmeyer, and U. Vishkin, "Solving NP-Hard Problems on Graphs That Are Almost Trees and an Application to Facility Location Problems," *Journal of the Association of Computing Machinery,* Vol. 31, 1984, 459–473.

59. Hakimi, S. L., "Optimal Locations of Switching Centers and the Absolute Centers and Medians of a Graph," *Operations Research,* Vol. 12, 1964, pp. 450–459.

60. Hakimi, S. L., "Optimum Distribution of Switching Centers in a Communication Network and Some Related Graph Theoretic Problems," *Operations Research,* Vol. 13, 1965, pp. 462–475.

61. Hakimi, S. L., and S. N. Maheshwari, "Optimum Locations of Centers in Networks," *Operations Research,* Vol. 20, 1972, pp. 967–973.

62. Hakimi, S. L., E. F. Smeichel, and J. G. Pierce, "On p-Centers in Networks," *Transportation Science,* Vol. 12, 1978, pp. 1–15.

63. Hakimi, S. L., "On Locating New Facilities in a Competitive Environment," *European Journal of Operational Research,* Vol. 12, 1983, pp. 29–35.

64. Halfin, S., "On Finding the Absolute and Vertex Centers of a Tree with Distances," *Transportation Science,* Vol. 8, 1974, pp. 75–77.

65. Halpern, J., "The Location of a Center-Median Convex Combination on an Undirected Tree," *Journal of Regional Science,* Vol. 16, 1976, pp. 237–245.

66. Halpern, J., "Finding Minimal Center-Median Convex Combinations (Cent-Dian) of a Graph," *Management Science,* Vol. 24, 1978, pp. 534–544.

67. Halpern, J., "A Simple Edge Elimination Criterion on a Search for the Center of a Graph," *Management Science,* Vol. 25, 1979, pp. 105–107.

68. Halpern, J., "Duality in the Cent-Dian of a Graph," *Operations Research,* Vol. 28, 1980, pp. 722–735.

69. Halpern, J., and O. Maimon, "Algorithms for the m-Center Problems: A Survey," *European Journal of Operations Research,* Vol. 10, 1982, pp. 90–99.

70. Halpern, J., and O. Maimon, "Accord and Conflicts among Several Objectives in Locational Decisions on Tree Networks," in J.-F. Thisse and H. G. Zoller, eds., *Locational Analysis of Public Facilities,* North-Holland Publishing Company, Amsterdam, 1983, pp. 301–314.

71. Handler, G. Y., "Minimax Location of a Facility in an Undirected Tree Graph," *Transportation Science,* Vol. 7, 1973, pp. 287–293.

72. Handler, G. Y., "Finding Two-Centers of a Tree: The Continuous Case," *Transportation Science,* Vol. 12, 1978, pp. 93–106.

73. Handler, G. Y., "Medi-centers of a Tree," *Transportation Science,* Vol. 19, No. 3, 1985, pp. 246–260.

74. Handler, G. Y., and M. Rozman, "The Continuous m-Center Problem on a Network," *Networks,* Vol. 15, 1985, pp. 191–204.

75. Handler, G. Y., "Complexity and Efficiency in Minimax Network Location," in N. Christofides, A. Mingozzi, P. Toth, and C. Sandi, eds., *Combinatorial Optimization,* John Wiley & Sons, Inc., New York, 1979.

76. Handler, G. Y., and P. B. Mirchandani, *Location on Networks,* The MIT Press, Cambridge, MA, 1979.

77. Hansen, P., and J.-F. Thisse, "Outcomes of Voting and Planning: Condorcet, Weber and Rawls Locations," *Journal of Public Economics,* Vol. 16, 1981, pp. 1–15.

78. Hansen, P., D. Peeters, and J.-F. Thisse, "Public Facility Location: A Selective Survey," in J.-F. Thisse and H. G. Zoller, eds., *Locational Analysis of Public Facilities,* North-Holland Publishing Company, Amsterdam, 1983.

79. Hansen, P., M. Labbé, D. Peeters, and J.-F. Thisse, "Single Facility Location on Networks," *Annals of Discrete Mathematics,* Vol. 31, 1987, pp. 113–145.

80. Hansen, P., D. Peeters, D. Richard, and J.-F. Thisse, "The Minisum and Minimax Location Problems Revisited," *Operations Research,* Vol. 33, 1985, pp. 1251–1265.

81. Hedetniemi, S. M., E. J. Cockayne, and S. T. Hedetniemi, "Linear Algorithms for Finding the Jordan Center and Path Center of a Tree," *Transportation Science,* Vol. 15, 1981, pp. 98–114.

82. Hooker, J. N., "Solving Nonlinear Single-Facility Network Location Problems," *Operations Research,* Vol. 34, 1986, pp. 732–743.

83. Horn, W. A., "Three Results for Trees, Using Mathematical Induction," *Journal of Research of the National Bureau of Standards,* Vol. 76B, 1972, pp. 39–43.

84. Hsu, W. L., and G. L. Nemhauser, "Easy and Hard Bottleneck Location Problems," *Discrete Applied Mathematics,* Vol. 1, 1979, pp. 209–215.

85. Hua, L. K., and others, "Applications of Mathematical Methods to Wheat Harvesting," *Chinese Mathematics,* Vol. 2, 1962, pp. 77–91.

86. Ja"rvinen, P., J. Fajala, and H. Sinervo, "A Branch-and-Bound Algorithm for Seeking the p-Median," *Operations Research,* Vol. 20, 1972, pp. 173–182.

87. Jordan, C., "Sur les assemblages des lignes," *Zeitschrift fuer Reine und Angewandte Mathematik,* Vol. 70, 1869, pp. 185–190.

88. Kariv, O., and S. L. Hakimi, "An Algorithmic Approach to Network Location Problems. Part 1: The p-Centers," *SIAM Journal of Applied Mathematics,* Vol. 37, 1979, pp. 513–538.

89. Kariv, O., and S. L. Hakimi, "An Algorithmic Approach to Network Location Problems. Part 2: The p-Medians," *SIAM Journal of Applied Mathematics,* Vol. 37, 1979, pp. 539–560.

90. Kolen, A., *Location Problems on Trees and in the Rectilinear Plane,* Stitchting Mathematisch Centrum, Amsterdam, 1982.

91. Kolen, A., "Equivalence between the Direct Search Approach and the Cut Approach to the Rectilinear Distance Location Problem," *Operations Research,* Vol. 29, 1981, pp. 616–620.

92. Kolen, A., "The Round-Trip p-Center and Covering Problem on a Tree," *Transportation Science,* Vol. 19, 1985, pp. 222–235.

93. Kolen, A., "Solving Covering Problems and the Uncapacitated Plant Location Problem on Trees," *European Journal of Operational Research,* Vol. 12, 1983, pp. 266–278.

94. Krarup, J., and P. Pruzan, "Selected Families of Location Problems," *Annals of Discrete Mathematics,* Vol. 5, 1979, pp. 327–387.

95. Labbé, M., "Essays in Network Location Theory," *Cahiers du Centre d'Etudes de Recherche Operationelle,* Vol. 27, 1985, pp. 5–130.

96. Labbé, M., "Outcomes of Voting and Planning in Single Facility Location Problems," *European Journal of Operational Research,* Vol. 20, 1985, pp. 299–313.

97. Levy, J., "An Extended Theorem for Location on a Network," *Operational Research Quarterly,* Vol. 18, 1972, pp. 433–442.

98. Lin, C. C., "On Vertex Addends in Minimax Location Problems," *Transportation Science,* Vol. 9, 1975, pp. 165–168.

99. Louveaux, F., J.-F. Thisse, and H. Beguin, "Location Theory and Transportation Costs," *Regional Science and Urban Economics,* Vol. 12, 1982, pp. 529–545.

100. Lowe, T. J., "Efficient Solutions in Multiobjective Tree Network Location Problems," *Transportation Science,* Vol. 12, 1979, pp. 298–316.

101. Mavrides, L. P., "An Indirect Method for the Generalized k-Median Problem Applied to Lock-Box Location," *Management Science,* Vol. 25, 1979, pp. 990–996.

102. McGinnis, L., "A Survey of Recent Results for a Class of Facilities Location Problems," *AIIE Transactions,* Vol. 9, 1977, pp. 11–18.

103. Megiddo, N., A. Tamir, E. Zemel, and R. Chandrasekaran, "An $O(n \log^2 n)$ Algorithm for the kth Longest Path in a Tree with Applications to Location Problems," *SIAM Journal of Computing,* Vol. 10, 1981, pp. 328–337.

104. Megiddo, N., "Linear-Time Algorithms for LP in R^3 and Related Problems," *SIAM Journal of Computing,* Vol. 12, 1983, pp. 759–776.

105. Megiddo, N., E. Zemel, and S. L. Hakimi, "The Maximum Coverage Location Problem," *SIAM Journal of Algebraic and Discrete Methods,* Vol. 4, 1983, pp. 253–261.

106. Meir, A., and J. W. Moon, "Relations between Packing and Covering Numbers of a Tree," *Pacific Journal of Mathematics,* Vol. 61, 1979, pp. 225–233.

107. Minieka, E., "The m-Center Problem," *SIAM Review,* Vol. 12, 1970, pp. 138–139.

108. Minieka, E., "The Centers and Medians of a Graph," *Operations Research,* Vol. 25, 1977, pp. 641–650.

109. Minieka, E., *Optimization Algorithms for Networks and Graphs,* Marcel Dekker, Inc., New York, 1978.

110. Minieka, E., "Conditional Centers and Medians of a Graph," *Networks,* Vol. 10, 1980, pp. 265–272.

111. Minieka, E., "A Polynomial Time Algorithm for Finding the Absolute Center of a Network," *Networks,* Vol. 11, 1981, pp. 351–355.

112. Minieka, E., "Radial Location Theory," *Networks,* Vol. 13, 1983, pp. 233–239.

113. Minieka, E., "Anti-centers and Anti-medians of a Network," *Networks,* Vol. 13, 1983, pp. 359–365.

114. Mirchandani, P. B., "Locational Decisions on Stochastic Networks," *Geographical Analysis,* Vol. 12, 1980, pp. 172–183.

115. Mirchandani, P. B., and A. R. Odoni, "Locating New Passenger Facilities on a Transportation Network," *Transportation Research,* Vol. 13B, 1979, pp. 113–122.

116. Mirchandani, P. B., and A. R. Odoni, "Locations of Medians on Stochastic Networks," *Transportation Science,* Vol. 13, 1979, pp. 85–97.

117. Mirchandani, P. B., and A. Oudjit, "Localizing 2-Medians on Probabilistic and Deterministic Tree Networks," *Networks,* Vol. 10, 1980, pp. 329–350.

118. Mirchandani, P. B., and R. L. Francis, eds., *Discrete Location Theory,* John Wiley & Sons, Inc., New York, 1990.

119. Mitchell, S. L., "Another Characterization of the Centroid of a Tree," *Discrete Mathematics,* Vol. 24, 1978, pp. 277–280.

120. Moon, D. I., and S. S. Chaudhry, "An Analysis of Network Location Problems with Distance Constraints," *Management Science,* Vol. 30, 1984, pp. 290–307.

121. Moreno, J. A., "A Correction to the Definition of Local Center," *European Journal of Operational Research,* Vol. 20, 1985, pp. 382–386.

122. Morgan, C. A., and P. J. Slater, "A Linear Algorithm for a Core of a Tree," *Journal of Algorithms,* Vol. 1, 1980, pp. 247–258.

123. Narula, S. C., V. I. Ogbu, and H. M. Samuelsson, "An Algorithm for the p-Median Problem," *Operations Research,* Vol. 25, 1977, pp. 709–712.

124. Picard, J. C., and H. D. Ratliff, "A Cut Approach to the Rectilinear Distance Location Problems," *Operations Research,* Vol. 26, 1978, pp. 422–434.

125. ReVelle, C., D. Marks, and J. C. Liebman, "An Analysis of Private and Public Sector Location Models," *Management Science,* Vol. 16, 1970, pp. 692–707.

126. Rose, D. J., R. E. Tarjan, and G. S. Lucker, "Algorithmic Aspects of Vertex Elimination in Graphs," *SIAM Journal of Computing,* Vol. 5, 1976, pp. 266–281.

127. Rosenthal, A., M. Hersey, J. Pino, and M. Coulter, "A Generalized Algorithm for Centrality Problems on Trees," *Proceedings of the Allerton Conference on Communication, Control, and Computing,* 1978, pp. 616–625.

128. Shier, D. R., "A Min-Max Theorem for p-Center Problems on a Tree," *Transportation Science,* Vol. 11, 1977, pp. 243–252.

129. Shier, D. R., and P. M. Daring, "Optimal Locations for a Class of Nonlinear Single Facility Location Problems on a Network," *Operations Research,* Vol. 31, 1983, pp. 292–303.

130. Slater, P. J., "Maxmin Facility Location," *Journal of Research of the National Bureau of Standards,* Vol. 79B, 1975, pp. 107–115.

131. Slater, P. J., "Central Vertices in a Graph," *Proceedings of the 7th Southeastern Conference on Combinatorics, Graph Theory, and Computing,* 1976, pp. 487–489.

132. Slater, P. J., "Center to Centroid in Graphs," *Journal of Graph Theory,* Vol. 2, 1978, pp. 209–222.

133. Slater, P. J., "Structure of the k-Centra in a Tree," *Proceedings of the 9th Southeastern Conference on Combinatorics, Graph Theory, and Computing,* 1978.

134. Slater, P. J., "Medians of Arbitrary Graphs," *Journal of Graph Theory,* Vol. 4, 1980, pp. 389–392.

135. Slater, P. J., "Locating Central Paths in a Graph," *Transportation Science,* Vol. 16, 1982, pp. 1–18.

136. Slater, P. J., *Some Definitions of Central Structures in a Graph,* Sandia Laboratories, Albuquerque, NM, 1980.

137. Slater, P. J., "On Locating a Facility to Service Areas within a Network," *Operations Research,* Vol. 29, 1981, pp. 527–531.

138. Slater, P. J., "The k-Nucleus of a Graph," *Networks,* Vol. 11, 1981, pp. 232–242.

139. Slater, P. J., "Locating Central Paths in a Graph," *Transportation Science,* Vol. 16, 1982, pp. 1–18.

140. Syslo, M. M., N. Deo, and J. S. Kowalik, *Discrete Optimization Algorithms with PASCAL Programs,* Prentice Hall, Englewood Cliffs, NJ, 1983.

141. Tamir, A., and E. Zemel, "Locating Centers on a Tree with Discontinuous Supply and Demand Regions," *Mathematics of Operations Research,* Vol. 7, 1982, pp. 183–197.

142. Tansel, B. C., R. L. Francis, and T. J. Lowe, "Binding Inequalities for Tree Network Location Problems with Distance Constraints," *Transportation Science,* Vol. 14, 1980, pp. 107–124.

143. Tansel, B. C., R. L. Francis, T. J. Lowe, and M.-L. Chen, "Duality and Distance Constraints for the Nonlinear p-Center Problem and Covering Problem on a Tree Network," *Operations Research,* Vol. 30, 1982, pp. 725–743.

144. Tansel, B. C., R. L. Francis, and T. J. Lowe, "A Bi-Objective Minimax Location Problem on a Tree Network," *Transportation Science,* Vol. 16, 1982, pp. 407–429.

145. Tansel, B. C., R. L. Francis, and T. J. Lowe, 1983, "Location on Networks: A Survey. Part I: The p-Center and p-Median Problems," *Management Science,* Vol. 29, 1983, pp. 482–497.

146. Tansel, B. C., R. L. Francis, and T. J. Lowe, "Location on Networks: A Survey. Part II: Exploiting Tree Network Structure," *Management Science,* Vol. 29, 1983, pp. 498–511.

147. Thisse, J.-F., and H. G. Zoller, eds., *Locational Analysis of Public Facilities,* North-Holland Publishing Company, Amsterdam, 1983.

148. Ting, S. S., "A Linear-Time Algorithm for Maxisum Facility Location on Tree Networks," *Transportation Science,* Vol. 18, 1984, pp.76–84.

149. Toregas, C., R. Swain, C. ReVelle, and L. Bergaman, "The Location of Emergency Service Facilities," *Operations Research,* Vol. 19, 1971, pp. 1363–1375.

150. Vincke, P., "Problèmes de localisation multicritères," *Cahiers du Centre d'Etudes de Recherche Operationelle,* Vol. 24, 1983, pp. 333–341.

151. Wendell, R. E., and A. P. Hurter, "Optimal Locations on a Network," *Transportation Science,* Vol. 7, 1973, pp. 18–33.

152. Wendell, R. E., and R. D. McKelvey, "New Perspectives in Competitive Location Theory," *European Journal of Operational Research,* Vol. 6, 1981, pp. 174–182.

153. Zelenka, B., "Medians and Peripherians of Trees," *Archivum Mathematicum (Brno),* 1968, pp. 87–95.

PROBLEMS

7.1. A tree is said to be in *canonical form* if each vertex except vertex v_j has exactly one adjacent vertex with a smaller index, called the *parent* of v_j, and denoted by $p(v_j)$. The following is a canonical form representation of the tree of Figure 7.1

Vertex	1	2	3	4	5	6
Parent	0	1	2	2	4	4
Arc length	0	6	4	4	6	8

By convention, the parent of vertex 1 is 0. Note that the arc lengths can also be easily recorded, so the length of the arc joining vertices 6 and 4 is 8, the length of the arc joining vertices 5 and 4 is 6, and so on. By convention, the arc joining vertices 1 and 0 has a length of 0.

(a) Does every vertex have a parent? Can a vertex have more than one parent? How would you define the "child" of a vertex? Does every vertex have a child?

(b) Explain why the following procedure will put a tree into canonical form. Initially, suppose that all the vertices of the tree are unnumbered. Choose any vertex and number it 1. Choose any vertex adjacent to 1 and number it 2. Choose any unnumbered vertex adjacent to 1 or 2, and number it 3. In general, supposing that

k vertices have been numbered, choose any unnumbered vertex adjacent to one of the vertices $1, 2, \ldots, k$, and number it $k + 1$. Repeat until all vertices are numbered.

(c) Can a graph with a cycle be put into canonical form? Why or why not?

7.2. The following is a canonical form representation of a tree, with the entries of the last row giving the vertex weights.

Vertex	1	2	3	4	5	6
Parent	0	1	2	2	4	4
Arc length	0	3	2	2	4	1
Weight	2	3	2	2	4	1

(a) Use the given tree to solve the 1-median problem by both the Chinese algorithm and the majority algorithm. Also compute the minimum objective function value.
(b) Use the given tree to solve the unweighted 1-center problem.
(c) Use the given tree to solve the weighted problem.
(d) In solving the 1-median problem, how much would the weight of vertex 5 have to increase for the vertex to be a unique 1-median, and why?
(e) Suppose for the given tree that the addends for the 1-center problem are taken to be the weights, while the weights are all 1. Solve the resulting 1-center problem with addends. Which vertex or vertices would be optimal among all vertices, and why?
(f) Explain how the tree illustrates the majority subtree property, and why.

7.3. Devise a *simple* modification of the majority algorithm that will compute the minimum objective function value while considering each arc length exactly once.

7.4. Devise a *simple* modification of the majority algorithm that will detect all alternative 1-medians, and explain why your modification is valid.

7.5. Can a tree ever have a cycle, in a trivial sense? Why or why not?

7.6. Prove that a sum of two convex functions is a convex function.

7.7. Prove that a maximum of two convex functions is a convex function.

7.8. Prove that the 1-median function on a tree is a convex function.

7.9. Prove that the 1-center function on a tree is a convex function.

7.10. Develop an example of the pairwise intersection property and of the neighborhood intersection property.

7.11. Develop an example of a 1-median problem with m nodes for which the Chinese algorithm requires m iterations, while the majority algorithm requires one iteration.

7.12. Develop a specialization of the majority algorithm appropriate for the case where the tree is in canonical form. If you omit the last j columns of the canonical form table, will column $m - j$ always be a tip of the corresponding tree?

7.13. Apply the pairwise neighborhood problem to the version of the 1-center problem for which the objective function is given by

$$g(y) = \max\{f_i(d(y, v_i)) : i = 1, \ldots, m\}$$

Each function f_i is continuous and strictly increasing and can be interpreted as a loss function for vertex i. Thus a 1-center minimizes the maximum loss. You may find it useful to know that a strictly increasing and continuous function always has an inverse function, which is also strictly increasing and continuous. The sum of two such functions is also strictly increasing and continuous.

7.14. For the covering problem shown in Figure 7.4, reduce each string length by 1 and solve the resulting cover problem, as well as the corresponding dual problem. Does reducing the string lengths cause more centers to be needed?

7.15. For the covering problem shown in Figure 7.4, reduce the length of string number 5 to 7 and solve the resulting cover problem as well as the corresponding dual problem. Does reducing the string lengths cause more centers to be needed?

7.16. Solve the vertex-restricted versions of the covering problems stated in Problems 7.14 and 7.15.

7.17. Apply Floyd's algorithm to the graph of Figure 7.7.

7.18. Solve the 1-, 2-, and 3-center problems for the tree of Figure 7.3 when the weights for vertices 1 to 4 are

(a) 1, 2, 1, and 2, respectively.
(b) 1, 2, 3, and 4, respectively.

7.19. In Problem 7.18, use NCENT to solve the 2-center problem.

7.20. Solve Problem 7.18 with the additional restriction that only vertex locations are allowed.

7.21. Solve Problem 7.19 with the additional restriction that only vertex locations are allowed.

7.22. Here is a path construction algorithm to find the unique path between any two vertices i and j of a tree in canonical form.

(0) Initially, define two subpaths consisting of i and j, respectively.
(1) Do the following until the two subpaths have the same last vertex index: extend by one arc the subpath with largest last index.
(2) Join the two subpaths to form the desired path.

Let the following tree be given in canonical form.

Vertex	1	2	3	4	5	6	7	8	9	10	11
Parent	0	1	1	1	4	1	4	5	5	3	7

(a) Make a drawing of the tree.
(b) Apply the algorithm to verify that the successive subpaths it constructs between vertices 3 and 8 are the following:

3 and 8; 3 and 8, 5; 3 and 8, 5, 4; 3 and 8, 5, 4, 1; 3, 1 and 8, 5, 4, 1

Hence the path between vertices 3 and 8 is as follows: 3, 1, 4, 5, 8. We can illustrate the idea of the algorithm by considering the construction of a path between vertices 9 and 11. Construct a path from vertex 9 to vertex 1. Then construct a path from vertex 11 to vertex 1. These two paths will include some duplicate arcs, namely (1, 4). By omitting the duplicate arcs we obtain the path 9, 5, 4, 7, 11 joining vertices 9 and 11. More generally, to construct a path between vertice i and j we could construct a path between vertices i and 1, and between vertices j and 1; on omitting duplicate arcs in the two paths and joining the resulting paths we get the path between vertices i and j.

(c) Apply the algorithm to construct the path between vertices 5 and 9.
(d) Apply the algorithm to construct the path between vertices 7 and 9.
(e) Apply the algorithm to construct the path between vertices 10 and 11.

7.23. Develop a proof of the piecewise linearity and concavity property.

7.24. Prove that the sum of concave functions is a concave function.

7.25. Prove that the 1-median function $f(x)$ is concave on each arc.

7.26. Is the 1-center function $g(y)$ concave on each arc? Why or why not?

7.27. Develop a general proof of the vertex optimality property for the n-median property.

7.28. Develop a general statement of the algorithm implicit in Figure 7.11 for solving the 1-center problem on a network with cycles.

7.29. Develop a proof of the VIP property for the absolute 1-center problem.

7.30. Develop a proof of the VIP property for the absolute n-center problem.

7.31. Suppose that we want to solve the absolute 1-center problem on a cyclic network for which each arc is *directed*.

(a) Prove that some vertex is an absolute 1-center.

(b) Generalize the result of part (a) for minimizing any function $h(y)$ that is nondecreasing in each of the distances $d(y, v_i)$.

7.32. Find all the intersection points for the graph of Figure 7.7.

7.33. This problem illustrates a quick and dirty way to construct a network map from an actual map. Part (a) can be done easily and conveys most of the insight. The remaining parts are time consuming but important in order to have an approach that works well in general.

Consider a hypothetical map, with cities 1 to 4 having coordinates of (0, 0), (4, 0), (4, 3), and (8, 0), respectively. Coordinates are specified in inches, and the map has a scale of 60 miles per inch. The arcs [1, 2], [1, 3], [2, 3], [2, 4], and [3, 4] represent road segments joining the indicated pairs of cities, and each length is *approximated* as the product of the map scale and the Euclidean distance between the associated coordinates (e.g., the length of arc [1, 2] is approximated as $60[(0-4)^2+(0-0)^2]^{1/2} = 240$ miles). Thus the arc lengths, listed in the same order as the arcs are 240, 300, 180, 240, and 300.

(a) Write a program that reads in all these data from an input file arranged in the following format, and computes the arc lengths:

```
60

4

1  0.00  0.00
2  4.00  0.00
3  4.00  3.00
4  8.00  0.00

1  2  3  -1
2  1  3  4  -1
3  1  2  4  -1
4  2  3  -1
```

Note that the map scale is listed first and the number of vertices second. Then follow the coordinates of the vertices, and information about the arcs (e.g., vertices 2 and 3 are adjacent to vertex 1, vertices 1, 3, and 4 are adjacent to vertex 2, and so on; the −1 is to denote the end of the line).

Approximating arc lengths by Euclidean distances underestimates actual distances, and a judicious increase of the map scale can reduce this underestimation. Arc information is intentionally duplicated to reduce the possibility of leaving out an arc by mistake. A more accurate approach to arc lengths is, of course, to use actual distances, but this requires a great deal more effort to obtain and input the data, resulting in an approach that is neither quick nor dirty. Print out the lengths in a matrix as follows ("inf" indicates infinity, i.e., no arc exists):

$$\begin{pmatrix} 0 & 240 & 300 & \text{inf} \\ 240 & 0 & 180 & 240 \\ 300 & 180 & 0 & 300 \\ \text{inf} & 240 & 300 & 0 \end{pmatrix}$$

(b) Generalize your program to handle an arbitrary number of vertices, m, while using the same basic input format and computational approach.

(c) Note that the matrix in part (a) is in the form needed for Floyd's algorithm. Add a code for Floyd's algorithm to your program of part (b) to compute a shortest distance between every pair of vertices. Store the distance matrix in memory, and write it to an output file. Round each computed distance to a nearest integer.

(d) Write a subroutine to display graphically—on a video monitor—the vertex locations and the arcs joining them. Represent each vertex location by a square or circle, and arcs by line segments joining associated vertex locations. If you have a color monitor, show vertices in green and arcs in yellow.

(e) Generalize your code of part (c) to solve vertex-restricted unweighted 1-center problems and unweighted 1-median problems. Show the solutions found in red on a color monitor [assuming that you have done part (d)].

(f) Generalize your code of part (c) to read in a name for each vertex (e.g., a city or town name, or the identity of an intersection). Write a procedure which, for any vertex number requested, writes the name to the screen and to an output file.

APPENDIX

Floyd's Algorithm for Computing Distances Between Every Pair of Vertices in a Network

Floyd's algorithm, for a network with nonnegative arc lengths and with m vertices, computes m tables successively, each being computed from the previous table. The entry in row i and column j of the last table, table m, is $d(v_i, v_j)$, the length of a shortest path from v_i to v_j. We compute the tables as follows.

Table 0. Let the entry in row i and column j of table 0 be the length of the arc from v_i to v_j if the arc exists, ∞ if the arc does not exist, and zero if $i = j$. Put row 1 of table 0 in the bottom margin of table 0, and column 1 of table 0 in the right margin of table 0.

Table p ($p = 1, \ldots, m$). Given that table $p - 1$ is computed, compute the entry in row i and column j of table p to be the minimum of

(a) The entry in row i and column j of table $p - 1$, and
(b) the sum of the row i right margin entry and column j bottom margin entry in table $p - 1$.

If $p < m$, put row $p + 1$ of table p in the bottom margin of table p, and put column $p + 1$ of table p in the right margin of table p.

We illustrate Floyd's algorithm below, using a network of five nodes, for which the arcs join vertices 1 and 2, 1 and 5, 2 and 3, 2 and 5, 3 and 4, 3 and 5, and 4 and 5; the arc lengths are, respectively, 10, 20, 14, 12, 10, 12, and 20. The example demonstrates a fact that is generally true; when table 0 is symmetric about the main diagonal, all subsequent tables will also be symmetric about the main diagonal, so that it is actually only necessary to compute the table entries above the main diagonal.

TABLE 0

	1	2	3	4	5	
1	0	10	∞	∞	20	0
2	10	0	14	∞	12	10
3	∞	14	0	10	12	∞
4	∞	∞	10	0	20	∞
5	20	12	12	20	0	20
	0	10	∞	∞	20	

TABLE 1

	1	2	3	4	5	
1	0	10	∞	∞	20	10
2	10	0	14	∞	12	0
3	∞	14	0	10	12	14
4	∞	∞	10	0	20	∞
5	20	12	12	20	0	12
	10	0	14	∞	12	

TABLE 2

	1	2	3	4	5	
1	0	10	24	∞	20	24
2	10	0	14	∞	12	14
3	24	14	0	10	12	0
4	∞	∞	10	0	20	10
5	20	12	12	20	0	12
	24	14	0	10	12	

TABLE 3

	1	2	3	4	5	
1	0	10	24	34	20	34
2	10	0	14	24	12	24
3	24	14	0	10	12	10
4	34	24	10	0	20	0
5	20	12	12	20	0	20
	34	24	10	0	20	

TABLE 4

	1	2	3	4	5	
1	0	10	24	34	20	20
2	10	0	14	24	12	12
3	24	14	0	10	12	12
4	34	24	10	0	20	20
5	20	12	12	20	0	0
	20	12	12	20	0	

TABLE 5

	1	2	3	4	5	
1	0	10	24	34	20	
2	10	0	14		12	
3	24	14	0	10	12	
4	34	24	10	0	20	
5	20	12	12	20	0	

The motivation for Floyd's algorithm is as follows. Let d_{ij}^0 denote the length L_{ij} of the arc from v_i to v_j if the arc exists, $+\infty$ if the arc does not exist, and 0 if $i = j$. Denote by $V_p = \{v_1, \ldots, v_p\}$ the set of the *first* p vertices for $p = 1, \ldots, m$. Define d_{ij}^p to be the length of a shortest path from v_i to v_j subject to the condition that every intermediate vertex in the path is in V_p. Any shortest path from v_i to v_j with its intermediate vertices in V_p either

(a) has all its intermediate vertices in V_{p-1}—and hence has length d_{ij}^{p-1}, or
(b) has v_p as one of its intermediate vertices and has all of its other intermediate vertices in V_{p-1}

and hence has length $d_{ip}^{p-1} + d_{pj}^{p-1}$. Thus d_{ij}^{p} is the minimum of d_{ij}^{p-1} and $d_{ip}^{p-1} + d_{pj}^{p-1}$. But d_{ij}^{p-1} is the entry in row i and column j of table $p - 1$, while d_{ip}^{p-1} is the row i right margin entry of table $p - 1$, and d_{pj}^{p-1} is the column j bottom margin entry of table $p - 1$. Hence the entry in row i and column j of table p is the minimum of

(a) the entry in row i and column j of table $p - 1$, and
(a) the entry in row i and column j of table $p - 1$, and
(b) the sum of the row i right margin entry of table $p - 1$ and the column j bottom margin entry of table $p - 1$.

Thus d_{ij}^{p} is the entry in row i and column j of table p.

The last table, table m, gives, for each row i and column j, d_{ij}^{m}, the length of a shortest path between vertices i and j, all of whose intermediate vertices are in V_m. But V_m is the set of *all* vertices, so we have $d(v_i, v_j) = d_{ij}^{m}$ for all i and j.

In closing, we remark that *the main virtue of Floyd's algorithm is its simplicity*, which makes it very easy to implement on a computer. There are other algorithms, including Dijkstra's, which are more efficient, and should be investigated if a very efficient algorithm is required.

Chapter 8

Cyclic Network Location Problems

8.1 INTRODUCTION

In this chapter we pick up the thread of Chapter 7 and study algorithms for solving three types of problems involving the location of two or more new facilities on a cyclic network: center problems, median problems, and a special case of median problems called the warehouse location problem. Recall that in Chapter 7 we already have introduced the n-center and n-median network location problems, developed techniques for solving 1-center and 1-median, and established characterizations of the optimal n-center and n-median solutions. You will need to review that material before continuing.

The problems and solution procedures discussed in this chapter represent something of a watershed in location analysis. Generally speaking, the location problems discussed in previous chapters are easy to solve. Granted, the algorithms may have seemed complicated, but the amount of computation they require is modest and does not increase too fast as the size of the problems increases. More precisely, the number of arithmetic operations required by any of these algorithms is bounded by (does not exceed) a polynomial function of the problem size. This is comforting, because it tends to indicate that if we can solve one instance of the problem, we probably can solve incrementally larger instances with acceptable increases in the level of effort required.

For the location problems considered in this chapter and the next, there are no known exact algorithms having a polynomial computing bound. What this means in practical terms is that we may be able to solve a particular instance of one of these problems, yet find that when a few more locations are added, it can no longer be solved with reasonable computing effort.

Lest you become discouraged, many realistic instances of the problems examined in these two chapters can be optimized, despite their inherent computational complexity. Furthermore, the modeling and exact algorithms presented are often helpful in developing approximation schemes. Nevertheless, to solve realistic versions of these problems, you *will* need computer implementations of the algorithms.

Developing computer implementations of the algorithms, collecting the necessary data, and performing the analysis quite often will be an expensive process. However, the cost will be justified whenever the penalty is large for making a wrong facility location decision. Having too many facilities or poorly located facilities can result in significant, ongoing excess costs. Furthermore, the cost to correct poor location decisions can be quite significant.

As we saw in Chapter 7, the optimal n-center and n-median solutions can be restricted to a finite set of points—vertices and intersection points for the n-center, and vertices for the n-median. Our formulations for these problems will associate with each candidate location a zero-one variable whose value in the solution indicates whether or not the location is used. Thus we shall formulate the n-center and n-median as *discrete optimization problems*. This alone should warn you that finding an optimal solution to one of these problems is likely to involve large amounts of computation.

The chapter begins with a generalization of the covering problem to networks other than trees. The covering problem will be formulated as a discrete optimization problem for which there are specialized solution techniques. As in the case of tree networks, we will be able to solve the n-center problem via a sequence of covering problems. Unfortunately, covering problems on general networks will be much harder to construct and solve than their counterparts for tree networks.

We shall formulate a generalized version of the n-median problem and discuss effective heuristics for its solution, as well as optimization procedures. Finally, we shall present solution procedures for the special case of the generalized n-median problem in which n and m are equal (i.e., there is no binding constraint on the number of medians). This special case is commonly referred to as the *warehouse location problem*. We shall see that there are very effective algorithms for solving the warehouse location problem; and in studying it, we shall learn a very important lesson in problem formulation.

In this chapter and the next, we shall use the terminology and some methodology of discrete optimization. Although we shall attempt to explain the basic concepts and methods within the location problem context, you should feel free to consult the excellent sources identified in the list of references for more details on basic concepts, or elaboration of more advanced topics.

8.2 THE COVERING PROBLEM ON CYCLIC NETWORKS

Recall from Section 7.5 that the covering problem involves locating the minimum number of new facilities, or centers, so that each of the existing facilities is within a specified distance of some center.

Example 8.1

The Consolidated Consumer Corporation serves retail stores in a 10-city region. CCC has built its business on a next-day delivery policy (i.e., if an order is received by 5 P.M. today, it will be delivered by 5 P.M. tomorrow). This means that CCC must have a warehouse within a 7-hour drive of each retail store. Currently, CCC operates warehouses in all 10 cities. If necessary, the warehouse in any city could be expanded to serve all the other cities.

To reduce operating costs, CCC wishes to consolidate as many of the warehouses as possible while maintaining the next-day delivery service. The network in Figure 8.1 gives the driving times between each city. What is the minimum number of warehouses that CCC must operate, and where should they be located?

As in Chapter 7, denote by $w_j D(X, v_j)$ the weighted distance from vertex v_j to its closest center, and let s_j be the maximum weighted distance allowed between v_j and its closest center. The covering problem is to determine X so that $|X|$ is minimized and $w_j D(X, v_j) \le s_j$ for all vertices. For this example we have

$$w_j = 1, \qquad j = 1, \ldots, 10$$

$$s_j = 7, \qquad j = 1, \ldots, 10$$

Our approach to solving the covering problem will be first to restate it as a discrete optimization problem in zero–one variables and then to use discrete optimization solution techniques. Although we first consider a vertex-restricted version of the problem, we shall see that the resulting discrete optimization model applies

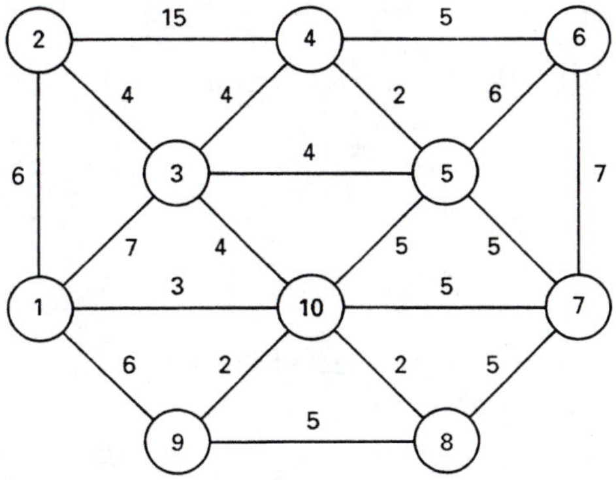

Figure 8.1 Example covering problem.

to the absolute-center version as well, and even to covering problems where there is no explicit network!

8.2.1 Covering Problem Formulation

For any vertex v_q, we easily can determine the set of potential vertices that can cover v_q [i.e., the set $J_q = \{v_j | w_q D(v_j, v_q) \leq s_q\}$]. In Example 8.1, $s_q = 7$ for each vertex, and the resulting *covering sets* are:

$$J_1 = \{1,2,3,8,9,10\}, \qquad J_2 = \{1,2,3\}$$
$$J_3 = \{2,3,4,5,8,9,10\}, \qquad J_4 = \{3,4,5,6,7,10\}$$
$$J_5 = \{3,4,5,6,7,8,9,10\}, \qquad J_6 = \{4,5,6,7\}$$
$$J_7 = \{4,5,6,7,8,9,10\}, \qquad J_8 = \{1,3,5,7,8,9,10\}$$
$$J_9 = \{1,3,5,7,8,9,10\}, \qquad J_{10} = \{1,3,4,5,7,8,9,10\}$$

A feasible solution to the covering problem is a set of vertices that has a nonempty intersection with each of the covering sets. For Example 8.1, a feasible solution is $X = \{1,3,4\}$. An optimal solution is one with the fewest vertices.

We now can give a formal statement of the covering problem as a discrete optimization problem:

(C) $$\text{minimize} \quad z = \sum_{j=1}^{m} x_j$$

subject to

$$\sum_{j=1}^{m} a_{ij} x_j \geq 1, \qquad i = 1, \ldots, m$$
$$x_j \in \{0,1\}, \qquad j = 1, \ldots, m \tag{8.1}$$

where

$$a_{ij} = \begin{cases} 1 & \text{if } j \in J_i \\ 0 & \text{otherwise} \end{cases}$$

$$x_j = \begin{cases} 1 & \text{if vertex } v_j \text{ is selected for a center} \\ 0 & \text{otherwise} \end{cases}$$

For Example 8.1, the matrix of a_{ij} values is

i \ j	1	2	3	4	5	6	7	8	9	10
1	1	1	0	0	0	0	0	1	1	1
2	1	1	1	0	0	0	0	0	0	0
3	1	1	1	1	1	0	0	1	1	1
4	0	0	1	1	1	1	1	0	0	1
5	0	0	1	1	1	1	1	1	1	1
6	0	0	0	1	1	1	1	0	0	0
7	0	0	0	1	1	1	1	1	1	1
8	1	0	1	0	1	0	1	1	1	1
9	1	0	1	0	1	0	1	1	1	1
10	1	0	1	1	1	0	1	1	1	1

Note that a_{ij} is a parameter in this model and depends on the particular values of s_q used.

Now suppose that we do not restrict the centers to be located at vertices. Can we still use a formulation like C to determine the minimum number of centers? The answer is: "Yes, but some additional work will be required."

Recall that the vertex and intersection point (VIP) property presented in Chapter 7 defines a finite set of points that must be considered for locating an absolute n-center, for $n \geq 1$. Therefore, in solving the covering problem for absolute centers, we can restrict center locations to vertex *and intersection* points. In terms of the model, C, this simply means that there are more variables, x_j. We adopt the convention that intersection points are indexed by $j = m + 1, \ldots, p$. Now J_q, the covering set for vertex v_q, includes all vertices *and intersection points* satisfying the distance constraint.

Upon reflection, you might observe that we do not require an explicit network nor explicit value of s_j. All we really need is to be able to specify the points to be served, the potential center locations, and the covering coefficients, a_{ij}, which need not be defined by distances. The following example illustrates an application of the covering location model, which does not involve location at all.

Example 8.2

The Consolidated Consumer Corporation purchases 10 items from vendors. Each of seven potential vendors can supply several of the items, but not all 10. CCC wants to select the minimum number of vendors required to supply all 10 items. How should vendors be chosen if there are no price differences between alternate vendors?

The vendor selection problem in this example can be formulated as follows:

$$a_{ij} = \begin{cases} 1 & \text{if vendor } j \text{ can supply item } i \\ 0 & \text{otherwise} \end{cases}$$

$$x_j = \begin{cases} 1 & \text{if vendor } j \text{ is selected} \\ 0 & \text{otherwise} \end{cases}$$

$$\text{minimize} \quad z = \sum_{j=1}^{7} x_j$$

subject to

$$\sum_{j=1}^{7} a_{ij} x_j \geq 1, \qquad i = 1, \ldots, m$$

$$x_j \in \{0, 1\}, \qquad j = 1, \ldots, 7$$

As a final note, it often is necessary or convenient to associate a cost, c_j, with decision variable x_j. For instance, c_j might be the cost to establish a center at location j, and we may want to minimize the total cost of centers rather than simply minimizing the number of centers.

This more general problem, with cost coefficients, is called the *set covering problem*. There is always a network associated with a set covering problem, and we will use the term *set covering location problem* (SCLP), or just *covering location problem*, to distinguish the special case in which the underlying network's vertices represent physical locations and its arcs represent travel distances or times.

The general set covering model has application in a number of diverse areas, such as product delivery, switching circuit design, truck dispatching, information retrieval, political districting, crew scheduling, and assembly line balancing, as well as facility location. The general set covering problem tends to be more difficult to solve than the covering location problem. You should keep this in mind when reading the following three sections, where we shall discuss techniques for solving the set covering problem, and consequently, solving the covering location problem. For the interested reader, the references cited at the end of the chapter will provide additional insight and details.

8.2.2 Reducing the Set Covering Problem

The set covering formulation for Example 8.1 has 10 variables and 10 constraints, and we could use any appropriate discrete optimization technique to solve it. However, it often is possible to eliminate some variables and constraints before applying the general solution procedure. In Example 8.1, observe that any vertex covered by v_2 is also covered by v_1. Furthermore, in this case we should always select v_1 instead of v_2, which implies that $x_2 = 0$ in an optimal solution.

In this section we will give several such *reduction rules* that may be applied to any set covering problem. The aim of the reduction rules is to detect infeasibility, to identify the optimal values for some of the decision variables, or to show that some of the constraints will be satisfied "automatically." By applying the reduction rules, the size of the problem in terms of variables and constraints often can be reduced, and in some cases the problem may be solved by the reduction rules alone.

The reduction rules use the set J_i, already defined, and another set, I_j, which is the set of all vertices that can be covered by a center at location j. Note that J_i corresponds to a *row* of the constraint coefficient matrix, and I_j corresponds to a *column*. The four reduction rules are stated and then briefly explained.

Row Infeasibility Rule. If for any row $k, J_k = \emptyset$, there is no feasible solution.

If there is no potential center location that can cover vertex v_k, there is no feasible solution to the covering problem. This test would only be useful in network covering problems where not all vertices are candidates for center locations, so it does not apply to Example 8.1.

Row Feasibility Rule. If for some row r and column c, $J_r = \{c\}$, $x_c = 1$ is required for feasibility. Rows $k \in I_c$ may be dropped.

If there is only one candidate center location that covers vertex r, it must be selected in any feasible solution. If in Example 8.1, $s_6 = 4$, there must be a center at vertex v_6, since no other vertex is within 4 hours.

Row Dominance Rule. If there are two different rows, r and q, such that $J_r \subseteq J_q$, row q may be dropped.

Any center that covers vertex r will also cover vertex q. For example, in the constraint coefficient table for Example 8.1, note that if v_2 is covered, v_1 also must be covered. Thus the constraint associated with v_1 can be deleted from the model, since it will be satisfied by any solution that satisfies the constraint associated with v_2.

Column Dominance Rule. If there are two distinct columns, s and t, such that $I_s \supseteq I_t$, and $c_s \leq c_t$, some optimal solution will have $x_t = 0$.

If the potential center location corresponding to x_s covers all the vertices covered by the potential center location associated with x_t, but incurs no greater cost, there is no reason to select x_t in an optimal solution. We can always do at least as well by selecting x_s. In our example problem, $I_1 \supseteq I_2$, so we can set $x_2 = 0$ and delete it from the formulation.

To apply these rules, simply cycle through them repetitively until no further reductions are possible. If the problem is not completely reduced (i.e., all variables are not specified), some additional solution procedure is required. For example 8.1 the first cycle yields

1. {Row infeasibility} no $J_k = \emptyset$
2. {Row feasibility} no J_r with only one element

3. {Row dominance}
   - row 2 dominates row 1
   - row 2 dominates row 3
   - row 4 dominates row 5
   - row 6 dominates row 4
   - row 6 dominates row 7
   - row 8 dominates row 9
   - row 8 dominates row 10

After the row dominance rule, the reduced matrix is

i \ j	1	2	3	4	5	6	7	8	9	10
2	1	1	1	0	0	0	0	0	0	0
6	0	0	0	1	1	1	1	0	0	0
8	1	0	1	0	1	0	1	1	1	1

4. {Column dominance}
   - column 1 dominates column 2
   - column 1 dominates column 3
   - column 5 dominates column 4
   - column 5 dominates column 6
   - column 5 dominates column 7
   - column 5 dominates column 8
   - column 5 dominates column 9
   - column 5 dominates column 10

After the column dominance rule, the reduced matrix is

i \ j	1	5
2	1	0
6	0	1
8	1	1

The second cycle of reduction rules yields:

1. Row infeasibility} no $J_k = \emptyset$
2. {Row feasibility} $J_2 = \{1\}$, so $x_1 = 1$ and row 8 is dropped
   $J_6 = \{5\}$, so $x_5 = 1$

For this particular problem instance, the reduction rules alone solve the covering location problem. Note that we need to record the results of the rules in order to be able to reconstruct the solution to the original problem.

Example 8.3

The Cascade Chemical Company is planning to install a system of state-of-the-art decontamination chambers in its pesticide manufacturing plant. Figure 8.2 summarizes the travel time in minutes between the several manned stations in the plant. Travel times reflect stairways, safety doors, and equipment congestion. The decontamination chambers are very expensive, so CCC wants to install the minimum number necessary. On the other hand, there must be a chamber within 6 minutes of each operator. How many chambers are required, and where should they be located?

Example 8.3 presents a covering problem that is not completely solved by the reduction rules. You should verify that when all possible reductions are completed, the reduced constraint coefficient matrix is

i \ j	1	2	4	5	6
1	1	1	0	1	1
2	1	1	1	0	0
4	0	1	1	1	1
5	1	0	1	1	0
6	1	0	1	0	1

for which no further reductions are possible.

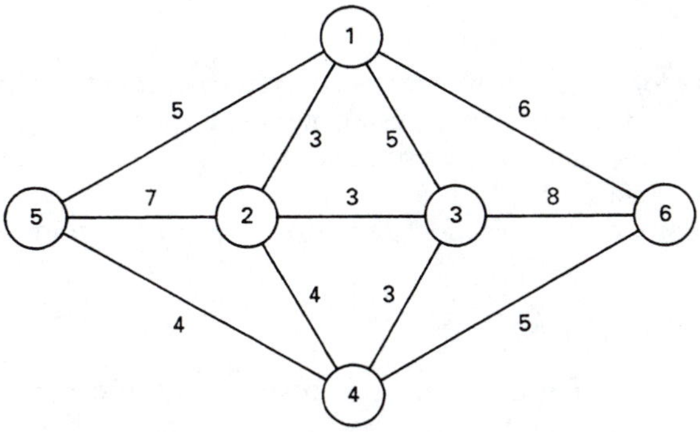

Figure 8.2 Covering problem network.

8.2.3 A Greedy Heuristic for Set Covering

In this section we present a simple but effective heuristic for the general set covering problem:

(SCP) $$\text{minimize} \quad z = \sum_{j=1}^{p} c_j x_j$$

subject to

$$\sum_{j=1}^{p} a_{ij}x_j \geq 1, \qquad i = 1, \ldots, m$$

$$x_j \in \{0, 1\}, \qquad j = 1, \ldots, p$$

For the covering location problem we may take the c_j's to be 1. The algorithm we present starts with all $x_j = 0$ and proceeds to select one x_j at a time to make nonzero, stopping when all constraints are satisfied. It is called a *greedy algorithm* because it always does the best it can at each sequential decision, even though this may not lead to the globally optimum solution.

In stating the greedy algorithm, we will use the following notation:

I = set of rows not yet covered; initially $I = \{1, \ldots, m\}$
J = set of variables chosen to be one; initially $J = \emptyset$

$k_j = |I \cap I_j| = \sum_{i \in I \cap I_j} a_{ij}$, the number of not yet covered rows that could be covered by x_j

$f(c_j, k_j)$ = "payoff" function, a measure of the immediate reward for selecting x_j

The greedy algorithm is

WHILE $I \neq \emptyset$ DO
(1) $f(c_v, k_v) = \text{minimum } \{f(c_j, k_j) | j \notin J\}$
(2) $J \leftarrow J \cup \{v\}$
(3) $I \leftarrow I \backslash (I \cap I_v)$
ENDWHILE

Steps (1) to (3) are repeated as long as any rows are not covered (constraints are not satisfied). Step (1) selects the next variable (center) using the payoff function $f(c_j, k_j)$. Step (2) adds the variable to the selected variable set, and step (3) deletes all newly covered rows. ($\leftarrow$ means "is replaced by.")

There is considerable latitude in the specification of the payoff function, but one good choice is

$$f(c_j, k_j) = \frac{c_j}{k_j}$$

which can be interpreted as the cost per additional row covered. If ties occur, use a largest k_j rule as the first tiebreaker, and the smallest j as the second tiebreaker. Note that if $k_j = 0$, there is no point in selecting x_j, so we may take $f(c_j, k_j)$ to be infinite in this case.

For Example 8.3, the greedy algorithm gives the following results:

1. $J = \emptyset$, $I = \{1, 2, 4, 5, 6\}$

j	1	2	4	5	6
$f(c_j, k_j)$	$\frac{1}{4}$	$\frac{1}{3}$	$\frac{1}{4}$	$\frac{1}{3}$	$\frac{1}{3}$

Select x_1.

2. $J = \{1\}$, $I = \{4\}$

j	2	4	5	6
$f(c_j, k_j)$	1	1	1	1

Select x_2.

The greedy algorithm is easy to implement and has been observed to give good results, at least for reasonably large problems in which $|I_j|$ is relatively small for all j. It is, however, only a heuristic, so it gives us an *upper bound* on the minimum number of centers required. To guarantee that we have the minimum number, techniques of discrete optimization will be required.

Two types of discrete optimization techniques have been used with some success for solving set covering problems in general as well as the covering location problem. They are *cutting planes* and *branch and bound*. Our presentation is intended to convey the concepts underlying these techniques and some simple procedures for using them to solve covering location problems. Additional information, more sophisticated algorithms, and details on implementation can be found in the references at the end of the chapter.

8.2.4 Cutting-Plane Approach

The cutting plane procedure for solving C is based on the linear programming problem:

(LRC)
$$\text{minimize} \quad f = \sum_{j=1}^{p} x_j$$

subject to

$$\sum_{j=1}^{p} a_{ij} x_j \geq 1, \qquad i = 1, \ldots, m$$

$$0 \leq x_j \leq 1, \qquad j = 1, \ldots, p$$

LRC is called a *relaxation* of C because any solution to C is feasible in LRC, but there are many solutions to LRC that are not feasible in C. In other words, the set of all feasible solutions to LRC includes the set of all feasible solutions to C, plus some additional solutions. Since the criterion or objective function is the same in both problems, the optimal solution for LRC provides a *lower bound* on the optimal solution for C (i.e., $f^* \leq z^*$). If we solve LRC, and the optimal solution also is feasible in C (i.e., all the x_j are either zero or 1), it must be optimal in C as well.

Note that LRC is a linear programming problem, so it can be solved with reasonable effort, even for large problems [e.g., hundreds of constraints (vertices in the network)]. Unfortunately, the optimal solution to LRC may be fractional. The basic concept in cutting plane procedures is to formulate a constraint that is satisfied by any feasible solution to C, but which is violated by the fractional optimal solution to LRC. If we augment the constraints set of LRC by this new constraint, called a *cut* or *cut constraint*, and solve LRC again, the new constraint "cuts off" the previous fractional optimal solution.

Of course, the new solution also may be fractional, so cutting-plane algorithms generally incorporate a method for generating a series of cut constraints, terminating when the linear programming solution of the augmented problem is feasible in the original discrete optimization problem. For the covering location problem, experience has shown that:

1. Often, LRC gives an optimal solution to C.
2. When the solution to LRC is fractional, a single cut of the form

$$\sum_{j=1}^{p} x_j \geq \lceil f^* \rceil$$

usually is sufficient, where $\lceil f^* \rceil$ is the smallest integer larger than f^*.

We hasten to add that this conclusion applies only to covering location problems. You should not assume that such a simple cutting-plane procedure will be successful for general set covering problems.

To illustrate this cutting-plane procedure, consider once again Example 8.3. We already have seen that the reduction rules reduce the covering problem to the following:

$$\text{minimize} \quad z = x_1 + x_2 + x_4 + x_5 + x_6$$

subject to

$$\begin{aligned}
x_1 + x_2 \qquad\quad + x_5 + x_6 &\geq 1 \\
x_1 + x_2 + x_4 \qquad\qquad\quad &\geq 1 \\
x_2 + x_4 + x_5 + x_6 &\geq 1 \\
x_1 \quad + x_4 + x_5 \qquad &\geq 1 \\
x_1 \quad + x_4 \qquad\quad + x_6 &\geq 1 \\
x_j \in \{0, 1\}, \quad j \in (1, 2, 4, 5, 6)&
\end{aligned}$$

An optimal solution to LRC is $f^* = 1.4$, $x_1 = 0.4$, $x_2 = 0.2$, $x_4 = 0.4$, $x_5 = 0.2$, and $x_6 = 0.2$. Since this solution is fractional, we would add the cut constraint:

$$x_1 + x_2 + x_4 + x_5 + x_6 \geq 2$$

An optimal solution to the augmented problem is $f^* = 2, x_1 = 1, x_4 = 1$, and all other variables zero; thus the simple cutting plane procedure does solve our example problem.

Obviously, we have discussed only the simplest type of cut constraint. If you wish to pursue the study of cutting planes, the references at the end of the chapter provide an introduction to cutting-plane theory in general as well as specific applications to set covering problems.

8.2.5 Branch-and-Bound Approach

While the cutting-plane procedure described here works well for the special case of covering location problems, we will need a different approach for solving other discrete location problems. That approach, called branch and bound, is the most widely used general methodology for solving discrete optimization problems.

The philosophy underlying branch and bound is best described by the phrase "divide and conquer." In the context of discrete optimization, this means that we break apart a difficult problem into two or more smaller problems which we hope will be easier to solve than the original problem. We may have to apply the divide-and-conquer strategy to the smaller problems as well, so it can be a recursive strategy. To put this philosophy into action, we need to specify how a problem is to be "broken apart," how we will "solve" the smaller problems, and how we will manage the process if it becomes recursive.

The most common method for managing the divide-and-conquer strategy is based on the concept of *enumeration*. For example, the solution to the covering problem, C, is a p-vector, $\mathbf{X}$, whose components are all either zero or 1, so there are 2^p potential solutions. Enumeration is a systematic way of listing all of these 2^p solution vectors and can be represented by an *enumeration tree*, such as the one in Figure 8.3 for the case $p = 3$.

Using Figure 8.3 for reference, we can make several general statements about enumeration trees for binary p-vectors. An enumeration tree has $p + 1$ *levels*, numbered $0, 1, \ldots, p$. On level i of the tree there are 2^i nodes, and each corresponds to a version of the original problem in which some variables have been specified, or *fixed*. These fixed variables and their values are determined by following the path from the node of interest back to node 1. At any node in the tree, variables that are not fixed are called *free* variables. Note that the total number of nodes having free variables is $2^0 + 2^1 + 2^2 + \cdots + 2^{p-1} = 2^p - 1$.

The downward branches from a node on level $k < p$ represent the *separation* of the corresponding problem into two smaller problems, accomplished by specifying the values of variable $x_{[k]}$. This is the "divide" part of the divide-and-conquer strategy, and $x_{[k]}$ is called the *branching* variable. The brackets around the index indicate that $x_{[k]}$ is the kth variable chosen to branch on, and $[k] \neq k$ is permitted. For example, we might branch on x_3 first, in which case $[1] = 3$.

At node i on level $k < p$, the fixed variables constitute a *partial solution*. The remaining problem in the free variables often is called a *candidate problem*. A candidate problem is "solved" if we can:

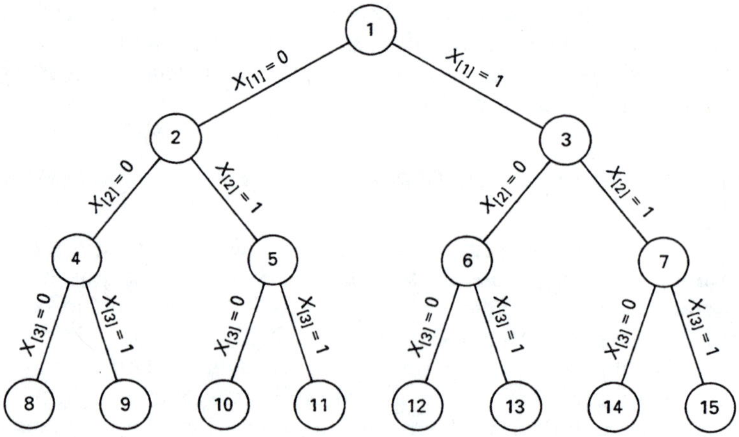

Figure 8.3 Example enumeration tree.

1. determine the optimal values for its free variables, or
2. determine that it has no feasible solution, or
3. prove that its optimal solution will be no better than some other solution that we already know, called an *incumbent solution*.

For conditions 2 and 3, we say that the candidate problem or node is *fathomed*. If one of these three conditions is satisfied, we do not need to construct explicitly the part of the enumeration tree below node i, which is the origin of the term *implicit enumeration*.

As an example of branch and bound, consider once more the covering problem for Example 8.2. After applying the reduction rules, we still have a problem with five zero–one variables and five constraints. This is the candidate problem associated with node 1 of the enumeration tree. We cannot reduce this candidate problem. We could attempt to solve it using the cutting-plane procedure; but for the moment, suppose that we do not have access to a linear programming optimizer. We can solve the problem using the reduction rules in conjunction with the divide-and-conquer strategy of branch and bound.

First, we must choose a branching variable. In general, the choice of the branching variable can have a large impact on how quickly we are able to find and verify an optimal solution, so it is worthwhile to spend some effort to discover a good rule to use. For simplicity, we will arbitrarily choose $x_{[1]} = x_2$. After branching, we have two new candidate problems; one with x_2 fixed to zero, and one with x_2 fixed to 1. Refer to these as CP2 and CP3, respectively.

Consider CP2, whose constraint coefficient matrix is

i \ j	1	4	5	6
1	1	0	1	1
2	1	1	0	0
4	0	1	1	1
5	1	1	1	0
6	1	1	0	1

Applying the reduction rules, we find that row dominance eliminates constraints 5 and 6, and column dominance eliminates column 6. The reduced constraint coefficient matrix is

i \ j	1	4	5
1	1	0	1
2	1	1	0
4	0	1	1

We must again choose a branching variable (apply the divide-and-conquer strategy to CP2). Suppose that we select $x_{[2]} = x_5$, and create two more candidate problems, CP4 and CP5, with $x_5 = 0$ and $x_5 = 1$, respectively. The corresponding enumeration tree is shown in Figure 8.4.

For CP4, the reduction rules lead us to drop row 2; and as a result, we must have both x_1 and $x_4 = 1$. All constraints are satisfied, and $z = 2$. We now have an incumbent solution.

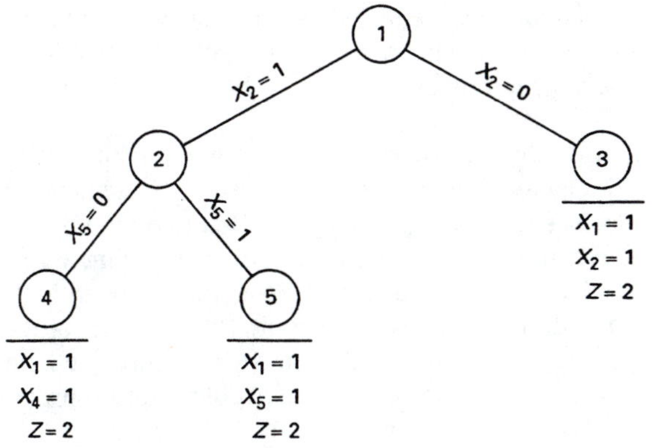

Figure 8.4 Enumeration tree for example problem.

We still have CP3 and CP5 to consider. For CP5, the constraint coefficient matrix is:

i \ j	1	4	6
2	1	1	0
6	1	1	1

because setting $x_5 = 1$ satisfied the other constraints. Now, the reduction rules solve the candidate problem, selecting $x_1 = 1$. The optimal solution to CP5 has $z = 2$, so is no improvement over the incumbent solution. For CP3, the constraint coefficient matrix is

i \ j	1	4	6
4	1	1	0
5	1	1	1

because setting $x_2 = 1$ satisfied the other constraints. Again, the reduction rules solve the candidate problem, setting $x_1 = 1$, with $z = 2$, which is no improvement over the incumbent solution. There are no remaining candidate problems to examine, so we are finished. Note that we explicitly considered only five of the $2^5 - 1 = 31$ possible candidate problems.

If you review the branch-and-bound solution to the example problem, you will observe that several times we had two or more candidate problems available to be examined. To specify a branch-and-bound procedure completely, we would need to state the rule to be used for selecting the next candidate problem to examine in this situation. There are two common strategies in which candidate problems are added to a *list* as they are created. A last-in, first-out (LIFO) rule selects the "newest" candidate problem on the list to examine next, while a first-in, first-out (FIFO) rule selects the "oldest" candidate problem on the list to examine next. LIFO rules tend to be used more often, since they generally require a much smaller list for the candidate problems.

Recall that the third way to solve a candidate problem is to show that its optimal solution will be no better than the incumbent solution. This usually is achieved by solving some relaxed problem, such as LRC, to obtain a lower bound on the optimal candidate problem solution value. If this lower bound is greater than the incumbent solution value, the candidate problem optimal solution also will be greater than the incumbent value. Thus we actually do not need to determine the optimal solution to the candidate problem. In the simple branch-and-bound procedure just illustrated, the number of variables fixed to one in the candidate problem would provide such a lower bound.

This operation is referred to as *bounding* and is a crucial element in the branch-and-bound method. As we shall see later, only slight changes in the formulation of the bounding problem (the relaxation) can make a great difference in the value of the lower bound, thus in the number of candidate problems that can be fathomed. Note that finding a good initial incumbent solution also is important.

This brief discussion of branch-and-bound methodology is intended only to familiarize you with the concepts and terminology and to give you a way to solve small covering location problems by hand. There is a rich and extensive literature on all aspects of branch and bound—the references cited at the end of the chapter will provide you with an entry point into that literature.

8.2.6 Other Covering Problems

The covering location problem formulation, (C), is sometimes referred to as the *total cover problem*, because it requires every vertex to be covered by a center. There are two variations of the total cover problem, the *partial cover problem* and the *generalized partial cover problem*, which address situations in which there are not enough centers, or new facilities, to guarantee that every vertex is covered.

Example 8.4

Don's Dippy Donuts' recently completed market survey in Greater Ashley, Arkansas, identified nine suburban developments that currently are underserved by high-carbohydrate fast-food outlets. Nine possible locations have been identified, and the company can obtain financing for two new stores. Experience in other major metropolitan areas in Arkansas indicates that a store's business from a suburban development falls off drastically if it is more than 10 minutes' driving time away. In the matrix given below, a coefficient of 1 indicates that the corresponding suburban development is within 10 minutes of the corresponding potential store site. Where should the two new stores be built to maximize the number of developments within 10 minutes of one of Don's stores?

Development	Potential Site (j)								
(i)	1	2	3	4	5	6	7	8	9
1	1	1	0	0	1	0	0	0	0
2	1	1	0	0	0	0	0	0	0
3	0	0	1	0	0	1	0	0	0
4	0	1	0	1	1	0	0	1	0
5	0	0	0	0	1	1	0	0	0
6	0	0	0	0	0	1	0	0	0
7	0	0	0	0	0	0	1	1	0
8	0	0	1	0	0	0	0	1	0
9	1	0	0	0	0	0	0	0	1

The problem in Example 8.4 is a partial cover problem, which can be stated generally as: How many vertices can be covered if only k centers are available? The mathematical formulation is

$$\text{PC}(k) \qquad \text{maximize} \quad z_{pc} = \sum_{i=1}^{m} \max\{a_{ij}x_j \,|\, j = 1, \ldots, p\}$$

subject to

$$\sum_{j=1}^{p} x_j \le k$$

$$x_j \in \{0, 1\}, \qquad j = 1, \ldots, p$$

where a_{ij} and x_j are defined as they were for C. Note that $a_{ij}x_j = 1$ only if vertex i is covered by a center at location j, and the max operator ensures that at most one covering is counted for each vertex. Thus z_{pc} is the number of vertices covered.

Example 8.5

Don's Darling Daycare recently completed a market survey in Lower Dunwoody which identified five major condominium developments with inadequate day-care service. The follow-up site selection survey revealed four suitable day-care facility sites, with travel times from the condominiums as shown in the following table:

Condo	Site (j)			
(i)	1	2	3	4
1	1	9	17	24
2	10	2	8	15
3	16	8	2	11
4	20	12	4	5
5	24	16	10	1

Based on the market survey, the number of families in each condominium that would sign up with DDD is 18, 25, 23, 21, and 40. DDD's corporate philosophy is to try to minimize the total trip time for all families using their facilities. If financing and staff are available for two new facilities, and families take their children to the nearest facility, which two sites should be selected?

The problem in Example 8.5 is a *generalized partial cover problem* and can be stated generally as: What is the minimum sum of distances traveled between vertices and their closest centers if only k centers are available? Letting t_{ij} be the total distance traveled between vertex v_i and center location j if the closest center to v_i is at location j, the generalized partial cover problem can be formulated as

$$\text{GPC}(k) \qquad \text{minimize} \quad \sum_{i=1}^{m} \min\{t_{ij} \,|\, j \in \theta(\mathbf{X})\}$$

subject to

$$1 \le \sum_{j=1}^{p} x_j \le k$$

$$x_j \in \{0, 1\}, \qquad j = 1, \ldots, p$$

where $\theta(\mathbf{X}) = \{j \mid x_j = 1\}$.

For Example 8.5, the condominiums correspond to vertices. The travel distance, t_{ij}, is measured in terms of travel time per day and computed as the product of trip time and number of families in the corresponding condominium.

You should think carefully about the difference between the partial cover model and the generalized partial cover model. For example, will the partial cover model necessarily provide the doughnut company with the maximum *amount* of business? Would the generalized partial cover model be an equitable method for locating emergency services, such as a fire station?

A number of specialized methods have been developed for solving PC(k) and GPC(k), some of which are cited in the references at the end of the chapter. However, we shall see later that both these problems are special cases of the generalized p-median problem, which is presented in Section 8.4. The methods described for that problem should also work well for these two special cases.

8.3 *n*-CENTER PROBLEM ON CYCLIC NETWORKS

The statement of the n-center problem is the same for both cyclic networks and tree networks; that is, determine a set of n points, $Y = \{y_j \mid j = 1, \ldots, n\}$, such that the function

$$g(Y) = \max\{w_i D(Y, v_i) \mid i = 1, \ldots, m\}$$

is minimized, where $D(Y, v_i)$ is the distance from vertex v_i to its nearest center. The difference between the two types of networks is that for trees, the distances $D(y_j, v_i)$ are determined by the unique path between y_j and v_i, whereas for a cyclic network, the path may not be unique, and typically, $D(y_j, v_i)$ must be found using an iterative procedure.

Example 8.6

Six communities in Eastaboga, Alabama, are planning a volunteer fire department with two stations. The problem facing the planning committee is the location of the two stations, which should minimize the maximum time required to respond to an alarm in any community. Figure 8.5 illustrates the travel times for the road network connecting the communities.

For this problem, $D(Y, v_i)$ will be in terms of travel time rather than distance, the weights, w_i, all will be 1 and $n = 2$. If there is a center at v_2, the distance to it

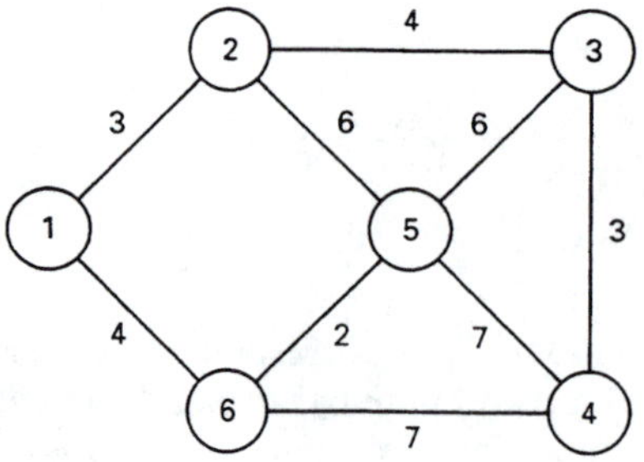

Figure 8.5 Network for n-center example.

from v_4 is 7, corresponding to the path (v_2, v_3, v_4). In this case the shortest path from v_2 to v_4 is determined easily by inspection. For large networks, however, a more formal computational algorithm is required. One particular shortest-path algorithm is presented in the appendix to Chapter 7.

The approach we shall describe for solving the n-center problem uses a discrete optimization model based on the VIP property presented in Chapter 7. Although this model does not lead directly to a solution procedure, it does have a useful relationship to the covering problem. We shall discuss this relationship, and see how it leads to an indirect approach to solving the n-center problem, analogous to the solution procedure for the case of tree networks.

8.3.1 Formulating the *n*-Center Problem

Because of the VIP property, there is a finite, although possibly very large set of potential center locations. For each potential location, we have a binary decision; either include it in the n-center or exclude it. Let $Q = \{q_1, q_2, \ldots, q_p\}$ be the set of all potential center locations, with the first m of them corresponding to vertices. Table 8.1 lists the distinct intersection point locations for the network of Figure 8.5. For even a small problem, there can be a great many intersection points.

Note that the intersection points, q_k, are given as a distance from v_i on the arc $[v_i, v_j]$, and each has an associated *bound value*, b_k. The bound value is simply the weighted distance from either one of the two vertices used to define the intersection point. For example, the entry in Table 8.1 for $k = 9$ corresponds to an intersection point on the arc $[v_1, v_2]$. The intersection point is defined by vertices v_5 and v_6, so it is located a distance of 2.5 from v_1 and its weighted distance to either v_5 or v_6 is 6.5.

The set S_k listed in Table 8.1 is the set of all vertices that would be covered by a center at q_k, with a weighted distance less than or equal to b_k. For example, a center at q_9 could cover every vertex except v_4, with a maximum weighted distance of 6.5.

The procedure for finding an n-center also will employ a set, $Q(z)$, defined for a particular value of z as

$$Q(z) = \{q_1, \ldots, q_m\} \cup \{q_j \mid j > m \text{ and } b_j \leq z\}$$

TABLE 8.1 DISTINCT INTERSECTION POINTS FOR FIGURE 8.5

k	arc $[v_i, v_j]$	s	t	q_k	b_k	S_k	$Q(5)$
7	(1, 2)	1	2	1.5	1.5	1, 2	*
8		6	3	1.5	5.5	1, 2, 3, 6	
9		6	5	2.5	6.5	1, 2, 3, 5, 6	
10	(1, 6)	1	6	2	2	1, 6	*
11		2	6	.5	3.5	1, 2, 6	*
12		1	5	3	3	1, 5, 6	*
13		2	5	1.5	4.5	1, 2, 5, 6	*
14		3	4	2	9	1, 2, 3, 4, 5, 6	
15	(2, 3)	2	3	2	2	2, 3	*
16		1	3	.5	3.5	1, 2, 3	*
17		2	4	3.5	3.5	2, 3, 4	*
18		1	4	2	5	1, 2, 3, 4	*
19	(2, 5)	2	5	3	3	2, 5	*
20		1	5	1.5	4.5	1, 2, 5	*
21		2	6	4	4	2, 5, 6	*
22		1	6	2.5	5.5	1, 2, 5, 6	
23		1	4	5	8	1, 2, 3, 4, 5, 6	
24		3	5	1	5	1, 2, 3, 5	*
25		3	6	2	6	1, 2, 3, 5, 6	
26	(3, 4)	3	4	1.5	1.5	3, 4	*
27		1	6	1.5	8.5	1, 2, 3, 4, 5, 6	
28	(3, 5)	3	5	3	3	3, 5	*
29		2	5	1	5	2, 3, 4, 5	*
30		2	6	2	6	2, 3, 4, 5, 6	
31		3	6	4	4	3, 5, 6	*
32		4	6	2.5	5.5	3, 4, 5, 6	
33	(4, 5)	4	5	3.5	3.5	4, 5	*
34		4	1	6.5	6.5	1, 2, 3, 4, 5, 6	
35		4	2	6.5	6.5	1, 2, 3, 4, 5, 6	
36		4	6	4.5	4.5	4, 5, 6	*
37		3	5	2	5	3, 4, 5	*
38		3	1	5	8	1, 2, 3, 4, 5, 6	
39	(4, 6)	4	6	3.5	3.5	4, 6	*
40		4	1	5.5	5.5	1, 4, 5, 6	
41		4	5	4.5	4.5	4, 5, 6	*
42		3	6	2	5	3, 4, 6	*
43		3	1	4	7	1, 3, 4, 5, 6	
44		3	5	3	6	3, 4, 5, 6	
45	(5, 6)	5	6	1	1	5, 6	*

that is, $Q(z)$ is the subset of Q obtained by excluding intersection points whose bound value exceeds z. Table 8.1 indicates the intersection points that would be included in $Q(z)$ for $z = 5$. It also will be convenient to let $d_{ij} = D(v_i, q_j)$.

The n-center problem now can be formulated as:

NCP(n) $\qquad$ minimize z_n

subject to

$$z_n \geq \min\{w_i d_{ij} \,|\, j \in \theta(\mathbf{X})\}, \qquad i = 1, \ldots, m \tag{8.2}$$

$$\sum_{j=1}^{p} x_j \leq n \tag{8.3}$$

$$x_j \in \{0, 1\}, \qquad j = 1, \ldots, p$$

where

$$x_j = \begin{cases} 1 & \text{if a center is at } q_j \\ 0 & \text{otherwise} \end{cases}$$

and

$$\theta(\mathbf{X}) = \{j \,|\, x_j = 1\}$$

The set $\theta(\mathbf{X})$ is the set of vertices and intersection points that are used in the solution. Constraints (8.2) force the objective function to be at least as large as the smallest weighted distance from a vertex to any center location. Constraint (8.3) ensures that at most n center locations are selected.

Although this is a mathematically precise statement of the problem, it is not a formulation for which we have ready-made general-purpose solution procedures. [For example, try to state a linear programming relaxation for NCP(n).] Therefore, our approach to solving NCP(n) is indirect and involves solving a sequence of related covering problems, for which we do have ready-made solution procedures.

8.3.2 Relationship between n-Center and Covering

Denote by $(z^*, \mathbf{X}^*)$ an optimal solution to NCP(n). Now consider the covering problem:

CP(z) $$\text{minimize} \quad n = \sum_{j=1}^{p} x_j$$

subject to

$$\sum_{j=1}^{p} a_{ij} x_j \geq 1, \qquad i = 1, \ldots, m$$

$$x_j \in \{0, 1\}, \qquad j = 1, \ldots, p$$

where

$$a_{ij} = \begin{cases} 1 & \text{if } w_i d_{ij} \leq z \\ 0 & \text{otherwise} \end{cases}$$

Suppose that z^1 and z^* are specific values for the parameter z, and the optimal solution to CP(z^1) is $(n^1, \mathbf{X}^1)$. The relationship between NCP(n) and CP(z) is summarized in the following claims, offered without proof:

Relationship between Covering and n-Center Problems

1. If $(z^*, \mathbf{X}^*)$ solves NCP(n), then $(n, \mathbf{X}^*)$ solves CP(z^*).
2. If $(z^*, \mathbf{X}^*)$ solves NCP(n) and $(n^1, \mathbf{X}^1)$ solves CP(z^1), then:
   (a) If $n^1 > n$, then $z^1 < z^*$.
   (b) If $n^1 \le n$, then $z^1 \ge z^*$.

It is important to study these claims carefully. Part 1 tells us that *if we knew* z^*, the optimal solution *value* for the n-center, we could use CP(z^*), the covering problem formulation, to determine $\mathbf{X}^*$, an optimal n-center. This is progress, since we at least have some methods for solving CP(z^*). Part 2 tells us that if we guess a value for z^*, call it z^1, and solve CP(z^1), we obtain either a lower bound on z^*, case (a), or a feasible solution and an upper bound on z^*, case (b).

8.3.3 Search Algorithm for n-Center

In essence, part 1 of the claim tells us that we can use the covering problem to solve the n-center problem, and part 2 shows us how to do it. We must *search* for z^*, using CP(z) to evaluate each trial value and reduce the search interval. The search procedure will be quite similar to the one used to find an n-center on a tree network.

Table 8.2 contains a quite general statement of an algorithm for solving the n-center problem for specified n. In this algorithm, $\underline{z}$ and $\bar{z}$ are, respectively, lower and upper bounds on z^* and ϵ is a "convergence tolerance parameter." For example, if $\epsilon = 0.01$, then when the algorithm terminates, the best solution found is guaranteed to have a value no greater than $(1.01)z^*$ (i.e., 1% greater than the minimum possible value). The bounds can be initialized in any convenient way [e.g., setting $\underline{z}$ to zero and $\bar{z}$ to the optimal solution value for NCP(1)].

Step 3 of the algorithm could be as simple as a bisection search, or it could involve searching over the set of bound values corresponding to the intersection points in $Q(\bar{z})$.

TABLE 8.2 GENERAL ALGORITHM FOR NCP(n)

1. Initialize $\underline{z}$, $\bar{z}$, and ϵ; $t \leftarrow 1$
2. WHILE $(\bar{z} - \underline{z})/z > \epsilon$
3. Select $z_t \in (\underline{z}, \bar{z})$
4. Solve CP(z_t), obtaining $n_t, \mathbf{X}_t$
   (let $n_t = \infty$ if there is no feasible solution)
5. IF $n_t > n$, THEN $z \leftarrow z_t$
   ELSE $\displaystyle \bar{z} \leftarrow \max_{i=1,\ldots,m} \left(\min_{j \in \theta(\mathbf{X}_t)} (w_i d_{ij}) \right)$
   $\bar{\mathbf{X}} \leftarrow \mathbf{X}_t$
6. $t \leftarrow t + 1$
7. ENDWHILE

Let us apply this algorithm to solve NCP(2) for Example 8.6. We first calculate the values $D(v_i, v_j)$ and obtain:

i \ j	1	2	3	4	5	6	Row maximum
1	0	3	7	10	6	4	10
2	3	0	4	7	6	7	7
3	7	4	0	3	6	8	8
4	10	7	3	0	7	7	10
5	6	6	6	7	0	2	7
6	4	7	8	7	2	0	8

From the distances, we conclude that an optimal vertex-restricted 1-center would be at either v_2 or v_5, with solution value of 7. From Table 8.1 we see that the optimal absolute 1-center is q_{34} (or q_{35}) on arc (v_4, v_5) at a distance of 6.5 from v_4, with solution value of 6.5. Thus we can initialize the upper bound $\bar{z}$, to a value of 6.5. Since the shortest arc has a length of 2, we can initialize $\underline{z} \leftarrow 1$. We will initialize $\epsilon \leftarrow 0$, since we want an optimal solution.

Because we have enumerated all the bound values, in Table 8.1, and the distances $D(v_i, v_j)$, we have at hand a set of values, one of which is equal to z^*. Therefore, we may restrict our trial values, z_t, to this set:

$$\{1, 1.5, 2, 3, 3.5, 4, 4.5, 5, 5.5, 6, 6.5\}$$

The details of the solution for the example are shown below.

Iteration 1: ($\underline{z} = 1, \bar{z} = 6.5$)

2. $(6.5 - 1)/6.5 > 0$, so do not terminate
3. Midpoint of search interval is 3.75; use next smallest value from set, which is 3.5; $z_1 \leftarrow 3.5$.
4. Coefficient matrix for CP(3.5) is

i \ j	1	2	3	4	5	6	7	10	11	12	15	16	17	19	26	28	33	39	45
1	1	1					1	1	1	1		1							
2	1	1					1		1		1	1	1	1					
3			1	1							1	1	1		1	1			
4			1	1									1		1		1	1	
5					1	1				1				1		1	1		1
6					1	1		1	1	1								1	1

Column dominance eliminates q_1, q_2, q_3, q_4, q_5, q_6, q_7, q_{10}, q_{15}, q_{26}, and q_{45}.

By inspection, q_{12} and q_{17} form a 2-center with value 3.5.

5. $n_1 \leftarrow 2, \theta(\mathbf{X}_1) = \{q_{12}, q_{17}\}$, since $n_1 \leq 2$, $\bar{z} \leftarrow 3.5$, $\overline{\mathbf{X}} \leftarrow \mathbf{X}_1$.
6. $t \leftarrow 2$.

Iteration 2: ($\underline{z} = 1, \bar{z} = 3.5$)

2. $(3.5 - 1)/3.5 > 0$; do not terminate.
3. Midpoint of search interval is 2.25; use next-largest value from set, which is 3; $z_2 \leftarrow 3$.
4. Coefficient matrix for CP(3) is

i \ j	1	2	3	4	5	6	7	10	12	15	19	26	28	45
1	1	1					1	1	1					
2	1	1					1			1	1			
3			1	1						1		1	1	
4			1	1								1		
5					1	1			1		1		1	1
6					1	1		1	1					1

Column dominance eliminates $q_1, q_2, q_4, q_5, q_6, q_{10}, q_{26}$, and q_{45}. By inspection $n_2 > 2$, since there are six rows and only one column can cover more than two rows, and it can cover only three.

5. $\underline{z} \leftarrow 3$.
6. $t \leftarrow 3$.

At this point, the algorithm terminates, since there are no possible final values between 3 and 3.5. You might reflect for a moment on the wisdom of setting $\epsilon \leftarrow 0$ when you do not have the list of bound values to use in guiding the search.

Now suppose that we want to solve NCP(3). In this particular instance, we already know that $z^* < 3.5$, so we can start with CP(3) from iteration 2 of the solution to CNP(2). We see by inspection that there is a solution with $n = 3$, using q_3, q_{12}, and one of (q_7, q_{15}, or q_{19}). Suppose that we select q_7 to complete the solution. Since vertex v_1 is covered by both q_7 and q_{12}, we assign it to the closest, which is q_7. Since we have removed one of the vertices covered by q_{12}, we need to see if there is a better location for the center serving the remaining vertices, v_5 and v_6. There is, and the resulting objective function value is 1.5.

8.3.4 Summary

The approach we have described for solving the n-center problem is based on the fundamental relationship between the n-center problem and the covering problem. A key element in understanding the solution procedure, therefore, is understanding this relationship. The algorithm itself is a straightforward exploitation of the relationship. It should also be clear from the example, however, that *implementing* the algorithm requires specifying many detailed steps and devising computational methods both for enumerating the intersection points and for solving the covering problems.

It is easy to devise poor methods for doing these steps in the algorithm. Success in solving realistic n-center problems depends not only on a sound solution strategy, which we have described, but also on an effective computer implementation, which we have not described. Computational techniques are beyond the scope of our presentation, but the interested reader will find some details described in the references at the end of the chapter.

8.4 *n*-MEDIAN PROBLEM ON CYCLIC NETWORKS

To complete our study of location problems on cyclic networks, we now turn our attention to the n-median problem. Recall from Chapter 7 that this problem can be stated formally as

$$\text{minimize} \quad f(Y) = \sum_{i=1}^{m} w_i D(Y, v_i)$$

where $Y = \{y_1, \ldots, y_n\}$ is a set of n medians, or locations on N, a cyclic network, and $D(Y, v_i)$ is the shortest distance between vertex v_i and its nearest median. Also, because of the vertex optimality property of Chapter 7 we need only consider vertices as potential median locations.

Example 8.7

Six major cotton farms in Muddy Flat, Mississippi, have organized a cooperative to provide ginning services by building two cotton gins. Aside from the costs of operating the cotton gins, which will not vary much over the region, the major cost consideration is transporting cotton to the gins. The members of the cooperative have agreed to share the transportation costs in proportion to the amount of cotton they send to be ginned. Therefore, minimizing total transportation costs is a key objective in locating the gins. The road network connecting the farms is shown in Figure 8.5, where the numbers on the arcs represent distance in miles. If each farm has offered a site, where should the gins be located?

In addition to demonstrating the versatility of Figure 8.5, this example presents a problem that fits the n-median model. The weights, w_i, can be determined from the number of truckloads of cotton from farm i, and the operating cost per mile for trucks used by farm i (the operating cost could be the same for all farms). Note that if the farmers had not agreed to share transportation costs, an n-center formulation might be more acceptable to them. Why?

8.4.1 Formulating the *n*-Median Problem

The specification of an optimal n-median has two components: (1) selecting vertices to be median locations, and (2) assigning vertices to medians. These are easily interpreted as decision variables; that is, let

$$y_j = \begin{cases} 1 & \text{if vertex } v_j \text{ is selected as a median} \\ 0 & \text{otherwise} \end{cases}$$

and let

$$x_{ij} = \begin{cases} 1 & \text{if vertex } v_i \text{ is assigned to a median at vertex } v_j \\ 0 & \text{otherwise} \end{cases}$$

We cannot have more than n medians, so

$$\sum_{j=1}^{m} y_j \leq n$$

Also, we can only assign vertices to medians that have been selected, that is,

$$x_{ij} \leq y_j \qquad \text{for all } i \text{ and } j$$

Finally, we need to ensure that every vertex is assigned to exactly one median:

$$\sum_{j=1}^{m} x_{ij} = 1 \qquad \text{for all } i$$

The constraints define a set of feasible alternatives, and an n-median is one of these alternatives that also achieves minimum total weighted distance. Treating weighted distance as a "cost," we have a standard discrete optimization problem formulation:

NMP(n)

$$\text{minimize} \quad z = \sum_{j=1}^{m} \sum_{i=1}^{m} w_i d_{ij} x_{ij}$$

subject to

$$\sum_{j=1}^{m} x_{ij} = 1, \qquad i = 1, \ldots, m$$

$$-x_{ij} + y_j \geq 0, \qquad i = 1, \ldots, m; \quad j = 1, \ldots, m$$

$$\sum_{j=1}^{m} y_j \leq n$$

$$y_j \in \{0, 1\}, \qquad j = 1, \ldots, m$$

$$x_{ij} \geq 0, \qquad i = 1, \ldots, m; \quad j = 1, \ldots, m$$

where

$$d_{ij} = d(v_i, v_j)$$

The requirements on the x_{ij}'s have been written as simple nonnegativity constraints. This is sufficient, at least for the case of nonnegative weighted distances, because once the y_j's are specified, it is never optimum for any x_{ij} to take on a value other than zero or 1.

Problem NMP(n) is a *mixed-integer programming problem*, since it has both discrete and continuous variables, and there are standard methods for solving such problems (e.g., linear programming–based branch and bound). As we shall see, there also are specialized and reasonably effective algorithms for solving NMP(n). Prior to discussing those algorithms, however, we would like to generalize our formulation.

Example 8.8

The cotton farmers of Muddy Flat (see Example 8.7) discovered, upon closer examination, that the costs to set up and operate a cotton gin would vary slightly by farm. A detailed analysis resulted in the following annual cost estimates for a 3-year planning period:

Farm	1	2	3	4	5	6
Annual cost	8000	8000	10,000	8000	9000	8000

In addition, for planning purposes, it has been determined that each farm will require ginning for 1000 truckloads per year, and the average vehicle cost is \$1 per mile. The problem now is to minimize the total cost, facility plus transportation, associated with the cotton ginning operation.

First, since weighted distance often is just a proxy for cost, we can generalize the criterion by converting it into a total cost model. Let

f_j = fixed or overhead cost associated with a median at v_j
c_{ij} = cost to serve v_i from a median at v_j

and rewrite the objective function as

$$\text{minimize} \quad z = \sum_{j=1}^{m} \left(f_j y_j + \sum_{i=1}^{m} c_{ij} x_{ij} \right)$$

There may be some vertex-to-median assignments, x_{ij}, that would not be allowed, for example, because of jurisdictional boundaries or simply due to distance. Let

$$a_{ij} = \begin{cases} 1 & \text{if vertex } i \text{ can be assigned to a median at vertex } j \\ 0 & \text{otherwise} \end{cases}$$

Note that if $a_{ij} = 0$, we can delete x_{ij} from the model. From a computational point of view, if there are many prohibited assignments, deleting these variables allows a more efficient implementation.

Finally, not every vertex may present a suitable median location, for example, due to space limitation, utilities availability, and so on. To incorporate this in the model, index the suitable locations by $j = 1, 2, \ldots, p$, where usually, $p < m$. We can now state the *generalized n-median problem:*

GMP(n)

$$\text{minimize} \quad z = \sum_{j=1}^{p} \left(f_j y_j + \sum_{i=1}^{m} c_{ij} x_{ij} \right)$$

subject to

$$\sum_{j=1}^{p} a_{ij}x_{ij} = 1, \qquad i = 1,\ldots,m$$

$$-x_{ij} + y_j \geq 0, \qquad i = 1,\ldots,m; \quad j = 1,\ldots,p$$

$$\sum_{j=1}^{p} y_j \leq n$$

$$y_j \in \{0,1\}, \qquad j = 1,\ldots,p$$

$$x_{ij} \geq 0, \qquad i = 1,\ldots,m; \quad j = 1,\ldots,p$$

By appropriately specifying the parameters f_j, c_{ij}, and a_{ij}, we obtain the simple n-median problem, NMP(n).

We stated in Section 8.2.5 that the partial cover problem and generalized partial cover problem could be formulated as special cases of GMP(n). To use GMP(n) for solving PC(k), first observe that maximizing the number of covered vertices is equivalent to minimizing the number of vertices covered by an "artificial" center. Define

$$x_{i,p+1} = \begin{cases} 1 & \text{if } v_i \text{ assigned to artificial center} \\ 0 & \text{otherwise} \end{cases}$$

$$a_{ij} = \begin{cases} 1 & \text{if } w_i d(v_i, v_j) \leq s_i \text{ or } j = p+1 \\ 0 & \text{otherwise} \end{cases}$$

$$f_j = \begin{cases} -(m+1) & \text{if } j = p+1 \\ 0 & \text{otherwise} \end{cases}$$

$$c_{ij} = \begin{cases} 1 & \text{if } j = p+1 \\ 0 & \text{otherwise} \end{cases}$$

$$n = k + 1$$

The optimal solution to this instance of GMP(n) will always select v_{p+1} for a median, whether or not it is "used." The remaining k centers will be selected so as to minimize the number of vertices assigned to the artificial center.

Solving GPC(k) is even easier; define

$$p = m$$

$$f_j = 0, \qquad j = 1,\ldots,p$$

$$c_{ij} = t_{ij}, \qquad i = 1,\ldots,m; \quad j = 1,\ldots,p$$

$$a_{ij} = 1, \qquad i = 1,\ldots,m; \quad j = 1,\ldots,p$$

$$n = k$$

In fact, if you have not been napping, you have already observed that GPC(k) is exactly the same formulation as GMP(n)!

As you might now suspect, our generalized n-median formulation can be used to model a great many facility location problems in addition to simple median and covering problems on cyclic networks. Furthermore, GMP(n) makes no direct reference to an underlying network, although one usually is present in a location problem context. Because of the generality of this model, we will refer henceforth to existing facilities (EF) or customers rather than vertices, and new facilities (NF) rather than centers or medians.

In presenting algorithms for GMP(n), we will assume that all $a_{ij} = 1$ (i.e., all EF to NF assignments are feasible). This assumption simplifies the algorithms, although they can be specialized to exploit sparsity if it is present in the problem. We always can prevent an assignment in an optimal solution by making the corresponding c_{ij} very large.

8.4.2 Lower Bounds for GMP(n)

Since GMP(n) is a mixed-integer programming problem, branch and bound is a natural strategy to employ in solving it. Recall from our earlier discussion of branch and bound that a key element in this approach is the development of lower bounds for the candidate problems. In this section we present lower bounding schemes for GMP(n), based on a particular relaxation. The relaxation is an especially nice one, since it provides not only very good lower bounds, but also gives us a mechanism for constructing good feasible solutions.

The relaxed problem is obtained from GMP(n) by associating a *multiplier*, u_i, with each of the assignment constraints (i.e., with each EF) and transforming them into a "penalty function." The resulting objective function is

$$\text{minimize} \quad \sum_{j=1}^{p}\left(f_j y_j + \sum_{i=1}^{m} c_{ij} x_{ij}\right) + \sum_{i=1}^{m} u_i\left(1 - \sum_{j=1}^{p} x_{ij}\right)$$

Rearranging terms gives us the relaxed problem:

$$\text{PD}(\mathbf{U}, n) \qquad \text{minimize} \quad z_D(\mathbf{U}) = \sum_{j=1}^{p}\left[f_j y_j + \sum_{i=1}^{m} (c_{ij} - u_i)x_{ij}\right] + \sum_{i=1}^{m} u_i$$

subject to

$$-x_{ij} + y_j \geq 0, \qquad i = 1, \ldots, m; \quad j = 1, \ldots, p$$

$$\sum_{j=1}^{p} y_j \leq n$$

$$y_j \in \{0, 1\}, \qquad j = 1, \ldots, p$$

$$x_{ij} \geq 0, \qquad i = 1, \ldots, m; \quad j = 1, \ldots, p$$

The u_i's are often called *Lagrangian multipliers* or *dual variables*, and PD($\mathbf{U}, n$) is called the *partial dual* or *Lagrangian relaxation* of GMP(n).

PD(U, n) has two very important features. First, since it is a relaxation of GMP(n), it always provides the required lower bound [i.e., $z_D(\mathbf{U}) \leq z^*$ for any choice of U]. If we are fortunate enough to find a multiplier vector for which the solution to PD(U, n) is feasible in GMP(n), this solution is also optimal.

The second important feature of PD(U, n) is that it is quite easily solved for a given multiplier vector, U. For each possible NF location, define

$$\rho_j(\mathbf{U}) = f_j + \sum_{i=1}^{m} \min(0, c_{ij} - u_i)$$

Now PD(U, n) may be rewritten as

$$z_D(\mathbf{U}) = \sum_{i=1}^{m} u_i + \min \sum_{j=1}^{p} \rho_j(\mathbf{U})y_j$$

subject to

$$\sum_{j=1}^{p} y_j \leq n$$

$$y_j \in \{0, 1\}, \qquad j = 1, \ldots, p$$

If you have trouble convincing yourself that this is an equivalent statement of PD(U, n), note that if $y_j = 0$, there is no contribution to $z_D(\mathbf{U})$, while if $y_j = 1$, $z_D(\mathbf{U})$ includes f_j and all values $(c_{ij} - u_i)$ that are negative.

To solve PD(U, n), sort the $\rho_j(\mathbf{U})$ in increasing order, discard the nonnegative values, and then select the n smallest remaining values or all values if there are fewer than n. The corresponding sites are selected for NF. For a particular U it may happen that all $\rho_j(\mathbf{U}) > 0$. In this case it is legitimate to select the smallest $\rho_j(\mathbf{U})$, which can be interpreted as adding the following redundant constraint to GMP(n):

$$\sum_{j=1}^{p} y_j \geq 1$$

Table 8.3 gives a precise statement of the algorithm for solving PD(U, n), and Table 8.4 presents the result obtained by applying the algorithm to the problem in Example 8.8 with particular values for the multipliers, and $n = 2$. In the algorithm, J^* is the set of NF locations selected.

Note that in this example, the solution to PD(U, n) is not feasible in GMP(n) since neither EF two nor three is assigned to a NF and both EF five and six are assigned twice. Also note, however, that using the set J^* from the solution, it is trivial to construct a feasible solution, simply by assigning each EF to the cheapest NF in J^*. For the example, the resulting feasible solution has a total cost of 41.

At this point, the perceptive reader will be asking: "Where did you get the multiplier vector, and is it a good one?" There are two strategies for obtaining a multiplier vector. One is to use an ad hoc procedure, for example,

$$u_i = \frac{1}{p} \sum_{j=1}^{p} c_{ij}$$

TABLE 8.3 ALGORITHM FOR SOLVING PD(**U**, n)

1. FOR $j = 1$ to p
   (a) Compute $\rho_j(\mathbf{U})$
   ENDFOR
2. $t \leftarrow 1$; $J^* \leftarrow \emptyset$
3. $\rho_{r(t)} \leftarrow \min_{j=1,\ldots,p} \{\rho_j(\mathbf{U})\}$
4. IF $\rho_{r(t)} < 0$, THEN
   WHILE $\rho_{r(t)} < 0$ and $|J^*| < n$
   (a) $J^* \leftarrow J^* \cup \{r(t)\}$
   (b) $t \leftarrow t + 1$
   (c) $\rho_{r(t)} \leftarrow \min_{j \notin J^*} \{\rho_j(\mathbf{U})\}$
   ENDWHILE
   $t \leftarrow t - 1$
   ELSE
   (d) $J^* \leftarrow \{r(1)\}$
   ENDIF
5. $z_D(\mathbf{U}) = \sum_{i=1}^{m} u_i + \sum_{j=1}^{t} \rho_{r(j)}$

TABLE 8.4 SAMPLE SOLUTION OF PD(**U**, n)

i \ j	1	2	3	4	5	6	u_i
1	0	3	7	10	6	4	4
2	3	0	4	7	6	7	4
3	7	4	0	3	6	8	5
4	10	7	3	0	7	7	8
5	6	6	6	7	0	2	7
6	4	7	8	7	2	0	7
f_j	8	8	10	8	9	9	
$\rho_j(\mathbf{U})$	−1	0	−1	−2	−4	−5	
$\rho_{r(1)}$					*		
$\rho_{r(2)}$						*	
	$z_D(\mathbf{U}) = 35 + (-9) = 26$						

Such procedures may work well for some problems, but in general are not robust. In the next section we will see how to obtain multiplier vectors as a by-product of a greedy heuristic, which is somewhat more appealing as it is a general method, even though it may not always result in good lower bounds.

A second strategy is to solve the *Lagrangian dual* problem,

$$\underset{\mathbf{U}}{\text{maximize}}\ z_D(\mathbf{U})$$

which requires searching in m-space for the optimal multiplier vector $\mathbf{U}^*$. If we find $\mathbf{U}^*$, we are guaranteed to have the best lower bound obtainable from PD($\mathbf{U}, n$).

Unfortunately, solving the Lagrangian dual may require a substantial computational effort. Very good results have been reported, however, using a search technique called *subgradient optimization*. The general idea is to increase u_i if the corresponding EF is not assigned in the solution to the current PD($\mathbf{U}, n$), and to decrease u_i if the corresponding EF is assigned to more than one facility. The resulting sequence of vectors can be shown to converge to $\mathbf{U}^*$ provided that certain technical requirements are satisfied. You should go to the references cited at the end of the chapter for more details.

8.4.3 Heuristics for GMP(n)

In this section we examine two types of heuristics applicable to the solution of the generalized n-median problem. The first is a greedy heuristic based on the Lagrangian relaxation, and the second is a local improvement or neighborhood search heuristic. Both approaches are fairly easy to implement and have been shown to give very good solutions, especially for large problems.

Consider the Lagrangian multiplier, u_i, associated with customer i in PD($\mathbf{U}, n$), and suppose that it represents an estimate of the cost of serving that customer. Then we can interpret $\rho_j(\mathbf{U})$ as a measure of the incremental cost of a facility at potential location j. If $c_{ij} < u_i$, customer i could be served at lower cost from site j. If $\rho_j(\mathbf{U}) > 0$, the incremental cost is positive; in other words, this site could not "return its fixed costs," so it is not a good one to use, given the current service costs estimates, $\mathbf{U}$. We shall refer to $\rho_j(\mathbf{U})$ as the site's *relative cost factor*.

How might we assign values to $\mathbf{U}$? Suppose that we have $k < n$ NF sites already selected, and site j is not one of them. Further suppose that each customer is assigned to one of the k sites. Site j can improve the solution (i.e., reduce total cost) only if it allows some customer to be served at a lower cost. Thus we could let u_i be the current cost to serve customer i, and if $\rho_j(\mathbf{U}) < 0$, we should add a NF at site j.

This is the motivation for the greedy algorithm stated in Table 8.5. The initial multiplier vector is based on the worst possible situation for each customer (i.e., highest possible service cost). Comparing Tables 8.3 and 8.5 reveals a striking similarity between the algorithm for solving PD($\mathbf{U}, n$) and the greedy algorithm for GMP(n). In fact, the only significant difference is that in the latter, the multipliers are revised each time the set J^* is augmented by adding another NF.

The greedy algorithm is illustrated in Table 8.6, using the data for Example 8.6. The solution tableau has four quadrants, defined by the heavy bold lines. The upper left-hand quadrant contains the problem data. The columns in the upper right-hand quadrant contain the multiplier vectors determined at each iteration. Similarly, the rows in the lower left-hand quadrant contain values of $\rho_j(\mathbf{U}^r)$ determined at each

TABLE 8.5 GREEDY ALGORITHM FOR GMP(n)

1. $J^* \leftarrow \emptyset$; $t \leftarrow 1$
2. FOR j = 1 TO m
   $u_i^1 \leftarrow$ maximum $\{c_{ij} | j = 1, \ldots, p\}$
   ENDFOR
3. $\rho_{r(1)} \leftarrow$ minimum $\{\rho_j(\mathbf{U}^1) | j = 1, \ldots, p\}$
4. IF $\rho_{r(1)} < 0$, THEN
   WHILE $\rho_{r(t)} < 0$ and $|J^*| < n$
   (a) $J^* \leftarrow J^* \cup \{r(t)\}$
   (b) IF $|J^*| < n$ THEN $t \leftarrow t - 1$
   (c) FOR i = 1 TO m
   $u_i^t \leftarrow$ minimum $(u_i^{t-1}, c_{ir(t-1)})$
   ENDFOR
   (d) $\rho_{r(t)} \leftarrow$ minimum $\{\rho_j(\mathbf{U}^t) | j \in J^*\}$
   ENDWHILE
   $t \leftarrow t - 1$
   ELSE
   (e) $J^* \leftarrow \{r(1)\}$
   ENDIF
5. $z_g \leftarrow \sum_{j \in J^*} f_j + \sum_{i=1}^{m} \min(c_{ij} | j \in J^*)$

TABLE 8.6 SOLUTION OF SAMPLE PROBLEM USING THE GREEDY ALGORITHM

i \ j	1	2	3	4	5	6	u_i^1	u_i^2	u_i^3
1	0	3	7	10	6	4	10	3	3
2	3	0	4	7	6	7	7	0	0
3	7	4	0	3	6	8	8	4	4
4	10	7	3	0	7	7	10	7	7
5	6	6	6	7	0	2	7	6	2
6	4	7	8	7	2	0	8	7	0
f_j	8	8	10	8	9	8	$z_g(\mathbf{U}^t)$	$z_d(\mathbf{U}^t)$	
$\rho_j(\mathbf{U}^1)$	−12	−15	−12	−8	−14	−14	35	7	
$\rho_j(\mathbf{U}^2)$	2	8	2	0	−2	−3	32	22	
$\rho_j(\mathbf{U}^3)$	5	8	2	0	7	8	32	16	

iteration. The two columns in the lower right-hand quadrant contain, in order, the feasible solution value, $z_g(\mathbf{U}^t)$, corresponding to J^* in each iteration, and the lower bound, $z_D(\mathbf{U}^t)$, corresponding to $\mathbf{U}^t$ obtained at each iteration:

$$z_g(\mathbf{U}^t) = \sum_{j \in J^*} f_j + \sum_{i=1}^{m} \min\{c_{ij} | j \in J^*\}$$

$$z_D(\mathbf{U}^t) = \sum_{i=1}^{m} u_i^t + \sum_{j \in J^*} \rho_j(\mathbf{U}^t)$$

Solutions in Table 4.6 are $J^* = \{2\}$, $\{2, 6\}$ and $\{2, 6\}$ for iterations 1, 2 and 3 respectively. Note that the greedy algorithm produces solutions for one NF, two NF, and so on, up to n NF. When the algorithm terminates, we may take the *largest* of the lower bounds, since $z_D(\mathbf{U}) \leq z^*$ for any $\mathbf{U}$. Even so, in our example, the lower bound is only 69% of the upper bound, so we cannot have much confidence in the quality of this solution. Recall that in the preceding section, we obtained a lower bound of 27 using an arbitrary multiplier vector, which indicates that the multiplier vectors from the greedy algorithm need not give good lower bounds.

The greedy algorithm allows us to obtain a feasible solution to GMP(n), and an upper bound, $\bar{z}$ on z^*, and also generates several multiplier vectors which we can use in PD($\mathbf{U}, n$) to obtain a lower bound, $\underline{z}$. When $\bar{z} > \underline{z}$, it is often possible to improve the upper bound by modifying the greedy solution. For example, you may verify that replacing site 2 by site 3 in the greedy solution leads to a total cost of 31.

The idea of modifying a given solution to GMP(n) to see if a better one can be found is appealing. However, if we take a completely ad hoc approach, we easily may find ourselves enumerating all the feasible solutions, which is not appealing! What we need is a systematic method for examining a subset of feasible solutions. Fortunately, we can define three kinds of modifications which will give us only a limited number of modified solutions to examine.

The three types of solution modifications or local search heuristics are called *drop*, *add*, and *swap*, and all three are guided by evaluating the site relative cost factors, $\rho_j(\mathbf{U})$, for specially defined vectors, $\mathbf{U}$. In presenting the three local search heuristics, we shall let J be the set of sites currently selected for a NF, and define

$$c_{i\alpha} = \min\{c_{ij} \mid j \in J\}$$

$$c_{i\beta} = \min\{c_{ij} \mid j \in J \text{ and } j \neq \alpha\}$$

Thus $c_{i\alpha}$ is the current cost to serve customer i and $c_{i\beta}$ would be the cost if the NF at site α were abandoned. Note both α and β depend on i.

Drop heuristic. Suppose, for the cotton gin location problem, that we have a trial solution with $J = \{1, 2, 5\}$. The NF at site 1 serves only customer 2, since one of the other sites has a lower variable cost for each of the other customers. Thus dropping site 1 from the solution would yield a net savings of $8 - 3 = 5$. This analysis can be formalized by defining customer service costs:

$$v_i^j = \begin{cases} c_{i\alpha} & \text{if } j \neq \alpha \\ c_{i\beta} & \text{if } j = \alpha \end{cases}$$

Then if $\rho_j(\mathbf{V}^j) > 0$, we may delete site j from the current solution and reduce the total cost.

Add heuristic. The drop heuristic seeks to reduce cost by deleting a NF, and the add heuristic seeks to reduce cost by adding a NF. In fact, the greedy algorithm itself is nothing more than repeated application of the add heuristic to a specific starting solution. The add heuristic uses customer service costs $v_i^j = c_{i\alpha}$ and if $\rho_j(\mathbf{V}^j) < 0$, the total cost will be reduced by adding site j.

Swap heuristic. Consider the solution $J = \{2, 5\}$, with total cost of 32, obtained from the greedy solution of Example 8.6. Observe that it is not profitable to drop site 2, nor is it profitable to add site 3. However, the total cost is reduced to 31 if we exchange site 2 for site 3. The swap heuristic permits improvements to the solution that cannot be obtained by a sequence of adding and dropping where each add and drop improves the solution value. To evaluate the simultaneous dropping of site r and adding of site s, define customer service costs:

$$v_i^s = c_{i\alpha},\ i = 1, \ldots, m$$

$$v_i^r = \begin{cases} \min(c_{is}, c_{i\beta}) & \text{if } \alpha = r \\ \min(c_{is}, c_{i\alpha}) & \text{if } \alpha \neq r \end{cases}$$

If $\rho_s(\mathbf{V}^s) < \rho_r(\mathbf{V}^r)$, the swap will reduce total cost. Note that $\rho_s(\mathbf{V}^s)$ is the cost for adding site s and $\rho_r(\mathbf{V}^r)$ is the cost for deleting site r, assuming that site s has been added.

The drop, add, and swap heuristics are conceptually simple and can be implemented quite easily. Table 8.7 gives a formal statement of a local improvement or neighborhood search algorithm for GMP(n). You should not infer that developing a good computer implementation of a local improvement algorithm is a trivial undertaking. In the first place, the algorithm in Table 8.7 is only one simple way to use the drop, add, and swap procedures, and many variations are possible. For example, the swap procedure might be applied to solutions with fewer than n sites already selected. Especially for large problems, you might not want to do an exhaustive search for the *best* drop, add, or swap. Also, for large problems, the organization of the data and the calculations will have an important impact on computational resources, such as random access memory and run time.

TABLE 8.7 LOCAL IMPROVEMENT ALGORITHM FOR GMP(n)

```
1. Initialize J ← {j}                              [Select some site]
2. Repeat
        Repeat
              Find best site to add and if          [Add phase]
              profitable, add it
        Until (no change in J or |J| = n)
        Repeat
              Find best site to drop and if         [Drop phase]
              profitable, drop it
        Until (no change in J)
   Until (no change in J)
3. Repeat
        Find best swap and if                       [Swap phase]
        profitable, make swap
   Until (no change in J)
```

For large instances of GMP(n) with n "not too large," say, less than 10, a combination of the greedy algorithm followed by local improvement has proven quite effective. Solutions within a few percent of optimum have been obtained with computation times of only a few seconds on a mainframe computer.

8.4.4 Optimizing GMP(n)

Heuristic solutions from the greedy algorithm and local improvement, while usually good, provide at best an upper bound, $\bar{z}$, on the optimal solution value z^*. Similarly, the Lagrangian relaxation, PD($\mathbf{U}, n$), perhaps embedded in a search procedure (e.g., subgradient optimization), provides only a lower bound $\underline{z}$ on the optimal value. If the gap, $\bar{z} - \underline{z}$, is too large to be acceptable, other methods will be required to solve the problem.

Since GMP(n) is a mixed-integer programming problem, branch and bound is a natural approach, using PD($\mathbf{U}, n$) to obtain the required lower bounds. The success of this approach (i.e., whether or not a given problem can be solved with reasonable effort) generally depends on the quality of the lower bound. Thus it probably is wise at least to try to solve the Lagrangian dual. The effort required usually is recouped manyfold through reducing the number of candidate problems that must be evaluated explicitly.

8.4.5 Summary

The generalized n-median formulation can be used to model a wide variety of facility location problems in addition to the n-median on a cyclic network. It has the form of a mixed-integer programming problem with a convenient mathematical structure that can be exploited easily in devising solution procedures. In discussing the solution procedures, such as Lagrangian relaxation and subgradient optimization, we have touched on topics that are outside the scope of this book, but which must be mentioned in any modern treatment of the problem.

GMP(n) represents a transition from network-based location problems to more general discrete optimization models of location problems. In fact, we have started to move away from the terms *arc*, *vertex*, *center*, and *median*, and toward terms like *customer*, *site*, *fixed cost*, and *variable cost*. These discrete optimization models permit a more detailed description of the problem (e.g., inclusion of site-specific costs) but at the same time, require more complex solution procedures and greater computational resources.

8.5 WAREHOUSE LOCATION PROBLEM

"Warehouse location" is the name given to one of the first widely studied discrete location problems. In this problem we have m customers, each with a known delivery requirement of r_i units per period. There are p potential warehouse locations, and the cost to establish a warehouse at location j is f_j dollars per period. The marginal cost to ship to customer i from a warehouse at site j is assumed to be v_{ij} dollars per unit per period, which includes variable costs for both handling and transportation. The problem is to determine which warehouse sites should be used, and the assignment of customers to warehouses, in order to minimize total cost per period.

This problem statement embodies some stringent assumptions about costs. Figure 8.6 illustrates the required form of facility cost as a function of throughput and Figure 8.7 illustrates the required form of transportation cost as a function of shipment size. Note that in both cases, the marginal costs are the same at all levels of activity (i.e., there are no scale economies in handling or transportation). Even with these simplified cost models, it may be a difficult task to estimate the cost parameters. For example, what is the length of a "period," and how many periods are to be considered? Is warehouse supervision a fixed cost or a variable cost? These and many other similar questions must be resolved to develop the cost models for a particular problem.

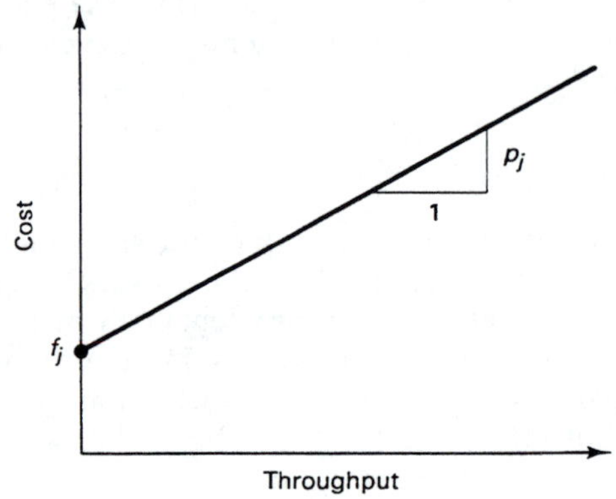

Figure 8.6 Warehouse j cost model.

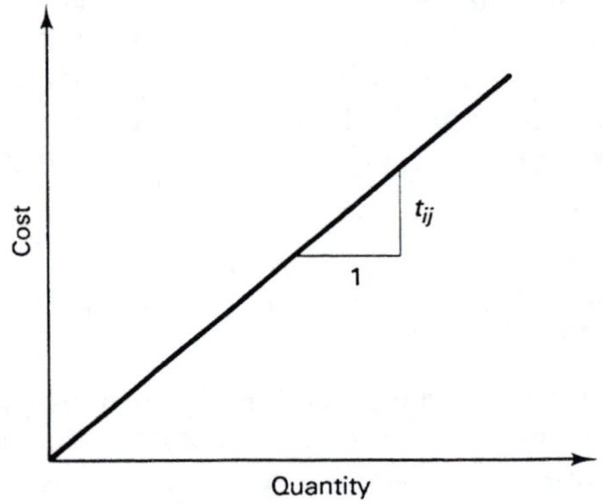

Figure 8.7 Transportation cost model.

The warehouse location problem can be formulated as a mixed-integer programming problem as follows:

(WLP) minimize $z = \sum_{j=1}^{p} \left(f_j y_j + \sum_{i=1}^{m} c_{ij} x_{ij} \right)$

subject to

$$\sum_{j=1}^{p} x_{ij} = 1, \qquad i = 1, \ldots, m \qquad [u_i] \tag{8.4}$$

$$-x_{ij} + y_j \geq 0, \qquad i = 1, \ldots, m; \quad j = 1, \ldots, p \qquad [w_{ij}] \tag{8.5}$$

$$0 \leq x_{ij} \leq 1, \qquad i = 1, \ldots, m; \quad j = 1, \ldots, p \tag{8.6}$$

$$y_j \in \{0, 1\}, \qquad j = 1, \ldots, p \tag{8.7}$$

where

$$c_{ij} = v_{ij} r_i$$

$$y_j = \begin{cases} 1 & \text{if site } j \text{ is selected} \\ 0 & \text{otherwise} \end{cases}$$

Note that $x_{ij} r_i$ is the amount shipped from site j to customer i (i.e., x_{ij} is the fraction of r_i shipped to customer i from site j). The constraints (8.6) and (8.7) could be relaxed to

$$x_{ij} \geq 0 \tag{8.6a}$$

$$y_j \in \{0, 1, 2, \ldots\} \tag{8.7a}$$

The assignment constraints (8.4) effectively limit any x_{ij} to a maximum value of 1, which would force the corresponding y_j to be at least 1. As long as $f_j > 0$, however, it would never be optimal for y_j to be any larger than 1. The treatment of bounds on the variables is important in developing the dual formation.

With this formulation, we see that WLP is a special case of GMP(n); one in which we have deleted the constraint limiting the number of facilities. Thus we could solve WLP by setting $n = p$ in the corresponding generalized n-median formulation.

The reason for studying WLP as a separate problem is twofold. First, in the context of the GMP(n) methodology, there is an elegant direct search algorithm for finding optimal or near-optimal Lagrangian multipliers. As a result, WLP is a very easy to solve discrete optimization problem. Second, from WLP, we shall learn a very important lesson about formulating discrete location problems.

8.5.1 The Linear Programming Dual of WLP

The linear programming relaxation of WLP could be used in a branch-and-bound algorithm. Suppose that instead of solving the relaxed problem, we solve its linear programming dual. Let u_i and w_{ij} be the dual variables; they are shown in brackets in the statement of WLP. The linear programming dual of the relaxed problem is

$$\text{(DWLP)} \qquad \text{minimize} \quad z_d = \sum_{i=1}^{m} u_i$$

subject to

$$\sum_{i=1}^{m} w_{ij} \le f_j, \qquad j = 1, \ldots, p \qquad [y_j] \tag{8.8}$$

$$u_i - w_{ij} \le c_{ij}, \qquad i = 1, \ldots, m; \quad j = 1, \ldots, p \qquad [x_{ij}] \tag{8.9}$$

$$w_{ij} \ge 0, \qquad i = 1, \ldots, m; \quad j = 1, \ldots, p$$

$$u_i \text{ is unrestricted,} \qquad i = 1, \ldots, m$$

Since the linear relaxation of WLP is the "dual" of DWLP, we have shown in brackets the "dual" variables from the relaxed WLP which correspond to each constraint in DWLP.

We are going to concentrate on solving DWLP. To do so, we first present a condensed dual formulation and a simple "ascent" algorithm for finding a good feasible dual solution. We then use the complementary slackness conditions from linear programming duality theory to construct a corresponding feasible solution to WLP. If our dual feasible solution is not optimal (i.e., the primal and dual solutions violate complementary slackness), the complementary solution will be the basis for an "adjustment" algorithm that tries to improve the dual solution.

To develop the condensed dual problem, first observe that the dual constraints (8.9) can be rewritten as

$$w_{ij} \ge u_i - c_{ij}$$

or, since w_{ij} is a nonnegative variable,

$$w_{ij} = \max(0, u_i - c_{ij})$$

Now, we can sum the w_{ij} over i to obtain

$$\sum_{i=1}^{m} w_{ij} = \sum_{i=1}^{m} \max(0, u_i - c_{ij}) \tag{8.10}$$

Substituting (8.10) into (8.8) gives

$$\sum_{i=1}^{m} \max(0, u_i - c_{ij}) \le f_j$$

Finally, one can rearrange this into an equivalent and slightly more familiar form:

$$\rho_j(\mathbf{U}) = f_j + \sum_{i=1}^{m} \min(0, c_{ij} - u_i) \ge 0 \tag{8.11}$$

Thus our condensed dual problem is

$$\text{maximize} \quad z_d = \sum_{i=1}^{m} u_i$$

subject to

$$\rho_j(\mathbf{U}) \geq 0, \qquad j = 1, \ldots, p$$

$$u_i \text{ unrestricted}, \qquad i = 1, \ldots, m$$

A feasible solution to the condensed dual problem provides a lower bound on the solution value for WLP. For example, the algorithms in Table 8.3 or 8.5 with $n = p$ would yield feasible dual solutions. However, the special structure of WLP allows us to do even better than these two general methods. Exploiting this special structure requires a good grasp of the application of complementary slackness theory from linear programming, so the remainder of Section 8.5 should be considered advanced material.

8.5.2 Complementary Slackness Relationships

Suppose that $(\mathbf{Y}^*, \mathbf{X}^*)$ is optimal in the relaxed WLP, and $(\mathbf{U}^*, \mathbf{W}^*)$ is optimal in DWLP. Then the following complementary slackness conditions are satisfied:

(1) IF $y_j^* \rho_j(\mathbf{U}^*) = 0$
THEN
IF $y_j^* = 1$, THEN $\rho_j(\mathbf{U}^*) = 0$ (8.12a)
AND IF $\rho_j(\mathbf{U}^*) \neq 0$, THEN $y_j^* = 0$ (8.12b)

(2) IF $x_{ij}^* (c_{ij} - u_i^* + w_{ij}^*) = 0$
THEN
IF $x_{ij}^* = 1$, THEN $w_{ij}^* - u_i^* + c_{ij} = 0$ (8.13a)
(OR EQUIVALENTLY, $u_i^* \geq c_{ij}$)
AND IF $c_{ij} - u_i^* + w_{ij}^* > 0$
(OR EQUIVALENTLY, $u_i^* < c_{ij}$) (8.13b)
THEN $x_{ij}^* = 0$

(3) IF $w_{ij}^* (y_j^* - x_{ij}^*) = 0$
THEN
IF $w_{ij}^* > 0$ (OR EQUIVALENTLY, $u_i^* > c_{ij}$) (8.14a)
THEN $x_{ij}^* = y_j^*$
AND IF $x_{ij}^* < y_j^*$, THEN $w_{ij}^* = 0$ (8.14b)
(OR EQUIVALENTLY, $u_{ij}^* \leq c_{ij}$)

Suppose that we are given a feasible dual solution, $\mathbf{U}^+$. If we can find a feasible primal solution, $(\mathbf{Y}^+, \mathbf{X}^+)$, and if $\mathbf{U}^+$ and $(\mathbf{Y}^+, \mathbf{X}^+)$ satisfy the complementary slackness condition, they are optimal solutions. In the following, we show how to use $\mathbf{U}^+$ and the complementary slackness conditions to *construct* $(\mathbf{Y}^+, \mathbf{X}^+)$, a feasible solution to WLP. The construction attempts to satisfy the complementary slackness conditions.

From (8.12) it is clear that y_j^+ can be nonzero only if $\rho_j(\mathbf{U}^+) = 0$. Therefore, define

$$J = \{j \mid \rho_j(\mathbf{U}^+) = 0\}$$

and require $y_j^+ = 0$ for $j \notin J$. We only want to have $y_j^+ = 1$ if there is at least one customer that "should" be assigned to a warehouse at site j. From (8.13) we can only have $x_{ij}^+ = 1$ if $u_i^+ > c_{ij}$. Furthermore, for every customer, we need to have at least one site selected for which $u_i^+ > c_{ij}$. Therefore, we need a set of sites J^+ satisfying

(i) $J^+ \subseteq J$ (8.15)

(ii) for every customer, i, there is at least one $j \in J^+$ for which $u_i^+ > c_{ij}$. (8.16)

In case for some customers there are two or more $j \in J^+$ satisfying (8.16), we need a method for selected the one to be used in $\mathbf{X}^+$. The best choice is the one that results in the smallest cost, so define:

$$c_{i\alpha(i)} = \min(c_{ij} \mid j \in J^+)$$

In other words, if a warehouse is opened at each site in J^+, customer i should be assigned to site $\alpha(i)$.

In summary, given $\mathbf{U}^+$ for which $J \neq \emptyset$ and there is a J^+ satisfying (8.15) and (8.16), $(\mathbf{Y}^+, \mathbf{X}^+)$ is constructed as follows:

$$y_j^+ = \begin{cases} 1 & \text{if } j \in J^+ \\ 0 & \text{otherwise} \end{cases} \tag{8.17}$$

$$x_{ij}^+ = \begin{cases} 1 & \text{if } j = \alpha(i) \\ 0 & \text{otherwise} \end{cases} \tag{8.18}$$

$$x_p^+ = \sum_{j \in J^+} f_j + \sum_{i=1}^{m} c_{i\alpha(i)} \tag{8.19}$$

In the next section we develop an algorithm that creates a feasible dual solution with the required structure.

To illustrate the construction procedure, consider once again Example 8.8, and suppose that $\mathbf{U}^+ = (4, 6, 4, 7, 6, 4)$. Table 8.8 displays the corresponding values of

TABLE 8.8 $\rho_j(\mathbf{U}^+)$ AND $u_i^+ - c_{ij}$ FOR EXAMPLE 8.8

i \ j	1	2	3	4	5	6	u_i^+
1	4	1	−3	−6	−2	0	4
2	3	6	2	−1	0	−1	6
3	−3	0	4	1	−2	−4	4
4	−3	0	4	7	0	−1	7
5	0	0	0	−1	6	4	6
6	0	−3	−4	−4	2	4	4
f_j	8	8	10	8	9	8	
$\rho_j(\mathbf{U}^+)$	1	1	0	0	1	0	

$\rho_j(\mathbf{U}^+)$ and $u^+ - c_{ij}$. From Table 8.8, we have $J = \{3, 4, 6\}$, and we could simply let $J^+ = J$. However, closer examination of Table 8.8 reveals that we also could choose $J^+ = \{3, 6\}$. For the latter choice of J^+, the primal solution is

$$\mathbf{Y}^+ = (0, 0, 1, 0, 0, 1)$$

$$x_{i3}^+ = \begin{cases} 1 & \text{for } i = 2, 3, 4 \\ 0 & \text{otherwise} \end{cases}$$

$$x_{i6}^+ = \begin{cases} 1 & \text{for } i = 1, 5, 6 \\ 0 & \text{otherwise} \end{cases}$$

Checking the complementary slackness conditions, we discover that they are all satisfied and conclude that we have optimal primal and dual solutions. Moreover, since $(\mathbf{Y}^+, \mathbf{X}^+)$ is feasible in WLP, it also is optimal in WLP. Note that $z_p^+ = z_d^+ = 31$.

We will not always be so fortunate! In the previous example, if we select $J^+ = \{4, 5, 6\}$, we get the following primal solution:

$$\mathbf{Y}^+ = (0, 0, 1, 1, 0, 1)$$

$$x_{i3}^+ = \begin{cases} 1 & \text{if } i = 2, 3 \\ 0 & \text{otherwise} \end{cases}$$

$$x_{i4}^+ = \begin{cases} 1 & \text{if } i = 4 \\ 0 & \text{otherwise} \end{cases}$$

$$x_{i6}^+ = \begin{cases} 1 & \text{if } i = 1, 5, 6 \\ 0 & \text{otherwise} \end{cases}$$

For this solution, checking the complementary slackness conditions reveals that (8.14) is violated since

$$y_3^+ = 1, \qquad x_{43}^+ = 0, \qquad \text{and} \qquad w_{43}^+ > 0$$

$$y_4^+ = 1, \qquad x_{34}^+ = 0, \qquad \text{and} \qquad w_{34}^+ > 0$$

Note that $z_p^+ = 36$, $z_d^+ = 31$, and $w_{34}^+ + w_{43}^+ = 5 = z_p^+ - z_d^+$.

In general, $(\mathbf{Y}^+, \mathbf{X}^+)$ will violate complementary slackness if $w_{ij}^+ > 0$ for a given i and more than one $j \in J^+$. This observation is formalized in the following:

Lemma. Suppose that we are given J^+ satisfying (8.15) and (8.16). Then

$$z_p^+ - z_d^+ = \sum_{i=1}^{m} \sum_{\substack{j \in J^+ \\ j \neq \alpha(i)}} w_{ij}^+$$

Note that if $z_p^+ = z_d^+$, then $(\mathbf{Y}^+, \mathbf{X}^+)$ is also an optimal solution to WLP.

To exploit these relationships, we need an algorithm for determining $\mathbf{U}^+$, the feasible dual solution. In addition, we need a rule for selecting J^+ if it is to be a strict subset of J.

8.5.3 Dual Ascent Algorithm

The objective function for DWLP is just the sum of the dual variables, u_i. It is intuitively appealing to make each of them as large as possible in order to maximize z_d. Suppose that we have an initial feasible dual solution, $\mathbf{U}$. If $u_i \geq c_{ij}$, there is a limit to how much larger u_i can be without making $\rho_j(\mathbf{U})$ negative, thus violating a dual constraint. Specifically, if Δ is a proposed increase in u_i, then to maintain dual feasibility we must have

$$\Delta \leq \min_j \{\rho_j(\mathbf{U}) \mid u_i > c_{ij}\} \equiv \Delta_1$$

This ensures that no constraint is violated by the new dual solution. Similarly, if initially $u_i < c_{ij}$, we can increase u_i by $c_{ij} - u_i$ without affecting the associated dual constraints. Thus

$$\Delta \leq \min_j \{c_{ij} - u_i \mid u_i < c_{ij}\} \equiv \Delta_2$$

This ensures no change in the dual slacks, $\rho_j(\mathbf{U})$. Thus $\Delta = \min(\Delta_1, \Delta_2)$ is an increase in u_i that will not cause any dual constraint to be violated.

This gives us a simple "dual ascent" algorithm for finding a good dual solution. Cycle through the customers, computing Δ for each one. For any customers, if $\Delta > 0$, increase the corresponding u_i by Δ. Stop when all customers have been considered without increasing any u_i.

The dual ascent algorithm is stated formally in Table 8.9. It is important to note that in the algorithm, whenever a dual variable, u_i^+, is increased, all the $\rho_j(\mathbf{U}^+)$ are recomputed *before* going to the next u_i^+. A tableau format for the dual ascent calculations is given in Table 8.10 and illustrated using the data for Example 8.8.

The tableau in Table 8.10(a) illustrates the initialization of $u_i^+ \leftarrow 0, i = 1, \ldots, m$. The tableau is completed by considering each row, computing Δ_1 and Δ_2,

TABLE 8.9 DUAL ASCENT ALGORITHM

```
FOR i ∈ I
    u_i^+ ← min (c_ij | j = 1,...,p)
END
WHILE I ≠ ∅
    FOR i ∈ I
      Δ ← min{min_j [ρ_j(U^+) | u_i^+ ≥ c_ij], min_j [c_ij − u_i^+ | u_i^+ < c_ij]}
      IF Δ > 0, THEN
            u_i^+ ← u_i^+ + Δ
      ELSE
            I ← I\{i}
      ENDIF
    ENDFOR
ENDWHILE
```

TABLE 8.10(a) INITIALIZED TABLEAU

	1	2	3	4	5	6	u_i^+	Δ_1	Δ_2	u_i^+
1	0	3	7	10	6	4	0			
2	3	0	4	7	6	7	0			
3	7	4	0	3	6	8	0			
4	10	7	3	0	7	8	0			
5	6	6	6	7	0	2	0			
6	4	7	8	8	2	0	0			
f_j	8	8	10	8	9	8				
$\rho_j(\mathbf{U})$	8	8	10	8	9	8				

TABLE 8.10(b) TABLEAU AFTER ONE PASS

i	1	2	3	4	5	6	u_i^+	Δ_1	Δ_2	u_i^+
1	0	3	7	10	6	4	0	8	3	3
2	3	0	4	7	6	7	0	8	3	3
3	7	4	0	3	6	8	0	10	3	3
4	10	7	3	0	7	8	0	8	3	3
5	6	6	6	7	0	2	0	9	2	2
6	4	7	8	8	2	0	0	8	2	2
f_j	8	8	10	8	9	8				16
$\rho_j(\mathbf{U})$	8	8	10	8	9	8				
	5									
		5								
			7							
				5						
					7					
						6				

and determining whether or not the corresponding dual variable can be increased. If it can, the $\rho_j(\mathbf{U}^+)$ are recomputed. Table 8.10(b) shows the result after the first pass through the dual variables. When a change to u_i^+ causes some $\rho_j(\mathbf{U}^+)$ to change, the new value is shown in the corresponding row of $\rho_j(\mathbf{U}^+)$ values. Note that since $\Delta > 0$ for every i, all the dual variables will be considered again on the second pass.

Table 8.10(c) shows the result after the second pass through the dual variables. Since $\rho_4(\mathbf{U}^+) = \rho_6(\mathbf{U}^+) = 0$, we can construct a primal solution with sites 4 and 6. The

TABLE 8.10(c) TABLEAU AFTER TWO PASSES

	1	2	3	4	5	6	u_i^+	Δ_1	Δ_2	u_i^+
1	0	3	7	10	6	4	3	5	1	4
2	3	0	4	7	6	7	3	4	1	4
3	7	4	0	3	6	8	3	5	1	4
4	10	7	3	0	7	8	3	4	4	7
5	6	6	6	7	0	2	2	6	4	6
6	4	7	8	8	2	0	2	2	2	4
f_j	8	8	10	8	9	8				29
$\rho_j(\mathbf{U})$	5	5	7	5	7	6				
	4	4								
	3	3								
			6	4				$J = \{4,6\}$		
			2	0						
					3	2				
					1	0				

TABLE 8.10(d) TABLEAU AFTER THREE PASSES

	1	2	3	4	5	6	u_i^+	Δ_1	Δ_2	u_i^+
1	0	3	7	10	6	4	4	0	—	4
2	3	0	4	7	6	7	4	2	2	6
3	7	4	0	3	6	8	4	0	—	4
4	10	7	3	0	7	8	7	0	—	7
5	6	6	6	7	0	2	6	0	—	6
6	4	7	8	8	2	0	4	0	—	4
f_j	8	8	10	8	9	8				31
$\rho_j(\mathbf{U})$	3	3	2	0	1	0				
	1	1	0							
								$J = \{3,4,6\}$		

corresponding $z_p^+ = 32$. Note that $z_d^+ = 29$. Again, we have $\Delta > 0$ for all the dual variables, so they all will be considered on the third pass.

The result of the third pass is shown in Table 8.10(d). Only u_2^+ was increased on this pass. We have $z_d = 31$ and $J = \{3, 4, 6\}$. Site 3 provides lower service cost than site 4 for every one except customer 4. Since $c_{43} - c_{44} = 3$, and $f_4 = 8$, we obtain a better primal solution with $J^+ = \{3, 6\}$, $z_p^+ = 31$. Since $z_p^+ = z_d^+$, we now have an optimal primal solution.

From this example, you might have concluded that the set J^+ should be as small as possible. If so, you were very perceptive. In fact, the Lemma in Section 8.5.2 provides a theoretical justification for including the fewest possible sites in J^+. Suppose that there is some site $j' \in J^+$, and $j' \neq \alpha(i)$ for any customer. Dropping j' from J^+ will not affect any of the x_{ij}^+ values, thus will not affect the variable costs. However, it will permit a site to be dropped resulting in a lower fixed cost.

A site is *essential* if it cannot be dropped from J^+. A site that can be dropped from J^+ is *inessential*. As the Lemma indicates, dropping inessential sites from J^+ may improve the primal solution value.

8.5.4 Dual Adjustment Algorithm

The dual ascent algorithm does not always terminate with an optimal dual solution, as the example in Table 8.11* illustrates. The final result is shown in the table. We have $J = \{1,3,5\}$, $J^+ = \{1,3,5\}$, $z_p^+ = 1840$, and $z_d^+ = 1535$. The violations of complementary slackness are:

$$w_{13}^+ = 30 \quad \text{but} \quad x_{13}^+ = 0 \quad \text{and} \quad y_3^+ = 1$$

$$w_{15}^+ = 40 \quad \text{but} \quad x_{15}^+ = 0 \quad \text{and} \quad y_5^+ = 1$$

$$w_{21}^+ = 60 \quad \text{but} \quad x_{21}^+ = 0 \quad \text{and} \quad y_1^+ = 1$$

$$w_{23}^+ = 50 \quad \text{but} \quad x_{23}^+ = 0 \quad \text{and} \quad y_3^+ = 1$$

$$w_{33}^+ = 40 \quad \text{but} \quad x_{33}^+ = 0 \quad \text{and} \quad y_3^+ = 1$$

$$w_{35}^+ = 40 \quad \text{but} \quad x_{35}^+ = 0 \quad \text{and} \quad y_5^+ = 1$$

$$w_{43}^+ = 45 \quad \text{but} \quad x_{43}^+ = 0 \quad \text{and} \quad y_3^+ = 1$$

TABLE 8.11 DUAL ASCENT NOT OPTIMUM

i	1	2	3	4	5	u_i^+	Δ_1	Δ_2
1	120	210	180	210	170	210	0	—
2	180	M	190	190	150	240	0	—
3	100	150	110	150	110	150	0	—
4	M	240	195	180	150	240	0	—
5	60	65	50	65	70	60	0	5
6	M	210	M	120	195	235	0	—
7	180	110	M	160	200	180	0	20
8	M	165	195	120	M	220	0	—
f_j	200	200	200	400	300	$z_d^+ = 1535$		
$\rho_j(\mathbf{U}^+)$	0	45	0	55	0	$J = \{1,3,5\}$ $J^+ = \{1,3,5\}$ $z_p^+ = 1840$		

* Table 8.11 is based on examples in [15] and [36].

The w_{ij}^+ for which the complementary slackness is violated sum up to 305, which is exactly $z_p^+ - z_d^+$, as required by the Lemma stated in Section 8.5.2.

Note that a better primal solution is obtained using only sites 3 and 5 with a total cost of 1720. This solution violates (8.16), and the corresponding complementary slackness condition as well. To illustrate the dual adjustment, we shall use the primal solution with sites 1, 3, and 5.

If we could correct all violations of complementary slackness, we would obtain an optimal dual solution. The *dual adjustment algorithm* given in Table 8.12 attempts to do this by reducing u_i for a customer having more than one $w_{ij}^+ > 0$. To ensure that z_d^+ does not get smaller, this reduction in one u_i^+ should allow us to increase at least one other dual variable by the same amount. The following additional notation is required for the dual adjustment algorithm:

$$J_i^* = \{j \in J^+ \mid u_i^+ \geq c_{ij}\}, \qquad i = 1, \ldots, m$$

$$J_i^+ = \{j \in J^+ \mid u_i^+ > c_{ij}\}, \qquad i = 1, \ldots, m$$

$$I_j^+ = \{i \mid J_i^* = \{j\}\}, \qquad j \in J^+$$

$$c_{i\beta(i)} = \min\{c_{ij} \mid j \in J^+, j \neq \alpha(i)\} \qquad \text{if } |J_i^+| > 1$$

J_i^* is the subset of warehouses to which customer i could be assigned without violating complementary slackness and J_i^+ is the subset of warehouses to which customer i must be assigned to satisfy complementary slackness. For a given warehouse, $j \in J^+$, the set I_j^+ represents the set of customers that can be assigned only to warehouse j in a complementary primal solution.

Complementary slackness condition (8.14) is violated if $|J_i^+| > 1$. We can reduce $|J_i^+|$ by making u_i^+ smaller, thus eliminating at least one violation of complementary slackness. Unfortunately, this also reduces z_d^+. In the example, $|J_1^+| = 3$, indicating that complementary slackness is violated for two constraints associated with customer 1. If we reduced u_1^+ to 180, then $|J_1^+| = 2$, and one complementary slackness violation has been eliminated. In reducing u_1^+ by 30, we will cause $\rho_1(\mathbf{U}^+)$, $\rho_3(\mathbf{U}^+)$, and $\rho_5(\mathbf{U}^+)$ to increase by 30 each. This, in turn, may allow us to increase other dual variables. In this case we can increase u_6^+ by 30 and increase u_7^+ by 15, which represents a net increase in z_d^+ of 15.

In general, reducing u_i^+ creates slack in the constraints $\rho_j(\mathbf{U}^+) \geq 0$ for $j \in J_i^+$. This slack can be used to increase some other dual variable, say u_k^+, provided that u_k^+ is not blocked by a binding constraint [i.e., $\rho_b(\mathbf{U}^+) = 0$ for some $b \in J_i^+$]. Since $|J_i^+| > 1$, we know that $u_i^+ > c_{i\alpha(i)}$ and $u_i^+ > c_{i\beta(i)}$. Thus reducing u_i^+ will create slack in the dual constraints for sites $\alpha(i)$ and $\beta(i)$.

A dual variable u_k^+ is a candidate to be increased following the decrease in u_i^+ if u_k^+ is not blocked by any constraint other than for sites $\alpha(i)$ or $\beta(i)$. Now if $I_{\alpha(i)}^+ \neq \emptyset$, then for any $k \in I_{\alpha(i)}^+$, we have $u_k < c_{kj}$ for $j \neq \alpha(i)$. In other words, the dual variables, u_k^+ for $k \in I_{\alpha(i)}^+$, are not blocked by any other dual constraints. Similarly, if $I_{\beta(i)}^+ \neq \emptyset$, the dual variables u_k^+ for $k \in I_{\beta(i)}^+$ are not blocked by any other dual constraints. Note that $I_{\alpha(i)}^+ \neq \emptyset$ or $I_{\beta(i)}^+ \neq \emptyset$ does not *guarantee* that z_d^+ will increase; it simply indicates that an increase is possible.

If we have $|J_i^+| > 2$, we might consider $I_{\gamma(i)}^+$ where $\gamma(i)$ is the third smallest cost site in J^+ for customer i. However, for simplicity, we consider only $I_{\alpha(i)}^+$ and $I_{\beta(i)}^+$, and the conditions for reducing u_i^+ in the dual adjustment algorithm are

$$|J_i^+| > 1 \quad \text{and} \quad I_{\alpha(i)}^+ \neq \emptyset \quad \text{or} \quad I_{\beta(i)}^+ \neq \emptyset$$

The dual adjustment algorithm is formally stated in Table 8.12.

To start the dual adjustment algorithm, we cycle through the row indices until we reach one for which $|J_i^+| > 1$. Starting from the dual ascent solution displayed in Table 8.11, we find that $J_1^+ = \{1,3,5\}$. We determine $\alpha(1) = 1$, $\beta(1) = 5$, $I_1^+ = \{7\}$, and $I_5^+ = \{6\}$. Since $I_1^+ \cup I_5^+$ is nonempty, we reduce u_1^+ to the next smallest c_{ij}, which is $c_{13} = 180$. This reduction of 30 in u_1^+ results in an increase of 30 in $\rho_1(\mathbf{U}^+)$, $\rho_3(\mathbf{U}^+)$, and $\rho_5(\mathbf{U}^+)$. With the altered $\mathbf{U}^+$ and $\rho(\mathbf{U}^+)$, we apply the dual ascent algorithm but consider the dual variables in the order $\{u_6, u_7, u_1, u_2, u_3, u_4, u_5, u_8\}$. The corresponding dual ascent computations are displayed in Table 8.13. Note that we have improved both the dual solution and the primal solution, and the remaining gap is only 55, or roughly 3%.

Having completed the dual ascent, we continue cycling through the row indices until we find $J_6^+ = \{2,5\}$. We determine $\alpha(6) = 5$, $\beta(6) = 2$, $I_5^+ = \{1,2\}$, and $I_2^+ = \{5,7,8\}$. Since u_6^+ is a candidate for adjustment, we find its next smallest c_{6j}, which is $c_{62} = 210$. The reduction of 55 in u_6^+ allows a corresponding increase in $\rho_2(\mathbf{U}^+)$, $\rho_4(\mathbf{U}^+)$, and $\rho_5(\mathbf{U}^+)$. We again apply the dual ascent algorithm, considering the dual variables in the order $\{u_1, u_2, u_5, u_7, u_8, u_6, u_3, u_4\}$. The dual ascent computations are displayed in Table 8.14.

TABLE 8.12 DUAL ADJUST ALGORITHM

```
FOR i = 1,...,m
    IF |J_i^+| > 1, THEN
        IF I_α(i)^+ ∪ I_β(i)^+ ≠ ∅, THEN
            Δ = max_j {c_ij | u_i^+ > c_ij}
            FOR j such that u_i^+ > c_ij
                ρ_j(U^+) ← ρ_j(U^+) + u_i^+ − Δ
            ENDFOR
            u_i^+ ← Δ
            I ← I_α(i)^+ ∪ I^+_β(i)
            perform dual ascent
            I ← I ∪ {i}
            perform dual ascent
            I ← {1,2,...,m}
            perform dual ascent
        ENDIF
    ENDIF
ENDFOR
```

TABLE 8.13 DUAL ASCENT AFTER FIRST DUAL ADJUSTMENT

	1	2	3	4	5	u_i^+	Δ_1	Δ_2	u_i^+
1	120	210	180	210	170	180	0	—	180
2	180	M	190	190	150	240	0	—	240
3	100	150	110	150	110	150	0	—	150
4	M	240	195	180	150	240	0	—	240
5	60	55	50	65	70	60	0	—	60
6	M	210	M	120	195	235	30	—	265
7	180	110	M	160	200	180	15	20	195
8	M	165	195	120	M	220	0	—	220
f_j	200	200	200	400	300	$z_d^+ = 1550$			
$\rho_j(\mathbf{U}^+)$	30	45	30	55	30	$J = \{2,5\}$			
	30	15	30	25	0	$J^+ = \{2,5\}$			
	15	0	30	10	0	$z_p^+ = 1605$			

TABLE 8.14 DUAL ASCENT AFTER SECOND DUAL ADJUSTMENT

	1	2	3	4	5	u_i^+	Δ_1	Δ_2	u_i^+
1	120	210	180	210	170	180	15	30	195
2	180	M	190	190	150	240	0	—	240
3	100	150	110	150	110	150	0	—	150
4	M	240	195	180	150	240	0	—	240
5	60	55	50	65	70	60	0	—	60
6	M	210	M	120	195	210	40	M	250
7	180	110	M	160	200	195	0	—	195
8	M	165	195	120	M	220	15	M	235
f_j	200	200	200	400	300	$z_d^+ = 1565$			
$\rho_j(\mathbf{U}^+)$	15	55	30	65	55	$J = \{1,2,3,5\}$			
	0	55	15	65	40	$J^+ = \{1,2\}$			
	0	40	0	50	40	$z_p^+ = 1580$			
	0	0	0	10	0				

Upon completing the dual ascent computations, we find that $J = \{1, 2, 3, 5\}$ and for every customer there are at least two sites in J for which $u_i^+ \geq c_{ij}$. Thus we need to decide which sites will be in J^+. If we start with all four sites and apply the drop heuristic, we will end up with $J^+ = \{1, 2\}$. At this point we have once again improved both the dual and primal solutions, and the remaining gap is only 15, or less than 1%.

Continuing the adjustment algorithm, we find that $J_7^+ = \{1, 2\}$, $\alpha(7) = 2$, $\beta(7) = 1$, $I_2^+ = \{4, 6, 8\}$, and $I_1^+ = \{1, 2\}$. The adjustment to u_7^+ is 15, resulting in an increase in $\rho_1(\mathbf{U}^+)$, $\rho_2(\mathbf{U}^+)$, and $\rho_4(\mathbf{U}^+)$. Again, we apply the dual ascent algorithm, considering the dual variables in the order $\{u_1^+, u_2^+, u_4^+, u_6^+, u_8^+, u_7^+, u_3^+, u_5^+\}$. The dual

TABLE 8.15 DUAL ASCENT AFTER THIRD DUAL ADJUSTMENT

	1	2	3	4	5	u_i^+	Δ_1	Δ_2	u_i^+
1	120	210	180	210	170	195	0	—	195
2	180	M	190	190	150	240	0	—	240
3	100	150	110	150	110	150	0	—	150
4	M	240	195	180	150	240	0	—	240
5	60	55	50	65	70	60	0	—	60
6	M	210	M	120	195	250	0	—	250
7	180	110	M	160	200	180	15	20	195
8	M	165	195	120	M	235	0	—	235
f_j	200	200	200	400	300	$z_d^+ = 1565$			
$\rho_j(\mathbf{U}^+)$	15 0	15 0	0 0	25 10	0 0	$J = \{1,2,3,5\}$ $J^+ = \{1,2\}$ $z_p^+ = 1580$			

ascent calculations are shown in Table 8.15. In this case the adjustment did not cause any change to the solution.

Finally, we consider u_8^+, and find that it is not a candidate for adjustment since $J_8^+ = \{2\}$. The dual adjustment algorithm terminates, having improved our solution significantly. Unfortunately, we still have a gap between the primal solution, or upper bound, and the dual solution, or lower bound. To guarantee that we have an optimal solution, we would have to resort to branch and bound. The recommended strategy is to branch on some site (y_j) involved in a violation of complementary slackness (i.e., choose a customer for whom $|J_i^+| > 1$ and then branch on $y_{\alpha(i)}^+$).

8.5.5 Solving WLP

The dual ascent and dual adjustment algorithms are due to Erlenkotter [15]. His computer program for solving WLP is one of the best available methods. His computational experience, taken in historical context, is truly remarkable. His program was able to solve a large and varied set of standard or "benchmark" test problems with solution times that were, quite often, an order of magnitude better than the previous best known results. Problems up to $m = 100, p = 100$ were solved, and only for the very largest problems did the solution times exceed 1 second (IBM 360/91, all FORTRAN code). If you want to solve a warehouse location problem, we recommend Erlenkotter's code!

8.5.6 Alternative Formulation

You may have observed that our formulation of WLP is not the most parsimonious in terms of constraints. A formulation that is mathematically equivalent is

(WLP′) $$\text{minimize} \quad \sum_{j=1}^{p} \left(f_j y_j + \sum_{i=1}^{m} c_{ij} x_{ij} \right)$$

subject to

$$\sum_{j=1}^{p} x_{ij} = 1, \qquad i = 1,\ldots,m$$

$$\sum_{i=1}^{m} x_{ij} - my_j \le 0, \qquad j = 1,\ldots,p$$

$$0 \le x_{ij} \le 1, \qquad i = 1,\ldots,m; \quad j = 1,\ldots,p$$

$$y_j \in \{0,1\}, \qquad j = 1,\ldots,p$$

WLP and WLP′ have the same optimal solution, but their linear relaxations are quite different. In fact, the relaxation of WLP′ can be solved by inspection, because in an optimal solution we have

$$\sum_{i=1}^{m} x_{ij}^* = my_j^*$$

Otherwise, we could decrease y_j^* and therefore reduce the objective function value. We can solve the latter equation for y_j and substitute for y_j in the objective function, obtaining

$$\text{minimize} \quad \sum_{j=1}^{p}\sum_{i=1}^{m}\left(\frac{f_j}{m} + c_{ij}\right)x_{ij}$$

subject to

$$\sum_{j=1}^{p} x_{ij} = 1, \qquad i = 1,\ldots,m$$

$$0 \le x_{ij} \le 1, \qquad i = 1,\ldots,m; \quad j = 1,\ldots,p$$

Using this condensed version of WLP ′, we would determine, for each customer, the warehouse with the smallest $f_j/m + c_{ij}$ and assign the customer to that warehouse. Let $\bar{x}_{ij}$ represent the resulting assignments. Then

$$\bar{y}_j = \frac{1}{m}\sum_{i=1}^{m} \bar{x}_{ij}$$

which may be fractional.

Applying this procedure to the example in Table 8.9 gives

$$\bar{x}_{11} = \bar{x}_{22} = \bar{x}_{33} = \bar{x}_{44} = \bar{x}_{55} = \bar{x}_{66} = 1$$

$$\bar{y}_1 = \bar{y}_2 = \bar{y}_3 = \bar{y}_4 = \bar{y}_5 = \bar{y}_6 = \frac{1}{6}$$

$$\bar{z} = 8.5$$

Since we already know that $z^* = 31$, it is evident that the linear relaxation of WLP′ provides a very poor lower bound. Moreover, it is not obvious how we can use it to obtain a feasible solution to WLP.

This example highlights an important lesson—the most parsimonious formulation may not be the best one to use, especially if it is the basis for a relaxation. For location problems, the disaggregated constraints, $-x_{ij} + y_j > 0$, are preferred to the aggregated constraint

$$\sum_{i=1}^{m} x_{ij} - my_j \leq 0$$

even though they dramatically increase the size of the constraint set. WLP has come to be known as the "strong LP formulation" and WLP′ as the "weak LP formulation," for obvious reasons. Although a number of solution procedures for warehouse location have been based on the weak LP formulation, we will not discuss them, because they are not competitive with methods based on the strong LP formulation.

8.5.7 Summary

The warehouse location problem, a classical problem in location theory, is a special case of the generalized n-median problem. We have seen that there is a very efficient, though complicated algorithm for optimizing the Lagrangian relaxation of the warehouse location problem, and consequently, that optimal solutions can be found, even for large instances of the problem. Finally, we have learned that equivalent mathematical formulation of a discrete location problem can lead to quite different results in the bounding phase of a branch-and-bound algorithm.

8.6 CONCLUSION

Often it is convenient to classify location problems as either *continuous* or *discrete*. Chapter 4 introduced continuous location problems, and this chapter introduced discrete location problems. As you by now have discovered, the discrete location problems, although still relatively easy to state, are somewhat more difficult to solve exactly than are their continuous counterparts.

Nevertheless, there are effective algorithms for solving many discrete location problems, provided that you have the necessary computing resources. As we have seen already, there are methods for solving some discrete cover, center, and median problems. As the references and reading guide indicate, discrete location problems are a very active area for research. Thus we should expect that better and better solution methods will become available for an ever-growing body of discrete location problems.

An important assumption in all the discrete location problems addressed in this chapter is that the interactions among the sites can be described by *simple models*. To be more precise, all the costs and constraints can be expressed in terms of the interactions between single pairs of locations.

The next and final chapter will "close the circle." We present in Chapter 9 a family of discrete location problems that allow more complex relationships. These location problems will also be useful in the context of facility layout, which is discussed in Chapters 2 and 3.

8.7 FURTHER READING

If you find that the problems discussed in this chapter interest you and you would like to study them in more depth, you are in luck! There is a vast, reasonably mature, and growing literature on these problems. To comprehend fully the more elegant solution procedures, you will need to spend some time studying the basics of discrete optimization.

In addition to the large number of very good textbooks on operations research in general, you will find it helpful to study one or more discrete optimization texts. As a first step, the classic text by Garfinkel and Nemhauser [24] is excellent. More recent texts by Papadimitriou and Steiglitz [50], Nemhauser and Woolsey [49], and Parker and Rardin [51] reflect the advances in the area in the last decade. If you are considering the development of computational procedures, you certainly should read the classic paper by Geoffrion and Marsten [27]. Geoffrion [26] and Fisher [18] are excellent introductions to Lagrangian relaxation.

With suitable references to the methodology, you are ready to start exploring the location literature. Fortunately, there are several very good general references: the bibliographies by Domschke and Drexl [12] and Francis and Goldstein [19]; the survey articles by Francis et al. [20], Halpern and Maimon [31], Krarup and Pruzan [37,38], McGinnis [43], and Tansel et al. [56]; and the books by Mirchandani and Francis [46], Francis and White [21], Handler and Mirchandani [32], and Minieka [45]. As you will soon discover, there are literally thousands of papers on location problems, so you can profit by careful study of the surveys and bibliographies.

Set covering is a classic problem in discrete optimization, and many researchers have developed general algorithms. Garfinkel and Nemhauser [23] devote a chapter to set covering, discussing applications, reductions, and optimization methods. Computational procedures are presented by Marsten [41], Etcheberry [17], and Balas and Ho [1]. The greedy algorithm presented in Section 8.2.3 is based on [1]. The partial covering location problem and the generalized partial covering location problem were first analyzed in the early 1970s by Curry and Skeith [11], ReVelle and Swain [52], and Shannon and Ignizio [54]. The covering location problem was formulated by Toregas et al. [57] for locating emergency facilities, and they developed the cutting-plane approach. White and Case [59] provide a review of covering formulations.

The n-center problem was formulated by Hakimi [31]. Virtually all algorithms for n-center rely on solving a sequence of covering problems. The procedure we described is based on the work of Minieka [44] and Garfinkel et al. [25]. Christofides and Viola [5] propose an alternative approach.

The n-median problem was formulated by Hakimi [29,30], who also provided the vertex optimality property. The greedy heuristic for the generalized n-median problem can be traced back at least to Kuehn and Hamburger [39] and was analyzed by Cornuejols et al. [6], who provided a theoretical basis for the heuristic. They also solved the Lagrangian dual to obtain lower bounds in a branch-and-bound algorithm. The strong linear programming formulation of GMP(n) was solved by ReVelle and Swain [52], who found that the linear relaxation almost always gave integer solutions. The difficulty with the strong LP formulation was the large number of constraints. A specialized linear programming algorithm developed by Schrage [53] provided some relief from this difficulty.

The warehouse location problem was studied initially by Kuehn and Hamburger [39], who proposed the basic drop, add, and swap heuristics. Efroymson and Ray [13], and later Khumawala [36], developed branch and bound algorithms based on the weak LP formulation. The strong LP formulation was used by Erlenkotter [15] to develop a computational procedure that probably is the most effective, widely available tool for solving these problems. Related results are presented in Guignard and Spielberg [28] and Bilde and Krarup [2]. An excellent summary of the mathematics of the uncapacitated location problem is given by Cornuejols et al. [9].

The discrete location problems introduced in this chapter are only the simplest versions. Many extensions and variations have been formulated and analyzed. In the next chapter we present some of these extensions, primarily with additional constraints on the capacity of the facilities and on the assignment of EF to NF. We shall not consider such variations as multiple commodities, multiple time periods, stochastic parameters, or the location of paths. These and many other interesting variations can be found in the surveys and bibliographies listed in the references. The book by Mirchandani and Francis is the best recent compendium of research and an excellent starting point for advanced study.

REFERENCES

1. Balas, E., and A. Ho, "Set Covering Algorithms Using Cutting Planes, Heuristics, and Subgradient Optimization: A Computational Study," *Mathematical Programming Study,* Vol. 12, 1980, pp. 37–60.
2. Bilde, O., and J. Krarup, "Sharp Lower Bounds and Efficient Algorithms for the Simple Plant Location Problem," *Annals of Discrete Mathematics,* Vol. 1, 1977, pp. 79–97.
3. Boffey, T. B., and J. Karkazis, "*P*-Medians and Multi-Medians," *Journal of the Operational Research Society,* Vol. 35, No. 1, January 1984, pp. 57–64.
4. Christofides, N., *Graph Theory: An Algorithmic Approach,* Academic Press, New York, 1975.
5. Christofides, N., and P. Viola, "The Optimum Location of Multi-centers on a Graph," *Operational Research Quarterly,* Vol. 22, No. 2, 1971, pp. 145–154.
6. Cornuejols, G., M. L. Fisher, and G. L. Nemhauser, "Location of Bank Accounts to Optimize Float: An Analytic Study of Exact and Approximate Algorithms," *Management Science,* Vol. 23, No. 8, April 1977, pp. 789–810.
7. Cornuejols, G., M. L. Fisher, and G. L. Nemhauser, "On the Uncapacitated Location Problem," *Annals of Discrete Mathematics,* Vol. 1, 1977, pp. 163–177.

8. Cornuejols, G., G. L. Nemhauser, and L. A. Wolsey, "Worst-Case and Probabilistic Analysis of Algorithms for a Location Problem," *Operations Research,* Vol. 28, No. 4, July/August 1980, pp. 847–858.
9. Cornuejols, G., G. L. Nemhauser, and L. A. Wolsey, "The Uncapacitated Facility Location Problem," in P. Mirchandani and R. L. Francis, eds., *Discrete Location Theory,* Wiley-Interscience, New York, 1990.
10. Cornuejols, G., and J. M. Thizy, "A Primal Approach to the Simple Plant Location Problem," *SIAM Journal on Algebraic and Discrete Methods,* Vol. 3, No. 4, December 1982, pp. 503–510.
11. Curry, G. L., and R. W. Skeith, "A Dynamic Programming Algorithm for Facility Location and Allocation," *AIIE Transactions,* Vol. 1, No. 2, 1969, pp. 133–138.
12. Domschke, W., and A. Drexl, eds., *Location and Layout Planning: An International Bibliography,* Springer-Verlag, New York, 1985.
13. Efroymson, M. A., and T. L. Ray, "A Branch and Bound Algorithm for Plant Location," *Operations Research,* Vol. 14, No. 3, May/June 1966, pp. 361–368.
14. Erlenkotter, D., "Facility Location with Price-Sensitive Demands: Private, Public, and Quasi-public," *Management Science,* Vol. 24, No. 4, December 1977, pp. 378–386.
15. Erlenkotter, D., "A Dual-Based Procedure for Uncapacitated Facility Location," *Operations Research,* Vol. 26, No. 6, November/December 1978, pp. 992–1009.
16. Erlenkotter, D., "A Comparative Study of Approaches to Dynamic Location Problems," *European Journal of Operations Research,* Vol. 6, No. 2, February 1981, pp. 133–143.
17. Etcheberry, J., "The Set Covering Problem: A New Implicit Enumeration Algorithm," *Operations Research,* Vol. 25, No. 5, September/October 1977, pp. 760–772.
18. Fisher, M. L., "The Lagrangian Relaxation Method for Solving Integer Programming Problems," *Management Science,* Vol. 27, No. 1, January 1981, pp. 1–18.
19. Francis, R. L., and J. M. Goldstein, "Location Theory: A Selective Bibliography," *Operations Research,* Vol. 22, No. 2, March/April 1974, pp. 400–410.
20. Francis, R. L., L. F. McGinnis, and J. A. White, "Locational Analysis," *European Journal of Operations Research,* Vol. 12, No. 3, 1983, pp. 220–252.
21. Francis, R. L., and J. A. White, *Facility Layout and Location: An Analytical Approach,* Prentice Hall, Englewood Cliffs, NJ, 1974.
22. Galvao, R. D., "A Dual-Bounded Algorithm for the p-Median Problem," *Operations Research,* Vol. 28, No. 5, September/October 1980, pp. 1112–1121.
23. Garfinkel, R. S., and G. L. Nemhauser, "Optimal Set Covering: A Survey," in A. M. Geoffrion, ed., *Perspectives on Optimization: A Collection of Expository Articles,* Adison-Wesley Publishing Company, Inc., Reading, MA, 1972.
24. Garfinkel, R. S., and G. L. Nemhauser, *Integer Programming,* John Wiley & Sons, Inc., New York, 1972.
25. Garfinkel, R. S., A. W. Neebe, and M. R. Rao, "The M-Center Problem: Minimum Facility Location," *Management Science,* Vol. 23, No. 10, 1977, pp. 1133–1142.
26. Geoffrion, A. M., "Lagrangian Relaxation for Integer Programming," *Mathematical Programming Study,* Vol. 2, 1974, pp. 82–114.
27. Geoffrion, A. M., and R. E. Marsten, "Integer Programming Algorithms: A Framework

and State of the Art Survey," *Management Science,* Vol. 18, No. 9, May 1972, pp. 465–491.

28. Guignard, M., and K. Spielberg, "Algorithms for Exploiting the Structure of the Simple Plant Location Problem," *Annals of Discrete Mathematics,* Vol. 1, 1977, pp. 247–272.
29. Hakimi, S. L., "Optimum Location of Switching Centers and the Absolute Centers and Medians of a Graph," *Operations Research,* Vol. 12, No. 3, May/June 1964, pp. 450–459.
30. Hakimi, S. L., "Optimum Distribution of Switching Centers in a Communication Network and Some Related Graph Theoretic Problems," *Operations Research,* Vol. 13, No. 3, May/June 1965, pp. 462–475.
31. Halpern, J., and O. Maimon, "Algorithms for the *M*-Center Problems: A Survey," *European Journal of Operations Research,* Vol. 10, No. 1, 1982, pp. 90–99.
32. Handler, G. Y., and P. B. Mirchandani, *Location on Networks,* The MIT Press, Cambridge, MA, 1979.
33. Held, M., P. Wolfe, and H. P. Crowder, "Validation of Subgradient Optimization," *Mathematical Programming,* Vol. 6, No. 1, 1974, pp. 62–88.
34. Hochbaum, D. S., "Heuristics for the Fixed Cost Median Problem," *Mathematical Programming,* Vol. 22, No. 2, 1982, pp. 148–162.
35. Kariv, O., and S. L. Hakimi, "An Algorithmic Approach to Network Location Problems. Part I: The p-Centers; Part II: The p-Medians," *SIAM Journal of Applied Mathematics,* Vol. 37, No. 3, December 1979, pp. 513–560.
36. Khumawala, B. M., "An Efficient Branch and Bound Algorithm for the Warehouse Location Problem," *Management Science,* Vol. 18, No. 12, August 1972, pp. 718–731.
37. Krarup, J., and P. M. Pruzan, "Selected Families of Location Problems. Part I: Center Problems; Part II: Median Problems," *Annals of Discrete Mathematics,* Vol. 5, 1979, pp. 327–387.
38. Krarup, J., and P. M. Pruzan, "The Simple Plant Location Problem: Survey and Synthesis," *European Journal of Operations Research,* Vol. 12, No. 1, 1983, pp. 36–81.
39. Kuehn, A. A., and M. J. Hamburger, "A Heuristic Program for Locating Warehouses," *Management Science,* Vol. 9, No. 4, 1963, pp. 643–666.
40. Lemke, C. E., H. M. Salkin, and K. Spielberg, "Set Covering by Single Branch Enumeration with Linear Programming Subproblems," *Operations Research,* Vol. 19, No. 4, July/August 1971, pp. 998–1022.
41. Marsten, R. E., "An Algorithm for Large Set Partitioning Problems," *Management Science,* Vol. 20, No. 5, January 1974, pp. 774–787.
42. Mavrides, L. P., "An Indirect Method for the Generalized k-Median Problem Applied to Lock-Box Location," *Management Science,* Vol. 25, No. 10, October 1979, pp. 990–996.
43. McGinnis, L. F., "A Survey of Recent Results for a Class of Facilities Location Problems," *AIIE Transactions,* Vol. 9, No. 1, March 1977, pp. 11–18.
44. Minieka, E., "The *M*-Center Problem," *SIAM Review,* Vol. 12, No. 1, January 1970, pp. 138–139.
45. Minieka, E., *Optimization Algorithms for Networks and Graphs,* Marcel Dekker, Inc., New York, 1978.

46. Mirchandani, P. B., and R. L. Francis, eds., *Discrete Location Theory,* John Wiley & Sons, Inc., New York, 1990.

47. Mulvey, J. M., and H. L. Crowder, "Cluster Analysis: An Application of Lagrangian Relaxation," *Management Science,* Vol. 25, No. 4, April 1979, pp. 329–340.

48. Narula, S. C., U. I. Ogbu, and H. M. Samuelson, "An Algorithm for the p-Median Problem," *Operations Research,* Vol. 25, No. 4, July/August 1977, pp. 709–713.

49. Nemhauser, G. L., and L. A. Wolsey, *Integer and Combinatorial Optimization,* Wiley-Interscience, New York, 1988.

50. Papadimitriou, C. H., and K. Steiglitz, *Combinatorial Optimization: Algorithms and Complexity,* Prentice Hall, Englewood Cliffs, NJ, 1982.

51. Parker, R. G., and R. L. Rardin, *Discrete Optimization,* Academic Press, 1988.

52. ReVelle, C., and R. Swain, "Central Facilities Location," *Geographical Analysis,* Vol. 2, No. 1, 1970, pp. 30–42.

53. Schrage, L., "Implicit Representation of Generalized Variable Upper Bounds in Linear Programming," *Mathematical Programming,* Vol. 14, No. 1, 1978, pp. 11–20.

54. Shannon, R. E., and J. P. Ignizio, "A Heuristic Programming Algorithm for Warehouse Location," *AIIE Transactions,* Vol. 2, No. 4, 1970, pp. 334–339.

55. Soland, R. M., "Optimal Plant Location with Concave Costs," *Operations Research,* Vol. 22, No. 2, March/April 1974, pp. 373–385.

56. Tansel, B. C., R. L. Francis, and T. J. Lowe, "Location on Networks: A Survey. Part I: The p-Center and the p-Median Problems," *Management Science,* Vol. 29, No. 4, April 1983, pp. 482–497.

57. Toregas, G., R. Swain, C. ReVelle, and L. Bergman, "The Location of Emergency Service Facilities," *Operations Research,* Vol. 19, No. 6, October 1971, pp. 1362–1373.

58. Whitaker, R. A., "A Fast Algorithm for the Greedy Interchange for Large-Scale Clustering and Median Problems," *Infor,* Vol. 21, No. 2, May 1983, pp. 95–108 (errata: *Infor,* Vol. 22, 1984, pp. 70–71).

59. White, J. A., and K. E. Case, "On Covering Problems and the Central Facility Location Problem," *Geographical Analysis,* Vol. 6, No. 3, 1974, pp. 281–293.

PROBLEMS

8.1. Consider the covering problems having the covering coefficient matrices displayed below. For each one, apply the reduction rules from Section 8.2.2. If the problem is not completely reduced, apply the greedy algorithm from Section 8.2.3. What is the optimum solution for each problem?

(a)
$$\mathbf{A} = \begin{pmatrix} 1 & 0 & 0 & 1 & 1 \\ 1 & 1 & 0 & 0 & 1 \\ 0 & 0 & 1 & 0 & 0 \\ 0 & 1 & 1 & 1 & 0 \\ 0 & 1 & 0 & 0 & 1 \end{pmatrix}$$

(b)
$$\mathbf{A} = \begin{pmatrix} 1 & 1 & 0 & 0 & 0 \\ 0 & 1 & 0 & 1 & 1 \\ 0 & 0 & 1 & 1 & 0 \\ 1 & 0 & 0 & 1 & 0 \\ 1 & 1 & 0 & 0 & 1 \end{pmatrix}$$

(c)

$$\mathbf{A} = \begin{pmatrix} 1 & 1 & 0 & 0 & 0 & 0 & 0 & 0 & 1 \\ 1 & 1 & 0 & 1 & 0 & 0 & 0 & 0 & 0 \\ 0 & 0 & 1 & 0 & 0 & 0 & 0 & 1 & 0 \\ 0 & 0 & 0 & 1 & 0 & 0 & 0 & 0 & 0 \\ 1 & 0 & 0 & 1 & 1 & 0 & 0 & 0 & 0 \\ 0 & 0 & 1 & 0 & 1 & 1 & 0 & 0 & 0 \\ 0 & 0 & 0 & 0 & 0 & 0 & 1 & 0 & 0 \\ 0 & 0 & 0 & 1 & 0 & 0 & 1 & 1 & 0 \\ 0 & 0 & 0 & 0 & 0 & 0 & 0 & 0 & 1 \end{pmatrix}$$

(d)

$$\mathbf{A} = \begin{pmatrix} 1 & 0 & 1 & 0 \\ 1 & 0 & 0 & 1 \\ 0 & 1 & 1 & 0 \\ 0 & 1 & 0 & 0 \end{pmatrix}$$

(e)

$$\mathbf{A} = \begin{pmatrix} 1 & 0 & 1 & 0 \\ 0 & 1 & 1 & 0 \\ 1 & 0 & 0 & 1 \\ 0 & 0 & 0 & 1 \end{pmatrix}$$

(f)

$$\mathbf{A} = \begin{pmatrix} 1 & 0 & 1 & 0 \\ 1 & 1 & 1 & 0 \\ 0 & 0 & 0 & 1 \\ 0 & 0 & 1 & 0 \\ 1 & 1 & 0 & 1 \end{pmatrix}$$

8.2. Make a list of the 25 largest cities in the continental United States and construct the corresponding distance matrix. From the distance matrix, construct the cover problems corresponding to coverage constraints of 1000 miles, 500 miles, 300 miles, and 100 miles. Solve the cover problems using the reduction rules and the greedy algorithm.

8.3. Repeat Problem 8.2, but use only the 12 largest cities. How different are the solutions? What would you expect if you were to solve the problem for the 50 largest cities?

8.4. Use the cutting-plane approach described in Section 8.2.4 to solve Problems 8.2 and 8.3.

8.5. Use the greedy algorithm from Section 8.4.3 with $n = p$ to solve the following warehouse location problems:

(a) $\mathbf{C} = \begin{pmatrix} 6 & 8 & 10 \\ 10 & 5 & 7 \\ 12 & 8 & 6 \\ 8 & 4 & 6 \\ 4 & 5 & 8 \end{pmatrix}$, $\quad f = (8 \quad 10 \quad 6)$

(b) $\mathbf{C} = \begin{pmatrix} 0 & 6 & 8 \\ 3 & 4 & 7 \\ 5 & 0 & 7 \\ 6 & 8 & 6 \\ 8 & 4 & 0 \end{pmatrix}$, $\quad f = (10 \quad 8 \quad 12)$

(c) $\mathbf{C} = \begin{pmatrix} 14 & 16 & 18 \\ 13 & 14 & 17 \\ 20 & 15 & 17 \\ 22 & 18 & 16 \\ 18 & 14 & 16 \end{pmatrix}$, $\quad f = (0 \quad 0 \quad 0)$

8.6. Apply the drop, add, and swap heuristics to the solutions found in Problem 8.5.

8.7. The n-center formulation may be said to give a "fair" or *equitable* solution, while the n-median formulation gives a least-cost, or *efficient* solution. In this context, is it possible for a solution to be both equitable and efficient? Explain your answer.

8.8. In the context of Problem 8.7, suppose that you are concerned with both equity and efficiency. One approach to this type of bi-criterion problem is to constrain one of the objectives and optimize the other. The trade-off between the two can then be analyzed by parametric analysis of the constrained objective. Explain how this might be done using the models and methods presented in this chapter for the n-center and n-median problems.

8.9. Apply the dual ascent and dual adjustment algorithms to find lower bounds and feasible solutions for the warehouse location problems from Problem 8.5.

8.10. Use the weak linear programming formulation to compute lower bounds for Problem 8.5. How do these lower bounds compare to those computed in Problem 8.9?

8.11. Solve the problem given in Example 8.7.

8.12. Solve the problem given in Example 8.8.

8.13. ReVelle and Swain [52] present an example problem having the following generalized cover coefficients:

i \ j	1	2	3	4	5	6	7	8	9	10
1	0	0.875	1.50	2.00	2.875	3.625	3.50	4.25	4.50	4.43
2	1.750	0	2.75	2.250	7.500	7.75	6.50	6.75	8.50	8.36
3	3.750	3.44	0	1.88	6.25	6.07	5.00	7.50	7.50	7.32
4	3.500	1.97	1.31	0	5.70	4.81	3.72	3.94	5.46	5.34
5	3.450	4.50	3.00	3.90	0	2.85	3.81	5.92	4.80	4.94
6	8.350	8.91	5.58	6.32	5.46	0	2.88	6.90	3.74	4.89
7	15.050	13.98	8.60	9.14	13.69	5.38	0	7.53	4.30	4.00
8	12.750	10.11	9.00	6.75	14.80	9.00	5.25	0	6.00	4.50
9	10.800	10.20	7.20	7.51	9.60	3.90	2.40	4.80	0	1.20
10	6.650	6.27	4.39	4.58	6.16	3.19	1.40	2.25	0.75	0

Given four available facilities, solve the generalized partial cover problem using the greedy algorithm from Section 8.4.3.

8.14. A logging company is cutting trees in four different areas of Ashley County, Arkansas. The logs are transported by truck to a site along a railroad where special loading equipment is located for loading the logs on railcars. The firm has identified five candidate loading sites and has equipment available for at most three loading sites. Based on the volume of material handling between logging areas and loading sites, the mileages traveled per day between combinations of logging areas and loading sites are summarized below. It is the policy of the firm that all loads of logs from a given logging area are transported to the same loading site. Determine the optimum allocation of loading equipment to sites and the assignment of logging areas to loading sites to minimize distance traveled.

	Candidate loading sites				
Logging area	1	2	3	4	5
1	0	100	250	150	400
2	40	50	500	300	30
3	400	200	50	350	160
4	100	150	60	10	250

8.15. In Problem 8.14, suppose that the cost of transporting the equipment to the sites, setting up the equipment, and returning the equipment to the equipment storage yard is given to be

Candidate loading sites	1	2	3	4	5
Setup and Transportation cost ($)	200	100	50	150	200

Furthermore, let the travel cost per mile traveled by logging trucks between the logging areas and loading sites be $0.50. Determine the optimum allocation of loading equipment to sites and the assignment of logging areas to loading sites to minimize total cost over a 10-day planning horizon.

8.16. The community of Snyder, Arkansas, wishes to locate municipal parks for use by the public. The residential area in the community has been approximated by the 15 grid squares shown in Figure P8.16. Each grid square has an area of 0.25 square mile. Potential park sites are available in eight of the grid squares. It is desired that a park be located within 1 mile of the center of each grid square. Assume rectilinear travel and that each park site is in the center of the indicated grid square. Determine the minimum number of park sites necessary to satisfy the desired maximum distance criteria.

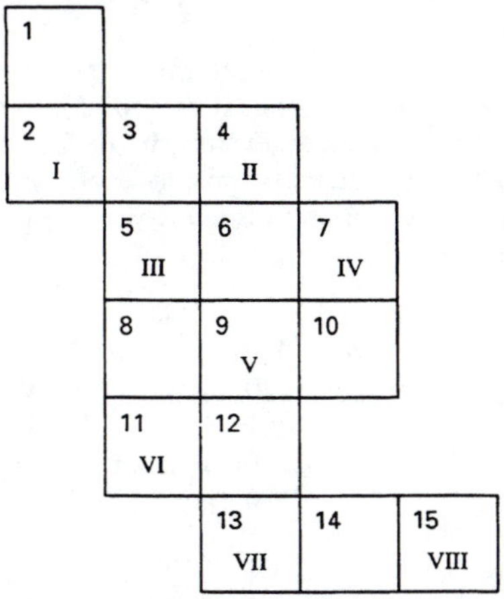

Figure P8.16

8.17. Solve Problem 8.16 when a maximum distance of 1.5 miles is used. Also, determine the locations and allocations of grid square to parks when a maximum of three parks is allowed.

8.18. In a rural section of North Carolina, health outreach clinics are to be located over an area coincident with 10 magisterial districts. Five potential sites are available for locating health outreach clinics. Distances between the centroids of the districts and the potential sites are tabulated, along with the population of each district. Determine the locations of the clinics that minimize the total distance traveled per unit time, based on a maximum of (a) one clinic, (b) two clinics, and (c) three clinics.

Magisterial district	Potential sites 1	2	3	4	5	Population
1	15	10	25	20	30	3000
2	30	10	20	20	35	4500
3	30	15	10	15	30	2500
4	25	0	15	10	25	5000
5	20	10	5	10	20	2000
6	30	20	5	10	20	4000
7	15	25	20	5	10	6000
8	0	25	30	20	15	7000
9	20	25	20	10	5	3000
10	10	30	30	20	10	5000

8.19. In Problem 8.18, suppose that a magisterial district is "covered" if a health outreach clinic is located within X miles of the centroid of the district. Determine the minimum number of clinics required to cover the districts when X equals (a) 5, (b) 10, (c) 15, and (d) 20 or more.

8.20. A study is being conducted to determine the optimum number and location of fire towers in a large national forest. It is desired that the minimum number of towers be used to provide a coverage of all tracts in the forest. The forest consists of 12 tracts to be included in the surveillance by rangers located in the fire towers. Eight locations have been selected as potential sites for fire towers due to their altitudes and visibility ranges. The coverage matrix is

$$\mathbf{A} = \begin{matrix} 1 & 1 & 0 & 0 & 0 & 0 & 0 & 0 \\ 0 & 1 & 0 & 0 & 0 & 0 & 0 & 0 \\ 0 & 1 & 1 & 0 & 0 & 0 & 0 & 0 \\ 1 & 1 & 0 & 1 & 0 & 0 & 0 & 0 \\ 0 & 1 & 1 & 1 & 0 & 1 & 0 & 0 \\ 0 & 0 & 1 & 1 & 1 & 1 & 0 & 1 \\ 1 & 0 & 1 & 1 & 1 & 1 & 0 & 0 \\ 1 & 0 & 0 & 1 & 0 & 0 & 0 & 0 \\ 0 & 0 & 0 & 1 & 1 & 1 & 1 & 1 \\ 1 & 0 & 0 & 0 & 1 & 0 & 1 & 0 \\ 0 & 0 & 0 & 0 & 1 & 1 & 1 & 1 \\ 0 & 0 & 0 & 0 & 1 & 1 & 1 & 1 \end{matrix}$$

Solve the problem by inspection and using the heuristic procedure.

8.21. Based on a maximum of three fire towers, determine the location of the towers that maximizes the coverage of tracts in Problem 8.20. Is the solution obtained a feasible solution to the total cover problem? Why or why not?

8.22. A textile plant has a number of automatic spinning machines that are assigned to operators who patrol the assigned area and repair any breaks that occur in the continuous filament fiber. The plant can be divided into 20 squares contained in a rectangle with a width of 4 and a length of 5. A patrol operator who is "based" in a square can patrol all machines in that square, as well as the eight adjacent squares. Formulate the problem of determining the minimum number of operators required to patrol the entire plant. Write out the corresponding cover matrix **A**.

8.23. The Republic of East Venutia is threatened by its neighbor, the Republic of West Venutia. Radar installations are to be located throughout East Venutia to provide protection against an attack by the West Venutians. Ten key areas of East Venutia have been designated as vital areas. Additionally, 10 sites have been designated as feasible sites for the radar installations. Determine the minimum number of radar installations to be located throughout the country to provide radar coverage for the 10 vital areas. Letting a_{ij} equal 1 (zero) if an installation at site i covers (does not cover) vital area j, the following covering coefficients are provided:

	Vital area									
Site	1	2	3	4	5	6	7	8	9	10
1	1	1	0	1	0	0	1	1	0	0
2	1	1	1	0	0	0	0	0	0	0
3	0	1	1	1	1	1	0	0	0	0
4	1	1	1	1	1	1	1	1	0	0
5	0	1	1	1	1	1	1	0	0	0
6	0	0	1	1	1	1	1	0	1	0
7	1	0	0	1	1	1	1	1	1	1
8	1	0	0	1	0	0	0	1	0	1
9	0	0	0	0	0	1	1	0	1	1
10	0	0	0	0	0	0	1	1	1	1

8.24. Consider the covering problem having the cover matrix

$$\mathbf{A} = \begin{matrix} 1 & 0 & 1 & 0 & 0 \\ 1 & 0 & 0 & 0 & 1 \\ 1 & 1 & 0 & 0 & 0 \\ 0 & 1 & 0 & 0 & 0 \\ 0 & 1 & 0 & 0 & 0 \\ 0 & 0 & 1 & 0 & 0 \\ 0 & 0 & 1 & 0 & 0 \\ 0 & 0 & 0 & 1 & 1 \\ 0 & 0 & 0 & 1 & 1 \end{matrix}$$

(a) Solve the total cover problem by enumerating all possible assignments.

(b) Solve the partial cover problem for the case of $k = 3$ by enumerating all possible assignments.

(c) Solve the partial cover problem for the case of $k = 3$ by using the heuristic solution procedure.

Chapter 9

Advanced Discrete Location Models

9.1 INTRODUCTION

Our treatment of discrete location problems in Chapter 8 was based on several key assumptions, including:

1. New facilities are not capacity constrained.
2. The location of one new facility does not directly affect the location of any other new facility.

There are practical problems in which these assumptions are violated. For example, in locating distribution warehouses, a particular location may have limited site space, limited transportation access, or a limited local labor pool—all factors that may limit the economically feasible throughput. Many production operations have distinct capacity limits, due to the processing equipment. For example, bottling and canning lines have rated capacities, so that the capacity of a bottling or canning plant is proportional to the number of lines. As we saw in Chapters 2 and 3, the relative location of activities is one of the keys to facilities design.

In this chapter we present models and solution methods for discrete location problems in which the new facilities have capacity constraints or there are interactions between the new facilities. These problems tend to be difficult to solve optimally, in both a theoretical and in a practical sense. The best of the optimization algorithms that have been developed to solve these problems generally involve specialized techniques and are beyond the scope of our presentation. Instead, we shall focus on useful heuristic methods, providing the basic concepts and references for exact solution methods.

The first problem we address is a generalization of the warehouse location problem in which the new facilities have limited capacities. We shall present two heuristic procedures: a generalization of the drop and add heuristics to the capacitated problem; and a heuristic that alternates between solving a particular warehouse location problem to choose sites, and a transportation problem to allocate new facility capacity to customer requirements. The latter heuristic method has been shown to give quite good results, with relatively modest computing requirements.

When there are binding capacity constraints, the heuristic methods may create solutions in which some customers' requirements are satisfied from two or more new facilities. In practice this situation, referred to as *dual sourcing*, may be undesirable because it prevents the clear definition of "customer zones" or "service areas." We present a formulation and a heuristic solution method for the problem in which single sourcing is required.

Finally, we address the problem of locating facilities when there are interactions between the facilities. This problem is commonly referred to as the *quadratic assignment problem* and is closely related to the block layout problems discussed in Chapters 2 and 3. The heuristic solution procedures for the quadratic assignment problem resemble those for computerized facility layout.

The notation and terminology used in this chapter follow from Chapter 8. We shall refer to existing facility, EF, and new facility, or NF, locations. You may wish to review Chapter 8 before proceeding.

9.2 CAPACITY-CONSTRAINED NEW FACILITIES

Example 9.1*

The Real Old Cola Company has experienced rapidly growing sales in the southeast region, and plans to open a series of regional bottling plants. Sales in eight cities account for 90% of ROC sales. A systematic analysis of potential sites has identified five acceptable plant locations. For each site the throughput capacity, in 1000 cases per week, and the weekly fixed operating costs in $100, are shown in Table 9.1. The projected weekly sales for each of the eight major sales regions and the variable cost per 1000 cases are also shown in Table 9.1. Because of major geographical features (rivers and mountains), certain sales region to plant location assignments were not considered. How many plants should ROC use, where should they be located, and how should the sales regions be assigned to plants to minimize the total annual costs?

The problem posed in Example 9.1 may be formulated as a mixed-integer linear programming problem, similar to the formulation for the warehouse location

* Based on a numerical example in [1].

TABLE 9.1 COSTS AND CAPACITIES FOR EXAMPLE 9.1

Sales region	Potential site 1	2	3	4	5	Demand
1	12	21	18	21	17	15
2	18	na	19	19	15	10
3	10	15	11	15	11	10
4	na	24	19.5	18	15	15
5	6	5.5	5	6.5	7	5
6	na	21	na	12	19.5	15
7	18	11	na	16	20	20
8	na	16.5	19.5	12	na	10
Capacity	25	20	20	50	35	
Fixed cost	100	70	60	110	80	

problem given in Section 8.5. Suppose that there are m customers, each with a known requirement of r_i units per period, and p potential NF locations, each with a known capacity of s_j units per period. Let y_j be a zero–one variable associated with the decision to locate a NF at site j, and let x_{ij} represent the fraction of requirement r_i satisfied from a facility at site j. The annualized cost of the facility at site j is f_j and the variable cost to produce r_i units at site j and ship it to customer i is c_{ij}. Then the optimal locations for NF and the optimal production/distribution plan can be determined by solving the following capacitated location problem:

CLP
$$Z = \min \sum_{j=1}^{p} f_j y_j + \sum_{j=1}^{p} \sum_{i=1}^{m} c_{ij} x_{ij} \tag{9.1}$$

$$\sum_{j=1}^{p} x_{ij} = 1, \qquad i = 1, \ldots, m \tag{9.2}$$

$$\sum_{i=1}^{m} r_i x_{ij} \leq s_j, \qquad j = 1, \ldots, p \tag{9.3}$$

$$x_{ij} - y_j \leq 0, \qquad i = 1, \ldots, m; \; j = 1, \ldots, p \tag{9.4}$$

$$x_{ij} \geq 0; \quad y_j \in \{0, 1\}, \qquad i = 1, \ldots, m; \; j = 1, \ldots, p$$

Formulation CLP differs from formulation WLP, given in Chapter 8, only in the addition of the *facility capacity constraint*, (9.3). The similarity between the two formulations motivates us to consider solution methods that were successful in solving WLP and attempt to generalize them for solving CLP. In the next two sections we consider a generalization of the add/drop/swap heuristic and a variant of the dual ascent/adjustment algorithm.

9.2.1 An Add/Drop Heuristic

Recall that in Chapter 8 we developed a set of reduction rules for the set covering problem, that is, a set of rules that could be used to determine the optimal values for some of the decision variables based on simple logical rules. For CLP we can develop rules that are similar in effect but require the solution of an optimization subproblem in their implementation.

When we *peg* a facility location variable in CLP, we specify the value that it must have in an optimal solution. Thus, for us to peg y_j to 1, we must know that the jth location *must* be used in any optimal solution. Similarly, for us to peg y_j to 0, we must know that the jth location *must never* be used in an optimal solution.

Let J_0 denote the set of all location variables that are pegged to 0, J_1 denote the set of all variables that are pegged to 1, and J_2 denote the set of all variables that are not pegged. In a complete solution to CLP, J_2 is empty, and in a partial solution, J_2 is nonempty. Now any partial or complete solution to CLP may be represented as a triple $\{J_0, J_1, J_2\}$.

Suppose that we have a partial solution in which y_j has not been pegged (i.e., $j \in J_2$). To peg y_j to 1, we must show that even if all the other sites represented in J_2 were used, it still would be beneficial to use site j. Let $z(J)$ be the optimal variable cost for a solution with facilities only at sites $j \in J$, and define

$$\Delta_j = z(J_1 \cup J_2 \backslash \{j\}) - z(J_1 \cup J_2) - f_j \tag{9.5}$$

We may interpret Δ_j as the minimum reduction in total cost for adding a facility at site j. Note that $z(J_1 \cup J_2 \backslash \{j\})$ is the minimum variable cost if site j is not used, since it allows every other site not already pegged to zero to be used. Since $z(J_1 \cup J_2)$ is the minimum variable cost, including site j, the difference represents the minimum reduction in variable cost from using site j. We must reduce the cost savings by the cost to open a facility at site j, f_j. Thus given a partial solution $\{J_0, J_1, J_2\}$, we compute Δ_j, and apply the

Δ-Rule: If $\Delta_j \geq 0$, then $y_j = 1$ in an optimal completion.

In a similar fashion, for site $j \in J_2$, in order to peg y_j to zero, we must show that even if no other facilities represented in J_2 are used, and J_1 is capacity feasible, it still would not be beneficial to use site j. Since $z(J_1)$ is greater than or equal to the minimum variable cost for any completion of the given partial solution, clearly the maximum variable cost savings for using site j is $z(J_1) - z(J_1 \cup \{j\})$. Thus we may compute

$$\Omega_j = z(J_1) - z(J_1 \cup \{j\}) - f_j \tag{9.6}$$

and apply the

Ω-Rule: If $\Omega_j \leq 0$, then $y_j = 0$ in an optimal completion.

We emphasize again that this pegging test may be used *only if* J_1 is capacity feasible.

Finally, we note that the values of $z(J)$ for any J can be obtained by solving a standard transportation problem with the destinations corresponding to the sites in J. In this transportation model, the cost coefficients are c_{ij}/r_i, the "capacities" are r_i, and the "demands" are s_j. If this formulation seems confusing to you, think of it as allocating the customer requirements to the open facilities, rather than as shipping goods "from" the facilities "to" the customers. If the total capacity for the sites represented in J is smaller than the total customer requirement, there is no feasible allocation, and we assign a very large value to $z(J)$.

To illustrate the application of the pegging rules, consider the data for Example 9.1. Initially, we have $J_0 = J_1 = \emptyset$, and $J_2 = \{1, 2, 3, 4, 5\}$, so the only rule that we can apply is the Δ-rule. Table 9.2 presents the standard transportation problem tableau

TABLE 9.2 TRANSPORTATION PROBLEM SOLUTION FOR **Y** = (1, 1, 1, 1, 1)

New facility sites

Sales areas		1	2	3	4	5	u_i	Dem.
	1	12 15	21	18	21	17	12	15
	2	18	M	19	19	15 10	15	10
	3	10 10	15	11	15	11	10	10
	4	M	24	19.5	18	15 15	15	15
	5	6	5.5	5 5	6.5	7	5	5
	6	M	21	M	12 15	19.5	12	15
	7	18	11 20	M	16	20	11	20
	8	M	16.5	19.5	12 10	M	12	10
	9	0 0	0 0	0 15	0 25	0 10	0	50
	v_j	0	0	0	0	0	c_{ij} x_{ij}	
	Cap.	25	20	20	50	35		

$z(\{1, 2, 3, 4, 5\}) = 1200$

with the optimal solution for $z(\{1, 2, 3, 4, 5\})$. Note that we have added a "dummy" customer whose requirement equals the slack capacity. Tables 9.3(a) to (e) present the tableaux for $z(J_1 \cup J_2 \backslash \{j\})$ for $j = 1, \ldots, 5$.

From the solutions in Table 9.2 and Table 9.3, we can compute the Δ values as:

1. $\Delta_1 = (1290 - 1200) - 100 = -10$
2. $\Delta_2 = (1300 - 1200) - 70 = 30$
3. $\Delta_3 = (1207.5 - 1200) - 60 = -52.5$
4. $\Delta_4 = (1407.5 - 1200) - 110 = 97.5$
5. $\Delta_5 = (1285 - 1200) - 80 = 5$

TABLE 9.3(a) TRANSPORTATION PROBLEM SOLUTION FOR **Y** = (0, 1, 1, 1, 1)

New facility sites

Sales areas	1	2	3	4	5	u_i	Dem.
1	12	21	18 5	21	17 10	18	15
2	18	M	19	19	15 10	16	10
3	10	15	11 10	15	11	11	10
4	M	24	19.5	18	15 15	16	15
5	6	5.5	5 5	6.5	7	5	5
6	M	21	M	12 15	19.5	12	15
7	18	11 20	M	16	20	11	20
8	M	16.5	19.5	12 10	M	12	10
9	0	0 0	0 0	0 25	0	0	25
v_j		0	0	0	0	c_{ij} x_{ij}	
Cap.	$\bar{0}$	20	20	50	35		

$z(\{2, 3, 4, 5\}) = 1290$

TABLE 9.3(b) TRANSPORTATION PROBLEM SOLUTION FOR **Y** = (1, 0, 1, 1, 1)

Sales areas	New facility sites: 1	2	3	4	5	u_i	Dem.
1	12 15	21	18	21	17	12	15
2	18	M	19	19	15 10	15	10
3	10 10	15	11	15	11	10	10
4	M	24	19.5	18	15 15	15	15
5	6	5.5	5 5	6.5	7	5	5
6	M	21	M	12 15	19.5	12	15
7	18	11	M	16 20	20	16	20
8	M	16.5	19.5	12 10	M	12	10
9	0 0	0	0 15	0 5	0 10	0	30
v_j	0		0	0	0	c_{ij} x_{ij}	
Cap.	25	0	20	50	35		

$z(\{1, 3, 4, 5\}) = 1300$

Applying the Δ-rule results in a new partial solution with $J_1 = \{2, 4, 5\}$. Even though the total capacity of J_1 is 105 and the total customer requirement is 100, we would need to solve the corresponding transportation problem to ensure that J_1 is capacity feasible, because of the infeasible allocations represented in the tableau by a cost of M.

Table 9.4 contains the transportation tableau for $z(J_1)$, which is capacity feasible, with a total variable cost of 1337.5. Since J_1 is capacity feasible, we can apply the Ω-rule to the remaining sites in J_2. Note that in doing so, we will again use the transportation problem solutions in Table 9.3. We compute the Ω-values as:

TABLE 9.3(c) TRANSPORTATION PROBLEM SOLUTION FOR **Y** = (1, 1, 0, 1, 1)

New facility sites

Sales areas	1	2	3	4	5	u_i	Dem.
1	12 15	21	18	21	17	12.5	15
2	18	M	19	19	15 10	15	10
3	10 10	15	11	15	11	10.5	10
4	M	24	19.5	18	15 15	15	15
5	6 0	5.5	5	6.5 5	7	6.5	5
6	M	21	M	12 15	19.5	12	15
7	18	11 20	M	16 0	20	16	20
8	M	16.5	19.5	12 10	M	12	10
9	0	0	0	0 20	0 10	0	30
v_j	0.5	5		0	0	c_{ij} x_{ij}	
Cap.	25	20	0	50	35		

$z(\{1, 2, 4, 5\}) = 1207.5$

1. $\Omega_1 = [1337.5 - 1207.5] - 100 = 30$
2. $\Omega_3 = [1337.5 - 1290] - 60 = -12.5$

Thus we are able to peg $y_3 = 0$.

Since we have successfully pegged a variable to zero, we can apply the Δ-rule again, computing

$$\Delta_1 = (1337.5 - 1207.5) - 100 = 30$$

Therefore, we can peg y_1 to 1. For the example problem, we have pegged all the

TABLE 9.3(d) TRANSPORTATION PROBLEM SOLUTION FOR **Y** = (1, 1, 1, 0, 1)

New facility sites

Sales areas	1	2	3	4	5	u_i	Dem.
1	12 / 15	21	18	21	17	13	15
2	18 / 0	M	19 / 5	19	15 / 5	19	10
3	10 / 10	15	11	15	11	11	10
4	M	24	19.5	18	15 / 15	19	15
5	6	5.5	5 / 5	6.5	7	5	5
6	M	21	M	12	19.5 / 15	23.5	15
7	18	11 / 20	M	16	20	14	20
8	M	16.5 / 0	19.5 / 10	12	M	19.5	10
9	0	0	0 / 0	0	0	0	0
v_j	1	3	0		4	c_{ij}	
Cap.	25	20	20	0	35		x_{ij}

$z(\{1, 2, 3, 5\}) = 1407.5$

variables, resulting in the optimal decision vector $(1, 1, 0, 1, 1)$, with cost $360 + 1207.5 = 1567.5$.

Example 9.2

The pegging tests will not always completely solve CLP. Consider the data displayed in Table 9.5. You should verify that the transportation problem solutions give

$$z(\{1, 2, 3, 4\}) = 35$$

$$z(\{2, 3, 4\}) = 65$$

$$z(\{1, 3, 4\}) = 50$$

TABLE 9.3(e) TRANSPORTATION PROBLEM SOLUTION FOR **Y** = (1, 1, 1, 1, 0)

New facility sites

Sales areas	1	2	3	4	5	u_i	Dem.
1	12 / 15	21	18	21	17	13	15
2	18 / 0	M	19	19 / 10	15	19	10
3	10 / 10	15	11	15	11	11	10
4	M	24	19.5	18 / 15	15	18	15
5	6	5.5	5 / 5	6.5	7	5	5
6	M	21	M	12 / 15	19.5	12	15
7	18	11 / 20	M	16	20	11	20
8	M	16.5	19.5	12 / 10	M	12	10
9	0	0 / 0	0 / 15	0 / 0	0	0	15
v_j	1	3	0	4			
Cap.	25	20	20	50	0		

Cell legend: c_{ij} (upper left) / x_{ij} (lower right)

$z(\{1, 2, 3, 4\}) = 1285$

$$z(\{1,2,4\}) = 35$$

$$z(\{1,2,3\}) = 35$$

Applying the Δ-rule we find that we can peg y_4 to 1, since site 4 has a zero fixed cost. None of the other sites can be pegged, and, since the partial solution is not capacity feasible, we cannot apply the Ω-rule.

As this example illustrates, the pegging rules may not completely solve CLP, just as the reduction rules in Chapter 8 did not completely solve the set covering problem. Thus we need a heuristic rule for completing the solution (i.e., for switch-

TABLE 9.4 TRANSPORTATION PROBLEM SOLUTION FOR **Y** = (1, 0, 1, 0, 0)

New facility sites

Sales areas	1	2	3	4	5	u_i	Dem.
1	12	21	18	21	17 / 15	20	15
2	18	M	19	19	15 / 10	18	10
3	10	15	11	15	11 / 10	14	10
4	M	24	19.5	18 / 15	15 / 0	18	15
5	6	5.5	5	6.5 / 5	7	6.5	5
6	M	21	M	12 / 15	19.5	12	15
7	18	11 / 20	M	16 / 0	20	16	20
8	M	16.5	19.5	12 / 10	M	12	10
9	0	0	0	0 / 5	0	0	5
v_j		5		0	3		
Cap.	0	20	0	50	35		

Cell legend: c_{ij} (upper left), x_{ij} (lower right)

$z(\{2, 4, 5\}) = 1337.5$

TABLE 9.5 DATA FOR EXAMPLE 9.2

	Variable costs for site:				
Customer	1	2	3	4	Requirement
1	1	2	3	4	15
2	0	3	1	3	15
3	3	1	2	3	15
Capacity	25	25	25	25	
Fixed cost	40	40	40	0	

ing sites from J_2 to either J_0 or J_1). We can use the Δ_j and Ω_j values as the basis for the corresponding drop and add heuristics.

To *add* a site to the solution (i.e., to switch j from J_2 to J_1), we need an indication that y_j "should" be 1 in an optimal solution. At this point we must make a heuristic decision about which site to add to the partial solution. A number of alternatives could be used. For example, the Δ_j values could be treated as a measure of the benefit of adding sites, and we could select the site with the largest Δ_j value. At least we could argue that this site is the "closest" to being required in an optimal solution. Using a similar argument, we might select the site with the largest Ω_j value, since it is the site that is "least likely" to be excluded in an optimal solution. A third alternative might be to select the site that has the smallest percent slack remaining in the solution to $z(J_1 \cup J_2)$ since it is heavily utilized in the lower bound solution.

Applying the largest-Δ heuristic, we would select site 1 to add to J_1. Now we solve the transportation problem with only sites 1 and 4, obtaining $z(\{1,4\}) = 70$. Since this solution is capacity feasible, we may apply the Ω-test for pegging sites 2 and 3. We compute

$$\Omega_2 = 70 - 35 - 40 = -5$$

$$\Omega_3 = 70 - 50 - 40 = -20$$

Since both values are negative, we are able to peg y_2 and y_3 to 0, obtaining the final solution $J_1 = \{1, 4\}$, with solution value 110.

Even though we were able to peg three of the variables, the last two successful pegging tests *followed a heuristic decision*. Thus our final solution is not guaranteed to be optimum. Note also that the Ω-test may not succeed in pegging, so that a drop heuristic might be needed. In a manner analogous to the add heuristic, we could decide to peg to zero the variable with the smallest Ω, the largest Δ, or the largest percent slack in the solution for $z(J_1 \cup J_2)$.

There are many possible ways to structure a heuristic algorithm using the Δ_j and Ω_j indices and the pegging tests that use them. The general logic that we followed in solving the example problem is described in the heuristic procedure ADD&DROP given in Table 9.6.

In the ADD&DROP heuristic stated in Table 9.6, lines 1 to 8 apply the Δ-rule, in which sites are pegged open. Similarly, lines 9 to 18 apply the Ω-rule, and returns "SUCCESS = true" if a site is pegged closed. The ADD&DROP heuristic is given in lines 19 to 37. After trying to peg sites open, the test in line 21 checks for capacity feasibility. If the currently open sites do not provide sufficient capacity, the SELECT in line 23 can use any of the rules mentioned to select sites to peg open. The WHILE loop in lines 28 to 36 tries to peg the remaining sites, first by closing, then by pegging open after closing one or more. Any one of the heuristic rules for selecting a site to close may be used in the SELECT in line 31. The Δ and Ω values are recalculated each time they are used, and the algorithm terminates with a feasible solution. You should recognize that there often are several possible ways to structure a heuristic algorithm, and we have given only one way to structure ADD&DROP.

TABLE 9.6 ADD&DROP HEURISTIC ALGORITHM

```
PROCEDURE PEGOPEN
FOR j ∈ J2
        IF Δj ≥ 0 THEN
                J2←J2\{j}
                J1←J1 ∪ {j}
        ENDIF
ENDFOR
END PEGOPEN

PROCEDURE PEGSHUT{SUCCESS}
SUCCESS←false
FOR j ∈ J2
        IF Ωj ≤ 0 THEN
                J2←J2\{j}
                J0←J0 ∪ {j}
                SUCCESS←true
        ENDIF
ENDFOR
END PEGSHUT

PROCEDURE ADD&DROP
PEGOPEN
IF s{J1} < r, THEN
        WHILE s(J1) < r
                SELECT j ∈ J2
                J2←J2\{j}
                J1←J1 ∪ {j}
        ENDWHILE
ENDIF
WHILE J2 ≠ ∅
        PEGSHUT{SUCCESS}
        IF(not.SUCCESS) THEN
                SELECT j ∈ J2
                J2←J2\{j}
                J0←J0 ∪ {j}
        ENDIF
        PEGOPEN
ENDWHILE
END ADD&DROP
```

9.2.2 A Ping-Pong Heuristic

The ADD&DROP heuristic, when applied to Examples 9.1 and 9.2, found very good solutions. Even though ADD&DROP may require solving many transportation problems, there are very effective computer programs for solving them, so the computational burden is not excessive. The major drawback to the ADD&DROP heuristic is that it provides no indication of the quality of the solution. Recall that in Section 8.5 we found that a lower bound on the optimum solution value of WLP could be determined from the dual formulation, DWLP, using the dual ascent and adjustment algorithms. In this section we will see how to use the results from Chapter 8 to obtain lower bounds for CLP.

From CLP we can construct two *subproblems*, one a restriction of CLP and one a relaxation of CLP. Suppose that we have values for the y_j variables, denoted by $\mathbf{Y}$. If we restrict the solution of CLP to these y_j's, we obtain the following *primal subproblem*:

$$(\mathrm{SP_Y}) \qquad ZP_{\mathrm{Y}} = \sum_{j=1}^{p} f_j y_j + \min \sum_{j=1}^{p} \sum_{i=1}^{m} c_{ij} x_{ij} \tag{9.7}$$

subject to

$$\sum_{j=1}^{p} x_{ij} = 1, \qquad i = 1, \ldots, m \qquad [u_i]$$

$$x_{ij} \le y_j, \qquad i = 1, \ldots, m;\; j = 1, \ldots, p \qquad [w_{ij}]$$

$$\sum_{i=1}^{m} r_i x_{ij} \le s_j y_j, \qquad j = 1, \ldots, p \qquad [v_j]$$

$$x_{ij} \ge 0, \qquad i = 1, \ldots, m; \quad j = 1, \ldots, p$$

Note that $\mathrm{SP_Y}$ has the form of a transportation problem and can be solved easily using standard methods, simply by specifying a zero capacity for those sites for which y_j is zero in $\mathbf{Y}$. When $\mathrm{SP_Y}$ has a feasible solution, it provides both a feasible solution and an upper bound for CLP.

Suppose that we have available a vector of dual variables $\mathbf{V} = (v_j \mid j = 1, \ldots, p)$. Then we can construct a Lagrangian relaxation of CLP or *dual subproblem* as follows:

$$(\mathrm{SD_V}) \qquad ZD_{\mathrm{V}} = \min \sum_{j=1}^{p} (f_j - s_j v_j) y_j + \sum_{j=1}^{p} \sum_{i=1}^{m} (c_{ij} + r_i v_j) x_{ij} \tag{9.8}$$

subject to

$$\sum_{j=1}^{p} x_{ij} = 1, \qquad i = 1, \ldots, m \qquad [u_i]$$

$$x_{ij} - y_j \le 0, \qquad i = 1, \ldots, m; \quad j = 1, \ldots, p \qquad [w_{ij}]$$

$$x_{ij} \ge 0, \qquad i = 1, \ldots, m; \quad j = 1, \ldots, p$$

$$y_j \in \{0, 1\}, \qquad j = 1, \ldots, p$$

Note that $\mathrm{SD_V}$ is equivalent to the formulation of WLP given in Chapter 8; thus we might consider solving $\mathrm{SD_V}$ using Erlenkotter's algorithm.

There is an interesting and valuable relationship between these two subproblems. SP_Y requires a set of values for the y_j's and can produce, in addition to the primal solution, a set of dual variables, v_j. SD_V requires a set of dual variables, v_j, and produces as a solution y_j's. Thus a natural approach to solving CLP is to "ping-pong" between these two subproblems, as illustrated in Figure 9.1. For this procedure to work successfully, we need to ensure that the $\mathbf{Y}$ produced by SD_V is feasible in CLP, so that SP_Y will have a feasible solution. We can ensure that the solution from SD_V has sufficient total capacity, by adding the following constraint:

$$\sum_{j=1}^{p} s_j y_j \geq \sum_{i=1}^{m} r_i \qquad [\lambda] \tag{9.9}$$

Of course, the algorithm for solving WLP must be modified to accommodate the additional constraint. This is discussed further below.

To illustrate the ping-ponging approach, we shall apply it to Example 9.2. We begin by solving SP_Y for $\mathbf{Y} = (1, 1, 1, 1)$, which is guaranteed to be capacity feasible. The transportation problem tableau is shown in Table 9.7, from which we obtain $z(\{1, 2, 3, 4\}) = 35$, $\mathbf{V} = (1, 0, 0, 0)$ and $ZP_Y = 120 + 35 = 155$.

Next, we must solve SD_V, with the capacity feasibility constraint (9.9). The method suggested for dealing with this complicating constraint is to dualize it as well as the capacity constraints. For a particular Lagrangian multiplier, λ, we obtain the following problem:

$$SD_{V,\lambda} \quad ZD_{V,\lambda} = \min \sum_{j=1}^{p} (f_j - s_j v_j - \lambda s_j) y_j + \sum_{j=1}^{p} \sum_{i=1}^{m} (c_{ij} + r_i v_j) x_{ij} + \lambda \sum_{i=1}^{m} r_i \tag{9.10}$$

subject to

$$\sum_{j=1}^{p} x_{ij} = 1, \qquad i = 1, \ldots, m$$

$$-x_{ij} - y_j \leq 1, \qquad i = 1, \ldots, m; \quad j = 1, \ldots, p$$

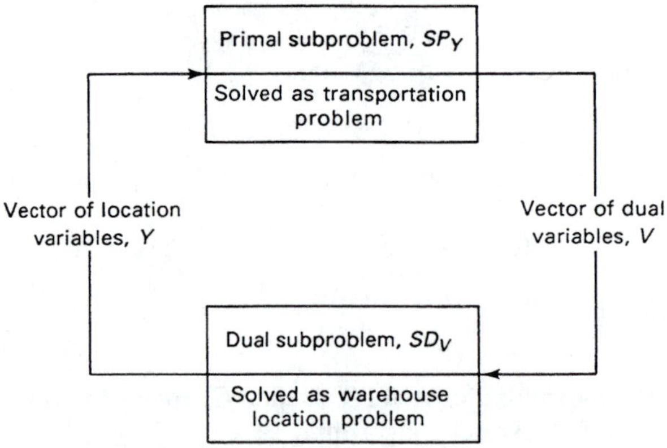

Figure 9.1 General organization of the ping-pong heuristic.

TABLE 9.7 TRANSPORTATION PROBLEM SOLUTION FOR Y = (1, 1, 1, 1)

New facility sites

	1	2	3	4	u_i	Dem.
1	1 / 15	2	3	3	2	15
2	0 / 10	3	1 / 5	3	1	15
3	3	1 / 15	2	3	1	15
4	0	0 / 10	0 / 20	0 / 25	0	55
v_j	1	0	0	0	c_{ij} / x_{ij}	
Cap.	25	25	25	25		

$z(\{1, 2, 3, 4\}) = 35$

$$x_{ij} \geq 0, \qquad i = 1, \ldots, m; \quad j = 1, \ldots, p$$

$$y_j \in \{0, 1\}, \qquad j = 1, \ldots, p$$

We may rewrite (9.10) as

$$ZD_{\mathrm{V},\lambda} = \lambda r + \min \sum_{j=1}^{p} \tilde{f}_j y_j + \sum_{j=1}^{p} \sum_{i=1}^{m} \bar{c}_{ij} x_{ij}$$

where

$$\tilde{f}_j = f_j - s_j v_j - \lambda s_j$$

$$\bar{c}_{ij} = c_{ij} + r_i v_{ij}$$

$$r = \sum_{i=1}^{m} r_i$$

Now $SD_{\mathrm{V},\lambda}$ is exactly a warehouse location problem as formulated in Chapter 8.

Suppose that we perform a simple linear search on λ, that is, we solve for

$$ZD'_{\mathrm{V}} = \max\{ZD_{\mathrm{V},\lambda} \mid \lambda\}$$

using Erlenkotter's algorithm to determine $ZD_{\mathrm{V},\lambda}$ for each value of λ. For any particular λ, the resulting solution either satisfies or violates the capacity feasibility constraint (9.9). If it satisfies (9.9), we have a feasible, though perhaps not optimal, solution to SD_{V}, which we then can use to construct a new primal subproblem, SP_{Y}.

In either case we have a lower bound on ZD_V, thus a lower bound on Z, the solution to CLP:

$$ZD_{V,\lambda} \leq ZD'_V \leq ZD_V \leq Z$$

Note that if

$$\lambda\left(-\sum_{j=1}^{p} s_j y_j + \sum_{i=1}^{m} r_i\right) = 0 \qquad (9.11)$$

for the solution to $SD_{V,\lambda}$, there is no gap between the lower and upper bounds, so we have an optimal solution to SD_V. The reason for this is that the difference between (9.8) and (9.10) is exactly the left side of (9.11).

Unfortunately, it is quite unlikely that (9.11) will hold, simply because of the difficulty of exactly matching capacity to requirements. In general, as λ is varied, we can expect the solution to $SD_{V,\lambda}$ to behave as shown in Figure 9.2; that is, as λ is increased, total capacity will tend to increase, but in "jumps." Thus we may not be able to find a value of λ such that (9.11) is satisfied in the solution to $SD_{V,\lambda}$.

Continuing our illustration, we take the dual variables from Table 9.7, $\mathbf{V} = (1, 0, 0, 0)$, and solve SD_V, assuming that $\lambda = 0$. The corresponding tableau for the dual ascent algorithm is shown in Table 9.8. Since $v_1 = 1$, we have computed modified variable and fixed costs for site 1 in the tableau. From Table 9.8 we can glean several helpful facts. First, the sum of the dual variables is 100, which provides a lower bound on the optimum value of SD_V, thus also providing a lower bound on the optimum solution to CLP. Second, if we take $J^+ = \{1, 2, 4\}$, we can construct a complementary primal solution for SD_V with a value of 115; thus the gap for SD_V is only 15. Finally, to satisfy complementary slackness, customer 2 can be assigned only to site 1 and customer 3 can be assigned only to site 2, so there is the possibility that the dual adjustment phase will improve the lower bound.

Table 9.9 presents the results of the dual adjustment phase, based on an initial reduction in u_1. Note that the lower bound on ZD_V has improved slightly, to a value of 105. Also, the complementary primal solution has a value of 105, so we have solved SP_V, obtaining $\mathbf{Y} = (1, 0, 0, 1)$, which is optimal for SP_V but not necessarily

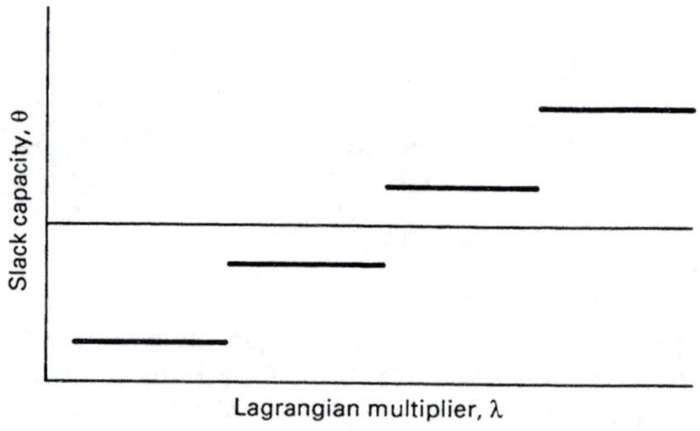

Figure 9.2 Slack capacity in $SD_{V,\lambda}$.

TABLE 9.8 DUAL ASCENT FOR EXAMPLE 9.2 WITH **V** = (1, 0, 0, 0)

	1	2	3	4	u_i
1	+ * 30	+ 30	45	45	(1) ~~30~~ 45
2	* 15	45	15	45	15
3	60	+ * 15	+ 30	45	(2) ~~15~~ 40
$p_j(U)$	~~15~~ 0 (1)	~~40~~ ~~25~~ (1) 0 (2)	~~40~~ 30 (2)	0	$z_d = 100$ $w_{12}^+ = 15$

Legend: ~~15~~ indicates a value that changed.
(1) indicates iteration when value changed.
+ indicates $w_{ij}^+ > 0$.
* indicates $x_{ij}^+ = 1$.

TABLE 9.9 DUAL ADJUSTMENT FOR EXAMPLE 9.2 WITH **V** = (1, 0, 0, 0)

	1	2	3	4	u_i
1	* 30	30	45	45	30
2	+ * 15	45	+ 15	45	(1) ~~15~~ 30
3	60	+ 15	+ 30	* 45	(2) ~~40~~ 45
$p_j(U)$	~~15~~ 0 (1)	~~15~~ ~~10~~ (2)	~~30~~ (1) ~~15~~ 10 (2)	0	$z_d = 105$ $J^+ = \{1, 4\}$ $z_p = 105$

Legend: ~~15~~ indicates a value that changed.
(1) indicates iteration when value changed.
+ indicates $w_{ij}^+ > 0$.
* indicates $x_{ij}^+ = 1$.

optimal for CLP. Note also that the solution to $SP_{V,0}$ is capacity feasible, so we need not search over λ.

Returning to SP_Y, we again solve the transportation problem, obtaining the final tableau shown in Table 9.10. Note that if $y_j = 0$, we simply set $s_j = 0$ in the transportation problem tableau. We have $ZP_Y = 110$ and $\mathbf{V} = (2, 0, 0, 0)$. At this point the gap between the upper and lower bounds for CLP is less than 5%.

Continuing the "ping-pong" heuristic, we prepare to solve the dual subproblem, SD_V, for the second time, using the dual variables $\mathbf{V} = (2, 0, 0, 0)$ obtained from the primal subproblem solution. As before, we assume that $\lambda = 0$, and compute the adjusted costs, $\bar{f}_j$ and $\bar{c}_{ij}$. Upon doing so, we discover that $\bar{f}_1 = -10$. If we simply substitute this value into the condensed dual formulation for the dual ascent algorithm, we obtain the obviously infeasible constraint:

$$\sum_{i=1}^{m} \max(0, u_i - c_{i1}) \leq -10$$

This apparent infeasibility is a result of an assumption we made in Chapter 8, namely, the assumption that $f_j \geq 0$ for all sites. Even if $f_j \geq 0$ in the original problem data for CLP, the adjusted costs, $\bar{f}_j$, may be negative, so we must revise the dual ascent and adjustment methods to accommodate negative fixed costs.

There are at least two alternative ways to deal with negative fixed costs. One is to go back to the original primal formulation of WLP and develop a new condensed

TABLE 9.10 PRIMAL SUBPROBLEM SOLUTION FOR EXAMPLE 9.2 USING $\mathbf{Y} = (1, 0, 0, 1)$

New facility sites

	1	2	3	4	u_i	Dem.
1	1 10	2	3	3 5	3	15
2	0 15	3	1	3	2	15
3	3	1	2	3 15	3	15
4	0	0	0	0 5	0	5
v_j	2			0	c_{ij} x_{ij}	
Cap.	25	0	0	25		

$z(\{1, 4\}) = 70$

dual formulation without the assumption of nonnegative fixed costs. Naturally, we would have to modify the dual ascent and dual adjustment algorithms.

The second approach, which we shall adopt, is to make the following key observation: If $f_j < 0$ in WLP, we may peg $y_j = 1$, even though there may be no shipments out of facility j in an optimal solution. As a result of this observation, we may be faced with solving a variant of WLP in which we have a set of facilities that are required to be "open," even though they may not be used.

Using our previous notation, let J_1 be the index set for sites having $\bar{f}_j < 0$, and let J_2 be the index set for the rest of the sites. Because there are no capacity constraints in $SD_{V,\lambda}$, the set J_1 represents a feasible solution by itself. If we define

$$\bar{c}_i(J_1) = \min\{\bar{c}_{ij} \mid j \in J_1\}$$

the cost associated with this feasible solution is

$$FC_{V,\lambda}(J_1) = \sum_{j \in J_1} \bar{f}_j + \sum_{i=1}^{m} \bar{c}_i(J_1) + \lambda \sum_{i=1}^{m} r_i$$

The location problem now is to decide which *additional* sites to use to maximize the cost savings relative to $FC_{V,\lambda}(J_1)$. This is a problem that we may solve using Erlenkotter's algorithm.

We shall define a *derived* dual subproblem, $SD_{V,\lambda}(J_1)$, as follows:

$$c'_{ij} = \bar{c}_{ij} - \bar{c}_i(J_1), \qquad i = 1, \ldots, m; \quad j \in J_2$$

$$c'_{i0} = 0, \qquad i = 1, \ldots, m$$

$$f'_j = \bar{f}_j, \qquad j \in J_2$$

$$f'_0 = 0$$

$$J'_2 = J_2 \cup \{0\}$$

Note that the index $j = 0$ corresponds to an artificial site or pseudofacility and represents the use of one of the sites in J_1. The derived dual subproblem is

$SD_{V,\lambda}(J_1)$
$$ZD_{V,\lambda}(J_1) = \min \sum_{j \in J'_2} f'_j y_j + \sum_{j \in J'_2} \sum_{i=1}^{m} c'_{ij} x_{ij}$$

subject to

$$\sum_{j \in J'_2} x_{ij} = 1, \qquad i = 1, \ldots, m$$

$$-x_{ij} + y_j \geq 0, \qquad i = 1, \ldots, m; \quad j \in J'_2$$

$$y_j \in \{0, 1\}, \qquad j \in J'_2$$

$$x_{ij} \geq 0, \qquad i = 1, \ldots, m; \quad j \in J'_2$$

$SD_{V,\lambda}(J_1)$ has exactly the same mathematical form as WLP and also has nonnegative fixed costs, thus may be solved using Erlenkotter's algorithm. From the solution to

$SD_{V,\lambda}(J_1)$ we may determine the solution to $SD_{V,\lambda}$ as

$$ZD_{V,\lambda} = FC_{V,\lambda}(J_1) + ZD_{V,\lambda}(J_1)$$

Applying this approach to our dual subproblem for $\mathbf{V} = (2, 0, 0, 0)$, we compute the modified costs shown in Table 9.11 and observe that $\bar{f}_1 < 0$. We set $J_1 = \{1\}$ and compute $\bar{c}_i(J_1)$ as $(45, 30, 75)$ and $FC_{V,\lambda}(J_1) = 140$. The dual ascent tableau for the derived problem is shown in Table 9.12. We see that $ZD_{V,\lambda}(J_1) = -35$ and conclude that using additional sites will reduce total cost. In Table 9.12, the complementary primal solution value is -35, so we have an optimal solution to $SD_{V,\lambda}$, $\mathbf{Y} = (1, 1, 0, 0)$, and $ZD_{V,\lambda} = 140 - 35 = 105$. Since this solution is capacity feasible, we do not need to search over λ, so we also have an optimal solution to SD_V.

TABLE 9.11 MODIFIED COSTS IN $SD_{V,\lambda}$ FOR EXAMPLE 9.2 USING $\mathbf{V} = (2, 0, 0, 0)$

	1	2	3	4
1	45	30	45	45
2	30	45	15	45
3	75	15	30	45
$\bar{f}_j$	−10	40	40	0

TABLE 9.12 DUAL ASCENT FOR THE DERIVED DUAL SUBPROBLEM OF EXAMPLE 9.2 WITH $\mathbf{V} = (2, 0, 0, 0)$

	1	2	3	4	u_i
1	0	+ * -15	0	0	(1) ~~-15~~ 0
2	0 *	15	+ -15	15	(2) ~~-15~~ 0
3	0	+ * -60	+ -45	-30	(3) (4) ~~-60~~ ~~-45~~ -35
$p_j(U)$	0	~~40~~ (1) ~~25~~ (3) ~~10~~ 0 (4)	~~40~~ (2) ~~25~~ (4) 15	0	$z_d = -35$ $z_p = -35$

Legend: ~~15~~ indicates a value that changed.

(1) indicates iteration when value changed.

+ indicates $w_{ij}^{+} > 0$.

* indicates $x_{ij}^{+} = 1$.

Returning to solve SP_Y for the third time, with $\mathbf{Y} = (1, 1, 0, 0)$, we obtain the optimal solution shown in Table 9.13, which has a total cost of 115. Note that the dual variables for this solution are $\mathbf{V} = (1, 0, 0, 0)$, or the same as for the first primal subproblem. If we go back to the dual subproblem with this dual vector, we will simply reproduce the earlier solution, so we terminate the ping-pong heuristic. We have a feasible solution with a cost of 110 and a lower bound of 105. Table 9.14 summarizes the iterations of the ping-pong approach for solving Example 9.2.

The ping-pong approach has a fundamental simplicity, which is illustrated in Figure 9.1. As we have just seen, however, there is a wealth of intricate detail in its implementation. Table 9.15 presents a general statement of the approach. Line 4 in procedure PING-PONG indicates solving the primal subproblem (e.g., using a transportation or general linear programming algorithm). The test in line 5 uses a set of

TABLE 9.13 PRIMAL SUBPROBLEM SOLUTION FOR EXAMPLE 9.2 USING $\mathbf{Y} = (1, 1, 0, 0)$

New facility sites

(each cell: c_{ij} / x_{ij}; columns 3 and 4 shaded)

	1	2	3	4	u_i	Dem.
1	1 / 10	2 / 5	3	3	2	15
2	0 / 15	3	1	3	1	15
3	3	1 / 15	2	3	1	15
4	0	0 / 5	0	0	0	5
v_j	1	0				
Cap.	25	25	0	0		

Key: c_{ij} (upper left of cell), x_{ij} (lower right of cell)

$z(\{1, 2\}) = 35$

TABLE 9.14 SUMMARY OF PING-PONG ITERATIONS FOR EXAMPLE 9.2

Primal subproblem (restriction)	Dual subproblem (relaxation)
$\mathbf{Y} = (1,1,1,1)$: $ZP_Y = 155$	$\mathbf{V} = (1,0,0,0)$: $ZD_{\mathbf{V}} = 105$
$\mathbf{V} = (1,0,0,0)$	$\mathbf{Y} = (1,0,0,1)$
$\mathbf{Y} = (1,0,0,1)$: $ZP_Y = 110$	$\mathbf{V} = (2,0,0,0)$: $ZD_{\mathbf{V}} = 105$
$\mathbf{V} = (2,0,0,0)$	$\mathbf{Y} = (1,1,0,0)$
$\mathbf{Y} = (1,1,0,0)$: $ZP_Y = 115$	STOP
$\mathbf{V} = (1,0,0,0)$	

TABLE 9.15 PROCEDURE PING-PONG

```
1.  PROCEDURE PING-PONG
2.  Initialize Y, 𝒴, 𝒱, LB, UB, DONE ← false
3.  WHILE NOT DONE
4.      CALL PRIMAL(Y: V, ZP_Y)
5.      IF V ∉ 𝒱, THEN
6.          IF ZP_Y < UB, THEN UB ← ZP_Y
7.          𝒱 ← 𝒱 ∪ {V}
8.          CALL DUAL(V: Y, ZD_V)
9.          IF Y ∉ 𝒴, THEN
10.             IF ZD_V > LB, THEN LB ← ZD_V
11.             𝒴 ← 𝒴 ∪ {Y}
12.         ELSE
13.             DONE ← true
14.         ENDIF
15.     ELSE
16.         DONE ← true
17.     ENDIF
18. ENDWHILE
19. END PING-PONG
```

previously determined primal solutions, **Y**, to detect the onset of cycling, so that the algorithm may be terminated. A similar test for repeated dual solutions is performed in line 9. The dual subproblem is solved at line 8. Note that if a search over λ is required and an optimum solution to SD_V is not guaranteed (e.g., because the dual ascent and adjustment phases are not followed by a branch-and-bound phase, even though a gap remains), the value returned as ZD_V should be the *lower bound*, not the feasible solution value.

9.2.3 Solving CLP

We have presented two heuristic approaches for solving the capacity-constrained location problem, CLP. The ADD&DROP heuristic is quite straightforward and should give good solutions but provides no indication of solution quality. The PING-PONG heuristic is fairly complicated, should give good solutions, and does provide lower bounds on the minimum total cost.

If you require an optimum solution to CLP, there are a number of solution procedures reported in the literature. In practice, some form of enumeration generally is required, although certain specific problem instances may be solved without enumeration. The Further Reading section at the end of the chapter will indicate key references for your study of exact methods.

9.3 CAPACITATED LOCATION WITH REQUIRED SINGLE SOURCING

In Section 9.2 we presented two approaches to solving location problems with limited capacity facility sites. These two approaches may produce a solution in which customers are dual sourced. For example, in Table 9.10, the requirements of customer 1 are satisfied partially by site 1 and partially by site 4.

In many settings, this type of service is undesirable. It creates potential confusion on the part of the customer and divides responsibility for service among the facilities. Thus we may want to require single sourcing for all customers. In this section we explore the potential impacts of this added constraint.

We discuss two approaches to the single-sourcing problem. One approach is to attempt to adjust the solution to CLP heuristically so that it satisfies the single-sourcing constraint. The second approach is to develop a mathematical formulation and solution procedure specifically for the single-sourcing problem.

9.3.1 Adjusting the CLP Solution

Suppose that the solution to CLP calls for n new facilities. The optimal allocation of the capacity at the n facilities to the m customer requirements is determined by solving a transportation problem. From the theory of linear programming, we know that there will be at most $m + n - 1$ nonzero x_{ij}'s in a basic feasible solution to this transportation problem. Therefore, the number of customers who are dual sourced cannot exceed $\min(m - 1, n - 1)$.

Obviously, when n is small (e.g., less than 5) and m is relatively large (e.g., 25 to 50) the number of dual-sourced customers in the solution of CLP will be relatively small. In this situation, a natural approach is to attempt to "adjust" the CLP solution so that it satisfies the single-sourcing constraint. We can identify two distinct scenarios in which this adjustment approach might be taken:

1. Capacity constraints are approximate (i.e., s_j is an *estimate* and actually could take on specific values that are somewhat larger or smaller without significantly affecting the cost coefficients); or
2. Capacity constraints are precise (i.e., s_j is based on technological or economic factors and cannot be changed without significantly affecting the cost coefficients).

In the first scenario, we may attempt to *adjust the capacities* to achieve single sourcing. In the second scenario, we may attempt to *adjust the capacity allocations* to achieve single sourcing. Each is discussed below.

Capacity adjustment. We shall consider two types of capacity adjustment: adding capacity to an NF and switching capacity from one NF to another. As we shall see, the dual variables from SP_Y can guide us in making these changes. In fact, we may reduce the total variable cost. Our objective, however, is not to reduce variable cost; rather, it is to achieve a "sensible" single-sourcing solution (i.e., one in which the sites selected in the solution to CLP are not adjusted "too much." It should be intuitively obvious that we could adjust the capacities in such a way that every customer is assigned to the NF with the smallest c_{ij}, in which case there would be no reason to solve SP_Y in the first place.

Example 9.3

The optimum solution to CLP yields the primal subproblem solution shown in Table 9.16. Customers 2, 4, and 5 are dual sourced. How should the capacities be adjusted to eliminate the dual sourcing?

Consider customer 2 first. The requirement of 20 is supplied from NF 1 and 2, each supplying 10. We can eliminate the dual sourcing of customer 2 by switching 10 units of capacity between NF 1 and 2. Which way should the capacity be switched (i.e., which NF should have its capacity increased and which decreased)? All other things being equal, we would want to switch the capacity so that variable costs do not increase.

Recall from the theory of linear programming that v_j may be interpreted as the value of an additional unit of capacity at NF j. Thus $v_1 - v_2 = 3$ would be the rate of change in variable cost as capacity is switched from NF 2 to NF 1. If we switch

TABLE 9.16 SP SOLUTION FOR EXAMPLE 9.2

i \ j	1	2	3	4	r_i	u_i
1	c_{ij} = 0, x_{ij} = 15	c_{ij} = 10	c_{ij} = 12	c_{ij} = 14	15	7
2	c_{ij} = 5, x_{ij} = 10	c_{ij} = 8, x_{ij} = 10	c_{ij} = 10	c_{ij} = 12	20	12
3	c_{ij} = 10	c_{ij} = 4, x_{ij} = 10	c_{ij} = 12	c_{ij} = 14	10	8
4	c_{ij} = 12	c_{ij} = 6, x_{ij} = 5	c_{ij} = 8, x_{ij} = 10	c_{ij} = 10	15	10
5	c_{ij} = 12	c_{ij} = 10	c_{ij} = 4, x_{ij} = 15	c_{ij} = 6, x_{ij} = 10	25	6
6	c_{ij} = 0	c_{ij} = 0	c_{ij} = 0	c_{ij} = 0, x_{ij} = 10	10	0
s_j	25	25	25	20		
v_j	7	4	2	0		

Key: each cell shows c_{ij} (upper left) and x_{ij} (lower right).

10 units of capacity, we obtain $s_1 = 35$ and $s_2 = 15$ (i.e., NF 2 now has less than half the capacity of NF 1, and variable cost is reduced by 30). Such a large imbalance might not be acceptable in practice; fortunately, this is only an academic exercise!

After switching capacity from NF 2 to NF 1, we have the transportation problem solution shown in Table 9.17. We observe that customer 4 is dual sourced, and using the same argument as before, we would switch 10 units of capacity from NF 3 to NF 2. The resulting transportation problem solution is shown in Table 9.18. Finally, to force single sourcing for customer 5, we would switch capacity from NF 4 to NF 3, obtaining the solution shown in Table 9.19.

In Table 9.19, NF 4 has a capacity of 10, which is allocated to the dummy customer. Thus we could eliminate NF 4 from the solution! In addition, the variable cost has decreased from the original CLP solution by a total of 70. On the other hand, the NF capacities are not as evenly balanced as they were initially.

In our example, we adjusted capacity so as to reduce the variable costs. There might be situations, however, in which the decision should be to adjust capacity so as to better balance the NF capacities, resulting in an increase in variable cost. As should be obvious by now, the capacity adjustment approach is ad hoc at best. We can summarize the general approach as follows:

1. Select a dual-sourced customer.

TABLE 9.17 CAPACITY ALLOCATION AFTER SWITCHING CAPACITY FROM NF 2 TO NF 1

i \ j	1	2	3	4	r_i	u_i
1	0 / 15	10	12	14	15	7
2	5 / 20	8 / 0	10	12	20	12
3	10	4 / 10	12	14	10	8
4	12	6 / 5	8 / 10	10	15	10
5	12	10	4 / 15	6 / 10	25	6
6	0	0	0	0 / 10	10	0
s_j	35	15	25	20		
v_j	7	4	2	0		

Key: each cell shows c_{ij} (upper left) / x_{ij} (lower right).

TABLE 9.18 CAPACITY ALLOCATION AFTER SWITCHING CAPACITY FROM NF 3 TO NF 2

i \ j	1	2	3	4	r_i	u_i
1	0 / 15	10	12	14	15	7
2	5 / 20	8 / 0	10	12	20	12
3	10	4 / 10	12	14	10	8
4	12	6 / 15	8 / 0	10	15	10
5	12	10	4 / 15	6 / 10	25	6
6	0	0	0	0 / 10	10	0
s_j	35	25	15	20	c_{ij} / x_{ij}	
v_j	7	4	2	0		

TABLE 9.19 CAPACITY ALLOCATION AFTER SWITCHING CAPACITY FROM NF 4 TO NF 3

i \ j	1	2	3	4	r_i	u_i
1	0 / 15	10	12	14	15	7
2	5 / 20	8 / 0	10	12	20	12
3	10	4 / 10	12	14	10	8
4	12	6 / 15	8 / 0	10	15	10
5	12	10	4 / 25	6 / 0	25	6
6	0	0	0	0 / 10	10	0
s_j	35	25	25	10	c_{ij} / x_{ij}	
v_j	7	4	2	0		

2. Select two supplying NFs.
3. If neither supplying NF has any slack, switch capacity between them, either to minimize cost or improve capacity balance.
4. If one of the supplying NFs has slack, either increase the capacity of the other, or decrease its capacity until the slack has been eliminated, which may lead to case 3.

After studying Tables 9.16 to 9.19, you may ask: "Why didn't we just switch capacity from NF 4 to NF 1 and complete the adjustment in one step?" In fact, this could have been done for Example 9.3. However, there is a potential problem in switching capacity between two NFs that do not serve a common customer. Consider the transportation problem solution shown in Table 9.20, which closely resembles the solution in Table 9.16 except that $s_2 = 20$ and $r_6 = 5$.

Now suppose that we try to switch capacity directly from NF 4 to NF 1, even though they do not serve a common customer. Using the tableau in Table 9.20, we might do this by setting $x_{61} = 5$ and $x_{64} = 0$, and looking for a pivot or flow change that would increase x_{64}. We would find the flow change sequence:

$$(x_{54} - \delta, x_{53} + \delta, x_{43} - \delta, x_{42} + \delta, x_{22} - \delta, x_{21} + \delta, x_{61} - \delta, x_{64} + \delta)$$

with a maximum allowable flow change of $\delta = 5$ units. Note, however, that one of

TABLE 9.20 CAPACITY ADJUSTMENT WITH DEGENERATE ASSIGNMENTS

i \ j	1	2	3	4	r_i	u_i
1	0 / 15	10	12	14	15	7
2	5 + / 10	8 − / 10	10	12	20	12
3	10	4 / 10	12	14	10	8
4	12	6 + / 0	8 − / 15	10	15	10
5	12	10	4 + / 15	6 − / 10	25	6
6	0 −	0	0	0 + / 5	5	0
s_j	25	20	25	20		
v_j	7	4	2	0		

(Cell key: c_{ij} / x_{ij})

the flows being increased is the degenerate flow, x_{42}. Thus, not only have we failed to eliminate any dual sourcing, we have added another!

There is an important general lesson here. The danger of an ad hoc procedure is that we may not be able to imagine all the situations to which it may be applied. Thus we may obtain quite unexpected results from ad hoc methods, especially if they are applied in an uncontrolled fashion. In this particular case, the root cause of the problem is degenerate variables in the basic feasible solution. We cannot know, a priori, whether or not a degenerate variable will be in the flow change sequence, and once we discover that it is, we have no way around the problem.

In conclusion, we suggest that capacity adjustment may be a good practical solution to the problem of dual sourcing. However, it should be applied cautiously, with the direction of adjustment carefully determined, considering both cost and capacity balance.

Capacity allocation adjustment. An allocation of capacity that satisfies the single-sourcing requirement has exactly m nonzero x_{ij}'s. Thus, given a single-sourcing solution, we can construct a basic feasible solution to the transportation problem, SP_Y, by appropriately specifying $n - 1$ degenerate basic variables. For example, Table 9.21 displays a basic feasible solution, which provides single

TABLE 9.21 SINGLE-SOURCING SOLUTION FOR EXAMPLE 9.2

i \ j	1	2	3	4	r_i	u_i
1	0 / 15	10	12	14	15	0
2	5 / 0	8	10	12 / 20	20	5
3	10	4 / 10	12	14	10	1
4	12	6 / 15	8 / 0	10	15	3
5	12	10	4 / 25	6 / 0	25	−1
6	0 / 10	0	0	0	10	0
s_j	25	25	25	20	c_{ij} / x_{ij}	
v_j	0	−3	−5	−7		

sourcing, for Example 9.3. Obviously, the solution in Table 9.21 is not optimum with respect to the transportation tableau.

Because the single-sourcing solutions, with additional degenerate capacity allocations, also are basic feasible solutions to SP_Y, it is tempting to try to adjust the allocations in the solution to CLP by pivoting operations in order to achieve single sourcing. For example, we can transform the solution in Table 9.16 into the solution in Table 9.21 in two pivots; on the first, x_{42} enters the solution, and on the second, x_{16} enters the solution.

There are two major difficulties with this approach. First, since we do not usually know the single-sourcing solution a priori, we would have to *enumerate* all possible pivots—in effect, searching for a single-sourcing basic feasible solution in the neighborhood of the given basic feasible solution for SP_Y. This search would be much like searching from a given chess position to try to find a winning "next" move.

The second major problem with this approach is that there may be *no* single-sourcing solution using the facilities selected in the solution to CLP. In this case we could enumerate all possible basic feasible solutions to SP_Y before we discovered that there is no single-sourcing solution.

As with the capacity adjustment approach, we conclude that although adjusting the capacity allocations is intuitively appealing, it has many technical limitations. One should use this approach with caution, recognizing the likelihood of failure.

9.3.2 Generalized Assignment Model

Adjusting the solution to CLP to achieve single sourcing, while intuitively appealing, has serious deficiencies as a general solution approach. In this section we discuss a more direct approach, which involves reformulating the problem to include the single-sourcing requirement explicitly.

The *generalized assignment problem* (GAP) is a well-known discrete optimization problem and can be stated generally as

$$\text{(GAP)} \qquad \text{minimize} \sum_{i \in I} \sum_{j \in J} c_{ij} x_{ij} \tag{9.12}$$

subject to

$$a_j \leq \sum_{i \in I} r_{ij} x_{ij} \leq b_j, \qquad j \in J \tag{9.13}$$

$$\sum_{j \in J} x_{ij} = 1, \qquad i \in I \tag{9.14}$$

$$x_{ij} \in \{0, 1\}, \qquad i \in I; \quad j \in J \tag{9.15}$$

The set I may be interpreted as an index set for *tasks*, and J may be interpreted as an index set for *resources*. Task i requires r_{ij} units of resource j if it is performed using resource j, with corresponding cost c_{ij}. The available amount of resource j is b_j, and at least a_j units of resource j must be used. Every task must be performed, and the objective is to minimize the total cost.

GAP models have been proposed for production planning problems, task assignment problems, districting problems, vehicle routing problems, and location problems. In particular, the GAP formulation may be used to model the capacitated location problem with single sourcing.

In the GAP formulation, each customer corresponds to a task and there are p additional tasks, or dummy customers, indexed $i = m + 1, \ldots, m + p$, with

$$r_{ij} = \begin{cases} s_j & \text{if } i = m + j \\ 0 & \text{otherwise} \end{cases}$$

If task $m + j$ is performed by resource j (i.e., if NF j has dummy customer $m + j$ assigned), no real customers can be assigned to NF j. Each NF site corresponds to a resource in GAP, and there is one additional resource, or dummy site, indexed $j = p + 1$. The capacities of the resources in GAP are

$$b_j = \begin{cases} s_j & \text{if } j = 1, \ldots, p \\ p & \text{if } j = p + 1 \end{cases}$$

and

$$r_{i,p+1} = \begin{cases} 0 & \text{for } i = 1, \ldots, m \\ 1 & \text{for } i = m + 1, \ldots, m + p \end{cases}$$

Note that if task $m + j$ is assigned resource $p + 1$ (i.e., dummy customer $m + j$ is assigned to the dummy site), resource j is available for other task assignments (i.e., NF site j is "selected"). Thus the cost of this assignment should be f_j. Table 9.22 presents the parameters for the GAP formulation of Example 9.3, where $f = (250, 250, 250, 200)$.

The reason for presenting the GAP formulation of the CLP with single sourcing is that GAP has been studied extensively, so general solution methods are available. We hasten to point out that GAP is a discrete optimization problem, and the available solution methods are essentially enumerative. The best solution methods are based on a particular Lagrangian relaxation of GAP, in which the assignment constraints, (9.14), are dualized. Letting $\lambda_i, i \in I$, be the Lagrangian multipliers, the relaxation is

$$\text{GAP}_\lambda \qquad \text{minimize} \sum_{i \in I} \sum_{j \in J} (c_{ij} - \lambda_i) x_{ij} + \sum_{i \in I} \lambda_i$$

subject to (9.13) and (9.15).

The key observation is that GAP_λ may be solved as a set of $p + 1$ independent binary knapsack problems, one for each resource, or NF site. This is good news, because even though the binary knapsack problem is yet another discrete optimization problem, there are exact branch-and-bound algorithms for solving it that are able to solve reasonably sized problems with modest computational requirements.

Suppose that we were to optimize GAP_λ with respect to λ. The theory of Lagrangian duality indicates that we would obtain a relatively good lower bound, because GAP_λ is a discrete optimization problem. Unfortunately, doing so would

TABLE 9.22 GAP FORMULATION FOR EXAMPLE 9.3

i \ j	1	2	3	4	5	λ_i
1	0 / 15	150 / 15	180 / 15	210 / 15	M	
2	100 / 20	160 / 20	200 / 20	240 / 20	M	
3	100 / 10	40 / 10	120 / 10	140 / 10	M	
4	180 / 15	90 / 15	120 / 15	150 / 15	M	
5	300 / 25	250 / 25	100 / 25	150 / 25	M	
6	0 / 25	M	M	M	250 / 1	
7	M	0 / 25	M	M	250 / 1	
8	M	M	0 / 25	M	250 / 1	
9	M	M	M	0 / 20	200 / 1	
b_j	25	25	25	20	4	c_{ij} / r_{ij}

require us to search in $m + p$ space, and for every trial point, we would have to solve $p + 1$ binary knapsack problems. This seems an excessive amount of work, so the best available algorithms use a set of rules for specifying $\boldsymbol{\lambda}$, perhaps with an ascent-type adjustment scheme for modifying $\boldsymbol{\lambda}$ to improve the lower bound.

To illustrate these methods, we will use Example 9.3. Table 9.22 gives the GAP data. For $i \leq m$, we take λ_i to be the *second smallest* c_{ij}, and for $k > m$, we take λ_k to be the negative of the sum of the negative values of $c_{ij} - \lambda_i$ in column k. The intuitive justification for these choices is as follows. If customer i is not assigned to its cheapest source, the difference between the smallest and the second smallest cost is the minimum penalty that customer i faces, in terms of variable cost. Similarly, if facility k is not used, the minimum variable cost penalty will be λ_k.

Table 9.23 displays the data for the corresponding problem, GAP$_\lambda$. Because the binary knapsack problems are so small for this example, we can solve them by inspection. Solutions also are indicated in Table 9.23. We can make the following observations:

TABLE 9.23 GAP$_\lambda$ SOLUTION: INITIAL

i \ j	1	2	3	4	5	λ_i
1	−150 / 15	0 / 15	30 / 15	60 / 15	M	150
2	−60 / 20	0 / 20	40 / 20	80 / 20	M	160
3	0 / 10	−60 / 10	20 / 10	180 / 10	M	100
4	60 / 15	−30 / 15	0 / 15	30 / 15	M	120
5	150 / 25	100 / 25	−50 / 25	0 / 25	M	150
6	−210 / (+) 25	M	M	M	40 / 1	210
7		−90 / (+) 25	M	M	160 / 1	90
8	M	M	−50 / (+) 25	M	200 / 1	50
9	M	M	M	0 / 20	200 / 1	0
b_j	25	25	25	20	4	$c_{ij} - \lambda_i$ / r_{ij}

Legend: (+): $x_{ij} = 1$

1. None of the facilities are available for customer assignment in the solution to GAP.
2. The lower bound from GAP$_\lambda$ is 680.
3. In a feasible solution, all facilities must be used.
4. A feasible assignment of customers to facilities is $x_{11} = x_{24} = x_{32} = x_{42} = x_{53} = 1$, with a total cost of 1420.

Thus GAP$_\lambda$ does not provide a feasible solution to GAP, nor is it obvious how to use the solution to GAP$_\lambda$ to obtain a feasible solution to GAP. (Recall that the Lagrangian relaxation of WLP provided a mechanism for constructing a feasible solution.) Moreover, we know that any feasible solution will have a cost at least as large as the sum of the fixed costs and the minimum possible assignment costs, which is 1280, so the lower bound is no better than 53% of the optimum solution.

TABLE 9.24 GAP$_\lambda$ SOLUTION: AFTER PEGGING

i \ j	1	2	3	4	λ_i
1	−150 / ⊕ 15	0 / 15	30 / 15	60 / 15	150
2	−60 / 20	0 / 20	40 / 20	80 / 20	160
3	0 / 10	−60 / ⊕ 10	20 / 10	180 / 10	100
4	60 / 15	−30 / ⊕ 15	0 / 15	30 / 15	120
5	150 / 25	100 / 25	−50 / ⊕ 25	0 / 25	150
b_j	25	25	25	20	$c_{ij} - \lambda_i$ / r_{ij}

Legend: ⊕: $x_{ij} = 1$

Now, you may be asking yourself, "Why should I be so enthusiastic over a relaxation that performs so poorly?" The answer is, "A slow horse is faster than no horse!" You may be able to improve the lower bound significantly by adjusting the multipliers. Furthermore, as the enumeration proceeds, the quality of the relaxation's performance may improve. To illustrate, suppose that we use our earlier observation that all the sites are required (i.e., we peg $x_{m+j,p+1} = 1$ for $j = 1, \ldots, m$). The corresponding problem is shown in Table 9.24. Now the lower bound has improved to 1340 and all the customers except customer 2 are assigned to a facility. If we make an ad hoc adjustment and increase λ_2 to 245, we obtain an improved lower bound of 1420 and a feasible solution. Thus the performance of the relaxation is much better when we have a capacity feasible solution.

We conclude this section by reiterating an earlier caution. The GAP formulation can be quite difficult to solve when there is little slack capacity. The Lagrangian relaxation, GAP$_\lambda$ may not give good lower bounds without some enumeration or a multiplier adjustment scheme, and does not appear to be very useful for generating feasible solutions. Thus GAP, although very similar to CLP in many regards, is quite different from the computational perspective. You should be certain that the GAP formulation is really necessary before you undertake to solve it!

9.4 INTERACTION BETWEEN NEW FACILITIES

In Chapter 4 we studied multifacility planar location problems in which there were interactions between the NFs, so that the locations of the NFs relative to one another had an impact on the location criterion. In the discrete location problems we have studied so far, however, the contribution of one location decision to the performance measure has not depended directly on any other location decision. Any interaction between location decisions has occurred as a result of constraints, such as a limit on the number of facilities.

In this section we examine a class of discrete location models that permit us to address certain interactions between facilities. Although at first blush this might seem to be a relatively minor extension of our models, we shall see that the resulting problems are notoriously difficult to solve optimally. Fortunately, there are a number of reasonably good heuristics for these problems.

Example 9.4

A manufacturing cell is being designed with manual material handling between workstations. Figure 9.3 presents a schematic illustration of the possible locations of the workstations along the aisle. Each workstation requires the same amount of space, so any assignment of workstations to locations is feasible. The objective of the location decisions is to minimize the total distance that material moves. Table 9.25(a) lists the material movement distances between locations, and Table 9.25(b) lists the material flow rates between workstations, as determined from process route sheets and planned production quantities. What is the optimum assignment of workstations to locations?

The problem in Example 9.4 can be formulated as an optimization problem as follows. Let r_{ij} be the rate of item movement from workstation i to workstation j, and let d_{rs} be the distance between locations r and s. Define

$$x_{ir} = \begin{cases} 1 & \text{if workstation } i \text{ is assigned to location } r \\ 0 & \text{otherwise} \end{cases}$$

Then the total distance that items move per unit time will be

$$z = \sum_{i=1}^{6} \sum_{j=1}^{6} \sum_{r=1}^{6} \sum_{s=1}^{6} r_{ij} d_{rs} x_{ir} x_{js}$$

Since every workstation must be assigned to a location and every location must have a workstation assigned to it, the complete formulation is

$$\text{minimize} \quad z = \sum_{i=1}^{6} \sum_{j=1}^{6} \sum_{r=1}^{6} \sum_{s=1}^{6} r_{ij} d_{rs} x_{ir} x_{js}$$

subject to

$$\sum_{r=1}^{6} x_{ir} = 1, \qquad i = 1, \ldots, 6$$

$$\sum_{i=1}^{6} x_{ir} = 1, \qquad r = 1, \ldots, 6$$

$$x_{ir} \in \{0, 1\}, \qquad i = 1, \ldots, 6; \quad r = 1, \ldots, 6$$

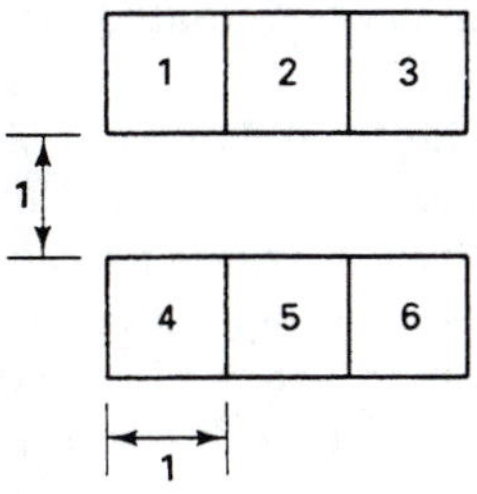

Figure 9.3 Location configuration for Example 9.4.

TABLE 9.25 DISTANCE MATRIX AND DIRECTED FLOWS FOR EXAMPLE 9.4

(a) Distance matrix, $d(k,h)$

	Site					
Site	1	2	3	4	5	6
1	0	1	2	1	2	3
2	1	0	1	2	1	2
3	2	1	0	3	2	1
4	1	2	3	0	1	2
5	2	1	2	1	0	1
6	3	2	1	2	1	0

(b) Direct flows

From facility:	To facility: 1	2	3	4	5	6
1	0	3	2	2	1	2
1	0	3	2	2	1	2
2	1	0	3	1	0	5
3	4	1	0	1	2	2
4	0	1	1	0	4	1
5	3	2	0	2	0	5
6	2	3	4	1	5	0

Note that the optimization criterion involves the products of pairs of location decision variables, so it has a *quadratic form*. Also note that the constraints are exactly the same as for the classical assignment problem. Thus, in its general form, this problem is called the *quadratic assignment problem* (QAP).

The QAP is a classic model in discrete optimization and is a suitable formulation for a wide variety of real problems, from backplane wiring in computers to the design of control panels or keyboards. Since the early 1960s, a tremendous amount of research effort has been directed toward optimization algorithms for QAP. Yet the QAP remains computationally intractable for problems with more than 15 to 20 facilities, and this situation has changed very little since the mid-1970s. One of the great ironies of discrete optimization is that QAP is so simple to state but so difficult to solve.

We focus our discussion of QAP on heuristic solution methods. It should be noted that there is solid theoretical evidence to indicate that heuristic methods should give good results, at least for large problems. As in Chapter 3, we shall see that there are two major classes of heuristics for QAP: *construction* and *improvement*. We shall discuss both approaches. Before discussing solution methods, however, it will be necessary to present a different formulation of the problem.

9.4.1 Formulation

In solving the discrete location problems considered up to now, we have worked directly with the location decision variables, x_{ir}. For the QAP, however, we shall find it more convenient to represent a solution as a *permutation*, $\mathbf{a}$. Let $a(i)$ represent the location to which workstation i has been assigned. Suppose that a solution to Example 9.4 is $\mathbf{a} = (2, 4, 5, 3, 1, 6)$. This solution is illustrated in Figure 9.4, where, for example, workstation 3 has been assigned to location 5.

Naturally, we are interested in the total cost of a solution, which we denote by $TC(\mathbf{a})$. In developing an expression for $TC(\mathbf{a})$, we note that the *direction* of material movement is really not important. Rather, we are concerned primarily with the flow *between workstations*, which can be represented as the sum of the i-to-j flow and the j-to-i flow, as shown in Table 9.26. In general, we denote the rate of flow between facilities i and j as a *weight*, w_{ij}. Note that the weight matrix shown in Table 9.26 is symmetric, which we will assume is true in general.

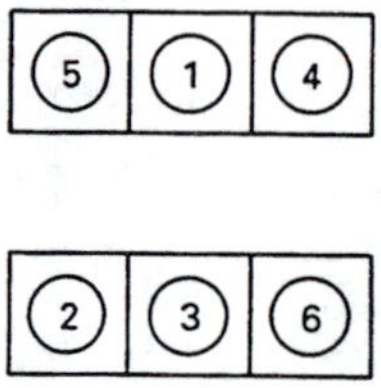

Figure 9.4 Solution $a = (2,4,5,3,1,6)$ for Example 9.4.

TABLE 9.26 SYMMETRIC WEIGHT MATRIX FOR EXAMPLE 9.4

From facility:	To facility: 1	2	3	4	5	6
1	0	4	6	2	4	4
2	4	0	4	2	2	8
3	6	4	0	2	2	6
4	2	2	2	0	6	2
5	4	2	2	6	0	10
6	4	8	6	2	10	0

We will denote the distance between locations r and s by $d(r,s)$. The distance matrix illustrated in Table 9.25(a) is symmetric, and for the moment, we will find it a notational convenience to assume that the distance matrix generally is symmetric. Although this seems like a sensible assumption, we shall subsequently present an example to show that the distances need not be symmetric.

With this notation, we can express the total cost of a solution to QAP as

$$TC(\mathbf{a}) = \sum_{1 \le i < j \le n} w_{ij}\, d(a(i), a(j)) \tag{9.16}$$

The total cost for the solution illustrated in Figure 9.4 is 114.

Notice that the weight matrix $\mathbf{W} = (w_{ij})$ shown in Figure 9.25 includes entries on and below the diagonal, as well as above the diagonal. Also, notice that $w_{ij} = w_{ji}$ and that all the diagonal terms are zero. It will be assumed in general that any weight matrix $\mathbf{W}$ has the same property. The reason for the assumption is basically one of computational convenience; for example, the total cost of an assignment is also given by

$$TC(\mathbf{a}) = \frac{1}{2}\sum_{i=1}^{n}\sum_{j=1}^{n} w_{ij}\, d(a(i), a(j)) \tag{9.17}$$

To establish that expressions (9.16) and (9.17) are identical, note that (9.17) may be written as

$$\frac{1}{2}\left[\sum_{1 \le i < j \le n} w_{ij}\, d(a(i), a(j)) + \sum_{n \ge i > j \ge 1} w_{ij}\, d(a(i), a(j)) + \sum_{i=1}^{n} w_{ii}\, d(a(i), a(i))\right]$$

Now, the last sum in the latter expression is just zero, while the second sum may be written, using a change of indices, as

$$\sum_{n \ge e > f \ge 1} w_{ef}\, d(a(e), a(f)) \tag{9.18}$$

By assumption, $w_{ef} = w_{fe}$ and $d(a(e), a(f)) = d(a(f), a(e))$, so (9.18) is the same as

$$\sum_{n \ge e > f \ge 1} w_{fe}\, d(a(f), a(e)) = \sum_{1 \le i < j \le n} w_{ij}\, d(a(i), a(j)) \tag{9.19}$$

It now follows from the last three expressions that (9.16) and (9.17) give the same total cost for an assignment. Note that the right side of (9.19) is obtained from the left side of (9.19) just by replacing f by i and e by j, and observing that the set of summation indices $\{e,f\colon n \ge e > f \ge 1\}$ is the same as the set of summation indices $\{i,j\colon 1 \le i < j \le n\}$; the only difference between the left and right sides of (9.19) is the order of summation.

Another, and different, total cost expression will be useful subsequently; given an assignment of facilities to locations, the total cost for facility k is given, for $k = 1, \ldots, n$, by

$$p_k(\mathbf{a}) = \sum_{1 \le i < k} w_{ik}\, d(a(i), a(k)) + \sum_{k < j \le n} w_{kj}\, d(a(k), a(j)) \tag{9.20}$$

With reference to the example, (9.20), for $k = 3$, is

$$p_3(\mathbf{a}) = w_{13}\, d(a(1), a(3)) + w_{23}\, d(a(2), a(3)) + w_{34}\, d(a(3), a(4)) + w_{35}\, d(a(3), a(5)) + w_{36}\, d(a(3), a(6)) \tag{9.21}$$

Notice that (9.21) may also be written, using the symmetry of the **D** and **W** matrices, as

$$w_{31}\, d(a(3), a(1)) + w_{32}\, d(a(3), a(2)) + w_{33}\, d(a(3), a(3)) + w_{34}\, d(a(3), a(4)) + w_{35}\, d(a(3), a(5)) + w_{36}\, d(a(3), a(6))$$

By simply generalizing the same approach, another equivalent expression for $p_k(\mathbf{a})$ is

$$p_k(\mathbf{a}) = \sum_{j=1}^{n} w_{kj}\, d(a(k), a(j)) \tag{9.22}$$

The notion of *pairwise interchanges* was introduced in Chapter 3. Pairwise interchanges also may be used in an attempt to improve a given solution for QAP, so a convenient format is needed for computing the impact of pairwise interchanges. Suppose that we have a given solution, **a**, and we wish to evaluate the impact of interchanging the locations of facilities u and v [i.e., we wish to evaluate the change in total cost, denoted $DTC_{uv}(\mathbf{a})$]. If we denote by $\mathbf{a}'$ the solution that results from interchanging facilities u and v in **a**, we can express $DTC_{uv}(\mathbf{a})$ as

$$DTC_{uv}(\mathbf{a}) = TC(\mathbf{a}) - TC(\mathbf{a}')$$

This approach is unduly complicated, however, since both the costs $TC(\mathbf{a})$ and $TC(\mathbf{a}')$ include a number of common terms, that is, all those not involving facilities u or v. Thus, to compute $TC(\mathbf{a}) - TC(\mathbf{a}')$, it is only necessary to subtract those terms involving facilities u or v in $TC(\mathbf{a}')$ from those terms involving facilities u or v in $TC(\mathbf{a})$. The sum of all terms involving facilities u or v in $TC(\mathbf{a})$ is given by

$$\sum_{\substack{i=1\\ \neq v}}^{n} w_{iu}\, d(a(i), a(u)) + \sum_{\substack{i=1\\ \neq u}}^{n} w_{iv}\, d(a(i), a(v)) + w_{uv}\, d(a(u), a(v)) \tag{9.23}$$

When the sites at which facilities u and v are located are interchanged, facility u will be located at site $a(v)$, and facility v will be located at site $a(u)$, so that the sum of all terms involving facilities u or v in $TC(\mathbf{a}')$ is given by

$$\sum_{\substack{i=1\\ \neq v}}^{n} w_{iu}\, d(a(i), a(v)) + \sum_{\substack{i=1\\ \neq u}}^{n} w_{iv}\, d(a(i), a(u)) + w_{uv}\, d(a(v), a(u)) \tag{9.24}$$

Thus it is only necessary to subtract (9.24) from (9.23) to compute $TC(\mathbf{a}) - TC(\mathbf{a}')$. Prior to carrying out the computation, it is convenient to obtain alternative expressions for (9.23) and (9.24). Note that (9.23) may be written, on adding

$$0 = w_{vu}\, d(a(v), a(u)) + w_{uv}\, d(a(u), a(v)) - w_{vu}\, d(a(v), a(u)) - w_{uv}\, d(a(u), a(v))$$

as

$$\sum_{i=1}^{n} w_{iu}\, d(a(i), a(u)) + \sum_{i=1}^{n} w_{iv}\, d(a(i), a(v)) - w_{vu}\, d(a(v), a(u)) \qquad (9.25)$$

Similarly, (9.24) may be written, on adding

$$0 = w_{vu}\, d(a(v), a(v)) + w_{uv}\, d(a(u), a(u))$$

as

$$\sum_{i=1}^{n} w_{iu}\, d(a(i), a(v)) + \sum_{i=1}^{n} w_{iv}\, d(a(i), a(u)) + w_{uv}\, d(a(v), a(u)) \qquad (9.26)$$

Thus the change in cost obtained by interchanging the location of facilities u and v for a given assignment, which will be denoted by $DTC_{uv}(\mathbf{a})$, is readily obtained by subtracting (9.26) from (9.25), yielding

$$DTC_{uv}(\mathbf{a}) = \sum_{i=1}^{n} (w_{iu} - w_{iv})[d(a(i), a(u)) - d(a(i), a(v))] - 2w_{uv}\, d(a(u), a(v)) \qquad (9.27)$$

Equation (9.27) will be illustrated for $u = 2$ and $v = 4$ in the example; $DTC_{24}(\mathbf{a})$ is equal to

$$\begin{aligned}
&(w_{12} - w_{14})[d(2,4) - d(2,3)] + (w_{22} - w_{24})[d(4,4) - d(4,3)] \\
&\quad + (w_{32} - w_{34})[d(5,4) - d(5,3)] + (w_{42} - w_{44})[d(3,4) - d(3,3)] \\
&\quad + (w_{52} - w_{54})[d(1,4) - d(1,3)] + (w_{62} - w_{64})[d(6,4) - d(6,3)] \\
&\quad - 2w_{24}\, d(4,3) \\
&= 2(1) - 2(-3) + 2(-1) + 2(3) - 4(-1) + 6(1) - 2(2)(3) = 10
\end{aligned}$$

Thus, when the locations of facilities 2 and 4 are interchanged, giving the new assignment

$$\mathbf{a}' = (2, 3, 5, 4, 1, 6)$$

then the value of $TC(\mathbf{a}')$ is $TC(\mathbf{a}) - DTC_{24}(\mathbf{a}) = 114 - 10 = 104$.

Several points about the expression $DTC_{uv}(\mathbf{a})$ are worth making. Note that in $DTC_{uv}(\mathbf{a})$ the terms $w_{1u} - w_{1v}, w_{2u} - w_{2v}, \ldots, w_{nu} - w_{nv}$ are just the differences of the entries in columns u and v of the matrix $\mathbf{W}$, computed first for the first row, then for the second row, and so forth. If the computation of $DTC_{uv}(\mathbf{a})$ is to be carried out for a large number of different assignments, it may be convenient to construct a table of these differences, rather than recompute them using the data in the $\mathbf{W}$ matrix each time. The terms

$$d(a(i), a(u)) - d(a(i), a(v), \quad i = 1, \ldots, n$$

are the differences between the entries in columns $a(u)$ and $a(v)$ of the $\mathbf{D}$ matrix, computed in the row order $a(1), a(2), \ldots, a(n)$. It is the computation of the terms $d(a(i), a(u)) - d(a(i), a(v))$ that constitutes most of the work in computing $DTC_{uv}(\mathbf{a})$, since the assignment **a** must constantly be made use of in computing these terms. Also, if a least total cost assignment, say $\mathbf{a}^*$, is found, then for the interchange of locations of *any* two facilities u and v, $DTC_{uv}(\mathbf{a}^*) \leq 0$, so that we have a necessary condition for a least total cost assignment. Unfortunately, the necessary condition is not in general sufficient, since there is no guarantee that an assignment which satisfies the necessary condition cannot be improved upon by considering, for example, three-way interchanges or four-way interchanges. Another point to be made is that many computational procedures that attempt to find a least-cost assignment do so by making pairwise interchanges, so that it is often convenient to use the expression for $DTC_{uv}(\mathbf{a})$ in these procedures.

9.4.2 Construction Heuristics

Construction heuristics, as the name implies, begin with the basic problem data and build up a solution, or permutation, in an iterative fashion. Initially, none of the NF will have been assigned to a location, which we indicate by the convention $a(i) = 0$. At any intermediate point in the construction process, we will have a set, F, of NF which have been assigned to a location. We will use the notation $\mathbf{a}(F)$ to indicate the set of locations that are assigned to the NFs in the set F. With this additional notation, we can state the general form of the construction heuristic as follows:

WHILE $|F| < n$
1. Select some $i \notin F$
2. Select some $r \notin \mathbf{a}(F)$
3. $a(i) \leftarrow r$
4. $F \leftarrow F \cup \{i\}$

ENDWHILE

A particular implementation of the general construction heuristic requires particular rules for performing steps 1 and 2, and there are many reasonable rules that might be used. We shall illustrate only one particular implementation, in which the selection in step 1 is made arbitrarily and the selection in step 2 is made to minimize the additional total cost for the partial solution.

Given a partial solution, $(F, \mathbf{a}(F))$, we can compute an associated total cost, $TC(\mathbf{a}(F))$, by ignoring all the NF not yet assigned locations. If we now augment the partial solution by assigning NF k to location r, the increment in total cost is

$$p_k(\mathbf{a}(F), r) = \sum_{j \in F} w_{kj} d(r, a(j))$$

Note that to compute an *increment* in total cost, we must have at least one NF already assigned (i.e., we must have $|F| \geq 1$). Thus we must specify an initial assignment for the construction heuristic. In the version of the construction heuristic given below the initial assignment is made arbitrarily.

Procedure CONSTRUCT **(W, D: a)**

0. Select $i \in \{1, 2, \ldots, n\}$ arbitrarily
   $a(i) \leftarrow 1$
   $F = \{i\}$
1. WHILE $|F| < n$
   1.1 Select $i \notin F$ arbitrarily
   1.2 $p_i(\mathbf{a}(F)) = \text{minimum } \{p_i(\mathbf{a}(F), r) \mid r \notin \mathbf{a}(F)\}$
   1.3 $a(i) \leftarrow k$
   $F \leftarrow F \cup \{i\}$
   ENDWHILE

One method for making the arbitrary selection is to introduce some randomization (i.e., to select at random from among the alternatives). This type of randomization results in a different solution being constructed each time the heuristic algorithm is executed. Thus one might choose to repeat the construction algorithm several times, taking the best of the solutions constructed.

To illustrate procedure CONSTRUCT, we shall apply it to Example 9.4, with the arbitrary order for considering the NF being (3, 4, 2, 5, 1, 6). Table 9.27 displays the computations required, and the results of each iteration are

(0) $a(3) \leftarrow 1$
(1) $a(4) \leftarrow 2$
(2) $a(2) \leftarrow 4$
(3) $a(5) \leftarrow 5$
(4) $a(1) \leftarrow 3$
(5) $a(6) \leftarrow 6$

The total cost for the solution is 108. [You should verify this by computing $TC(\mathbf{a})$ yourself.]

Figure 9.5 illustrates the placement of the selected NF on each iteration of the heuristic. Observe that each NF added to a partial solution is assigned to a location adjacent to a previously assigned location. Thus one might suspect that the choice of the initial location could have a significant impact on the quality of the final solution. If the NF selected in step 0 has large weights, it probably should be assigned to a location that has a relatively large number of relatively small distances to other locations. On the other hand, if it has relatively small weights, perhaps it should be assigned to a less desirable location (i.e., one that has a relatively small number of small distances to other locations).

Reasoning in this way, we can construct an alternative solution to Example 9.4. We might conclude, for example, that workstations 5 and 6 should be assigned to locations 2 and 5, since these are "central" locations, and workstations 5 and 6 have the largest total flows. Next, we might conclude that workstations 2 and 3 should be located as close as possible to workstation 6, since the corresponding weights are the

TABLE 9.27 COMPUTATIONS FOR CONSTRUCTION HEURISTIC

Iteration 1: $i = 3$; $a(3) \leftarrow 1$		
Iteration 2: $i = 4$		
r	$p_i(\mathbf{a}(F), r)$	k
2	$w_{43} d_{21} = (2)(1) = 1$	*
3	$w_{43} d_{31} = (2)(2) = 4$	
4	$w_{43} d_{41} = (2)(1) = 2$	
5	$w_{43} d_{51} = (2)(2) = 4$	
6	$w_{43} d_{61} = (2)(3) = 6$	
	$a(4) \leftarrow 2$	
Iteration 3: $i = 2$		
r	$p_i(\mathbf{a}(F), r)$	k
3	$w_{42} d_{32} + w_{32} d_{31} = (2)(1) + (4)(2) = 10$	
4	$w_{42} d_{42} + w_{32} d_{41} = (2)(2) + (4)(1) = 8$	*
5	$w_{42} d_{52} + w_{32} d_{51} = (2)(1) + (4)(2) = 10$	
6	$w_{42} d_{62} + w_{32} d_{61} = (2)(2) + (4)(3) = 14$	
	$a(2) \leftarrow 4$	
Iteration 4: $i = 5$		
r	$p_i(\mathbf{a}(F), r)$	k
3	$w_{25} d_{34} + w_{45} d_{32} + w_{35} d_{31} = (2)(3) + (6)(1) + (2)(2) = 16$	
5	$w_{25} d_{54} + w_{45} d_{52} + w_{35} d_{51} = (2)(1) + (6)(1) + (2)(2) = 12$	*
6	$w_{25} d_{64} + w_{45} d_{62} + w_{35} d_{61} = (2)(2) + (6)(2) + (2)(3) = 22$	
	$a(5) \leftarrow 5$	
Iteration 5: $i = 1$		
r	$p_i(\mathbf{a}(F), r)$	k
3	$w_{51} d_{35} + w_{21} d_{34} + w_{41} d_{32} + w_{31} d_{31} = (4)(3) + (4)(3) + (2)(1) + (6)(2) = 34$	*
6	$w_{51} d_{65} + w_{21} d_{64} + w_{41} d_{62} + w_{31} d_{61} = (4)(1) + (4)(2) + (2)(2) + (6)(3) = 34$	
	$a(1) \leftarrow 3$	
Iteration 6: $i = 6$; $a(i) \leftarrow 6$.		

largest for unassigned workstations. Finally, we might conclude that workstation 1 should be located adjacent to workstation 3, since w_{13} is the largest of the weights for the remaining unassigned workstations. Figure 9.6 illustrates this solution, which has a total cost of 92.

Our intuitive analysis of the weights and the distances between locations leads us to a solution that is almost 15% better than the one from procedure CONSTRUCT. Naturally, we become quite excited at this result, and attempt to capture our intuitive analysis in a formal algorithm. As an exercise (in futility, perhaps) you should attempt to formalize the logic leading to the solution in Figure 9.6. Your formal algorithm should be one that may be applied to *any* problem, not just one with the locations configured as in Example 9.4.

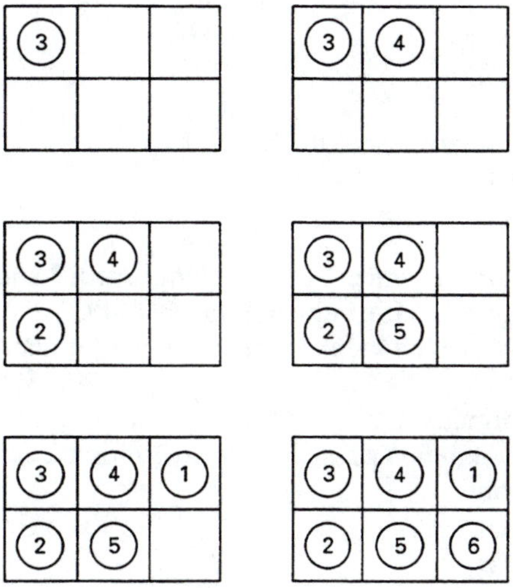

Figure 9.5 Sequence of assignments by the construction heuristic.

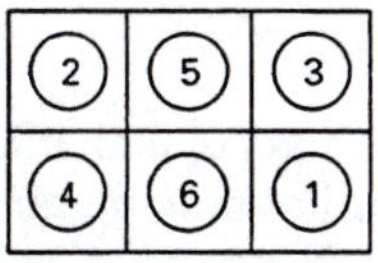

Figure 9.6 Better solution for Example 9.4.

As a final note on procedure CONSTRUCT, the selection in step 1.2 considers only the interactions with NFs already assigned to locations. There does not appear to be an obvious and effective way to consider the interactions with other NFs not yet located. One difficulty is that even if a way could be found to estimate the magnitude of the distance-weighted interactions, it is not clear that minimizing the incremental cost would be wise, since it might simply postpone the assignment of a critical NF until there were no more "good" locations for it.

As our discussion should have shown you by now, there are many possible variations of the basic construction heuristic. We encourage you to develop some of these yourself and apply them to some small problems.

9.4.3 Improvement Heuristics

The basic notion behind improvement heuristics is to modify a given solution so that the total cost is reduced. When the solution is represented as a permutation, the simplest possible modification is to select two elements of the permutation and interchange their values. This type of modification is the basis for *pairwise interchange improvement heuristics*. Since we already have developed an expression for the resulting change in total cost, $DTC_{uv}(\mathbf{a})$, we can state the general form of a pairwise interchange improvement heuristic as follows:

```
LOOP (a0:a)
0. a ← a0
1. Select a pair of facilities, (u, v)
2. Evaluate DTCuv(a)
3. Decide whether or not to make the interchange
4. Decide whether or not to continue
ENDLOOP
```

In LOOP, $\mathbf{a}_0$ is the initial solution, which is to be "improved," and $\mathbf{a}$ is the final solution. Developing an effective specific implementation of LOOP requires careful implementation of steps 1 to 4, which means that there can be many variations of the basic algorithm.

Generally speaking, a formally organized method is desirable for selecting the pair of facilities in step 1, although it is possible to use a randomized selection rule. Thus a nested loop of the form

```
FOR i = 1 TO n − 1
        FOR j = i + 1 TO n
        ENDFOR
ENDFOR
```

will serve to enumerate all the pairs of activities. If none of the $n(n - 1)/2$ pairs has a positive $DTC_{ij}(\mathbf{a})$, the algorithm may be terminated.

Based on the nested loop structure, we can easily identify three different strategies that may be followed in step 3 of LOOP. First, we could interchange u and v whenever $DTC_{uv}(\mathbf{a}) > 0$ (i.e., we could perform the interchange inside the "FOR j" loop). At least early in the procedure, we might expect to find many such interchanges, or we might find that many of the interchanges result in only small changes to $TC(\mathbf{a})$. Thus a second alternative is interchange u and v only when $DTC_{uv}(\mathbf{a})$ is the largest value for facility u (i.e., we could perform the interchange inside the "FOR i" loop but outside the "FOR j" loop). Carrying this strategy one step further, we could evaluate all possible pairs, and choose the one with the largest $DTC_{uv}(\mathbf{a})$ value (i.e., we could perform the interchange outside the "FOR i" loop).

If we insist on the best possible pairwise interchange, we get a *steepest-descent pairwise-interchange* heuristic (SDPI), which is stated in Table 9.28. If we are given the solution $\mathbf{a} = (2, 4, 5, 3, 1, 6)$ for Example 9.4, and apply the algorithm in Table 9.28, we obtain the $DTC_{ij}(\mathbf{a})$ values shown in Table 9.29 on the first pass through the WHILE loop. The best interchange is $(u, v) = (2, 6)$ [or $(4, 5)$, since there is a tie], resulting in a decrease of 16 in the total cost. On the second pass through the WHILE loop, there is no improvement, so the algorithm terminates with the solution $\mathbf{a} = (2, 6, 5, 3, 1, 4)$, which is illustrated in Figure 9.7.

Our example solution using SDPI highlights a potential problem with pairwise interchange algorithms in general—they may be "trapped" by a bad solution. Note that the initial solution for our example had workstations 5 and 6 assigned to locations 1 and 4, which are at one end of the cell. Because of the high flow rate

TABLE 9.28 STEEPEST DESCENT PAIRWISE INTERCHANGE HEURISTIC

```
Procedure SDPI(a0, D, W: a)
    a ← a0
    DONE ← false
    WHILE (NOT.DONE)
        DONE ← true
        max ← 0
        FOR i = 1 TO n − 1
            FOR j = 1 TO n − 1
                IF DTCij (a) > max, THEN
                    max ← DTCij (a)
                    u ← i
                    v ← j
                    DONE ← false
                ENDIF
            ENDFOR
        ENDFOR
        IF max > 0 THEN
            temp ← a(u)
            a(u) ← a(v)
            a(v) ← temp
        ENDIF
    ENDWHILE
END SDPI
```

TABLE 9.29 FIRST-PASS RESULTS FROM SDPI FOR EXAMPLE 9.3

(i,j)	$DTC_{ij}(\mathbf{a})$
(1,2)	0
(1,3)	−4
(1,4)	−2
(1,5)	8
(1,6)	12
(2,3)	0
(2,4)	10
(2,5)	−4
(2,6)	16
(3,4)	−4
(3,5)	8
(3,6)	10
(4,5)	16
(4,6)	−4
(5,6)	6

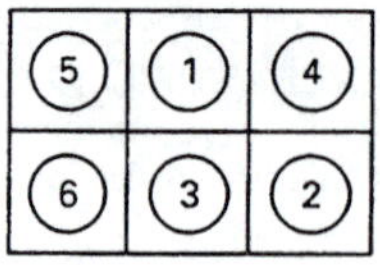

Figure 9.7 Best pairwise interchange solution to Example 9.4.

between workstations 5 and 6, any *pairwise interchange* that moves one but not the other results in an increase in total cost. Thus it is not possible, using SDPI, to reach the solution in Figure 9.6, which has a better value.

To overcome this problem with SDPI, one might consider interchanging three or more facilities simultaneously. Clearly, a steepest-descent algorithm based on k-tuple interchanges will always give a solution that is at least as good as the solution from a steepest descent $(k - 1)$-tuple interchange algorithm. The price for this potential improvement in solution quality is a computational one. The number of k-tuples that must be evaluated on each pass through the WHILE loop grows quite rapidly as k increases, as indicated by the following:

n	$k = 2$	$k = 3$	$k = 4$
5	15	20	15
10	45	120	210
20	190	1140	4845

In addition, when evaluating the interchange of a k-tuple, there are $(k - 1)!$ possible ways to perform the interchange. Thus the amount of calculation grows much faster than the figures above would indicate, when k-tuple interchanges are used.

Despite its computational requirements, the steepest-descent pairwise interchange heuristic remains popular because of the quality of the solutions produced. A number of variants have been developed, all in an attempt to overcome either the computational burden of SDPI, or the shortcomings of the pairwise interchanges. Many of these variants are described in the papers cited at the end of this chapter. One of the more interesting recent developments is described in the next section.

9.4.4 A Simulation Heuristic for QAP

Simulated annealing is an approach to solving difficult combinatorial problems that has gained considerable attention in recent years. The approach is based on a Monte Carlo model used to study the relationship between atomic structure, entropy, and temperature during the annealing of a sample of material. In the context of QAP, simulated annealing resembles the pairwise interchange heuristic, with one very important difference: If an interchange *increases* total cost, it still may be accepted, with some predetermined probability. Thus there is at least some chance that an unlucky choice for an intermediate solution, such as our example in Figure 9.7, will not cause the search to be trapped at a suboptimal solution.

In the simulated annealing method, there are three key concepts. The first is referred to as "temperature" and essentially is a parameter that controls the probability that a cost-increasing interchange will be accepted. During the course of the simulated annealing, the temperature will be reduced periodically, reducing the probability of accepting a cost-increasing interchange. The second key concept is "equilibrium" or a condition in which it is unlikely that further significant changes in the solution will occur with additional sampling. For example, if a large number of interchanges have been attempted at a given temperature without finding a better solution, it seems unlikely that additional sampling will be productive.

The third key concept is the "annealing schedule," which defines the set of temperatures to be used and how many interchanges to consider (or accept) before reducing the temperature. If there are too few temperatures, or not enough interchanges are attempted at each temperature, there is a greater likelihood of stopping with a suboptimal solution.

Using these concepts, and our earlier terminology and notation, we may give a very generic statement of the simulated annealing approach to solving QAP as

```
FOR i = 1 TO r DO
        EQUIL ← false
        WHILE NOT.EQUIL DO
                SELECT (u, v)
                IF DTC_uv(a) > 0, THEN
                        a(u)→a(v)
                ELSE
                        SELECT (x)
                        IF x < exp(−DTC_uv(a)/t_i), THEN
                                a(u)↔a(v)
                ENDIF
                CHECK (EQUIL)
        ENDWHILE
ENDFOR
```

In this statement of the method, t_i is the temperature, r is the number of different temperatures in the annealing schedule, and the notation "$a(u) \leftrightarrow a(v)$" indicates the interchange of u and v. The operation SELECT indicates randomly specifying an appropriate value, and the operation CHECK tests appropriate conditions to see if equilibrium has been attained.

There have been some promising results with the simulated annealing approach. However, those instances in which solutions found using simulated annealing are better than those found with more traditional heuristics usually involve a great deal of experimentation in order to find a good annealing schedule and appropriate tests for equilibrium. Unfortunately, at the time of this writing, there do not appear to be good rules for specifying these parameters a priori.

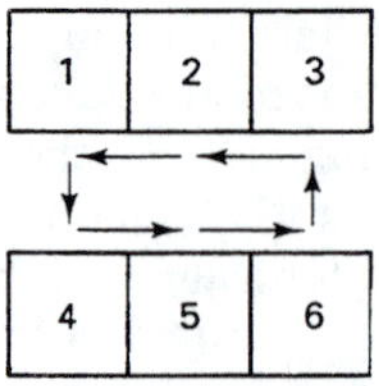

Figure 9.8 AGVS guidepath for Example 9.4.

TABLE 9.30 TRAVEL DISTANCES FOR EXAMPLE 9.4

From facility:	To facility: 1	2	3	4	5	6
1	0	5	4	1	2	3
2	1	0	5	2	3	4
3	2	1	0	3	4	5
4	5	4	3	0	1	2
5	4	3	2	5	0	1
6	3	2	1	4	5	0

9.4.5 Summary

Our development of heuristics for QAP has been based on an assumption of symmetric distances. Many real problems may not satisfy this assumption, as the following example demonstrates.

Example 9.5

Suppose that the cell described in Example 9.4 is being redesigned so that the material movement can be done using an automated guided vehicle. The guide path for the automated guided vehicle will be a simple loop, as illustrated in Figure 9.8. Because the vehicle travels in one direction only, the material movement distances are not symmetric. For example, the distance from location 1 to location 4 is only 1, whereas the distance from location 4 to location 1 is 5. Table 9.30 gives the distances between all pairs of locations. What is the assignment of workstations to locations that minimizes the total material movement distance?

You should convince yourself that the assumption of symmetric distances is not essential to the methods presented here for solving QAP. Rather, the assumption merely simplifies the notation and some of the computations.

9.5 FURTHER READING

In this chapter we have barely scratched the surface of an enormous body of literature on advanced discrete location problems. Although we do not intend to provide a complete listing of all the important references, we do want to provide some selected references for the interested student.

The capacitated location problem has been the focus of research on both exact and heuristic methods since the 1960s. One of the first exact algorithms, by Sa [38], was based on a weak linear programming formulation. Davis and Ray [12] employed the strong formulation, but the computational tools at the time prevented them from solving large problems. More recently, the problem has been attacked with good success by Nauss [34], Geoffrion and McBride [19], and Van Roy [40].

Because, for many years, exact algorithms could not solve realistic problems, there was a great deal of interest in heuristics for CLP. The essential ideas of adding, dropping, and interchanging can be found in the early work of Kuehn and Hamburger [30], Cooper [11], and Feldman et al. [15]. Khumawala [26] extended these ideas to solve capacitated problems, but used approximations to the optimal transportation problem solutions. Jacobsen [25] gives a detailed account of these heuristic methods. Our discussion of the ping-pong heuristic is based on the work of Van Roy [40], who also describes in much more detail how to use the method in an optimization framework.

The GAP model has been studied by a number of researchers, including DeMaio and Roveda [13], El Shaieb [14], Srinivasan and Thompson [39], and Ross and Soland [36]. Ross and Soland [37] described the formulation of the single-sourcing version of CLP as a GAP model, and our discussion of solution methods is based on their presentation. More recently, Klastorin [28] has reported success with a subgradient optimization-based heuristic, and Fisher et al. [17] have developed an effective multiplier adjustment method for improving the Lagrangian lower bound.

The QAP model has been, perhaps, one of the most extensively studied problems in all of discrete optimization. It was described very early by Koopmans and Beckman [29] in the context of location economics, and soon after a large number of other applications were described. Initially, many variations of the branch-and-bound method were proposed (e.g., by Gilmore [20], Gavett and Plyter [18], and many others). For a historical perspective on QAP, see Lawler [31], Hanan and Kurtzberg [22], and Burkard [5]. Lest you underestimate the difficulty of solving QAP, note that after all the effort to develop exact methods, general-purpose algorithms still are not able to routinely solve problems with more than about 15 locations.

Naturally, heuristic methods have played a central role in the study of QAP. Although there has been some work on construction heuristics (see, e.g., Gilmore [20] and Burkard and Derigs [6]), it is generally thought that improvement heuristics give better results for the effort required. The interchange notion, which is the basis for the SDPI algorithm, is found in the CRAFT algorithm of Armour and Buffa [2], and there have been a vast number of variants (e.g., Nugent et al. [35], Vollmann et al. [41], Hillier [23], and Hillier and Connors [24], to mention only a few).

There has been some recent interest in *hybrid* methods, which use some sort of truncated optimization method to establish a "good" initial solution, followed by improvement methods. Examples of this approach are found in Bazaraa and Sherali [3] and Burkard and Stratmann [9].

Simulated annealing is another approach of current interest. The method is based on a model proposed by Metropolis et al. [33], and its application to optimization was described by Kirkpatrick et al. [27]. Golden and Skiscim [21] and Wilhelm and Ward [42] described extensive computational studies of the method applied to traveling salesman, n-median, and quadratic assignment problems. The application to QAP was also discussed by Burkard and Rendl [8].

A final area of investigation into QAP is represented by the work of Burkard and Fincke [7]. They showed that under certain assumptions, as a QAP model grows large, the relative error of a heuristic algorithm grows small. Thus there is reason to hope that good heuristics will provide very good solutions when applied to large problems (e.g., the location of devices in a large-scale integrated circuit).

We must confess that the great interest in the models described in this chapter is probably because they have applications beyond location decision making. In fact, as you explore the literature, you will find interesting research in a number of diverse areas, including computer science, economics, mathematics, and statistical mechanics.

REFERENCES

1. Akinc, U., and B. M. Khumawala, "An Efficient Branch and Bound Algorithm for the Capacitated Warehouse Location Problem," *Management Science,* Vol. 23, No. 6, 1977, pp. 585–594.
2. Armour, G. C., and E. S. Buffa, "A Heuristic Algorithm and Simulation Approach to the Relative Location of Facilities," *Management Science,* Vol. 9, No. 2, 1963, pp. 294–309.
3. Bazaraa, M. S., and M. D. Sherali, "Bender's Partitioning Scheme Applied to a New Formulation of the Quadratic Assignment Problem," *Naval Research Logistics Quarterly,* Vol. 27, No. 1, 1980, pp. 29–41.
4. Bonomi, E., and J. Lutton, "The *N*-City Travelling Salesman Problem: Statistical Mechanics and the Metropolis Algorithm," *SIAM Review,* Vol. 26, No. 4, 1984, pp. 551–568.
5. Burkard, R. E., "Quadratic Assignment Problems," *European Journal of Operations Research,* Vol. 15, No. 3, 1984, pp. 283–289.
6. Burkard, R. E., and U. Derigs, *Assignment and Matching Problems: Solution Methods with FORTRAN-Programes,* Springer-Verlag, Berlin, 1980.
7. Burkard, R. E., and U. Fincke, "The Asymptotic Probabilistic Behaviour of Quadratic Sum Assignment Problems," *Zeitschrift fur Operations Research,* Vol. 27, 1983, pp. 73–81.
8. Burkard, R. E., and F. Rendl, "A Thermodynamically Motivated Simulation Procedure for Combinatorial Optimization Problems," *European Journal of Operations Research,* Vol. 17, 1984, pp. 169–174.
9. Burkard, R. E., and K. H. Stratmann, "Numerical Investigations on Quadratic Assignment Problems," *Naval Research Logistics Quarterly,* Vol. 27, No. 1, 1978, pp. 109–120.

10. Conway, R. W., and W. L. Maxwell, "A Note on the Assignment of Facility Locations," *Journal of Industrial Engineering,* Vol. 12, No. 1, 1961, pp. 34–36.

11. Cooper, L., "Heuristic Methods for Location-Allocation Problems," *SIAM Review,* Vol. 6, No. 1, 1964, pp. 37–53.

12. Davis, P. S., and T. L. Ray, "A Branch-Bound Algorithm for the Capacitated Facilities Location Problem," *Naval Research Logistics Quarterly,* Vol. 16, No. 3, 1969, pp. 331–334.

13. DeMaio, A., and C. Roveda, "An All Zero-One Algorithm for a Class of Transportation Problems," *Operations Research,* Vol. 19, No. 6, 1971, pp. 1406–1418.

14. El Shaieb, A. M., "A New Algorithm for Locating Sources among Destinations," *Management Science,* Vol. 20, No. 2, 1973, pp. 221–231.

15. Feldman, E., F. A. Lehrer, and T. L. Ray, "Warehouse Location under Continuous Economies of Scale," *Management Science,* Vol. 12, No. 9, 1966, pp. 670–684.

16. Fisher, M. L., and R. Jaikumar, "A Generalized Assignment Heuristic for Vehicle Routing," *Networks,* Vol. 11, No. 2, 1981, pp. 109–124.

17. Fisher, M. L., R. Jaikumar, and L. Van Wassenhove, "A Multiplier Adjustment Method for the Generalized Assignment Problem," *Management Science,* Vol. 32, No. 9, 1986, pp. 1095–1103.

18. Gavett, J. W., and N. V. Plyter, "The Optimal Assignment of Facilities to Locations by Branch and Bound," *Operations Research,* Vol. 14, No. 2, 1966, pp. 210–232.

19. Geoffrion, A. M., and R. McBride, "Lagrangean Relaxation Applied to Capacitated Facility Location Problems," *AIIE Transactions,* Vol. 10, No. 1, 1978, pp. 40–47.

20. Gilmore, P. C., "Optimal and Suboptimal Algorithms for the Quadratic Assignment Problem," *SIAM Journal,* Vol. 10, No. 2, 1962, pp. 305–313.

21. Golden, B. L., and C. C. Skiscim, "Using Simulated Annealing to Solve Routing and Location Problems," *Naval Research Logistics Quarterly,* Vol. 33, No. 2, 1986, pp. 261–279.

22. Hanan, M., and J. Kurtzberg, "A Review of the Placement and Quadratic Assignment Problems," *SIAM Review,* Vol. 14, No. 2, 1972, pp. 324–342.

23. Hillier, F. S., "Quantitative Tools for Plant Layout Analysis," *Journal of Industrial Engineering,* Vol. 14, No. 1, 1963, pp. 33–40.

24. Hillier, F. S., and M. M. Connors, "Quadratic Assignment Problem Algorithms and the Location of Indivisible Facilities," *Management Science,* Vol. 13, No. 1, 1966, pp. 42–57.

25. Jacobsen, S. K., "Heuristics for the Capacitated Plant Location Model," *European Journal of Operational Research,* Vol. 12, No. 3, 1983, pp. 253–261.

26. Khumawala, B. M., "An Efficient Heuristic Procedure for the Capacitated Warehouse Location Problem," *Naval Research Logistics Quarterly,* Vol. 21, No. 4, 1974, pp. 609–623.

27. Kirkpatrick, S., C. D. Gelatti, Jr., and M. P. Vecchi, "Optimization by Simulated Annealing," *Science,* Vol. 220, 1983, pp. 671–680.

28. Klastorin, T. D., "An Effective Subgradient Algorithm for the Generalized Assignment Problem," *Computers and Operations Research,* Vol. 6, No. 3, 1979, pp. 155–164.

29. Koopmans, T. C., and M. J. Beckman, "Assignment Problems and the Location of Economic Activities," *Econometrica,* Vol. 25, No. 1, 1957, pp. 53–76.

30. Kuehn, A. A., and M. J. Hamburger, "A Heuristic Program for Locating Warehouses," *Management Science,* Vol. 9, No. 4, 1963, pp. 643–666.

31. Lawler, E. L., "The Quadratic Assignment Problem," *Management Science,* Vol. 9, No. 4, 1963, pp. 586–599.

32. Martello, S., and P. Toth, "An Algorithm for the Generalized Assignment Problem," in J. P. Brans, ed., *Operational Research '81,* North-Holland Publishing Company, Amsterdam, 1981.

33. Metropolis, N., A. Rosenbluth, M. Rosenbluth, A. Teller, and E. Teller, "Equation of State Calculation by Fast Computing Machines," *Journal of Chemical Physics,* Vol. 21, 1953, pp. 1087–1092.

34. Nauss, R. M., "An Improved Algorithm for the Capacitated Facility Location Problem," *Journal of the Operational Research Society,* Vol. 29, No. 12, 1978, pp. 1195–1201.

35. Nugent, C. E., T. E. Vollmann, and J. Ruml, "An Experimental Comparison of Techniques for the Assignment of Facilities to Locations," *Operations Research,* Vol. 16, No. 1, 1968, pp. 150–173.

36. Ross, G. T., and R. M. Soland, "A Branch and Bound Algorithm for the Generalized Assignment Problem," *Mathematical Programming,* Vol. 8, 1975, pp. 91–103.

37. Ross, G. T., and R. M. Soland, "Modeling Facility Location Problems as Generalized Assignment Problems," *Management Science,* Vol. 24, No. 3, 1977, pp. 345–357.

38. Sa, G., "Branch-and-Bound and Approximate Solutions to the Capacitated Plant-Location Problem," *Operations Research,* Vol. 17, No. 6, 1969, pp. 1005–1016.

39. Srinivasan, V., and G. L. Thompson, "An Algorithm for Assigning Uses to Sources in a Special Class of Transportation Problems," *Operations Research,* Vol. 21, No. 1, 1973, pp. 284–295.

40. Van Roy, T. J., "A Cross Decomposition Algorithm for Capacitated Facility Location," *Operations Research,* Vol. 34, No. 1, 1986, pp. 145–163.

41. Vollmann, T. E., C. E. Nugent, and R. L. Zartler, "A Computerized Model for Office Layout," *Journal of Industrial Engineering,* Vol. 19, No. 7, 1968, pp. 321–329.

42. Wilhelm, M. R., and T. L. Ward, "Solving Quadratic Assignment Problems by Simulated Annealing," *IIE Transactions,* Vol. 19, No. 1, 1987, pp. 107–119.

PROBLEMS

9.1. For the data given below, calculate the Ω_i and Δ_i values for each site.

$$\mathbf{C} = \begin{pmatrix} 15 & 20 & 40 \\ 15 & 20 & 40 \\ 40 & 20 & 15 \\ 40 & 20 & 15 \end{pmatrix}, \qquad \mathbf{R} = \begin{pmatrix} 10 \\ 15 \\ 10 \\ 20 \end{pmatrix}$$

$$\mathbf{S} = (25 \quad 30 \quad 25)$$

$$\mathbf{F} = (200 \quad 200 \quad 150)$$

Determine a feasible solution using the ADD & DROP heuristic from Table 9.16. State explicitly how you make the "select" decisions in lines 22 and 30.

9.2. Since at least two sites must be selected in a feasible solution to Problem 9.1, your calculation of the Ω_i and Δ_i values provides all the information needed to enumerate the feasible solutions completely. Use complete enumeration to determine the optimum solution. Is it different from the solution you found using ADD & DROP?

9.3. Use the allocation adjustment approach to enforce single sourcing in the optimum solution to Problem 9.1. Find the best single-sourcing solution to Problem 9.1 using total enumeration. What is the difference in total cost?

9.4. For the data given below, calculate the Ω_i and Δ_i values for each site.

$$\mathbf{C} = \begin{pmatrix} 5 & 6 & 10 & 11 \\ 8 & 5 & 5 & 8 \\ 11 & 9 & 5 & 4 \\ 6 & 11 & 10 & 5 \end{pmatrix}, \qquad \mathbf{R} = \begin{pmatrix} 10 \\ 10 \\ 10 \\ 10 \end{pmatrix}$$

$$\mathbf{S} = (20 \quad 20 \quad 20 \quad 20)$$

$$\mathbf{F} = (10 \quad 10 \quad 10 \quad 10)$$

Determine a feasible solution using the ADD & DROP heuristic from Table 9.16. State explicitly how you make the "select" decisions in lines 22 and 30. Remember to recalculate the Ω_i and Δ_i values each time a variable is pegged.

9.5. Suppose that the fixed costs in Problem 9.4 are 20 instead of 10. Does the solution change?

9.6. For the data given below, calculate the Ω_i and Δ_i values for each site.

$$\mathbf{C} = \begin{pmatrix} 0 & 6 & 8 \\ 3 & 4 & 7 \\ 5 & 0 & 7 \\ 6 & 8 & 6 \end{pmatrix}, \qquad \mathbf{R} = \begin{pmatrix} 15 \\ 15 \\ 10 \\ 20 \end{pmatrix}$$

$$\mathbf{S} = (30 \quad 25 \quad 35)$$

$$\mathbf{F} = (100 \quad 80 \quad 120)$$

Determine a feasible solution using the ADD & DROP heuristic from Table 9.16. State explicitly how you make the "select" decisions in lines 22 and 30.

9.7. For the data given below, calculate the Ω_i and Δ_i values for each site.

$$\mathbf{C} = \begin{pmatrix} 6 & 8 & 10 \\ 10 & 5 & 7 \\ 12 & 8 & 6 \\ 8 & 4 & 6 \end{pmatrix}, \qquad \mathbf{R} = \begin{pmatrix} 5 \\ 5 \\ 5 \\ 5 \end{pmatrix}$$

$$\mathbf{S} = (10 \quad 10 \quad 10 \quad 20)$$

$$\mathbf{F} = (80 \quad 100 \quad 60)$$

Determine a feasible solution using the ADD & DROP heuristic from Table 9.16. State explicitly how you make the "select" decisions in lines 22 and 30.

9.8. In Problems 8.14 and 8.15 you were asked to analyze the location of railcar loading sites for a logging operation. The logging company has now realized that there may be a congestion problem at the loading sites if too many trucks are assigned to load railcars at a single site. Below we summarize the logging data in terms of truck trips per day and truck unloading capacity at the railcar loading sites.

$$\mathbf{C} = \begin{pmatrix} 0 & 1.00 & 2.50 & 3.50 & 4.00 \\ 0.20 & 0.25 & 2.50 & 1.60 & 0.15 \\ 2.00 & 1.00 & 0.25 & 1.75 & 0.80 \\ 0.50 & 0.75 & 0.30 & 0.05 & 1.25 \end{pmatrix}, \qquad \mathbf{R} = \begin{pmatrix} 50 \\ 100 \\ 100 \\ 100 \end{pmatrix}$$

$$\mathbf{S} = (150 \quad 150 \quad 150 \quad 150 \quad 150)$$

$$\mathbf{F} = (200 \quad 100 \quad 50 \quad 150 \quad 200)$$

Each railcar loading site can handle at most 150 truckloads per day. Calculate the Ω_i and Δ_i values for each site, and use the ADD & DROP heuristic to determine which sites should be used and how many trucks to send from each logging area to each site.

9.9. Use the heuristic solution you developed in Problem 9.8 as a starting point for the ping-pong heuristic. (*Hint*: You may need to recompute the dual variables in the transportation problem, forcing the dual variable corresponding to the "excess requirement" to be zero.)

9.10. Solve Problem 9.8 assuming a capacity at each railcar loading site of only 125 truckloads per day. What is the impact on total cost of the smaller capacities?

9.11. After reviewing your solution for the problem with railcar loading capacities of 125 truckloads per day, the logging company expressed concern about traffic control when one logging area must send trucks to more than one railcar loading site. Develop a single-sourcing solution without changing the site capacities, by reallocating the railcar loading capacity to the logging areas (i.e., by modifying the x_{ij} rather than the capacities). Can you find a better solution by trial and error?

9.12. Write a computer program for the steepest-descent pairwise interchange procedure.

9.13. Compute the values of $DTC_{ij}(\mathbf{a})$ shown in Table 9.29 and show all work.

9.14. Verify, by actually computing the total cost each way, that expressions (9.16) and (9.17) give the same total cost for Example 9.3.

9.15. Given the assignment (2, 4, 3, 1) and the matrices

$$\mathbf{D} = \begin{pmatrix} 0 & 1 & 1 & 2 \\ 1 & 0 & 2 & 1 \\ 1 & 2 & 0 & 1 \\ 2 & 1 & 1 & 0 \end{pmatrix}, \qquad \mathbf{W} = \begin{pmatrix} 0 & 12 & 4 & 2 \\ 12 & 0 & 2 & 3 \\ 4 & 2 & 0 & 12 \\ 2 & 3 & 12 & 0 \end{pmatrix}$$

(a) Find the total cost of the given assignment.

(b) Find the least cost assignment by total enumeration.

(c) Starting with the given assignment, complete one iteration of the SDPI algorithm (i.e., complete both the inner FOR loops).

9.16. The following **D** and **W** matrices are given:

$$\mathbf{D} = \begin{pmatrix} 0 & 1 & 1 & 2 & 3 \\ 1 & 0 & 2 & 1 & 2 \\ 1 & 2 & 0 & 1 & 2 \\ 2 & 1 & 1 & 0 & 1 \\ 3 & 2 & 2 & 1 & 0 \end{pmatrix}, \qquad \mathbf{W} = \begin{pmatrix} 0 & 5 & 2 & 4 & 1 \\ 5 & 0 & 3 & 0 & 2 \\ 2 & 3 & 0 & 0 & 0 \\ 4 & 0 & 0 & 0 & 5 \\ 1 & 2 & 0 & 5 & 0 \end{pmatrix}$$

(a) Starting with the assignment $(1, 4, 3, 5, 2)$, apply the SDPI procedure until at least two interchanges have been made, and compute the total cost for each assignment obtained.

(b) Apply procedure CONSTRUCT to generate an initial assignment. What rule did you use for selecting the next facility to locate (i.e., step 1.1 in the procedure)?

9.17. A classical lower bound for TC(**a**) is obtained as follows. Consider all the distance matrix elements above the diagonal and sort them in nondecreasing order. Consider all the weight matrix elements above the diagonal and sort them into nonincreasing order. Now take the inner product of these two vectors. The value of the inner product is a lower bound on the total solution value. The lower bound for problem 15 would be

$$(1, 1, 1, 1, 2, 2) \times (12, 12, 4, 3, 2, 2)^T = 39$$

Can you prove that this will always yield a lower bound?

9.18. Calculate the lower bound described in Problem 9.17 for the data of Example 9.3.

9.19. Given the assignment $(1, 3, 4, 2)$ and the matrices

$$\mathbf{D} = \begin{pmatrix} 0 & 1 & 2 & 3 \\ 1 & 0 & 1 & 2 \\ 2 & 1 & 0 & 1 \\ 3 & 2 & 1 & 0 \end{pmatrix}, \qquad \mathbf{W} = \begin{pmatrix} 0 & 12 & 4 & 2 \\ 12 & 0 & 2 & 3 \\ 4 & 2 & 0 & 12 \\ 2 & 3 & 12 & 0 \end{pmatrix}$$

(a) Apply the SDPI procedure until no further improvements are possible.

(b) Calculate the lower bound using the method described in Problem 9.17.

(c) Find a least-cost solution by total enumeration. Show all work.

(d) Compare the values obtained in steps (a) to (c).

9.20. List 10 applications of the quadratic assignment problem formulation. Clearly identify the facilities, the sites, and the weights.

9.21. Use the computer program obtained in Problem 9.12 to solve the quadratic assignment problem for which the data are

$$\mathbf{D} = \begin{pmatrix} 0 & 1 & 2 & 3 & 1 & 2 & 3 & 4 \\ 1 & 0 & 1 & 2 & 2 & 1 & 2 & 3 \\ 2 & 1 & 0 & 1 & 3 & 2 & 1 & 2 \\ 3 & 2 & 1 & 0 & 4 & 3 & 2 & 1 \\ 1 & 2 & 3 & 4 & 0 & 1 & 2 & 3 \\ 2 & 1 & 2 & 3 & 1 & 0 & 1 & 2 \\ 3 & 2 & 1 & 2 & 2 & 1 & 0 & 1 \\ 4 & 3 & 2 & 1 & 3 & 2 & 1 & 0 \end{pmatrix}, \qquad \mathbf{W} = \begin{pmatrix} 0 & 5 & 2 & 4 & 1 & 0 & 0 & 6 \\ 5 & 0 & 3 & 0 & 2 & 2 & 2 & 0 \\ 2 & 3 & 0 & 0 & 0 & 0 & 0 & 5 \\ 4 & 0 & 0 & 0 & 5 & 2 & 2 & 10 \\ 1 & 2 & 0 & 5 & 0 & 10 & 0 & 0 \\ 0 & 2 & 0 & 2 & 10 & 0 & 5 & 1 \\ 0 & 2 & 0 & 2 & 0 & 5 & 0 & 10 \\ 6 & 0 & 5 & 10 & 0 & 1 & 10 & 0 \end{pmatrix}$$

The sites for this problem are arranged as shown in Figure P9.21.

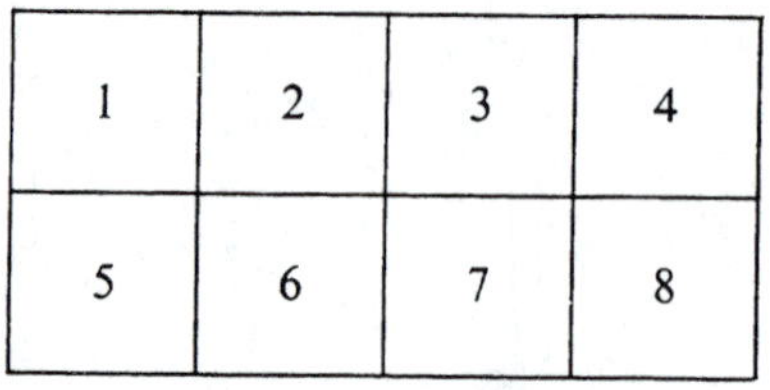

Figure P9.21

The optimum solution is known to have a total cost of 107. Note the effect of the initial assignment on the final solution by employing a number of different initial assignments.

9.22. Five manufacturing departments are to be assigned among the six sites shown in Figure P9.22. Four products are to be processed through the five departments according to the processing sequences shown, with the indicated frequencies of movement between departments.

Product	Processing sequence	Material-handling volume (loads/day)
1	A, B, C, D, E	20
2	A, C, B, C, D, E	30
3	A, D, E	15
4	A, C, D, B, E	40

(a) Develop the **D** and **W** matrices, assuming rectilinear travel between the centroids of the sites.
(b) Use procedure CONSTRUCT to generate an initial assignment.
(c) Calculate the lower bound as described in Problem 9.17.
(d) Apply procedure SDPI until no further improvements are possible.

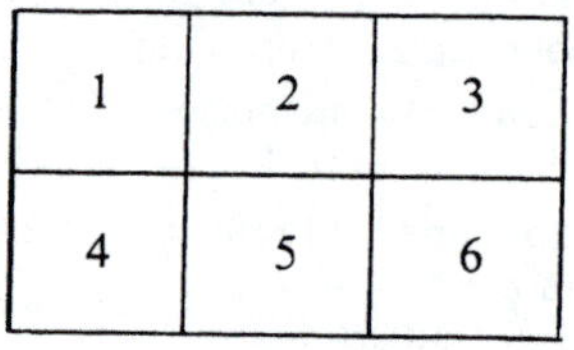

Figure P9.22

9.23. Solve Problem 9.22 assuming the facilities are to be arranged in a single row rather than in a double row.

9.24. O. R. Mann is designing his house so that distance traveled is minimized. Using standard room modules of size 12 by 12 ft, an eight-room house (including the garage) is to be designed. Based on the **W** matrix below and site arrangement shown in Figure P9.24, develop a house design using each of the heuristic algorithms. Evaluate each on the basis of practical limitations.

1	2	3	4
5	6	7	8
9	10	11	12
13	14	15	16

Site Arrangement

Figure P9.24

$\mathbf{W} =$

	L.R.	B.R. 1	B.R. 2	B.R. 3	Den	D.R.	Kit.	Gar.
L.R.	0	2	1	1	5	4	3	2
B.R. 1	2	0	2	2	6	3	4	1
B.R. 2	1	2	0	3	5	4	4	0
B.R. 3	1	2	3	0	4	3	5	0
Den	5	6	5	4	0	4	6	10
D.R.	4	3	4	3	4	0	8	0
Kit.	3	4	4	5	6	8	0	12
Gar.	2	1	0	0	10	0	12	0

(Note: The number of sites is greater than the number of rooms!)

Index

Q

R

Second Edition

FACILITY LAYOUT AND LOCATION: AN ANALYTICAL APPROACH

Richard L. Francis

Leon F. McGinnis, Jr.

John A. White

Providing a comprehensive introduction to quantitative methods for facility layout and location, this text is directed at senior and graduate level students in industrial engineering, manufacturing systems, management science, and operations research curricula. Problems of facility layout and location are treated together because of the similarity between arranging the space in a single facility and arranging a system of facilities. An introduction to the field's issues and literature is included, along with the basic tools and methodologies.

The Second Edition revises over half of the text to provide material reflecting the most current developments. Chapters contain explanations of what layout and location problems are, how to collect data, and show how to model and solve such problems.

Features include:

- basic design and layout approaches and problem definitions
- extensive references and homework problem sets
- chapters on network, discrete, and multi-objective location problems
- material on discrete optimization, networks, graphs, computer-aided layout, and storage system design
- guides for further reading
- various figures, tables, and numerical examples in each chapter

ISBN 0-13-29923
9 780132 992312

Prentice Hall
Englewood Cliffs, NJ 07632